Adolf Grünbaum

THE THIRTEEN BOOKS

OF

EUCLID'S ELEMENTS

Ἐὰν ἐν κύκλῳ δύο εὐθεῖαι τέμνωσιν ἀλλήλας μὴ διὰ τοῦ κέντρου οὖσαι, οὐ τέμνουσιν ἀλλήλας δίχα. Ἔστω κύκλος ὁ
ΑΒΓΔ καὶ ἐν αὐτῷ δύο εὐθεῖαι αἱ ΑΓ ΒΔ τεμνέτωσαν ἀλλήλας
κατὰ τὸ Ε μὴ διὰ τοῦ κέντρου οὖσαι. λέγω ὅτι οὐ τέμνουσιν
ἀλλήλας δίχα. εἰ γὰρ δυνατόν, τεμνέτωσαν ἀλλήλας δίχα,
ὥστε ἴσην εἶναι τὴν μὲν ΑΕ τῇ ΕΓ, τὴν δὲ ΒΕ τῇ ΕΔ· καὶ
εἰλήφθω τὸ κέντρον τοῦ ΑΒΓΔ κύκλου καὶ ἔστω τὸ Ζ, καὶ ἐπεζεύχθω ἡ ΖΕ. ἐπεὶ οὖν εὐθεῖά τις διὰ τοῦ κέντρου ἡ ΖΕ εὐθεῖάν
τινα τὴν ΑΓ διχατομεῖ, καὶ πρὸς ὀρθὰς αὐτὴν τέμνει·
ὀρθὴ ἄρα ἐστὶν ἡ ὑπὸ ΖΕΑ. πάλιν ἐπεὶ εὐθεῖά τις ἡ ΖΕ εὐθεῖάν
τινα τὴν ΒΔ διχατομεῖ, καὶ πρὸς ὀρθὰς
αὐτὴν τέμνει· ὀρθὴ ἄρα ἡ ὑπὸ
ΖΕΒ. ἐδείχθη δὲ καὶ ἡ ὑπὸ ΖΕΑ ὀρθή·
ἴση ἄρα ἡ ὑπὸ ΖΕΑ τῇ ὑπὸ ΖΕΒ, ἡ
ἐλάσσων τῇ μείζονι· ὅπερ ἐστὶν ἀδύνατον. οὐκ ἄρα αἱ ΑΓ ΒΔ τέμνουσιν ἀλλήλας δίχα. Ἐὰν ἄρα ἐν κύκλῳ δύο

εὐθεῖαι τέμνωσιν ἀλλήλας καὶ τὰ ἑξῆς. ὅπερ ἔδει δεῖξαι. ∴
Ἐὰν δύο κύκλοι τέμνωσιν ἀλλήλους, οὐκ ἔσται αὐτῶν τὸ αὐτὸ κέντρον. δύο γὰρ κύκλοι οἱ ΑΒΓ ΓΔΗ τεμνέτωσαν ἀλλήλους κατὰ τὰ Β Γ σημεῖα. λέγω ὅτι οὐκ ἔσται αὐτῶν τὸ
αὐτὸ κέντρον. εἰ γὰρ δυνατόν, ἔστω τὸ Ε καὶ ἐπεζεύχθω ἡ ΕΓ
καὶ διήχθω ἡ ΕΖΗ ὡς ἔτυχεν· καὶ ἐπεὶ τὸ Ε σημεῖον κέντρον
ἐστὶ τοῦ ΑΒΓ κύκλου, ἴση ἐστὶν ἡ ΕΓ τῇ ΕΖ. πάλιν ἐπεὶ τὸ
Ε σημεῖον κέντρον ἐστὶ τοῦ ΓΔΗ κύκλου, ἴση ἐστὶν ἡ ΕΓ τῇ
ΕΗ. ἐδείχθη δὲ ἡ ΕΓ καὶ τῇ ΕΖ ἴση. καὶ ἡ ΕΖ ἄρα τῇ ΕΗ

THE THIRTEEN BOOKS OF EUCLID'S ELEMENTS

TRANSLATED FROM THE TEXT OF HEIBERG

WITH INTRODUCTION AND COMMENTARY

BY

Sir THOMAS L. HEATH,

K.C.B., K.C.V.O., F.R.S.,

SC.D. CAMB., HON. D.SC. OXFORD

HONORARY FELLOW (SOMETIME FELLOW) OF TRINITY COLLEGE CAMBRIDGE

SECOND EDITION

REVISED WITH ADDITIONS

VOLUME I

INTRODUCTION AND BOOKS I, II

DOVER PUBLICATIONS, INC.

NEW YORK

PREFACE

" THERE never has been, and till we see it we never shall believe that there can be, a system of geometry worthy of the name, which has any material departures (we do not speak of *corrections* or *extensions* or *developments*) from the plan laid down by Euclid." De Morgan wrote thus in October 1848 (*Short supplementary remarks on the first six Books of Euclid's Elements* in the *Companion to the Almanac* for 1849); and I do not think that, if he had been living to-day, he would have seen reason to revise the opinion so deliberately pronounced sixty years ago. It is true that in the interval much valuable work has been done on the continent in the investigation of the first principles, including the formulation and classification of axioms or postulates which are necessary to make good the deficiencies of Euclid's own explicit postulates and axioms and to justify the further assumptions which he tacitly makes in certain propositions, content apparently to let their truth be inferred from observation of the figures as drawn ; but, once the first principles are disposed of, the body of doctrine contained in the recent text-books of elementary geometry does not, and from the nature of the case cannot, show any substantial differences from that set forth in the *Elements*. In England it would seem that far less of scientific value has been done; the efforts of a multitude of writers have rather been directed towards producing alternatives for Euclid which shall be more suitable, that is to say, easier, for schoolboys. It is of course not surprising that, in these days of short cuts, there should have arisen a movement to get rid of Euclid and to substitute a "royal road to geometry"; the marvel is that a book which was not written for schoolboys but for grown men (as all internal evidence shows, and in particular the essentially theoretical character of the work and its aloofness from anything of the nature of "practical" geometry) should have held its own as a school-book for so long. And now that Euclid's proofs and arrangement are no longer required from candidates at examinations there has been a rush of competitors anxious to be first in the field with a new text-book on the more "practical" lines which now find so much favour. The natural desire of each teacher who writes such a text-book is to give prominence to some special nostrum which he has found successful with pupils. One result is, too often, a loss of a due sense of proportion; and, in any case, it is inevitable that there should be great diversity of treatment. It was with reference to such a danger that Lardner wrote in 1846 : "Euclid once superseded, every teacher would esteem his own work the best, and every school would have its own class book. All that rigour and exactitude which have so long excited the admiration of men of science would be at an end. These very words would lose all definite meaning. Every school would have a different standard; matter of assumption in one being matter of demonstration in another; until, at length, GEOMETRY, in the ancient sense of the word, would be altogether frittered away or be only considered as a particular application of Arithmetic and Algebra." It is, perhaps, too early yet to prophesy what will be the ultimate outcome of the new order of things; but it would at least seem possible that history will repeat itself and that, when chaos has come again in geometrical teaching, there will be a return to Euclid more or less complete for the purpose of standardising it once more.

But the case for a new edition of Euclid is independent of any controversies as to how geometry shall be taught to schoolboys. Euclid's work will live long after all the text-books of the present day are superseded and forgotten. It is one of the noblest monuments of antiquity; no mathematician worthy of the name can afford not to know Euclid, the real Euclid as distinct from any revised or rewritten versions which will serve for schoolboys or engineers. And, to know Euclid, it is necessary to know his language, and, so far as it can be traced, the history of the "elements" which he collected in his immortal work.

This brings me to the *raison d'être* of the present edition. A new translation from the Greek was necessary for two reasons. First, though some time has elapsed since the appearance of Heiberg's definitive text and prolegomena, published between 1883 and 1888, there has not been, so far as I know, any attempt to make a faithful translation from it into English even of the Books which are commonly read. And, secondly, the other Books, VII. to X. and XIII., were not included by Simson and the editors who followed him, or apparently in any English translation since Williamson's (1781—8), so that they are now practically inaccessible to English readers in any form.

In the matter of notes, the edition of the first six Books in Greek and Latin with notes by Camerer and Hauber (Berlin, 1824—5) is a perfect mine of information. It would have been practically impossible to make the notes more exhaustive at the time when they were written. But the researches of the last thirty or forty years into the history of mathematics (I need only mention such names as those of Bretschneider, Hankel, Moritz Cantor, Hultsch, Paul Tannery, Zeuthen, Loria, and Heiberg) have put the whole subject upon a different plane. I have endeavoured in this edition to take account of all the main results of these researches up to the present date. Thus, so far as the geometrical Books are concerned, my notes are intended to form a sort of dictionary of the history of elementary geometry, arranged according to subjects; while the notes on the arithmetical Books VII.—IX. and on Book X. follow the same plan.

I desire to express here my thanks to my brother, Dr R. S. Heath, Vice-Principal of Birmingham University, for suggestions on the proof sheets and, in particular, for the reference to the parallelism between Euclid's definition of proportion and Dedekind's theory of irrationals, to Mr R. D. Hicks for advice on a number of difficult points of translation, to Professor A. A. Bevan for help in the transliteration of Arabic names, and to the Curators and Librarian of the Bodleian Library for permission to reproduce, as frontispiece, a page from the famous Bodleian MS. of the *Elements*. Lastly, my best acknowledgments are due to the Syndics of the Cambridge University Press for their ready acceptance of the work, and for the zealous and efficient coöperation of their staff which has much lightened the labour of seeing the book through the Press.

November, 1908.

PREFACE TO THE SECOND EDITION

I LIKE to think that the exhaustion of the first edition of this work furnishes a new proof (if such were needed) that Euclid is far from being defunct or even dormant, and that, so long as mathematics is studied, mathematicians will find it necessary and worth while to come back again and again, for one purpose or another, to the twenty-two-centuries-old book which, notwithstanding its imperfections, remains the greatest elementary textbook in mathematics that the world is privileged to possess.

The present edition has been carefully revised throughout, and a number of passages (sometimes whole pages) have been rewritten, with a view to bringing it up to date. Some not inconsiderable additions have also been made, especially in the Excursuses to Volume I, which will, I hope, find interested readers.

Since the date of the first edition little has happened in the domain of geometrical teaching which needs to be chronicled. Two distinct movements however call for notice.

The first is a movement having for its object the mitigation of the difficulties (affecting in different ways students, teachers and examiners) which are found to arise from the multiplicity of the different textbooks and varying systems now in use for the teaching of elementary geometry. These difficulties have evoked a widespread desire among teachers for the establishment of an agreed sequence to be generally adopted in teaching the subject. One proposal to this end has already been made: but the chance of the acceptance of an agreed sequence has in the meantime been prejudiced by a second movement which has arisen in other quarters.

I refer to the movement in favour of reviving, in a modified form, the proposal made by Wallis in 1663 to replace Euclid's Parallel-Postulate by a Postulate of Similarity (as to which see pp. 210—11 of Volume I of this work). The form of Postulate now suggested is an assumption that "Given one triangle, there can be constructed, on any arbitrary base, another triangle equiangular with (or similar to) the given triangle." It may perhaps be held that this assumption has the advantage of not referring, in the statement of it, to the fact that a straight line is of unlimited length; but, on the other hand, as is well known, Saccheri showed (1733) that it involves more than is necessary to enable Euclid's Postulate to be proved. In any case it would seem certain that a scheme based upon the proposed Postulate, if made scientifically sound, must be more difficult than the procedure now generally followed. This being so, and having regard to the facts (1) that the difference between the suggested Postulate and that of Euclid is in effect so slight and (2) that the historic interest of Euclid's Postulate is so great, I am of opinion that the proposal is very much to be deprecated.

T. L. H.

December 1925.

CONTENTS

VOLUME I.

INTRODUCTION.

Prop 20, p. 286-7
str line shorter than
△

INTRODUCTION.

CHAPTER I.

EUCLID AND THE TRADITIONS ABOUT HIM.

As in the case of the other great mathematicians of Greece, so in Euclid's case, we have only the most meagre particulars of the life and personality of the man.

Most of what we have is contained in the passage of Proclus' summary relating to him, which is as follows[1]:

"Not much younger than these (sc. Hermotimus of Colophon and Philippus of Medma) is Euclid, who put together the Elements, collecting many of Eudoxus' theorems, perfecting many of Theaetetus', and also bringing to irrefragable demonstration the things which were only somewhat loosely proved by his predecessors. This man lived[2] in the time of the first Ptolemy. For Archimedes, who came immediately after the first (Ptolemy)[3], makes mention of Euclid: and, further, they say that Ptolemy once asked him if there was in geometry any shorter way than that of the elements, and he answered that there was no royal road to geometry[4]. He is then younger than the pupils of Plato but older than Eratosthenes and Archimedes; for the latter were contemporary with one another, as Eratosthenes somewhere says."

This passage shows that even Proclus had no direct knowledge of Euclid's birthplace or of the date of his birth or death. He proceeds by inference. Since Archimedes lived just after the first

---

[1] Proclus, ed. Friedlein, p. 68, 6—20.

[2] The word γέγονε must apparently mean "flourished," as Heiberg understands it (*Litterargeschichtliche Studien über Euklid*, 1882, p. 26), not "was born," as Hankel took it: otherwise part of Proclus' argument would lose its cogency.

[3] So Heiberg understands ἐπιβαλὼν τῷ πρώτῳ (sc. Πτολεμαίῳ). Friedlein's text has καί between ἐπιβαλὼν and τῷ πρώτῳ; and it is right to remark that another reading is καὶ ἐν τῷ πρώτῳ (without ἐπιβαλών) which has been translated "in his first *book*," by which is understood *On the Sphere and Cylinder* I., where (1) in Prop. 2 are the words "let *BC* be made equal to *D by the second* (proposition) *of the first* of Euclid's (books)," and (2) in Prop. 6 the words "For these things are handed down in the Elements" (without the name of Euclid). Heiberg thinks the former passage is referred to, and that Proclus must therefore have had before him the words "by the second of the first of Euclid": a fair proof that they are genuine, though in themselves they would be somewhat suspicious.

[4] The same story is told in Stobaeus, *Ecl.* (II. p. 228, 30, ed. Wachsmuth) about Alexander and Menaechmus. Alexander is represented as having asked Menaechmus to teach him geometry concisely, but he replied: "O king, through the country there are royal roads and roads for common citizens, but in geometry there is one road for all."

Ptolemy, and Archimedes mentions Euclid, while there is an anecdote about *some* Ptolemy and Euclid, *therefore* Euclid lived in the time of the first Ptolemy.

We may infer then from Proclus that Euclid was intermediate between the first pupils of Plato and Archimedes. Now Plato died in 347/6, Archimedes lived 287–212, Eratosthenes *c.* 284–204 B.C. Thus Euclid must have flourished *c.* 300 B.C., which date agrees well with the fact that Ptolemy reigned from 306 to 283 B.C.

It is most probable that Euclid received his mathematical training in Athens from the pupils of Plato; for most of the geometers who could have taught him were of that school, and it was in Athens that the older writers of elements, and the other mathematicians on whose works Euclid's *Elements* depend, had lived and taught. He may himself have been a Platonist, but this does not follow from the statements of Proclus on the subject. Proclus says namely that he was of the school of Plato and in close touch with that philosophy[1]. But this was only an attempt of a New Platonist to connect Euclid with his philosophy, as is clear from the next words in the same sentence, " for which reason also he set before himself, as the end of the whole Elements, the construction of the so-called Platonic figures." It is evident that it was only an idea of Proclus' own to infer that Euclid was a Platonist because his *Elements* end with the investigation of the five regular solids, since a later passage shows him hard put to it to reconcile the view that the construction of the five regular solids was the end and aim of the *Elements* with the obvious fact that they were intended to supply a foundation for the study of geometry in general, " to make perfect the understanding of the learner in regard to the whole of geometry[2]." To get out of the difficulty he says[3] that, if one should ask him what was the aim (σκοπός) of the treatise, he would reply by making a distinction between Euclid's intentions (1) as regards the subjects with which his investigations are concerned, (2) as regards the learner, and would say as regards (1) that " the whole of the geometer's argument is concerned with the cosmic figures." This latter statement is obviously incorrect. It is true that Euclid's *Elements* end with the construction of the five regular solids; but the planimetrical portion has no direct relation to them, and the arithmetical no relation at all; the propositions about them are merely the conclusion of the stereometrical division of the work.

One thing is however certain, namely that Euclid taught, and founded a school, at Alexandria. This is clear from the remark of Pappus about Apollonius[4]: " he spent a very long time with the pupils of Euclid at Alexandria, and it was thus that he acquired such a scientific habit of thought."

It is in the same passage that Pappus makes a remark which might, to an unwary reader, seem to throw some light on the

---

[1] Proclus, p. 68, 20, καὶ τῇ προαιρέσει δὲ Πλατωνικός ἐστι καὶ τῇ φιλοσοφίᾳ ταύτῃ οἰκεῖος.
[2] *ibid.* p. 71, 8.   [3] *ibid.* p. 70, 19 sqq.
[4] Pappus, VII. p. 678, 10—12, συσχολάσας τοῖς ὑπὸ Εὐκλείδου μαθηταῖς ἐν Ἀλεξανδρείᾳ πλεῖστον χρόνον, ὅθεν ἔσχε καὶ τὴν τοιαύτην ἕξιν οὐκ ἀμαθῆ.

personality of Euclid.   He is speaking about Apollonius' preface
to the first book of his *Conics*, where he says that Euclid had not
completely worked out the synthesis of the "three- and four-line
locus," which in fact was not possible without some theorems first
discovered by himself.   Pappus says on this[1]: "Now Euclid—
regarding Aristaeus as deserving credit for the discoveries he had
already made in conics, and without anticipating him or wishing to
construct anew the same system (such was his scrupulous fairness and
his exemplary kindliness towards all who could advance mathematical
science to however small an extent), being moreover in no wise con-
tentious and, though exact, yet no braggart like the other [Apollonius]
—wrote so much about the locus as was possible by means of the
conics of Aristaeus, without claiming completeness for his demonstrations."   It is however evident, when the passage is examined in its
context, that Pappus is not following any tradition in giving this
account of Euclid: he was offended by the terms of Apollonius'
reference to Euclid, which seemed to him unjust, and he drew a
fancy picture of Euclid in order to show Apollonius in a relatively
unfavourable light.

Another story is told of Euclid which one would like to believe true.
According to Stobaeus[2], " some one who had begun to read geometry
with Euclid, when he had learnt the first theorem, asked Euclid, ' But
what shall I get by learning these things ?' Euclid called his slave
and said 'Give him threepence, since he must make gain out of what
he learns.'"

In the middle ages most translators and editors spoke of Euclid
as Euclid *of Megara*.   This description arose out of a confusion
between our Euclid and the philosopher Euclid of Megara who lived
about 400 B.C.   The first trace of this confusion appears in Valerius
Maximus (in the time of Tiberius) who says[3] that Plato, on being
appealed to for a solution of the problem of doubling the cubical
altar, sent the inquirers to "Euclid the geometer."   There is no doubt
about the reading, although an early commentator on Valerius
Maximus wanted to correct "Eucliden" into "*Eudoxum*," and this
correction is clearly right.   But, if Valerius Maximus took Euclid the
geometer for a contemporary of Plato, it could only be through
confusing him with Euclid of Megara.   The first specific reference to
Euclid as Euclid of Megara belongs to the 14th century, occurring in
the ὑπομνηματισμοί of Theodorus Metochita (d. 1332) who speaks of
"Euclid of Megara, the Socratic philosopher, contemporary of Plato,"
as the author of treatises on plane and solid geometry, data, optics
etc.: and a Paris MS. of the 14th century has "Euclidis philosophi
Socratici liber elementorum."   The misunderstanding was general
in the period from Campanus' translation (Venice 1482) to those of
Tartaglia (Venice 1565) and Candalla (Paris 1566).   But one
Constantinus Lascaris (d. about 1493) had already made the proper

[1] Pappus, VII. pp. 676, 25—678, 6. Hultsch, it is true, brackets the whole passage
pp. 676, 25—678, 15, but apparently on the ground of the diction only.
[2] Stobaeus, *l.c.*   [3] VIII. 12, ext. 1.

distinction by saying of our Euclid that "he was different from him of Megara of whom Laertius wrote, and who wrote dialogues"[1]; and to Commandinus belongs the credit of being the first translator[2] to put the matter beyond doubt : " Let us then free a number of people from the error by which they have been induced to believe that our Euclid is the same as the philosopher of Megara " etc.

Another idea, that Euclid was born at Gela in Sicily, is due to tne same confusion, being based on Diogenes Laertius' description[3] of the philosopher Euclid as being "of Megara, or, according to some, of Gela, as Alexander says in the Διαδοχαί."

In view of the poverty of Greek tradition on the subject even as early as the time of Proclus (410–485 A.D.), we must necessarily take *cum grano* the apparently circumstantial accounts of Euclid given by Arabian authors ; and indeed the origin of their stories can be explained as the result (1) of the Arabian tendency to romance, and (2) of misunderstandings.

We read[4] that " Euclid, son of Naucrates, grandson of Zenarchus[5], called the author of geometry, a philosopher of somewhat ancient date, a Greek by nationality domiciled at Damascus, born at Tyre, most learned in the science of geometry, published a most excellent and most useful work entitled the foundation or elements of geometry, a subject in which no more general treatise existed before among the Greeks : nay, there was no one even of later date who did not walk in his footsteps and frankly profess his doctrine.   Hence also Greek, Roman and Arabian geometers not a few, who undertook the task of illustrating this work, published commentaries, scholia, and notes upon it, and made an abridgment of the work itself.   For this reason the Greek philosophers used to post up on the doors of their schools the well-known notice : ' Let no one come to our school, who has not first learned the elements of Euclid.' "   The details at the beginning of this extract cannot be derived from Greek sources, for even Proclus did not know anything about Euclid's father, while it was not the Greek habit to record the names of grandfathers, as the Arabians commonly did.   Damascus and Tyre were no doubt brought in to gratify a desire which the Arabians always showed to connect famous Greeks in some way or other with the East.   Thus Naṣīraddīn, the translator of the *Elements*, who was of Ṭūs in Khurāsān, actually makes Euclid out to have been "Thusinus" also[6].   The readiness of the Arabians to run away with an idea is illustrated by the last words

[1] Letter to Fernandus Acuna, printed in Maurolycus, *Historia Siciliae*, fol. 21 r. (see Heiberg, *Euklid-Studien*, pp. 22—3, 25).

[2] Preface to translation (Pisauri, 1572).

[3] Diog. L. II. 106, p. 58 ed. Cobet.

[4] Casiri, *Bibliotheca Arabico-Hispana Escurialensis*, I. p. 339.   Casiri's source is al-Qifṭī (d. 1248), the author of the *Ta'rīkh al-Ḥukamā*, a collection of biographies of philosophers, mathematicians, astronomers etc.

[5] The *Fihrist* says "son of Naucrates, the son of Berenice (?) " (see Suter's translation in *Abhandlungen zur Gesch. d. Math.* VI. Heft, 1892, p. 16).

[6] The same predilection made the Arabs describe Pythagoras as a pupil of the wise Salomo, Hipparchus as the exponent of Chaldaean philosophy or as the Chaldaean, Archimedes as an Egyptian etc. (Ḥājī Khalfa, *Lexicon Bibliographicum*, and Casiri).

of the extract. Everyone knows the story of Plato's inscription over the porch of the Academy: "let no one unversed in geometry enter my doors"; the Arab turned geometry into *Euclid's* geometry, and told the story of Greek philosophers in general and "*their* Academies."

Equally remarkable are the Arabian accounts of the relation of Euclid and Apollonius[1]. According to them the *Elements* were originally written, not by Euclid, but by a man whose name was Apollonius, a carpenter, who wrote the work in 15 books or sections[2]. In the course of time some of the work was lost and the rest became disarranged, so that one of the kings at Alexandria who desired to study geometry and to master this treatise in particular first questioned about it certain learned men who visited him and then sent for Euclid who was at that time famous as a geometer, and asked him to revise and complete the work and reduce it to order. Euclid then re-wrote it in 13 books which were thereafter known by his name. (According to another version Euclid composed the 13 books out of commentaries which he had published on two books of Apollonius on conics and out of introductory matter added to the doctrine of the five regular solids.) To the thirteen books were added two more books, the work of others (though some attribute these also to Euclid) which contain several things not mentioned by Apollonius. According to another version Hypsicles, a pupil of Euclid at Alexandria, offered to the king and published Books XIV. and XV., it being also stated that Hypsicles had "discovered" the books, by which it appears to be suggested that Hypsicles had edited them from materials left by Euclid.

We observe here the correct statement that Books XIV. and XV. were not written by Euclid, but along with it the incorrect information that Hypsicles, the author of Book XIV., wrote Book XV. also.

The whole of the fable about Apollonius having preceded Euclid and having written the *Elements* appears to have been evolved out of the preface to Book XIV. by Hypsicles, and in this way; the Book must in early times have been attributed to Euclid, and the inference based upon this assumption was left uncorrected afterwards when it was recognised that Hypsicles was the author. The preface is worth quoting:

"Basilides of Tyre, O Protarchus, when he came to Alexandria and met my father, spent the greater part of his sojourn with him on account of their common interest in mathematics. And once, when

---

[1] The authorities for these statements quoted by Casiri and Ḥājī Khalfa are al-Kindī's tract *de instituto libri Euclidis* (al-Kindī died about 873) and a commentary by Qāḍīzāde ar-Rūmī (d. about 1440) on a book called *Ashkāl at-ta' sīs* (fundamental propositions) by Ashraf Shamsaddīn as-Samarqandī (*c.* 1276) consisting of elucidations of 35 propositions selected from the first books of Euclid. Naṣīraddīn likewise says that Euclid cut out two of 15 books of elements then existing and published the rest under his own name. According to Qāḍīzāde the king heard that there was a celebrated geometer named Euclid at *Tyre*: Naṣīraddīn says that he sent for Euclid of Ṭūs.

[2] So says the *Fihrist*. Suter (*op. cit.* p. 49) thinks that the author of the *Fihrist* did not suppose Apollonius *of Perga* to be the writer of the *Elements*, as later Arabian authorities did, but that he distinguished another Apollonius whom he calls "a carpenter." Suter's argument is based on the fact that the *Fihrist's* article on Apollonius (of Perga) says nothing of the *Elements*, and that it gives the three great mathematicians, Euclid, Archimedes and Apollonius, in the correct chronological order.

examining the treatise written by Apollonius about the comparison between the dodecahedron and the icosahedron inscribed in the same sphere, (showing) what ratio they have to one another, they thought that Apollonius had not expounded this matter properly, and accordingly they emended the exposition, as I was able to learn from my father. And I myself, later, fell in with another book published by Apollonius, containing a demonstration relating to the subject, and I was greatly interested in the investigation of the problem. The book published by Apollonius is accessible to all— for it has a large circulation, having apparently been carefully written out later—but I decided to send you the comments which seem to me to be necessary, for you will through your proficiency in mathematics in general and in geometry in particular form an expert judgment on what I am about to say, and you will lend a kindly ear to my disquisition for the sake of your friendship to my father and your goodwill to me."

The idea that Apollonius preceded Euclid must evidently have been derived from the passage just quoted. It explains other things besides. Basilides must have been confused with βασιλεύς, and we have a probable explanation of the "Alexandrian king," and of the "learned men who visited" Alexandria. It is possible also that in the "Tyrian" of Hypsicles' preface we have the origin of the notion that Euclid was born in Tyre. These inferences argue, no doubt, very defective knowledge of Greek: but we could expect no better from those who took the *Organon* of Aristotle to be "instrumentum musicum pneumaticum," and who explained the name of Euclid, which they variously pronounced as *Uclides* or *Icludes*, to be compounded of *Ucli* a key, and *Dis* a measure, or, as some say, geometry, so that *Uclides* is equivalent to the *key of geometry*!

Lastly the alternative version, given in brackets above, which says that Euclid made the *Elements* out of commentaries which he wrote on two books of Apollonius on conics and prolegomena added to the doctrine of the five solids, seems to have arisen, through a like confusion, out of a later passage[1] in Hypsicles' Book XIV.: "And this is expounded by Aristaeus in the book entitled 'Comparison of the five figures,' and by Apollonius in the second edition of his comparison of the dodecahedron with the icosahedron." The "doctrine of the five solids" in the Arabic must be the "Comparison of the five figures" in the passage of Hypsicles, for nowhere else have we any information about a work bearing this title, nor can the Arabians have had. The reference to the *two books* of Apollonius on *conics* will then be the result of mixing up the fact that Apollonius wrote a book on conics with the *second edition* of the other work mentioned by Hypsicles. We do not find elsewhere in Arabian authors any mention of a commentary by Euclid on Apollonius and Aristaeus: so that the story in the passage quoted is really no more than a variation of the fable that the *Elements* were the work of Apollonius.

---

[1] Heiberg's Euclid, vol. v. p. 6.

CHAPTER II.

EUCLID'S OTHER WORKS.

In giving a list of the Euclidean treatises other than the *Elements*, I shall be brief: for fuller accounts of them, or speculations with regard to them, reference should be made to the standard histories of mathematics[1].

I will take first the works which are mentioned by Greek authors.

1. The *Pseudaria*.

I mention this first because Proclus refers to it in the general remarks in praise of the *Elements* which he gives immediately after the mention of Euclid in his summary. He says[2]: "But, inasmuch as many things, while appearing to rest on truth and to follow from scientific principles, really tend to lead one astray from the principles and deceive the more superficial minds, he has handed down methods for the discriminative understanding of these things as well, by the use of which methods we shall be able to give beginners in this study practice in the discovery of paralogisms, and to avoid being misled. This treatise, by which he puts this machinery in our hands, he entitled (the book) of Pseudaria, enumerating in order their various kinds, exercising our intelligence in each case by theorems of all sorts, setting the true side by side with the false, and combining the refutation of error with practical illustration. This book then is by way of cathartic and exercise, while the Elements contain the irrefragable and complete guide to the actual scientific investigation of the subjects of geometry."

The book is considered to be irreparably lost. We may conclude however from the connexion of it with the *Elements* and the reference to its usefulness for beginners that it did not go outside the domain of elementary geometry[3].

---

[1] See, for example, Loria, *Le scienze esatte nell' antica Grecia*, 1914, pp. 245—268; T. L. Heath, *History of Greek Mathematics*, 1921, I. pp. 421—446. Cf. Heiberg, *Litterargeschichtliche Studien über Euklid*, pp. 36—153; *Euclidis opera omnia*, ed. Heiberg and Menge, Vols. VI.—VIII.

[2] Proclus, p. 70, 1—18.

[3] Heiberg points out that Alexander Aphrodisiensis appears to allude to the work in his commentary on Aristotle's *Sophistici Elenchi* (fol. 25 *b*): "Not only those (ἔλεγχοι) which do not start from the principles of the science under which the problem is classed...but also those which do start from the proper principles of the science but in some respect admit a paralogism, e.g. the *Pseudographemata* of Euclid." Tannery (*Bull. des sciences math. et astr.* 2ᵉ Série, VI., 1882, 1ᵉʳᵉ Partie, p. 147) conjectures that it may be from this treatise that the same commentator got his information about the quadratures of the circle by Antiphon and

2. The *Data*.

The *Data* (δεδομένα) are included by Pappus in the *Treasury of Analysis* (τόπος ἀναλυόμενος), and he describes their contents[1]. They are still concerned with elementary geometry, though forming part of the introduction to higher analysis. Their form is that of propositions proving that, if certain things in a figure are given (in magnitude, in species, etc.), something else is given. The subject-matter is much the same as that of the planimetrical books of the *Elements*, to which the *Data* are often supplementary. We shall see this later when we come to compare the propositions in the *Elements* which give us the means of solving the general quadratic equation with the corresponding propositions of the *Data* which give the solution. The *Data* may in fact be regarded as elementary exercises in analysis.

It is not necessary to go more closely into the contents, as we have the full Greek text and the commentary by Marinus newly edited by Menge and therefore easily accessible[2].

3. The book *On divisions (of figures)*.

This work (περὶ διαιρέσεων βιβλίον) is mentioned by Proclus[3]. In one place he is speaking of the conception or definition (λόγος) of *figure*, and of the divisibility of a figure into others differing from it in kind; and he adds: "For the circle is divisible into parts unlike in definition or notion (ἀνόμοια τῷ λόγῳ), and so is each of the rectilineal figures; this is in fact the business of the writer of the Elements in his Divisions, where he divides given figures, in one case into like figures, and in another into unlike[4]." "Like" and "unlike" here mean, not "similar" and "dissimilar" in the technical sense, but "like" or "unlike *in definition* or *notion*" (λόγῳ): thus to divide a triangle into triangles would be to divide it into "like" figures, to divide a triangle into a triangle and a quadrilateral would be to divide it into "unlike" figures.

The treatise is lost in Greek but has been discovered in the Arabic. First John Dee discovered a treatise *De divisionibus* by one Muhammad Bagdadinus[5] and handed over a copy of it (in Latin) in 1563 to Commandinus, who published it, in Dee's name and his own, in 1570[6]. Dee did not himself translate the tract from the Arabic; he

---

Bryson, to say nothing of the lunules of Hippocrates. I think however that there is an objection to this theory so far as regards Bryson; for Alexander distinctly says that Bryson's quadrature did *not* start from the proper principles of geometry, but from some principles more general.

[1] Pappus, VII. p. 638.

[2] Vol. VI. in the Teubner edition of *Euclidis opera omnia* by Heiberg and Menge. A translation of the *Data* is also included in Simson's Euclid (though naturally his text left much to be desired).

[3] Proclus, p. 69, 4.   [4] *ibid.* 144, 22—26.

[5] Steinschneider places him in the 10th c. H. Suter (*Bibliotheca Mathematica*, IV₃, 1903, pp. 24, 27) identifies him with Abū (Bekr) Muh. b. 'Abdalbāqī al-Baġdādī, Qāḍī (Judge) of Māristān (*circa* 1070-1141), to whom he also attributes the *Liber judei* (? judicis) *super decimum Euclidis* translated by Gherard of Cremona.

[6] *De superficierum divisionibus liber Machometo Bagdadino adscriptus, nunc primum Ioannis Dee Londinensis et Federici Commandini Urbinatis opera in lucem editus*, Pisauri, 1570, afterwards included in Gregory's Euclid (Oxford, 1703).

found it in Latin in a MS. which was then in his own possession but was about 20 years afterwards stolen or destroyed in an attack by a mob on his house at Mortlake[1]. Dee, in his preface addressed to Commandinus, says nothing of his having *translated* the book, but only remarks that the very illegible MS. had caused him much trouble and (in a later passage) speaks of "the actual, very ancient, copy from which I *wrote out*..." (in ipso unde descripsi vetustissimo exemplari). The Latin translation of this tract from the Arabic was probably made by Gherard of Cremona (1114–1187), among the list of whose numerous translations a "liber divisionum" occurs. The Arabic original cannot have been a direct translation from Euclid, and probably was not even a direct adaptation of it; it contains mistakes and unmathematical expressions, and moreover does not contain the propositions about the division of a circle alluded to by Proclus. Hence it can scarcely have contained more than a fragment of Euclid's work.

But Woepcke found in a MS. at Paris a treatise in Arabic on the division of figures, which he translated and published in 1851[2]. It is expressly attributed to Euclid in the MS. and corresponds to the description of it by Proclus. Generally speaking, the divisions are divisions into figures of the same kind as the original figures, e.g. of triangles into triangles; but there are also divisions into "unlike" figures, e.g. that of a triangle by a straight line parallel to the base. The missing propositions about the division of a circle are also here: "to divide into two equal parts a given figure bounded by an arc of a circle and two straight lines including a given angle" and "to draw in a given circle two parallel straight lines cutting off a certain part of the circle." Unfortunately the proofs are given of only four propositions (including the two last mentioned) out of 36, because the Arabic translator found them too easy and omitted them. To illustrate the character of the problems dealt with I need only take one more example: "To cut off a certain fraction from a (parallel-) trapezium by a straight line which passes through a given point lying inside or outside the trapezium but so that a straight line can be drawn through it cutting both the parallel sides of the trapezium." The genuineness of the treatise edited by Woepcke is attested by the facts that the four proofs which remain are elegant and depend on propositions in the *Elements*, and that there is a lemma with a true Greek ring: "to apply to a straight line a rectangle equal to the rectangle contained by *AB, AC and deficient by a square.*" Moreover the treatise is no fragment, but finishes with the words "end of the treatise," and is a well-ordered and compact whole. Hence we may safely conclude that Woepcke's is not only Euclid's own work but the whole of it. A restoration of the work, with proofs, was attempted by Ofterdinger[3], who however does not give Woepcke's props. 30, 31, 34, 35, 36. We have now a satisfactory restoration, with ample notes

---

[1] R. C. Archibald, *Euclid's Book on the Division of Figures with a restoration based on Woepcke's text and on the Practica geometriae of Leonardo Pisano*, Cambridge, 1915, pp. 4—9.

[2] *Journal Asiatique*, 1851, p. 233 sqq.

[3] L. F. Ofterdinger, *Beiträge zur Wiederherstellung der Schrift des Euklides über die Theilung der Figuren*, Ulm, 1853.

and an introduction, by R. C. Archibald, who used for the purpose Woepcke's text and a section of Leonardo of Pisa's *Practica geometriae* (1220)[1].

4. The *Porisms*.

It is not possible to give in this place any account of the controversies about the contents and significance of the three lost books of Porisms, or of the important attempts by Robert Simson and Chasles to restore the work. These may be said to form a whole literature, references to which will be found most abundantly given by Heiberg and Loria, the former of whom has treated the subject from the philological point of view, most exhaustively, while the latter, founding himself generally on Heiberg, has added useful details, from the mathematical side, relating to the attempted restorations, etc.[2] It must suffice here to give an extract from the only original source of information about the nature and contents of the *Porisms*, namely Pappus[3]. In his general preface about the books composing the *Treasury of Analysis* (τόπος ἀναλυόμενος) he says :

"After the Tangencies (of Apollonius) come, in three books, the Porisms of Euclid, [in the view of many] a collection most ingeniously devised for the analysis of the more weighty problems, [and] although nature presents an unlimited number of such porisms[4], [they have added nothing to what was written originally by Euclid, except that some before my time have shown their want of taste by adding to a few (of the propositions) second proofs, each (proposition) admitting of a definite number of demonstrations, as we have shown, and Euclid having given one for each, namely that which is the most lucid. These porisms embody a theory subtle, natural, necessary, and of considerable generality, which is fascinating to those who can see and produce results].

"Now all the varieties of porisms belong, neither to theorems nor problems, but to a species occupying a sort of intermediate position [so that their enunciations can be formed like those of either theorems or problems], the result being that, of the great number of geometers, some regarded them as of the class of theorems, and others of problems, looking only to the form of the proposition. But that the ancients knew better the difference between these three things is clear from the definitions. For they said that a theorem is that which is proposed with a view to the demonstration of the very thing proposed, a problem that which is thrown out with a view to the construction of the very thing proposed, and a porism that which is proposed with a view to the producing of the very thing proposed. [But this definition of the porism was changed by the more recent writers who could not produce everything, but used these elements

---

[1] There is a remarkable similarity between the propositions of Woepcke's text and those of Leonardo, suggesting that Leonardo may have had before him a translation (perhaps by Gherard of Cremona) of the Arabic tract.

[2] Heiberg, *Euklid-Studien*, pp. 56—79, and Loria, *op. cit.*, pp. 253—265.

[3] Pappus, ed. Hultsch, VII. pp. 648—660. I put in square brackets the words bracketed by Hultsch.

[4] I adopt Heiberg's reading of a comma here instead of a full stop.

and proved only the fact that that which is sought really exists, but did not produce it[1] and were accordingly confuted by the definition and the whole doctrine. They based their definition on an incidental characteristic, thus: A porism is that which falls short of a locus-theorem in respect of its hypothesis[2]. Of this kind of porisms loci are a species, and they abound in the Treasury of Analysis; but this species has been collected, named and handed down separately from the porisms, because it is more widely diffused than the other species]. But it has further become characteristic of porisms that, owing to their complication, the enunciations are put in a contracted form, much being by usage left to be understood; so that many geometers understand them only in a partial way and are ignorant of the more essential features of their contents.

"[Now to comprehend a number of propositions in one enunciation is by no means easy in these porisms, because Euclid himself has not in fact given many of each species, but chosen, for examples, one or a few out of a great multitude[3]. But at the beginning of the first book he has given some propositions, to the number of ten, of one species, namely that more fruitful species consisting of loci.] Consequently, finding that these admitted of being comprehended in one enunciation, we have set it out thus:

If, in a system of four straight lines[4] which cut each other two and two, three points on one straight line be given while the rest except one lie on different straight lines given in position, the remaining point also will lie on a straight line given in position[5].

---

[1] Heiberg points out that Props. 5—9 of Archimedes' treatise *On Spirals* are porisms in this sense. To take Prop. 5 as an example, *DBF* is a tangent to a circle with centre *K*. It is then possible, says Archimedes, to draw a straight line *KHF*, meeting the circumference in *H* and the tangent in *F*, such that

$$FH : HK < (\text{arc } BH) : c,$$

where *c* is the circumference of *any* circle. To prove this he assumes the following construction. *E* being any straight line greater than *c*, he says: let *KG* be parallel to *DF*, "and let the line *GH* equal to *E* be placed *verging* to the point *B*." Archimedes must of course have known how to effect this construction, which requires conics. But that it is *possible* requires very little argument, for if we draw any straight line *BHG* meeting the circle in *H* and *KG* in *G*, it is obvious that as *G* moves away from *C*, *HG* becomes greater and greater and may be made as great as we please. The "later writers" would no doubt have contented themselves with this consideration without actually *constructing HG*.

[2] As Heiberg says, this translation is made certain by a preceding passage of Pappus (p. 648, 1—3) where he compares two enunciations, the latter of which "falls short of the former in *hypothesis* but goes beyond it in *requirement*." E.g. the first enunciation requiring us, given three circles, to draw a circle touching all three, the second may require us, given only *two* circles (one less datum), to draw a circle touching them and *of a given size* (an extra requirement).

[3] I translate Heiberg's reading with a full stop here followed by πρὸς ἀρχῇ δὲ ὅμως [πρὸς ἀρχὴν (δεδομένον) Hultsch] τοῦ πρώτου βιβλίου....

[4] The four straight lines are described in the text as (the sides) ὑπτίου ἢ παρυπτίου, i.e. sides of two sorts of quadrilaterals which Simson tries to explain (see p. 120 of the *Index Graecitatis* of Hultsch's edition of Pappus).

[5] In other words (Chasles, p. 23; Loria, p. 256), if a triangle be so deformed that each of its sides turns about one of three points in a straight line, and two of its vertices lie on two straight lines given in position, the third vertex will also lie on a straight line.

"This has only been enunciated of four straight lines, of which not more than two pass through the same point, but it is not known (to most people) that it is true of any assigned number of straight lines if enunciated thus:

If any number of straight lines cut one another, not more than two (passing) through the same point, and all the points (of intersection situated) on one of them be given, and if each of those which are on another (of them) lie on a straight line given in position—

or still more generally thus:

if any number of straight lines cut one another, not more than two (passing) through the same point, and all the points (of intersection situated) on one of them be given, while of the other points of intersection in multitude equal to a triangular number a number corresponding to the side of this triangular number lie respectively on straight lines given in position, provided that of these latter points no three are at the angular points of a triangle (*sc.* having for sides three of the given straight lines)—each of the remaining points will lie on a straight line given in position[1].

"It is probable that the writer of the Elements was not unaware of this but that he only set out the principle; and he seems, in the case of all the porisms, to have laid down the principles and the seed only [of many important things], the kinds of which should be distinguished according to the differences, not of their hypotheses, but of the results and the things sought. [All the hypotheses are different from one another because they are entirely special, but each of the results and things sought, being one and the same, follow from many different hypotheses.]

"We must then in the first book distinguish the following kinds of things sought:

"At the beginning of the book[2] is this proposition:

I. '*If from two given points straight lines be drawn meeting on a straight line given in position, and one cut off from a straight line given in position (a segment measured) to a given point on it, the other will also cut off from another (straight line a segment) having to the first a given ratio.*'

"Following on this (we have to prove)

II. that such and such a point lies on a straight line given in position;

III. that the ratio of such and such a pair of straight lines is given;"

etc. etc. (up to XXIX.).

"The three books of the porisms contain 38 lemmas; of the theorems themselves there are 171."

---

[1] Loria (p. 256, *n.* 3) gives the meaning of this as follows, pointing out that Simson was the discoverer of it: "If a complete *n*-lateral be deformed so that its sides respectively turn about *n* points on a straight line, and $(n-1)$ of its $n(n-1)/2$ vertices move on as many straight lines, the other $(n-1)(n-2)/2$ of its vertices likewise move on as many straight lines: but it is necessary that it should be impossible to form with the $(n-1)$ vertices any triangle having for sides the sides of the polygon."

[2] Reading, with Heiberg, τοῦ βιβλίου [τοῦ ς' Hultsch].

Pappus further gives lemmas to the *Porisms* (pp. 866—918, ed. Hultsch).

With Pappus' account of Porisms must be compared the passages of Proclus on the same subject. Proclus distinguishes two senses in which the word πόρισμα is used. The first is that of *corollary* where something appears as an incidental result of a proposition, obtained without trouble or special seeking, a sort of bonus which the investigation has presented us with[1]. The other sense is that of Euclid's *Porisms*[2]. In this sense[3] "*porism* is the name given to things which are sought, but need some finding and are neither pure bringing into existence nor simple theoretic argument. For (to prove) that the angles at the base of isosceles triangles are equal is a matter of theoretic argument, and it is with reference to things existing that such knowledge is (obtained). But to bisect an angle, to construct a triangle, to cut off, or to place—all these things demand the making of something; and to find the centre of a given circle, or to find the greatest common measure of two given commensurable magnitudes, or the like, is in some sort between theorems and problems. For in these cases there is no bringing into existence of the things sought, but finding of them, nor is the procedure purely theoretic. For it is necessary to bring that which is sought into view and exhibit it to the eye. Such are the porisms which Euclid wrote, and arranged in three books of Porisms."

Proclus' definition thus agrees well enough with the first, "older," definition of Pappus. A porism occupies a place between a theorem and a problem: it deals with something already *existing*, as a theorem does, but has to *find* it (e.g. the centre of a circle), and, as a certain operation is therefore necessary, it partakes to that extent of the nature of a problem, which requires us to construct or produce something not previously existing. Thus, besides III. 1 of the *Elements* and X. 3, 4 mentioned by Proclus, the following propositions are real porisms: III. 25, VI. 11—13, VII. 33, 34, 36, 39, VIII. 2, 4, X. 10, XIII. 18. Similarly in Archimedes *On the Sphere and Cylinder* I. 2—6 might be called porisms.

The enunciation given by Pappus as comprehending ten of Euclid's propositions may not reproduce the *form* of Euclid's enunciations; but, comparing the result to be proved, that certain points lie on straight lines given in position, with the *class* indicated by II. above, where the question is of such and such a point lying on a straight line given in position, and with other classes, e.g. (V.) that such and such a line is given in position, (VI.) that such and such a line verges to a given point, (XXVII.) that there exists a given point such that straight lines drawn from it to such and such (circles) will contain a triangle given in species, we may conclude that a usual form of a porism was " to prove that it is possible to find a point with such and such a property"

---

[1] Proclus, pp. 212, 14; 301, 22.
[2] *ibid.* p. 212, 12. "The term porism is used of certain problems, like the *Porisms* written by Euclid."
[3] *ibid.* pp. 301, 25 sqq.

or "a straight line on which lie all the points satisfying given conditions" etc.

Simson defined a porism thus: "Porisma est propositio in qua proponitur demonstrare rem aliquam, vel plures datas esse, cui, vel quibus, ut et cuilibet ex rebus innumeris, non quidem datis, sed quae ad ea quae data sunt eandem habent relationem, convenire ostendendum est affectionem quandam communem in propositione descriptam[1]."

From the above it is easy to understand Pappus' statement that *loci* constitute a large class of porisms. A *locus* is well defined by Simson thus: "Locus est propositio in qua propositum est datam esse demonstrare, vel invenire lineam aut superficiem cuius quodlibet punctum, vel superficiem in qua quaelibet linea data lege descripta, communem quandam habet proprietatem in propositione descriptam."

Heiberg cites an excellent instance of a *locus* which is a *porism*, namely the following proposition quoted by Eutocius[2] from the *Plane Loci* of Apollonius:

"Given two points in a plane, and a ratio between unequal straight lines, it is possible to draw, in the plane, a circle such that the straight lines drawn from the given points to meet on the circumference of the circle have (to one another) a ratio the same as the given ratio."

A difficult point, however, arises on the passage of Pappus, which says that a porism is "that which, in respect of its hypothesis, falls short of a locus-theorem" ($\tau o \pi \iota \kappa o \hat{v} \, \theta \epsilon \omega \rho \acute{\eta} \mu a \tau o \varsigma$). Heiberg explains it by comparing the porism from Apollonius' *Plane Loci* just given with Pappus' enunciation of the same thing, to the effect that, if from two given points two straight lines be drawn meeting in a point, and these straight lines have to one another a given ratio, the point will lie on either a straight line or a circumference of a circle given in position. Heiberg observes that in this latter enunciation something is taken into the hypothesis which was not in the hypothesis of the enunciation of the porism, viz. "that the ratio of the straight lines is the same." I confess this does not seem to me satisfactory: for there is no real difference between the enunciations, and the supposed difference in hypothesis is very like playing with words. Chasles says: "*Ce qui constitue le porisme est ce qui manque à l'hypothèse d'un théorème local* (en d'autres termes, le porisme est inférieur, par l'hypothèse, au théorème local; c'est-à-dire que quand quelques parties d'une proposition locale n'ont pas dans l'énoncé la détermination qui leur est propre, cette proposition cesse d'être regardée comme un théorème et devient un porisme)." But the subject still seems to require further elucidation.

While there is so much that is obscure, it seems certain (1) that the *Porisms* were distinctly part of higher geometry and not of elementary

---

[1] This was thus expressed by Chasles: "Le porisme est une proposition dans laquelle on demande de démontrer qu'une chose ou plusieurs choses sont *données*, qui, ainsi que l'une quelconque d'une infinité d'autres choses non données, mais dont chacune est avec des choses données dans une même relation, ont une certaine propriété commune, décrite dans la proposition."

[2] Commentary on Apollonius' *Conics* (vol. II. p. 180, ed. Heiberg).

geometry, (2) that they contained propositions belonging to the modern theory of transversals and to projective geometry. It should be remembered too that it was in the course of his researches on this subject that Chasles was led to the idea of *anharmonic ratios*.

Lastly, allusion should be made to the theory of Zeuthen[1] on the subject of the porisms. He observes that the only porism of which Pappus gives the complete enunciation, " If from two given points straight lines be drawn meeting on a straight line given in position, and one cut off from a straight line given in position (a segment measured) towards a given point on it, the other will also cut off from another (straight line a segment) bearing to the first a given ratio," is also true if there be substituted for the first given straight line a conic regarded as the " locus with respect to four lines," and that this extended porism can be used for completing Apollonius' exposition of that locus. Zeuthen concludes that the *Porisms* were in part by-products of the theory of conics and in part auxiliary means for the study of conics, and that Euclid called them by the same name as that applied to corollaries because they were corollaries with respect to conics. But there appears to be no evidence to confirm this conjecture.

5.　The *Surface-loci* (τόποι πρὸς ἐπιφανείᾳ).

The two books on this subject are mentioned by Pappus as part of the *Treasury of Analysis*[2]. As the other works in the list which were on plane subjects dealt only with straight lines, circles, and conic sections, it is *a priori* likely that among the loci in this treatise (loci which are surfaces) were included such loci as were cones, cylinders and spheres. Beyond this all is conjecture based on two lemmas given by Pappus in connexion with the treatise.

(1)　The first of these lemmas[3] and the figure attached to it are not satisfactory as they stand, but a possible restoration is indicated by Tannery[4]. If the latter is right, it suggests that one of the loci contained all the points on the elliptical parallel sections of a cylinder and was therefore an oblique circular cylinder. Other assumptions with regard to the conditions to which the lines in the figure may be subject would suggest that other loci dealt with were cones regarded as containing all points on particular elliptical parallel sections of the cones[5].

(2)　In the second lemma Pappus states and gives a complete proof of the focus-and-directrix property of a conic, viz. that *the locus of a point whose distance from a given point is in a given ratio to its distance from a fixed line is a conic section, which is an ellipse, a parabola or a hyperbola according as the given ratio is less than, equal to, or greater than unity*[6]. Two conjectures are possible as to the application of this theorem in Euclid's *Surface-loci*. (*a*) It may have been used to prove that the locus of a point whose distance from a given straight

---

[1] *Die Lehre von den Kegelschnitten im Altertum*, chapter VIII.
[2] Pappus, VII. p. 636.　　　[3] *ibid*. VII. p. 1004.
[4] *Bulletin des sciences math. et astron.*, 2ᵉ Série, VI. 149.
[5] Further particulars will be found in *The Works of Archimedes*, pp. lxii—lxiv, and in Zeuthen, *Die Lehre von den Kegelschnitten*, p. 425 sqq.
[6] Pappus, VII. pp. 1006—1014, and Hultsch's Appendix, pp. 1270—3.

line is in a given ratio to its distance from a given plane is a certain cone. (*b*) It may have been used to prove that the locus of a point whose distance from a given point is in a given ratio to its distance from a given plane is the surface formed by the revolution of a conic about its major or conjugate axis[1]. Thus Chasles may have been correct in his conjecture that the *Surface-loci* dealt with surfaces of revolution of the second degree and sections of the same[2].

6. The *Conics*.

Pappus says of this lost work: "The four books of Euclid's Conics were completed by Apollonius, who added four more and gave us eight books of Conics[3]." It is probable that Euclid's work was lost even by Pappus' time, for he goes on to speak of "Aristaeus, who wrote the *still extant* five books of Solid Loci connected with the conics." Speaking of the relation of Euclid's work to that of Aristaeus on conics regarded as loci, Pappus says in a later passage (bracketed however by Hultsch) that Euclid, regarding Aristaeus as deserving credit for the discoveries he had already made in conics, did not (try to) anticipate him or construct anew the same system. We may no doubt conclude that the book by Aristaeus on solid loci preceded Euclid's on conics and was, at least in point of originality, more important. Though both treatises dealt with the same subject-matter, the object and the point of view were different; had they been the same, Euclid could scarcely have refrained, as Pappus says he did, from attempting to improve upon the earlier treatise. No doubt Euclid wrote on the general theory of conics as Apollonius did, but confined himself to those properties which were necessary for the analysis of the *Solid Loci* of Aristaeus. The *Conics* of Euclid were evidently superseded by the treatise of Apollonius.

As regards the contents of Euclid's *Conics*, the most important source of our information is Archimedes, who frequently refers to propositions in conics as well known and not needing proof, adding in three cases that they are proved in the "elements of conics" or in "the conics," which expressions must clearly refer to the works of Aristaeus and Euclid[4]

Euclid still used the old names for the conics (sections of a right-angled, acute-angled, or obtuse-angled cone), but he was aware that an ellipse could be obtained by cutting a cone in any manner by a plane not parallel to the base (assuming the section to lie wholly between the apex of the cone and its base) and also by cutting a cylinder. This is expressly stated in a passage from the *Phaenomena* of Euclid about to be mentioned[5].

7. The *Phaenomena*.

This is an astronomical work and is still extant. A much inter-

---

[1] For further details see *The Works of Archimedes*, pp. lxiv, lxv, and Zeuthen, *l. c.*
[2] *Aperçu historique*, pp. 273—4.   [3] Pappus, VII. p. 672.
[4] For details of these propositions see my *Apollonius of Perga*, pp. xxxv, xxxvi.
[5] *Phaenomena*, ed. Menge, p. 6: "If a cone or a cylinder be cut by a plane not parallel to the base, the section is a section of an acute-angled cone, which is like a shield (θυρεός)."

polated version appears in Gregory's Euclid. An earlier and better recension is however contained in the MS. Vindobonensis philos. Gr. 103, though the end of the treatise, from the middle of prop. 16 to the last (18), is missing. The book, now edited by Menge[1], consists of propositions in *spheric* geometry. Euclid based it on Autolycus' work περὶ κινουμένης σφαίρας, but also, evidently, on an earlier textbook of *Sphaerica* of exclusively mathematical content. It has been conjectured that the latter textbook may have been due to Eudoxus[2].

8.   The *Optics*.

This book needs no description, as it has been edited by Heiberg recently[3], both in its genuine form and in the recension by Theon. The *Catoptrica* published by Heiberg in the same volume is not genuine, and Heiberg suspects that in its present form it may be Theon's. It is not even certain that Euclid wrote *Catoptrica* at all, as Proclus may easily have had Theon's work before him and inadvertently assigned it to Euclid[4].

9.   Besides the above-mentioned works, Euclid is said to have written the *Elements of Music*[5] (αἱ κατὰ μουσικὴν στοιχειώσεις). Two treatises are attributed to Euclid in our MSS. of the *Musici*, the κατατομὴ κανόνος, *Sectio canonis* (the theory of the intervals), and the εἰσαγωγὴ ἁρμονική (introduction to harmony)[6]. The first, resting on the Pythagorean theory of music, is mathematical, and the style and diction as well as the form of the propositions mostly agree with what we find in the *Elements*. Jan thought it genuine, especially as almost the whole of the treatise (except the preface) is quoted *in extenso*, and Euclid is twice mentioned by name, in the commentary on Ptolemy's *Harmonica* published by Wallis and attributed by him to Porphyry. Tannery was of the opposite opinion[7]. The latest editor, Menge, suggests that it may be a redaction by a less competent hand from the genuine Euclidean *Elements of Music*. The second treatise is not Euclid's, but was written by Cleonides, a pupil of Aristoxenus[8].

Lastly, it is worth while to give the Arabians' list of Euclid's works. I take this from Suter's translation of the list of philosophers and mathematicians in the *Fihrist*, the oldest authority of the kind that we possess[9]. "To the writings of Euclid belong further [in addition to the *Elements*]: the book of Phaenomena; the book of

[1] *Euclidis opera omnia*, vol. VIII., 1916, pp. 2—156.
[2] Heiberg, *Euklid-Studien*, p. 46; Hultsch, *Autolycus*, p. xii; A. A. Björnbo, *Studien über Menelaos' Sphärik* (*Abhandlungen zur Geschichte der mathematischen Wissenschaften*, XIV. 1902), p. 56 sqq.
[3] *Euclidis opera omnia*, vol. VII. (1895).
[4] Heiberg, Euclid's *Optics, etc.* p. 1.   [5] Proclus, p. 69, 3.
[6] Both treatises edited by Jan in *Musici Scriptores Graeci*, 1895, pp. 113—166, 167—207, and by Menge in *Euclidis opera omnia*, vol. VIII., 1916, pp. 157—183, 185—223.
[7] *Comptes rendus de l'Acad. des inscriptions et belles-lettres*, Paris, 1904, pp. 439—445. Cf. *Bibliotheca Mathematica*, VI₃, 1905-6, p. 225, note 1.
[8] Heiberg, *Euklid-Studien*, pp. 52—55; Jan, *Musici Scriptores Graeci*, pp. 169—174.
[9] H. Suter, *Das Mathematiker-Verzeichniss im Fihrist* in *Abhandlungen zur Geschichte der Mathematik*, VI., 1892, pp. 1—87 (see especially p. 17). Cf. Casiri, I. 339, 340, and Gartz, *De interpretibus et explanatoribus Euclidis Arabicis*, 1823, pp. 4, 5.

Given Magnitudes [*Data*]; the book of Tones, known under the name of Music, not genuine; the book of Division, emended by Thābit; the book of Utilisations or Applications [*Porisms*], not genuine; the book of the Canon; the book of the Heavy and Light; the book of Synthesis, not genuine; and the book of Analysis, not genuine."

It is to be observed that the Arabs already regarded the book of Tones (by which must be meant the εἰσαγωγὴ ἁρμονική) as spurious. The book of Division is evidently the book on *Divisions* (*of figures*). The next book is described by Casiri as "liber de utilitate suppositus." Suter gives reason for believing the *Porisms* to be meant[1], but does not apparently offer any explanation of why the work is supposed to be spurious. The book of the Canon is clearly the κατατομὴ κανόνος. The book on "the Heavy and Light" is apparently the tract *De levi et ponderoso*, included in the Basel Latin translation of 1537, and in Gregory's edition. The fragment, however, cannot safely be attributed to Euclid, for (1) we have nowhere any mention of his having written on mechanics, (2) it contains the notion of specific gravity in a form so clear that it could hardly be attributed to anyone earlier than Archimedes[2]. Suter thinks[3] that the works on Analysis and Synthesis (said to be spurious in the extract) may be further developments of the *Data* or *Porisms*, or may be the interpolated proofs of *Eucl.* XIII. 1—5, divided into *analysis* and *synthesis*, as to which see the notes on those propositions.

---

[1] Suter, *op. cit.* pp. 49, 50. Wenrich translated the word as "utilia." Suter says that the nearest meaning of the Arabic word as of "porism" is *use, gain* (Nutzen, Gewinn), while a further meaning is explanation, observation, addition: a gain arising out of what has preceded (cf. Proclus' definition of the porism in the sense of a corollary).

[2] Heiberg, *Euklid-Studien*, pp. 9, 10.   [3] Suter, *op. cit.* p. 50.

CHAPTER III.

GREEK COMMENTATORS ON THE *ELEMENTS* OTHER THAN PROCLUS.

THAT there was no lack of commentaries on the *Elements* before the time of Proclus is evident from the terms in which Proclus refers to them; and he leaves us in equally little doubt as to the value which, in his opinion, the generality of them possessed. Thus he says in one place (at the end of his second prologue)[1]:

"Before making a beginning with the investigation of details, I warn those who may read me not to expect from me the things which have been dinned into our ears *ad nauseam* (διατεθρύληται) by those who have preceded me, viz. lemmas, cases, and so forth. For I am surfeited with these things and shall give little attention to them. But I shall direct my remarks principally to the points which require deeper study and contribute to the sum of philosophy, therein emulating the Pythagoreans who even had this common phrase for what I mean 'a figure and a platform, but not a figure and sixpence[2].'"

In another place[3] he says: "Let us now turn to the elucidation of the things proved by the writer of the Elements, selecting the more subtle of the comments made on them by the ancient writers, while cutting down their interminable diffuseness, giving the things which are more systematic and follow scientific methods, attaching more importance to the working-out of the real subject-matter than to the variety of cases and lemmas to which we see recent writers devoting themselves for the most part."

At the end of his commentary on Eucl. I. Proclus remarks[4] that the commentaries then in vogue were full of all sorts of confusion, and contained no account of *causes*, no dialectical discrimination, and no philosophic thought.

These passages and two others in which Proclus refers to "the commentators[5]" suggest that these commentators were numerous. He does not however give many names; and no doubt the only important commentaries were those of Heron, Porphyry, and Pappus.

---

[1] Proclus, p. 84, 8.
[2] i.e. we reach a certain height, use the platform so attained as a base on which to build another stage, then use that as a base and so on.
[3] Proclus, p. 200, 10.   [4] *ibid.* p. 432, 15.   [5] *ibid.* p. 289, 11; p. 328, 16.

I. Heron.

Proclus alludes to Heron twice as Heron *mechanicus*[1], in another place[2] he associates him with Ctesibius, and in the three other passages[3] where Heron is mentioned there is no reason to doubt that the same person is meant, namely Heron of Alexandria. The date of Heron is still a vexed question. In the early stages of the controversy much was made of the supposed relation of Heron to Ctesibius. The best MS. of Heron's *Belopoeica* has the heading Ἥρωνος Κτησιβίου βελοποιϊκά, and an anonymous Byzantine writer of the tenth century, evidently basing himself on this title, speaks of Ctesibius as Heron's καθηγητής, "master" or "teacher." We know of two men of the name of Ctesibius. One was a barber who lived in the time of Ptolemy Euergetes II, i.e. Ptolemy VII, called Physcon (died 117 B.C.), and who is said to have made an improved water-organ[4]. The other was a mechanician mentioned by Athenaeus as having made an elegant drinking-horn in the time of Ptolemy Philadelphus (285–247 B.C.)[5]. Martin[6] took the Ctesibius in question to be the former and accordingly placed Heron at the beginning of the first century B.C., say 126–50 B.C. But Philo of Byzantium[7], who repeatedly mentions Ctesibius by name, says that the first mechanicians had the advantage of being under kings who loved fame and supported the arts. Hence our Ctesibius is more likely to have been the earlier Ctesibius who was contemporary with Ptolemy II Philadelphus.

But, whatever be the date of Ctesibius, we cannot safely conclude that Heron was his immediate pupil. The title "Heron's (edition of) Ctesibius's Belopoeica" does not, in fact, justify any inference as to the interval of time between the two works.

We now have better evidence for a *terminus post quem*. The *Metrica* of Heron, besides quoting Archimedes and Apollonius, twice refers to "the books about straight lines (chords) in a circle" (ἐν τοῖς περὶ τῶν ἐν κύκλῳ εὐθειῶν). Now we know of no work giving a Table of Chords earlier than that of Hipparchus. We get, therefore, at once, 150 B.C. or thereabouts as the *terminus post quem*. But, again, Heron's *Mechanica* quotes a definition of "centre of gravity" as given by "Posidonius, a Stoic": and, even if this Posidonius lived before Archimedes, as the context seems to imply, it is certain that another work of Heron's, the *Definitions*, owes something to Posidonius of Apamea or Rhodes, Cicero's teacher (135–51 B.C.). This brings Heron's date down to the end of the first century B.C., at least.

We have next to consider the relation, if any, between Heron and Vitruvius. In his *De Architectura*, brought out apparently in 14 B.C., Vitruvius quotes twelve authorities on *machinationes* including Archytas

[1] Proclus, p. 305, 24; p. 346, 13.
[2] *ibid.* p. 41, 10.
[3] *ibid.* p. 196, 16; p. 323, 7; p. 429, 13.
[4] Athenaeus, *Deipno-Soph.* iv., c. 75, p. 174 *b—c.*.
[5] *ibid.* xi., c. 97, p. 497 *b—c.*
[6] Martin, *Recherches sur la vie et les ouvrages d'Héron d'Alexandrie*, Paris, 1854, p. 27.
[7] Philo, *Mechan. Synt.*, p. 50, 38, ed. Schöne.

(second), Archimedes (third), Ctesibius (fourth) and Philo of Byzantium (sixth), but does not mention Heron. Nor is it possible to establish inter-dependence between Vitruvius and Heron; the differences between them seem on the whole more numerous and important than the resemblances (e.g. Vitruvius uses 3 as the value of π, while Heron always uses the Archimedean value $3\frac{1}{7}$). The inference is that Heron can hardly have written earlier than the first century A.D.

The most recent theory of Heron's date makes him later than Claudius Ptolemy the astronomer (100–178 A.D.). The arguments are mainly these. (1) Ptolemy claims as a discovery of his own a method of measuring the distance between two places (as an arc of a great circle on the earth's surface) in the case where the places are neither on the same meridian nor on the same parallel circle. Heron, in his *Dioptra*, speaks of this method as of a thing generally known to experts. (2) The dioptra described in Heron's work is a fine and accurate instrument, much better than anything Ptolemy had at his disposal. (3) Ptolemy, in his work Περὶ ῥοπῶν, asserted that water with water round it has no weight and that the diver, however deep he dives, does not feel the weight of the water above him. Heron, strangely enough, accepts as true what Ptolemy says of the diver, but is dissatisfied with the explanation given by "some," namely that it is because water is uniformly heavy—this seems to be equivalent to Ptolemy's dictum that water in water has no weight—and he essays a different explanation based on Archimedes. (4) It is suggested that the Dionysius to whom Heron dedicated his *Definitions* is a certain Dionysius who was *praefectus urbi* in 301 A.D.

On the other hand Heron was earlier than Pappus, who was writing under Diocletian (284–305 A.D.), for Pappus alludes to and draws upon the works of Heron. The net result, then, of the most recent research is to place Heron in the third century A.D. and perhaps little earlier than Pappus. Heiberg[1] accepts this conclusion, which may therefore, perhaps, be said to hold the field for the present[2].

That Heron wrote a systematic commentary on the *Elements* might be inferred from Proclus, but it is rendered quite certain by references to the commentary in Arabian writers, and particularly in an-Nairīzī's commentary on the first ten Books of the *Elements*. The *Fihrist* says, under Euclid, that "Heron wrote a commentary on this book [the *Elements*], endeavouring to solve its difficulties[3]"; and under Heron, "He wrote: the book of explanation of the obscurities in Euclid[4]...." An-Nairīzī's commentary quotes Heron by name very frequently, and often in such a way as to leave no doubt that the author had Heron's work actually before him. Thus the extracts are given in the first person, introduced by "Heron says" ("Dixit Yrinus"

---

[1] *Heronis Alexandrini opera*, vol. v. (Teubner, 1914), p. ix.
[2] Fuller details of the various arguments will be found in my *History of Greek Mathematics*, 1921, vol. II., pp. 298–306.
[3] *Das Mathematiker-Verzeichniss im Fihrist* (tr. Suter), p. 16.
[4] *ibid.* p. 22.

or "Heron"); and in other places we are told that Heron "says nothing," or "is not found to have said anything," on such and such a proposition. The commentary of an-Nairīzī is in part edited by Besthorn and Heiberg from a Leiden MS. of the translation of the *Elements* by al-Ḥajjāj with the commentary attached[1]. But this MS. only contains six Books, and several pages in the first Book, which contain the comments of Simplicius on the first twenty-two definitions of the first Book, are missing. Fortunately the commentary of an-Nairīzī has been discovered in a more complete form, in a Latin translation by Gherardus Cremonensis of the twelfth century, which contains the missing comments by Simplicius and an-Nairīzī's comments on the first ten Books. This valuable work has recently been edited by Curtze[2].

Thus from the three sources, Proclus, and the two versions of an-Nairīzī, which supplement one another, we are able to form a very good idea of the character of Heron's commentary. In some cases observations given by Proclus without the name of their author are seen from an-Nairīzī to be Heron's; in a few cases notes attributed by Proclus to Heron are found in an-Nairīzī without Heron's name; and, curiously enough, one alternative proof (of I. 25) given as Heron's by Proclus is introduced by the Arab with the remark that he has not been able to discover who is the author.

Speaking generally, the comments of Heron do not seem to have contained much that can be called important. We find

(1) A few general notes, e.g. that Heron would not admit more than three axioms.

(2) Distinctions of a number of particular *cases* of Euclid's propositions according as the figure is drawn in one way or in another.

Of this class are the different cases of I. 35, 36, III. 7, 8 (where the chords to be compared are drawn on *different* sides of the diameter instead of on the same side), III. 12 (which is not Euclid's, but Heron's own, adding the case of external contact to that of internal contact in III. 11), VI. 19 (where the triangle in which an additional line is drawn is taken to be the *smaller* of the two), VII. 19 (where he gives the particular case of *three* numbers in continued proportion, instead of four proportionals).

(3) Alternative proofs. Of these there should be mentioned (*a*) the proofs of II. 1—10 "without a figure," being simply the algebraic forms of proof, easy but uninstructive, which are so popular nowadays, the proof of III. 25 (placed after III. 30 and starting from the *arc* instead of the chord), III. 10 (proved by III. 9), III. 13 (a proof preceded by a lemma to the effect that a straight line cannot meet a circle in more than two points). Another class of alternative proof is

---

[1] *Codex Leidensis* 399, 1. *Euclidis Elementa ex interpretatione al-Hadschdschadschii cum commentariis al-Narizii.* Five parts carrying the work to the end of Book IV. were issued in 1893, 1897, 1900, 1905 and 1910 respectively.

[2] *Anaritii in decem libros priores elementorum Euclidis commentarii ex interpretatione Gherardi Cremonensis...edidit* Maximilianus Curtze (Teubner, Leipzig, 1899).

(*b*) that which is intended to meet a particular *objection* (ἔνστασις) which had been or might be raised to Euclid's construction. Thus in certain cases he avoids *producing* a particular straight line, where Euclid produces it, in order to meet the objection of any one who should deny our right to assume that there is *any space available*[1]. Of this class are Heron's proofs of I. 11, I. 20, and his note on I. 16. Similarly on I. 48 he supposes the right-angled triangle which is constructed to be constructed on the *same* side of the common side as the given triangle is. A third class (*c*) is that which avoids *reductio ad absurdum*. Thus, instead of indirect proofs, Heron gives direct proofs of I. 19 (for which he requires, and gives, a preliminary lemma), and of I. 25.

(4) Heron supplies certain *converses* of Euclid's propositions, e.g. converses of II. 12, 13, VIII. 27.

(5) A few additions to, and extensions of, Euclid's propositions are also found. Some are unimportant, e.g. the construction of isosceles and scalene triangles in a note on I. 1, the construction of *two* tangents in III. 17, the remark that VII. 3 about finding the greatest common measure of three numbers can be applied to as many numbers as we please (as Euclid tacitly assumes in VII. 31). The most important extension is that of III. 20 to the case where the angle at the circumference is greater than a right angle, and the direct deduction from this extension of the result of III. 22. Interesting also are the notes on I. 37 (on I. 24 in Proclus), where Heron proves that two triangles with two sides of one equal to two sides of the other and with the included angles supplementary are equal, and compares the areas where the sum of the two included angles (one being supposed greater than the other) is less or greater than two right angles, and on I. 47, where there is a proof (depending on preliminary lemmas) of the fact that, in the figure of the proposition, the straight lines *AL*, *BK*, *CF* meet in a point. After IV. 16 there is a proof that, in a regular polygon with an even number of sides, the bisector of one angle also bisects its opposite, and an enunciation of the corresponding proposition for a regular polygon with an odd number of sides.

Van Pesch[2] gives reason for attributing to Heron certain other notes found in Proclus, viz. that they are designed to meet the same sort of points as Heron had in view in other notes undoubtedly written by him. These are (*a*) alternative proofs of I. 5, I. 17, and I. 32, which avoid the *producing* of certain straight lines, (*b*) an alternative proof of I. 9 avoiding the construction of the equilateral triangle on the side of *BC* opposite to *A*; (*c*) partial converses of I. 35—38, starting from the equality of the areas and the fact of the parallelograms or triangles being in the same parallels, and proving that the bases are the same or equal, may also be Heron's. Van Pesch further supposes that it was in Heron's commentary that the proof by Menelaus of I. 25 and the proof by Philo of I. 8 were given.

---

[1] Cf. Proclus, 275, 7 εἰ δὲ λέγοι τις τόπον μὴ εἰδέναι..., 289, 18 λέγει οὖν τις ὅτι οὐκ ἔστι τόπος....

[2] *De Procli fontibus*, Lugduni-Batavorum, 1900.

The last reference to Heron made by an-Naīrīzī occurs in the note on VIII. 27, so that the commentary of the former must at least have reached that point.

II.   Porphyry.

The Porphyry here mentioned is of course the Neo-Platonist who lived about 232–304 A.D.   Whether he really wrote a systematic commentary on the *Elements* is uncertain.   The passages in Proclus which seem to make this probable are two in which he mentions him (1) as having demonstrated the necessity of the words "not on the same side" in the enunciation of I. 14[1], and (2) as having pointed out the necessity of understanding correctly the enunciation of I. 26, since, if the particular injunctions as to the sides of the triangles to be taken as equal are not regarded, the student may easily fall into error[2]. These passages, showing that Porphyry carefully analysed Euclid's *enunciations* in these cases, certainly suggest that his remarks were part of a systematic commentary.   Further, the list of mathematicians in the *Fihrist* gives Porphyry as having written "a book on the Elements."   It is true that Wenrich takes this book to have been a work by Porphyry mentioned by Suidas and Proclus (*Theolog. Platon.*), περὶ ἀρχῶν libri II.[3]

There is nothing of importance in the notes attributed to Porphyry by Proclus.

(1)   Three alternative proofs of I. 20, which avoid *producing* a side of the triangle, are assigned to Heron and Porphyry without saying which belonged to which.   If the first of the three was Heron's, I agree with van Pesch that it is more probable that the two others were both Porphyry's than that the second was Heron's and only the third Porphyry's.   For they are similar in character, and the third uses a result obtained in the second[4].

(2)   Porphyry gave an alternative proof of I. 18 to meet a childish objection which is supposed to require the part of AC equal to AB to be cut off from CA and not from AC.

Proclus gives a precisely similar alternative proof of I. 6 to meet a similar supposed objection; and it may well be that, though Proclus mentions no name, this proof was also Porphyry's, as van Pesch suggests[5].

Two other references to Porphyry found in Proclus cannot have anything to do with commentaries on the *Elements*.   In the first a work called the Συμμικτά is quoted, while in the second a philosophical question is raised.

III.   Pappus.

The references to Pappus in Proclus are not numerous; but we have other evidence that he wrote a commentary on the *Elements*. Thus a scholiast on the definitions of the *Data* uses the phrase "as

---

[1] Proclus, pp. 297, 1—298, 10.   [2] *ibid.* p. 352, 13, 14 and the pages preceding.
[3] *Fihrist* (tr. Suter), p. 9, 10 and p. 45 (note 5).
[4] Van Pesch, *De Procli fontibus*, pp. 129, 130.   Heiberg assigned them as above in his *Euklid-Studien* (p. 160), but seems to have changed his view later. (See Besthorn-Heiberg, *Codex Leidensis*, p. 93, note 2.)
[5] Van Pesch, *op. cit.* pp. 130—1.

Pappus says at the beginning of his (commentary) on the 10th (book) of Euclid[1]." Again in the *Fihrist* we are told that Pappus wrote a commentary to the tenth book of Euclid in two parts[2]. Fragments of this still survive in a MS. described by Woepcke[3], Paris. No. 952. 2 (supplément arabe de la Bibliothèque impériale), which contains a translation by Abū 'Uthmān (beginning of 10th century) of a Greek commentary on Book X. It is in two books, and there can now be no doubt that the author of the Greek commentary was Pappus[4]. Again Eutocius, in his note on Archimedes, *On the Sphere and Cylinder* I. 13, says that Pappus explained in his commentary on the *Elements* how to inscribe in a circle a polygon similar to a polygon inscribed in another circle; and this would presumably come in his commentary on Book XII., just as the problem is solved in the second scholium on Eucl. XII. 1. Thus Pappus' commentary on the *Elements* must have been pretty complete, an additional confirmation of this supposition being forthcoming in the reference of Marinus (a pupil and follower of Proclus) in his preface to the *Data* to "the commentaries of Pappus on the book[5]."

The actual references to Pappus in Proclus are as follows:

(1) On the Postulate (4) that all right angles are equal, Pappus is quoted as saying that the converse, viz. that all angles equal to a right angle are right, is not true[6], since the angle included between the arcs of two semicircles which are equal, and have their diameters at right angles and terminating at one point, is equal to a right angle, but is not a right angle.

(2) On the axioms Pappus is quoted as saying that, in addition to Euclid's axioms, others are on record as well (συναναγράφεσθαι) about unequals added to equals and equals added to unequals[7]; these, says Proclus, follow from the Euclidean axioms, while others given by Pappus are involved by the definitions, namely those which assert that "all parts of the plane and of the straight line coincide with one another," that "a point divides a straight line, a line a surface, and a surface a solid," and that "the infinite is (obtained) in magnitudes both by addition and diminution[8]."

---

[1] Euclid's *Data*, ed. Menge, p. 262.   [2] *Fihrist* (tr. Suter), p. 22.

[3] *Mémoires présentés à l'académie des sciences*, 1856, XIV. pp. 658—719.

[4] Woepcke read the name of the author, in the title of the first book, as *B . los* (the dot representing a missing vowel). He quotes also from other MSS. (e.g. of the *Ta'rīkh al-Hukamā* and of the *Fihrist*) where he reads the name of the commentator as *B . lis*, *B . n . s* or *B . l . s.* Woepcke takes this author to be Valens, and thinks it possible that he may be the same as the astrologer Vettius Valens. This Heiberg (*Euklid-Studien*, pp. 169, 170) proves to be impossible, because, while one of the MSS. quoted by Woepcke says that "*B . n . s*, le *Roûmi*" (late-Greek) was later than Claudius Ptolemy and the *Fihrist* says "*B . l . s*, le *Roûmi*" wrote a commentary on Ptolemy's *Planisphaerium*, Vettius Valens seems to have lived under Hadrian, and must therefore have been an *elder* contemporary of Ptolemy. But Suter shows (*Fihrist*, p. 22 and p. 54, note 92) that *Banos* is only distinguished from *Babos* by the position of a certain dot, and *Balos* may also easily have arisen from an original *Babos* (there is no P in Arabic), so that Pappus must be the person meant. This is further confirmed by the fact that the *Fihrist* gives this author and Valens as the subjects of two separate paragraphs, attributing to the latter astrological works only.

[5] Heiberg, *Euklid-Studien*, p. 173; Euclid's *Data*, ed. Menge, pp. 256, lii.

[6] Proclus, pp. 189, 190.   [7] *ibid.* p. 197, 6—10.

[8] *ibid.* p. 198, 3—15.

(3) Pappus gave a pretty proof of I. 5. This proof has, I think, been wrongly understood; on this point see my note on the proposition.

(4) On I. 47 Proclus says[1]: "As the proof of the writer of the Elements is manifest, I think that it is not necessary to add anything further, but that what has been said is sufficient, since indeed those who have added more, like Heron and Pappus, were obliged to make use of what is proved in the sixth book, without attaining any important result." We shall see what Heron's addition consisted of; what Pappus may have added we do not know, unless it was something on the lines of his extension of I. 47 found in the *Synagoge* (IV. p. 176, ed. Hultsch).

We may fairly conclude, with van Pesch[2], that Pappus is drawn upon in various other passages of Proclus where he quotes no authority, but where the subject-matter reminds us of other notes expressly assigned to Pappus or of what we otherwise know to have been favourite questions with him. Thus:

1. We are reminded of the curvilineal angle which is equal to but not a right angle by the note on I. 32 to the effect that the converse (that a figure with its interior angles together equal to two right angles is a triangle) is not true unless we confine ourselves to rectilineal figures. This statement is supported by reference to a figure formed by four semicircles whose diameters form a square, and one of which is turned inwards while the others are turned outwards. The figure forms two angles "equal to" right angles in the sense described by Pappus on Post. 4, while the other curvilineal angles are not considered to be angles at all, and are left out in summing the internal angles. Similarly the allusions in the notes on I. 4, 23 to curvilineal angles of which certain moon-shaped angles (μηνοειδεῖς) are shown to be "equal to" rectilineal angles savour of Pappus.

2. On I. 9 Proclus says[3] that "Others, starting from the Archimedean spirals, divided any given rectilineal angle in any given ratio." We cannot but compare this with Pappus IV. p. 286, where the spiral is so used; hence this note, including remarks immediately preceding about the conchoid and the quadratrix, which were used for the same purpose, may very well be due to Pappus.

3. The subject of isoperimetric figures was a favourite one with Pappus, who wrote a recension of Zenodorus' treatise on the subject[4]. Now on I. 35 Proclus speaks[5] about the paradox of parallelograms having equal area (between the same parallels) though the two sides between the parallels may be of any length, adding that of parallelograms with equal perimeter the rectangle is greatest if the base be given, and the square greatest if the base be not given etc. He returns to the subject on I. 37 about triangles[6]. Compare[7] also his note on I. 4. These notes may have been taken from Pappus.

---

[1] Proclus, p. 429, 9—15.
[2] Van Pesch, *De Procli fontibus*, p. 134 sqq.   [3] Proclus, p. 272, 10.
[4] Pappus, v. pp. 304—350; for Zenodorus' own treatise see Hultsch's Appendix, pp. 1189 —1211.
[5] Proclus, pp. 396—8.   [6] *ibid.* pp. 403—4.   [7] *ibid.* pp. 236—7.

4. Again, on I. 21, Proclus remarks on the paradox that straight lines may be drawn from the base to a point within a triangle which are (1) together greater than the two sides, and (2) include a less angle, provided that the straight lines may be drawn from points in the base other than its extremities. The subject of straight lines satisfying condition (1) was treated at length, with reference to a variety of cases, by Pappus[1], after a collection of "paradoxes" by Erycinus, of whom nothing more is known. Proclus gives Pappus' first case, and adds a rather useless proof of the possibility of drawing straight lines satisfying condition (2) *alone*, adding that "the proposition stated has been proved by me without using the parallels of the commentators[2]." By "the commentators" Pappus is doubtless meant.

5. Lastly, the "four-sided triangle," called by Zenodorus the "hollow-angled,"[3] is mentioned in the notes on I. Def. 24—29 and I. 21. As Pappus wrote on Zenodorus' work in which the term occurred[4], Pappus may be responsible for these notes.

IV. Simplicius.

According to the *Fihrist*[5], Simplicius the Greek wrote "a commentary to the beginning of Euclid's book, which forms an introduction to geometry." And in fact this commentary on the definitions, postulates and axioms (including the postulate known as the Parallel-Axiom) is preserved in the Arabic commentary of an-Nairīzī[6]. On two subjects this commentary of Simplicius quotes a certain "Aganis," the first subject being the definition of an angle, and the second the definition of parallels and the parallel-postulate. Simplicius gives word for word, in a long passage placed by an-Nairīzī after I. 29, an attempt by "Aganis" to prove the parallel-postulate. It starts from a definition of parallels which agrees with Geminus' view of them as given by Proclus[7], and is closely connected with the definition given by Posidonius[8]. Hence it has been assumed that "Aganis" is none other than Geminus, and the historical importance of the commentary of Simplicius has been judged accordingly. But it has been recently shown by Tannery that the identification of "Aganis" with Geminus is practically impossible[9]. In the translation of Besthorn-Heiberg Aganis is called by Simplicius in one place "philosophus Aganis," in another "magister noster Aganis," in Gherard's version he is "socius Aganis" and "socius noster Aganis." These expressions seem to leave no doubt that Aganis was a contemporary and friend, if not master, of Simplicius; and it is impossible to suppose that Simplicius (fl. about 500 A.D.) could have used them of a man who lived four and

---

[1] Pappus, III. pp. 104—130.   [2] Proclus, p. 328, 15.
[3] Proclus, p. 165, 24; cf. pp. 328, 329.   [4] See Pappus, ed. Hultsch, pp. 1154, 1206.
[5] *Fihrist* (tr. Suter), p. 21.
[6] An-Nairīzī, ed. Besthorn-Heiberg, pp. 9—41, 119—133, ed. Curtze, pp. 1—37, 65—73. The *Codex Leidensis*, from which Besthorn and Heiberg's edition is taken, has unfortunately lost some leaves, so that there is a gap from Def. 1 to Def. 23 (parallels). The loss is, however, made good by Curtze's edition of the translation by Gherard of Cremona.
[7] Proclus, p. 177, 21.   [8] *ibid.* p. 176, 7.
[9] *Bibliotheca Mathematica*, II₃, 1900, pp. 9—11.

a half centuries before his time. A phrase in Simplicius' word-for-word quotation from Aganis leads to the same conclusion. He speaks of people who objected "even in ancient times" (iam antiquitus) to the use by geometers of this postulate. This would not have been an appropriate phrase had Geminus been the writer. I do not think that this difficulty can be got over by Suter's suggestion[1] that the passages in question may have been taken out of *Heron's* commentary, and that an-Nairīzī may have forgotten to name the author; it seems clear that Simplicius is the person who described "Aganis." Hence we are driven to suppose that Aganis was not Geminus, but some unknown contemporary of Simplicius[2]. Considerable interest will however continue to attach to the comments of Simplicius so fortunately preserved.

Proclus tells us that one Aegaeas (? Aenaeas) of Hierapolis wrote an epitome of the *Elements*[3]; but we know nothing more of him or of it.

---

[1] *Zeitschrift für Math. u. Physik*, XLIV., hist.-litt. Abth. p. 61.

[2] The above argument seems to me quite insuperable. The other arguments of Tannery do not, however, carry conviction to my mind. I do not follow the reasoning based on Aganis' definition of an angle. It appears to me a pure assumption that Geminus would have seen that Posidonius' definition of parallels was not admissible. Nor does it seem to me to count for much that Proclus, while telling us that Geminus held that the postulate ought to be proved and warned the unwary against hastily concluding that two straight lines approaching one another must necessarily meet (cf. a curve and its asymptote), gives no hint that Geminus did try to prove the postulate. It may well be that Proclus omitted Geminus' "proof" (if he wrote one) because he preferred Ptolemy's attempt which he gives (pp. 365—7).

[3] Proclus, p. 361, 21.

CHAPTER IV.

PROCLUS AND HIS SOURCES[1].

IT is well known that the commentary of Proclus on Eucl. Book I.
is one of the two main sources of information as to the history of
Greek geometry which we possess, the other being the *Collection* of
Pappus. They are the more precious because the original works of
the forerunners of Euclid, Archimedes and Apollonius are lost, having
probably been discarded and forgotten almost immediately after the
appearance of the masterpieces of that great trio.

Proclus himself lived 410–485 A.D., so that there had already
passed a sufficient amount of time for the tradition relating to the
pre-Euclidean geometers to become obscure and defective. In this
connexion a passage is quoted from Simplicius[2] who, in his account
of the quadrature of certain lunes by Hippocrates of Chios, while
mentioning two authorities for his statements, Alexander Aphrodisiensis (about 220 A.D.) and Eudemus, says in one place[3], "As
regards Hippocrates of Chios we must pay more attention to Eudemus,
since he was nearer the times, being a pupil of Aristotle."

The importance therefore of a critical examination of Proclus'
commentary with a view to determining from what original sources
he drew need not be further emphasised.

Proclus received his early training in Alexandria, where Olympio-
dorus was his instructor in the works of Aristotle, and mathematics
was taught him by one Heron[4] (of course a different Heron from the
"*mechanicus* Hero" of whom we have already spoken). He afterwards went to Athens where he was imbued by Plutarch, and by
Syrianus, with the Neo-Platonic philosophy, to which he then devoted

[1] My task in this chapter is made easy by the appearance, in the nick of time, of the
dissertation *De Procli fontibus* by J. G. van Pesch (Lugduni-Batavorum, Apud L. van
Nifterik, MDCCCC). The chapters dealing directly with the subject show a thorough
acquaintance on the part of the author with all the literature bearing on it; he covers
the whole field and he exercises a sound and sober judgment in forming his conclusions.
The same cannot always be said of his only predecessor in the same inquiry, Tannery
(in *La Géométrie grecque*, 1887), who often robs his speculations of much of their value
through his proneness to run away with an idea; he does so in this case, basing most of his
conclusions on an arbitrary and unwarranted assumption as to the significance of the words
οἱ περί τινα (e.g. Ἥρωνα, Ποσειδώνιον etc.) as used in Proclus.

[2] Simplicius on Aristotle's *Physics*, ed. Diels, pp. 54—69.

[3] *ibid.* p. 68, 32.

[4] Cf. Martin, *Recherches sur la vie et les ouvrages d'Héron d'Alexandrie*, pp. 240—2.

heart and soul, becoming one of its most prominent exponents. He speaks everywhere with the highest respect of his masters, and was in turn regarded with extravagant veneration by his contemporaries, as we learn from Marinus his pupil and biographer. On the death of Syrianus he was put at the head of the Neo-Platonic school. He was a man of untiring industry, as is shown by the number of books which he wrote, including a large number of commentaries, mostly on the dialogues of Plato. He was an acute dialectician, and pre-eminent among his contemporaries in the range of his learning[1]; he was a competent mathematician; he was even a poet. At the same time he was a believer in all sorts of myths and mysteries and a devout worshipper of divinities both Greek and Oriental.

Though he was a competent mathematician, he was evidently much more a philosopher than a mathematician[2]. This is shown even in his commentary on Eucl. I., where, not only in the Prologues (especially the first), but also in the notes themselves, he seizes any opportunity for a philosophical digression. He says himself that he attaches most importance to "the things which require deeper study and contribute to the sum of philosophy[3]"; alternative proofs, cases, and the like (though he gives many) have no attraction for him; and, in particular, he attaches no value to the addition of Heron to I. 47[4], which is of considerable mathematical interest. Though he esteemed mathematics highly, it was only as a handmaid to philosophy. He quotes Plato's opinion to the effect that "mathematics, as making use of hypotheses, falls short of the non-hypothetical and perfect science[5]"..."Let us then not say that Plato excludes mathematics from the sciences, but that he declares it to be secondary to the one supreme science[6]." And again, while "mathematical science must be considered desirable in itself, though not with reference to the needs of daily life," "if it is necessary to refer the benefit arising from it to something else, we must connect that benefit with intellectual knowledge ($\nu o \epsilon \rho \grave{\alpha} \nu \gamma \nu \hat{\omega} \sigma \iota \nu$), to which it leads the way and is a propaedeutic, clearing the eye of the soul and taking away the impediments which the senses place in the way of the knowledge of universals ($\tau \hat{\omega} \nu \ \ddot{o} \lambda \omega \nu$)[7]."

We know that in the Neo-Platonic school the younger pupils learnt mathematics; and it is clear that Proclus taught this subject, and that this was the origin of the commentary. Many passages show him as a master speaking to scholars. Thus "we have illustrated

---

[1] Zeller calls him "Der Gelehrte, dem kein Feld damaligen Wissens verschlossen ist."

[2] Van Pesch observes that in his commentaries on the *Timaeus* (pp. 671—2) he speaks as no real mathematician could have spoken. In the passage referred to the question is whether the sun occupies a middle place among the planets. Proclus rejects the view of Hipparchus and Ptolemy because "ὁ θεουργός" (sc. the Chaldean, says Zeller) thinks otherwise, "whom it is not lawful to disbelieve." Martin says rather neatly, " Pour Proclus, les Éléments d'Euclide ont l'heureuse chance de n'être contredits ni par les Oracles chaldaïqués, ni par les spéculations des pythagoriciens anciens et nouveaux......"

[3] Proclus, p. 84, 13.   [4] *ibid.* p. 429, 12.
[5] *ibid.* p. 31, 20.   [6] *ibid.* p. 32, 2.
[7] *ibid.* p. 27, 27 to 28, 7; cf. also p. 21, 25, pp. 46, 47.

and made plain all these things in the case of the first problem, but it is necessary that *my hearers* should make the same inquiry as regards the others as well[1]," and " I do not indicate these things as a merely incidental matter but as preparing *us* beforehand for the doctrine of the Timaeus[2]." Further, the pupils whom he was addressing were *beginners* in mathematics ; for in one place he says that he omits " for the present" to speak of the discoveries of those who employed the curves of Nicomedes and Hippias for trisecting an angle, and of those who used the Archimedean spiral for dividing an angle in any given ratio, because these things would be too difficult for beginners ($\delta\upsilon\sigma\theta\epsilon\omega\rho\dot{\eta}\tau\upsilon\upsilon\varsigma$ $\tau\hat{o}\hat{\iota}\varsigma$ $\epsilon\dot{\iota}\sigma\alpha\gamma\upsilon\mu\acute{\epsilon}\nu\upsilon\iota\varsigma$)[3]. Again, if his pupils had not been beginners, it would not have been necessary for Proclus to explain what is meant by saying that sides subtend certain angles[4], the difference between *adjacent* and *vertical* angles[5] etc., or to exhort them, as he often does, to work out other particular cases for themselves, for practice ($\gamma\upsilon\mu\nu\alpha\sigma\acute{\iota}\alpha\varsigma$ $\acute{\epsilon}\nu\epsilon\kappa\alpha$)[6].

The commentary seems then to have been founded on Proclus' lectures to beginners in mathematics. But there are signs that it was revised and re-edited for a larger public ; thus he gives notice in one place[7] "to those who shall come upon" his work ($\tau\hat{o}\hat{\iota}\varsigma$ $\grave{\epsilon}\nu\tau\epsilon\upsilon\xi\upsilon$-$\mu\acute{\epsilon}\nu\upsilon\iota\varsigma$). There are also passages which could not have been understood by the beginners to whom he lectured, e.g. passages about the cylindrical helix[8], conchoids and cissoids[9]. These passages may have been added in the revised edition, or, as van Pesch conjectures, the explanations given in the lectures may have been much fuller and more comprehensible to beginners, and they may have been shortened on revision.

In his comments on the propositions of Euclid, Proclus generally proceeds in this way : first he gives explanations regarding Euclid's proofs, secondly he gives a few different cases, mainly for the sake of practice, and thirdly he addresses himself to refuting objections raised by cavillers to particular propositions. The latter class of note he deems necessary because of "sophistical cavils" and the attitude of the people who rejoiced in finding paralogisms and in causing annoyance to scientific men[10]. His commentary does not seem to have been written for the purpose of correcting or improving Euclid. For there are very few passages of mathematical content in which Proclus can be supposed to be propounding anything of his own ; nearly all are taken from the works of others, mostly earlier commentators, so that, for the purpose of improving on or correcting Euclid, there was no need for his commentary at all. Indeed only in one place does he definitely bring forward anything of his own to get over a difficulty which he finds in Euclid[11]; this is where he tries to

---

[1] Proclus, p. 210, 18.
[3] *ibid.* p. 272, 12.
[5] *ibid.* p. 298, 14.
[7] *ibid.* p. 84, 9.
[9] *ibid.* p. 113.
[11] *ibid.* pp. 368—373.

[2] *ibid.* p. 384, 2.
[4] *ibid.* p. 238, 12.
[6] Cf. p. 224, 15 (on I. 2).
[8] *ibid.* p. 105.
[10] *ibid.* p. 375, 8.

prove the parallel-postulate, after first giving Ptolemy's attempt and then pointing out objections to it. On the other hand, there are a number of passages in which he extols Euclid; thrice[1] also he supports Euclid against Apollonius where the latter had given proofs which he considered better than Euclid's (I. 10, 11, and 23).

Allusion must be made to the debated question whether Proclus continued his commentaries beyond Book I. His intention to do so is clear from the following passages. Just after the words above quoted about the trisection etc. of an angle by means of certain curves he says, "For we may perhaps more appropriately examine these things on the third book, where the writer of the Elements bisects a given circumference[2]." Again, after saying that of all parallelograms which have the same perimeter the square is the greatest "and the rhomboid least of all," he adds: "But this we will prove in another place; for it is more appropriate to the (discussion of the) hypotheses of the second book[3]." Lastly, when alluding (on I. 45) to the squaring of the circle, and to Archimedes' proposition that any circle is equal to the right-angled triangle in which the perpendicular is equal to the radius of the circle and the base to its perimeter, he adds, "But of this elsewhere[4]"; this may imply an intention to treat of the subject on Eucl. XII., though Heiberg doubts it[5]. But it is clear that, at the time when the commentary on Book I. was written, Proclus had not yet begun to write on the other Books and was uncertain whether he would be able to do so: for at the end he says[6], "For my part, if I should be able to discuss the other books[7] in the same manner, I should give thanks to the gods; but, if other cares should draw me away, I beg those who are attracted by this subject to complete the exposition of the other books as well, following the same method, and addressing themselves throughout to the deeper and better defined questions involved" (τὸ πραγματειῶδες πανταχοῦ καὶ εὐδιαίρετον μεταδιώκοντας).

There is in fact no satisfactory evidence that Proclus did actually write any more commentaries than that on Book I.[8] The contrary view receives support from two facts pointed out by Heiberg, viz. (1) that the scholiast's copy of Proclus was not so much better than our

---

[1] Proclus, p. 280, 9; p. 282, 20; pp. 335, 336.
[2] ibid. p. 272, 14.
[3] ibid. p. 398, 18.
[4] ibid. p. 423, 6.
[5] Heiberg, Euklid-Studien, p. 165, note.
[6] Proclus, p. 432, 9.
[7] The words in the Greek are: εἰ μὲν δυνηθείημεν καὶ τοῖς λοιποῖς τὸν αὐτὸν τρόπον ἐξελθεῖν. For ἐξελθεῖν Heiberg would read ἐπεξελθεῖν.
[8] True, a Vatican ms. has a collection of scholia on Books I. (extracts from the extant commentary of Proclus), II., V., VI., X. headed Εἰς τὰ Εὐκλείδου στοιχεῖα προλαμβανόμενα ἐκ τῶν Πρόκλου·σποράδην καὶ κατ' ἐπιτομήν. Heiberg holds that this title itself suggests that the authorship of Proclus was limited to the scholia on Book I.; for προλαμβανόμενα ἐκ τῶν Πρόκλου suits extracts from Proclus' prologues, but hardly scholia to later Books. Again, a certain scholium (Heiberg in Hermes XXXVIII., 1903, p. 341, No. 17) purports to quote words from the end of "a scholium of Proclus" on X. 9. The words quoted are from the scholium X. No. 62, one of the Scholia Vaticana. But none of the other, older, sources connect Proclus' name with X. No. 62; it is probable therefore that a Byzantine, who had in his ms. of Euclid the collection of Schol. Vat. and knew that those on Book I. came from Proclus, himself attached Proclus' name to the others.

MSS. as to suggest that the scholiast had further commentaries of Proclus which have vanished for us[1]; (2) that there is no trace in the scholia of the notes which Proclus promised in the passages quoted above.

Coming now to the question of the sources of Proclus, we may say that everything goes to show that his commentary is a compilation, though a compilation "in the better sense" of the term[2]. He does not even give us to understand that we shall find in it much of his own; "let us," he says, "now turn to the exposition of the theorems proved by Euclid, selecting the more subtle of the comments made on them by the ancient writers, and cutting down their interminable diffuseness...[3]": not a word about anything of his own. At the same time, he seems to imply that he will not necessarily on each occasion quote the source of each extract from an earlier commentary; and, in fact, while he quotes the name of his authority in many places, especially where the subject is important, in many others, where it is equally certain that he is not giving anything of his own, he mentions no authority. Thus he quotes Heron by name six times; but we now know, from the commentary of an-Nairīzī, that a number of other passages, where he mentions no name, are taken from Heron, and among them the not unimportant addition of an alternative proof to I. 19. Hence we can by no means conclude that, where no authority is mentioned, Proclus is giving notes of his own. The presumption is generally the other way; and it is often possible to arrive at a conclusion, either that a particular note is not Proclus' own, or that it is definitely attributable to someone else, by applying the ordinary principles of criticism. Thus, where the note shows an unmistakable affinity to another which Proclus definitely attributes to some commentator by name, especially when both contain some peculiar and distinctive idea, we cannot have much doubt in assigning both to the same commentator[4]. Again, van Pesch finds a criterion in the *form* of a note, where the explanation is so condensed as to be only just intelligible; the note is that in which a converse of I. 32 is proved[5] the proposition namely that a rectilineal figure which has all its interior angles together equal to two right angles is a triangle.

It is not safe to attribute a passage to Proclus himself because he uses the first person in such expressions as "I say" or "I will prove" —for he was in the habit of putting into his own words the substance of notes borrowed from others—nor because, in speaking of an

---

[1] While one class of scholia (Schol. Vat.) have some better readings than our MSS. of Proclus have, and partly fill up the gaps at I. 36, 37 and I. 41—43, the other class (Schol. Vind.) derive from an inferior Proclus MS. which also had the same lacunae.

[2] Knoche, *Untersuchungen über des Proklus Diadochus Commentar zu Euklid's Elementen* (1862), p. 11.

[3] Proclus, p. 200, 10—13.

[4] Instances of the application of this criterion will be found in the discussion of Proclus' indebtedness to the commentaries of Heron, Porphyry and Pappus.

[5] Van Pesch attributes this converse and proof to Pappus, arguing from the fact that the proof is followed by a passage which, on comparison with Pappus' note on the postulate that all right angles are equal, he feels justified in assigning to Pappus. I doubt if the evidence is sufficient.

objection raised to a particular proposition, he uses such expressions as "perhaps someone may object" (ἴσως δ' ἄν τινες ἐνσταῖεν...): for sometimes other words in the same passage.indicate that the objection had actually been taken by someone[1]. Speaking generally, we shall not be justified in concluding that Proclus is stating something new of his own unless he indicates this himself in express terms.

As regards the form of Proclus' references to others by name, van Pesch notes that he very seldom mentions the particular *work* from which he is borrowing. If we leave out of account the references to Plato's dialogues, there are only the following references to books: the *Bacchae* of Philolaus[2], the *Symmikta* of Porphyry[3], Archimedes *On the Sphere and Cylinder*[4], Apollonius *On the cochlias*[5], a book by Eudemus on *The Angle*[6], a whole book of Posidonius directed against Zeno of the Epicurean sect[7], Carpus' *Astronomy*[8], Eudemus' *History of Geometry*[9], and a tract by Ptolemy on the parallel-postulate[10].

Again, Proclus does not always indicate that he is quoting something at second-hand. He often does so, e.g. he quotes Heron as the authority for a statement about Philippus, Eudemus as attributing a certain theorem to Oenopides etc.; but he says on I. 12 that "Oenopides first investigated this problem, thinking it useful for astronomy" when he cannot have had Oenopides' work before him.

It has been said above that Proclus was in the habit of stating in his own words the substance of the things which he borrowed. We are prepared for this when we find him stating that he will select the best things from ancient commentaries and "cut short their interminable diffuseness," that he will "briefly describe" (συντόμως ἱστορῆσαι) the other proofs of I. 20 given by Heron and Porphyry and also the proofs of I. 25 by Menelaus and Heron. But the best evidence is of course to be found in the passages where he quotes works still extant, e.g. those of Plato, Aristotle and Plotinus. Examination of these passages shows great divergences from the original; even where he purports to quote textually, using the expressions "Plato says," or "Plotinus says," he by no means quotes word for word[11]. In fact, he seems to have had a positive distaste for quoting textually from other works. He cannot conquer this even when quoting from Euclid; he says in his note on I. 22, "we will follow the words of the geometer" but fails, nevertheless, to reproduce the text of Euclid unchanged[12].

We now come to the sources themselves from which Proclus drew

---

[1] Van Pesch illustrates this by an objection refuted in the note on I. 9, p. 273, 11 sqq. After using the above expression to introduce the objection, Proclus uses further on (p. 273, 25) the term "they say" (φασίν).

[2] Proclus, p. 22, 15. [3] *ibid.* p. 56, 25.
[4] *ibid.* p. 71, 18. [5] *ibid.* p. 105, 5.
[6] *ibid.* p. 125, 8. [7] *ibid.* p. 200, 2.
[8] *ibid.* p. 241, 19. [9] *ibid.* p. 352, 15.
[10] *ibid.* p. 362, 15.

[11] See the passages referred to by van Pesch (p. 70). The most glaring case is a passage (p. 21, 19) where he quotes Plotinus, using the expression "Plotinus says......" Comparison with Plotinus, *Ennead.* I. 3, 3, shows that *very few* words are those of Plotinus himself; the rest represent Plotinus' views in Proclus' own language.

[12] Proclus, p. 330, 19 sqq

in writing his commentary. Three have already been disposed of, viz. Heron, Porphyry and Pappus, who had all written commentaries on the *Elements*[1]. We go on to

Eudemus, the pupil of Aristotle, who, among other works, wrote a history of arithmetic, a history of astronomy, and a history of geometry. The importance of the last mentioned work is attested by the frequent use made of it by ancient writers. That there was no other history of geometry written after the time of Eudemus seems to be proved by the remark of Proclus in the course of his famous summary: "Those who compiled histories bring the development of this science up to this point. *Not much younger than these is Euclid*[2]...." The loss of Eudemus' history is one of the gravest which fate has inflicted upon us, for it cannot be doubted that Eudemus had before him a number of the actual works of earlier geometers, which, as before observed, seem to have vanished completely when they were superseded by the treatises of Euclid, Archimedes and Apollonius. As it is, we have to be thankful for the fragments from Eudemus which such writers as Proclus have preserved to us.

I agree with van Pesch[3] that there is no sufficient reason for doubting that the work of Eudemus was accessible to Proclus at first hand. For the later writers Simplicius and Eutocius refer to it in terms such as leave no room for doubt that *they* had it before them. I have already quoted a passage from Simplicius' account of the lunes of Hippocrates to the effect that Eudemus must be considered the best authority since he lived nearer the times[4]. In the same place Simplicius says[5], "I will set out what Eudemus says word for word (κατὰ λέξιν λεγόμενα), adding only a little explanation in the shape of reference to Euclid's Elements *owing to the memorandum-like style of Eudemus* (διὰ τὸν ὑπομνηματικὸν τρόπον τοῦ Εὐδήμου) who sets out his explanations in the abbreviated form usual with ancient writers. Now in the second book of the history of geometry he writes as follows[5]." It is not possible to suppose that Simplicius would have written in this way about the style of Eudemus if he had merely been copying certain passages second-hand out of some other author and had not the original work itself to refer to. In like manner, Eutocius speaks of the paralogisms handed down in connexion with the attempts of Hippocrates and Antiphon to square the circle[6], "with which I imagine that those are accurately acquainted who have examined (ἐπεσκεμμένους) the geometrical history of Eudemus and know the Ceria Aristotelica." How could the contemporaries of Eutocius have *examined* the work of Eudemus unless it was still extant in his time?

The passages in which Proclus quotes Eudemus by name as his authority are as follows:

(1) On I. 26 he says that Eudemus in his history of geometry

---

[1] See pp. 20 to 27 above.
[2] Proclus, p. 68, 4—7.
[3] *De Procli fontibus*, pp. 73—75.
[4] See above, p. 29.
[5] Simplicius, *loc. cit.*, ed. Diels, p. 60, 27.
[6] Archimedes, ed. Heiberg, vol. III. p. 228.

referred this theorem to Thales, inasmuch as it was necessary to Thales' method of ascertaining the distance of ships from the shore[1].

(2) Eudemus attributed to Thales the discovery of Eucl. I. 15[2], and

(3) to Oenopides the problem of I. 23[3].

(4) Eudemus referred the discovery of the theorem in I. 32 to the Pythagoreans, and gave their proof of it, which Proclus reproduces[4].

(5) On I. 44 Proclus tells us[5] that Eudemus says that "these things are ancient, being discoveries of the Pythagorean muse, the application (παραβολή) of areas, their exceeding (ὑπερβολή) and their falling short (ἔλλειψις)." The next words about the appropriation of these terms (parabola, hyperbola and ellipse) by later writers (i.e. Apollonius) to denote the conic sections are of course not due to Eudemus.

Coming now to notes where Eudemus is not named by Proclus, we may fairly conjecture, with van Pesch, that Eudemus was really the authority for the statements (1) that Thales first proved that a circle is bisected by its diameter[6] (though the proof by *reductio ad absurdum* which follows in Proclus cannot be attributed to Thales[7]), (2) that "Plato made over to Leodamas the analytical method, by means of which *it is recorded* (ἱστόρηται) that the latter too made many discoveries in geometry[8]," (3) that the theorem of I. 5 was due to Thales, and that for equal angles he used the more archaic expression "similar" angles[9], (4) that Oenopides first investigated the problem of I. 12, and that he called the perpendicular the *gnomonic* line (κατὰ γνώμονα)[10], (5) that the theorem that only three sorts of polygons can fill up the space round a point, viz. the equilateral triangle, the square and the regular hexagon, was Pythagorean[11]. Eudemus may also be the authority for Proclus' description of the two methods, referred to Plato and Pythagoras respectively, of forming right-angled triangles in whole numbers[12].

We cannot attribute to Eudemus the beginning of the note on I. 47 where Proclus says that "if we listen to those who like to recount ancient history, we may find some of them referring this theorem to Pythagoras and saying that he sacrificed an ox in honour of his discovery[13]." As such a sacrifice was contrary to the Pythagorean tenets, and Eudemus could not have been unaware of this, the story cannot rest on his authority. Moreover Proclus speaks as though he were not certain of the correctness of the tradition; indeed,

---

[1] Proclus, p. 352, 14—18.
[2] *ibid.* p. 299, 3.
[3] *ibid.* p. 333, 5.
[4] *ibid.* p. 379, 1—16.
[5] *ibid.* p. 419, 15—18.
[6] *ibid.* p. 157, 10, 11.
[7] Cantor (*Gesch. d. Math.* I₃, p. 221) points out the connexion between the *reductio ad absurdum* and the analytical method said to have been discovered by Plato. Proclus gives the proof by *reductio ad absurdum* to meet an imaginary critic who desires a *mathematical* proof; possibly Thales may have been satisfied with the argument in the same sentence which mentions Thales, "the cause of the bisection being the unswerving course of the straight line through the centre."
[8] Proclus, p. 211, 19—23.
[9] *ibid.* p. 250, 20.
[10] *ibid.* p. 283, 7—10.
[11] *ibid.* pp. 304, 11—305, 3.
[12] *ibid.* pp. 428, 7—429, 9.
[13] *ibid.* p. 426, 6—9.

so far as the story of the sacrifice is concerned, the same thing is told of Thales in connexion with his discovery that the angle in a semi-circle is a right angle[1], and Plutarch is not certain whether the ox was sacrificed on the discovery of I. 47 or of the problem about application of areas[2]. Plutarch's doubt suggests that he knew of no evidence for the story beyond the vague allusion in the distich of Apollodorus "Logisticus" (the "calculator") cited by Diogenes Laertius also[3]; and Proclus may have had in mind this couplet with the passages of Plutarch.

We come now to the question of the famous historical summary given by Proclus[4]. No one appears to maintain that Eudemus is the author of even the early part of this summary in the form in which Proclus gives it. It is, as is well known, divided into two distinct parts, between which comes the remark, "Those who compiled histories[5] bring the development of this science up to this point. Not much younger than these is Euclid, who put together the Elements, collecting many of the theorems of Eudoxus, perfecting many others by Theaetetus, and bringing to irrefragable demonstration the things which had only been somewhat loosely proved by his predecessors." Since Euclid was later than Eudemus, it is impossible that Eudemus can have written this. Yet the style of the summary after this point does not show any such change from that of the former portion as to suggest different authorship. The author of the earlier portion recurs frequently to the question of the origin of the *elements* of geometry in a way in which no one would be likely to do who was not later than Euclid; and it must be the same hand which in the second portion connects Euclid's *Elements* with the work of Eudoxus and Theaetetus[6].

If then the summary is the work of one author, and that author not Eudemus, who is it likely to have been? Tannery answers that it is Geminus[7]; but I think, with van Pesch, that he has failed to show why it should be Geminus rather than another. And certainly the extracts which we have from Geminus' work suggest that the sort of topics which it dealt with was quite different; they seem rather to have been general questions of the *content* of mathematics, and even Tannery admits that historical details could only have come incidentally into the work[8].

Could the author have been Proclus himself? Circumstances

<hr/>

[1] Diogenes Laertius, I. 24, p. 6, ed. Cobet.
[2] Plutarch, *non posse suaviter vivi secundum Epicurum*, 11; *Symp.* VIII, 2.
[3] Diog. Laert. VIII. 12, p. 207, ed. Cobet:
     Ἡνίκα Πυθαγόρης τὸ περικλεὲς εὕρετο γράμμα,
          κεῖν᾽ ἐφ᾽ ὅτῳ κλεινὴν ἤγαγε βουθυσίην.
See on this subject Tannery, *La Géométrie grecque*, p. 105.
[4] Proclus, pp. 64—70.
[5] The plural is well explained by Tannery, *La Géométrie grecque*, pp. 73, 74. No doubt the author of the summary tried to supplement Eudemus by means of any other histories which threw light on the subject. Thus e.g. the allusion (p. 64, 21) to the Nile recalls Herodotus. Cf. the expression in Proclus, p. 64, 19, παρὰ τῶν πολλῶν ἱστόρηται.
[6] Tannery, *La Géométrie grecque*, p. 75.
[7] *ibid.* pp. 66—75.   [8] *ibid.* p. 19.

which seem to suggest this possibility are (1) that, as already stated, the question of the origin of the *Elements* is kept prominent, (2) that there is no mention of Democritus, whom Eudemus would not be likely to have ignored, while a follower of Plato would be likely enough to do him the injustice, following the example of Plato who was an opponent of Democritus, never once mentions him, and is said to have wished to burn all his writings[1], and (3) the allusion at the beginning to the "inspired Aristotle" (ὁ δαιμόνιος Ἀριστοτέλης)[2], though this may easily have been inserted by Proclus in a quotation made by him from someone else. On the other hand there are considerations which suggest that Proclus himself was *not* the writer. (1) The style of the whole passage is not such as to point to him as the author. (2) If he wrote it, it is hardly conceivable that he would have passed over in silence the discovery of the analytical method, the invention of Plato to which he attached so much importance[3].

There is nothing improbable in the conjecture that Proclus quoted the summary from a compendium of Eudemus' history made by some later writer: but as yet the question has not been definitely settled. All that is certain is that the early part of the summary must have been made up from scattered notices found in the great work of Eudemus.

Proclus refers to another work of Eudemus besides the history, viz. a book on *The Angle* (βιβλίον περὶ γωνίας)[4]. Tannery assumes that this must have been part of the history, and uses this assumption to confirm his idea that the history was arranged according to *subjects*, not according to chronological order[5]. The phraseology of Proclus however unmistakably suggests a separate work; and that the history was *chronologically* arranged seems to be clearly indicated by the remark of Simplicius that Eudemus "also counted Hippocrates among the more ancient writers" (ἐν τοῖς παλαιοτέροις)[6].

The passage of Simplicius about the lunes of Hippocrates throws considerable light on the style of Eudemus' history. Eudemus wrote in a memorandum-like or summary manner (τὸν ὑπομνηματικὸν τρόπον τοῦ Εὐδήμου)[7] when reproducing what he found in the ancient writers; sometimes it is clear that he left out altogether proofs or constructions of things by no means easy[8].

Geminus.

The discussions about the date and birthplace of Geminus form a whole literature, as to which I must refer the reader to Manitius and Tittel[9]. Though the name looks like a Latin name (Gemĭnus), Mani-

[1] Diog. Laertius, IX. 40, p. 237, ed. Cobet.   [2] Proclus, p. 64, 8.
[3] Proclus, p. 211, 19 sqq.; the passage is quoted above, p. 36.
[4] *ibid.* p. 125, 8.   [5] Tannery, *La Géométrie grecque*, p. 26.
[6] Simplicius, ed. Diels, p. 69, 23.   [7] *ibid.* p. 60, 29.
[8] Cf. Simplicius, p. 63, 19 sqq.; p. 64, 25 sqq.; also Usener's note "de supplendis Hippocratis quas omisit Eudemus constructionibus" added to Diels' preface, pp. xxiii—xxvi.
[9] Manitius, *Gemini elementa astronomiae* (Teubner, 1898), pp. 237—252; Tittel, art. "Geminos" in Pauly-Wissowa's *Real-Encyclopädie der classischen Altertumswissenschaft*, vol. VII., 1910.

tius concluded that, since it appears as Γεμῖνος in all Greek MSS. and as Γεμεῖνος in some inscriptions, it is Greek and possibly formed from γεμ as Ἐργῖνος is from ἐργ and Ἀλεξῖνος from ἀλεξ (cf. also Ἰκτῖνος, Κρατῖνος). Tittel is equally positive that it is Geminus and suggests that Γεμῖνος is due to a *false* analogy with Ἀλεξῖνος etc. and Γεμεῖνος wrongly formed on the model of Ἀντωνεῖνος, Ἀγριππεῖνα. Geminus, a Stoic philosopher, born probably in the island of Rhodes, was the author of a comprehensive work on the classification of mathematics, and also wrote, about 73–67 B.C., a not less comprehensive commentary on the meteorological textbook of his teacher Posidonius of Rhodes.

It is the former work in which we are specially interested here. Though Proclus made great use of it, he does not mention its title, unless we may suppose that, in the passage (p. 177, 24) where, after quoting from Geminus a classification of lines which never meet, he says, "these remarks I have selected from the φιλοκαλία of Geminus," φιλοκαλία is a title or an alternative title. Pappus however quotes a work of Geminus "on the classification of the mathematics" (ἐν τῷ περὶ τῆς τῶν μαθημάτων τάξεως)[1], while Eutocius quotes from "the sixth book of the doctrine of the mathematics" (ἐν τῷ ἕκτῳ τῆς τῶν μαθημάτων θεωρίας)[2]. Tannery[3] pointed out that the former title corresponds well enough to the long extract[4] which Proclus gives in his first prologue, and also to the fragments contained in the *Anonymi variae collectiones* published by Hultsch at the end of his edition of Heron[5]; but it does not suit most of the other passages borrowed by Proclus. The correct title was therefore probably that given by Eutocius, *The Doctrine*, or *Theory, of the Mathematics*; and Pappus probably refers to one particular portion of the work, say the first Book. If the sixth Book treated of conics, as we may conclude from Eutocius, there must have been more Books to follow, because Proclus has preserved us details about higher curves, which must have come later. If again Geminus finished his work and wrote with the same fulness about the other branches of mathematics as he did about geometry, there must have been a considerable number of Books altogether. At all events it seems to have been designed to give a complete view of the whole science of mathematics, and in fact to be a sort of encyclopaedia of the subject.

I shall now indicate first the certain, and secondly the probable, obligations of Proclus to Geminus, in which task I have only to follow van Pesch, who has embodied the results of Tittel's similar inquiry also[6]. I shall only omit the passages as regards which a case for attributing them to Geminus does not seem to me to have been made out.

First come the following passages which must be attributed to Geminus, because Proclus mentions his name:

    (1)   (In the first prologue of Proclus[7]) on the division of mathe-

---

[1] Pappus, ed. Hultsch, p. 1026, 9.   [2] Apollonius, ed. Heiberg, vol. II. p. 170.
[3] Tannery, *La Géométrie grecque*, pp. 18, 19.   [4] Proclus, pp. 38, 1—42, 8.
[5] Heron, ed. Hultsch, pp. 246, 16—249, 12.
[6] Van Pesch, *De Procli fontibus*, pp. 97—113. The dissertation of Tittel is entitled *De Gemini Stoici studiis mathematicis* (1895).
[7] Proclus, pp. 38, 1—42, 8, except the allusion in p. 41, 8—10, to Ctesibius and Heron and

matical sciences into arithmetic, geometry, mechanics, astronomy, optics, geodesy, canonic (science of musical harmony), and logistic (apparently arithmetical problems);

(2) (in the note on the definition of a straight line) on the classification of lines (including curves) as simple (straight or circular) and mixed, composite and incomposite, uniform (ὁμοιομερεῖς) and non-uniform (ἀνομοιομερεῖς), lines "about solids" and lines produced by cutting solids, including conic and spiric sections[1];

(3) (in the note on the definition of a plane surface) on similar distinctions extended to surfaces and solids[2];

(4) (in the note on the definition of parallels) on lines which *do not meet* (ἀσύμπτωτοι) but which are not on that account parallel, e.g. a curve and its asymptote, showing that the property of *not meeting* does not make lines parallel—a favourite observation of Geminus—and, incidentally, on *bounded* lines or those which *enclose a figure* and those which do not[3];

(5) (in the same note) the definition of parallels given by Posidonius[4];

(6) on the distinction between postulates and axioms, the futility of trying to prove axioms, as Apollonius tried to prove Axiom 1, and the equal incorrectness of assuming what really requires proof, "as Euclid did in the fourth postulate [equality of right angles] and in the fifth postulate [the parallel-postulate][5]";

(7) on Postulates 1, 2, 3, which Geminus makes depend on the idea of a straight line being described by the motion of a point[6];

(8) (in the note on Postulate 5) on the inadmissibility in geometry of an argument which is merely plausible, and the danger in this particular case owing to the existence of lines which do converge *ad infinitum* and yet never meet[7];

(9) (in the note on I. 1) on the subject-matter of geometry, theorems, problems and διορισμοί (conditions of possibility) for problems[8];

(10) (in the note on I. 5) on a generalisation of I. 5 by Geminus through the substitution for the rectilineal base of "one uniform line (curve)," by means of which he proved that the only "uniform lines"

---

their pneumatic devices (θαυματοποιϊκή), as regards which Proclus' authority may be Pappus (VIII. p. 1024, 24—27) who uses very similar expressions. Heron, even if not later than Geminus, could hardly have been included in a historical work by him. Perhaps Geminus may have referred to Ctesibius only, and Proclus may have inserted "and Heron" himself.

[1] Proclus, pp. 103, 21—107, 10; pp. 111, 1—113, 3.

[2] *ibid.* pp. 117, 14—120, 12, where perhaps in the passage pp. 117, 22—118, 23 we may have Geminus' own words.

[3] *ibid.* pp. 176, 18—177, 25; perhaps also p. 175. The note ends with the words "These things too we have selected from Geminus' Φιλοκαλία for the elucidation of the matters in question." Tannery (p. 27) takes these words coming at the end of the commentary on the definitions as referring to the whole of the portion of the commentary dealing with the definitions. Van Pesch properly regards them as only applying to the note on *parallels*. This seems to me clear from the use of the word *too* (τοσαῦτα καί).

[4] Proclus, p. 176, 5—17.

[5] *ibid.* pp. 178—182, 4; pp. 183, 14—184, 10; cf. p. 188, 3—11.

[6] *ibid.* p. 185, 6—25.

[7] *ibid.* p. 192, 5—29.   [8] *ibid.* pp. 200, 21—202, 25.

(alike in all their parts) are a straight line, a circle, and a cylindrical helix[1];

(11)　(in the note on I. 10) on the question whether a line is made up of indivisible parts (ἀμερῆ), as affecting the problem of bisecting a given straight line[2];

(12)　(in the note on I. 35) on *topical*, or *locus*-theorems[3], where the illustration of the equal parallelograms described between a hyperbola and its asymptotes may also be due to Geminus[4].

Other passages which may fairly be attributed to Geminus, though his name is not mentioned, are the following:

(1)　in the prologue, where there is the same allusion as in the passage (8) above to a remark of Aristotle that it is equally absurd to expect scientific proofs from a rhetorician and to accept mere plausibilities from a geometer[5];

(2)　a passage in the prologue about the subject-matter, methods, and bases of geometry, the latter including axioms and postulates[6];

(3)　another on the definition and nature of *elements*[7];

(4)　a remark on the Stoic use of the term axiom for every simple statement (ἀπόφανσις ἁπλῆ)[8];

(5)　another discussion on theorems and problems[9], in the middle of which however there are some sentences by Proclus himself[10].

(6)　another passage, in connexion with Def. 3, on lines including or not including a figure (with which cf. part of the passage (4) above)[11];

(7)　a classification of different sorts of angles according as they are contained by simple or mixed lines (or curves)[12];

(8)　a similar classification of figures[13], and of plane figures[14];

(9)　Posidonius' definition of a *figure*[15];

(10)　a classification of triangles into seven kinds[16];

(11)　a note distinguishing lines (or curves) producible indefinitely or not so producible, whether forming a figure or not forming a figure (like the "single-turn spiral")[17];

(12)　passages distinguishing different sorts of problems[18], different sorts of theorems[19], and two sorts of converses (complete and partial)[20];

(13)　the definition of the term "porism" as used in the title of Euclid's *Porisms*, as distinct from the other meaning of "corollary"[21];

(14)　a note on the Epicurean objection to I. 20 as being obvious even to an ass[22];

(15)　a passage on the properties of parallels, with allusions to

---

[1] Proclus, p. 251, 2—11.

[2] *ibid.* pp. 277, 25—279, 11.

[3] *ibid.* pp. 394, 11—395, 2 and p. 395, 13—21.

[4] *ibid.* p. 395, 8—12.

[5] *ibid.* pp. 33, 21—34, 1.

[6] *ibid.* pp. 57, 9—58, 3.

[7] *ibid.* pp. 72, 3—75, 4.

[8] *ibid.* p. 77, 3—6.

[9] *ibid.* pp. 77, 7—78, 13, and 79, 3—81, 4.

[10] *ibid.* pp. 78, 13—79, 2.

[11] *ibid.* pp. 102, 22—103, 18.

[12] *ibid.* pp. 126, 7—127, 16.

[13] *ibid.* pp. 159, 12—160, 9.

[14] *ibid.* pp. 162, 27—164, 6.

[15] *ibid.* p. 143, 5—11.

[16] *ibid.* p. 168, 4—12.

[17] *ibid.* p. 187, 19—27.

[18] *ibid.* pp. 220, 7—222, 14; also p. 330, 6—9.

[19] *ibid.* pp. 244, 14—246, 12.

[20] *ibid.* pp. 252, 5—254, 20.

[21] *ibid.* pp. 301, 21—302, 13.

[22] *ibid.* pp. 322, 4—323, 3.

Apollonius' *Conics*, and the curves invented by Nicomedes, Hippias and Perseus[1];

(16)   a passage on the parallel-postulate regarded as the converse of I. 17[2].

Of the authors to whom Proclus was indebted in a less degree the most important is **Apollonius of Perga**. Two passages allude to his *Conics*[3], one to a work on irrationals[4], and two to a treatise *On the cochlias* (apparently the cylindrical helix) by Apollonius[5]. But more important for our purpose are six references to Apollonius in connexion with elementary geometry.

(1)   He appears as the author of an attempt to explain the idea of a line (possessing length but no breadth) by reference to daily experience, e.g. when we tell someone to measure, merely, the length of a road or of a wall[6]; and doubtless the similar passage showing how we may in like manner get a notion of a surface (without depth) is his also[7].

(2)   He gave a new general definition of an angle[8].

(3)   He tried to prove certain axioms[9], and Proclus gives his attempt to prove Axiom 1, word for word[10].

Proclus further quotes:

(4)   Apollonius' solution of the problem in Eucl. I. 10, avoiding Euclid's use of I. 9[11],

(5)   his solution of the problem in I. 11, differing only slightly from Euclid's[12], and

(6)   his solution of the problem in I. 23[13].

Heiberg[14] conjectures that Apollonius departed from Euclid's method in these propositions because he objected to solving problems of a more general, by means of problems of a more particular, character. Proclus however considers all three solutions inferior to Euclid's; and his remarks on Apollonius' handling of these elementary matters generally suggest that he was nettled by criticisms of Euclid in the work containing the things which he quotes from Apollonius, just as we conclude that Pappus was offended by the remarks of Apollonius about Euclid's incomplete treatment of the "three- and four-line locus[15]." If this was the case, Proclus can hardly have got his information about these things at second-hand; and there seems to be no reason to doubt that he had the actual work of Apollonius before him. This work may have been the treatise mentioned by Marinus in the words "Apollonius in his general treatise" ('Απολλώνιος ἐν τῇ καθόλου πραγματείᾳ)[16]. If the notice in the *Fihrist*[17] stating, on the authority of Thābit b. Qurra, that

---

[1] Proclus, pp. 355, 20—356, 16.
[3] *ibid.* p. 71, 19; p. 356, 8, 6.
[5] *ibid.* pp. 105, 5, 6, 14, 15.
[7] *ibid.* p. 114, 20—25.
[9] *ibid.* p. 183, 13, 14.
[11] *ibid.* pp. 279, 16—280, 4.
[13] *ibid.* pp. 335, 16—336, 5.
[15] See above, pp. 2, 3.
[17] *Fihrist*, tr. Suter, p. 19.

[2] *ibid.* p. 364, 9—12; pp. 364, 20—365, 4.
[4] *ibid.* p. 74, 23, 24.
[6] *ibid.* p. 100, 5—19.
[8] *ibid.* p. 123, 15—19 (cf. p. 124, 17, p. 125, 17).
[10] *ibid.* pp. 194, 25—195, 5.
[12] *ibid.* p. 282, 8—19.
[14] *Philologus*, vol. XLIII. p. 489.
[16] *Marinus in Euclidis Data*, ed. Menge, p. 234, 16.

Apollonius wrote a tract on the parallel-postulate be correct, it may have been included in the same work. We may conclude generally that, in it, Apollonius tried to remodel the beginnings of geometry, reducing the number of axioms, appealing, in his definitions of lines, surfaces etc., more to experience than to abstract reason, and substituting for certain proofs others of a more general character.

The probabilities are that, in quoting from the tract of **Ptolemy** in which he tried to prove the parallel-postulate, Proclus had the actual work before him. For, after an allusion to it as "a certain book[1]" he gives two long extracts[2], and at the beginning of the second indicates the title of the tract, "in the (book) about the meeting of straight lines produced from (angles) less than two right angles," as he has very rarely done in other cases.

Certain things from **Posidonius** are evidently quoted at second-hand, the authority being Geminus (e.g. the definitions of *figure* and *parallels*); but besides these we have quotations from a separate work which he wrote to controvert Zeno of Sidon, an Epicurean who had sought to destroy the whole of geometry[3]. We are told that Zeno had argued that, even if we admit the fundamental principles (ἀρχαί) of geometry, the deductions from them cannot be proved without the admission of something else as well, which has not been included in the said principles[4]. On I. 1 Proclus gives at some length the arguments of Zeno and the reply of Posidonius as regards this proposition[5]. In this case Zeno's "something else" which he considers to be assumed is the fact that two straight lines cannot have a common segment, and then, as regards the "proof" of it by means of the bisection of a circle by its diameter, he objects that it has been assumed that two *circumferences* (arcs) of circles cannot have a common part. Lastly, he makes up, for the purpose of attacking it, another supposed "proof" of the fact that two straight lines cannot have a common part. Proclus appears, more than once, to be quoting the actual words of Zeno and Posidonius; in particular, two expressions used by Posidonius about "the acrid Epicurean" (τὸν δριμὺν Ἐπικούρειον)[6] and his "misrepresentations" (Ποσειδώνιός φησι τὸν Ζήνωνα συκοφαντεῖν)[7]. It is not necessary to suppose that Proclus had the original work of Zeno before him, because Zeno's arguments may easily have been got from Posidonius' reply; but he would appear to have quoted direct from the latter at all events.

The work of **Carpus** *mechanicus* (a treatise on astronomy) quoted from by Proclus[8] must have been accessible to him at first-hand, because a portion of the extract from it about the relation of theorems and problems[9] is reproduced word for word. Moreover, if he were not using the book itself, Proclus would hardly be in a position to question whether the introduction of the subject of theorems and problems

[1] Proclus, p. 191, 23.　　　[2] *ibid.* pp. 362, 14—363, 18; pp. 365, 7—367, 27.
[3] *ibid.* p. 200, 1—3.　　　　　　[4] *ibid.* pp. 199, 11—200, 1.
[5] *ibid.* pp. 214, 18—215, 13; pp. 216, 10—218, 11.
[6] *ibid.* p. 216, 21.　　　　　[7] *ibid.* p. 218, 1.
[8] *ibid.* pp. 241, 19—243, 11.　　[9] *ibid.* pp. 242, 22—243, 11.

was opportune in the place where it was found (εἰ μὲν κατὰ καιρὸν ἦ μή, παρείσθω πρὸς τὸ παρόν)[1].

It is of course evident that Proclus had before him the original works of Plato, Aristotle, Archimedes and Plotinus, as well as the Σύμμικτά of Porphyry and the works of his master Syrianus (ὁ ἡμέτερος καθηγεμών)[2], from whom he quotes in his note on the definition of an angle. Tannery also points out that he must have had before him a group of works representing the Pythagorean tradition on its mystic, as distinct from its mathematical, side, from Philolaus downwards, and comprising the more or less apocryphal ἱερὸς λόγος of Pythagoras, the Oracles (λόγια), and Orphic verses[3].

Besides quotations from writers whom we can identify with more or less certainty, there are many other passages which are doubtless quoted from other commentators whose names we do not know. A list of such passages is given by van Pesch[4], and there is no need to cite them here.

Van Pesch also gives at the end of his work[5] a convenient list of the books which, as the result of his investigation, he deems to have been accessible to and directly used by Proclus. The list is worth giving here, on the same ground of convenience. It is as follows:

Eudemus : *history of geometry.*
Geminus : *the theory of the mathematical sciences.*
Heron : *commentary on the Elements of Euclid.*
Porphyry :   „   „   „
Pappus :   „   „   „
Apollonius of Perga : a work relating to elementary geometry.
Ptolemy : *on the parallel-postulate.*
Posidonius : a book controverting Zeno of Sidon.
Carpus : *astronomy.*
Syrianus : a discussion on the *angle.*
Pythagorean philosophical tradition.
Plato's works.
Aristotle's works.
Archimedes' works.
Plotinus : *Enneades.*

Lastly we come to the question what passages, if any, in the commentary of Proclus represent his own contributions to the subject. As we have seen, the *onus probandi* must be held to rest upon him who shall maintain that a particular note is original on the part of Proclus. Hence it is not enough that it should be impossible to point to another writer as the probable source of a note; we must have a *positive* reason for attributing it to Proclus. The criterion must therefore be found either (1) in the general terms in which Proclus points out the deficiencies in previous commentaries and indicates the respects in which his own will differ from them, or (2) in specific expressions used by him in introducing particular notes which may

---

[1] Proclus, p. 241, 21, 22.
[3] Tannery, *La Géométrie grecque*, pp. 25, 26.
[4] Van Pesch, *De Procli fontibus*, p. 139.
[2] *ibid.* p. 123, 19.
[5] *ibid.* p. 155.

indicate that he is giving his own views. Besides indicating that he paid more attention than his predecessors to questions requiring deeper study (τὸ πραγματειῶδες) and " pursued clear distinctions" (τὸ εὐδιαίρετον μεταδιώκοντας)[1]—by which he appears to imply that his predecessors had confused the different departments of their commentaries, viz. lemmas, cases, and objections (ἐνστάσεις)[2]—Proclus complains that the earlier commentators had failed to indicate the ultimate grounds or *causes* of propositions[3]. Although it is from Geminus that he borrowed a passage maintaining that it is one of the proper functions of geometry to inquire into causes (τὴν αἰτίαν καὶ τὸ διὰ τί)[4], yet it is not likely that Geminus dealt with Euclid's propositions one by one ; and consequently, when we find Proclus, on I. 8, 16, 17, 18, 32, and 47[5], endeavouring to explain *causes*, we have good reason to suppose that the explanations are his own.

Again, his remarks on certain things which he quotes from Pappus can scarcely be due to anyone else, since Pappus is the latest of the commentators whose works he appears to have used. Under this head, come

(1) his objections to certain new axioms introduced by Pappus[6],

(2) his conjecture as to how Pappus came to think of his alternative proof of I. 5[7],

(3) an addition to Pappus' remarks about the curvilineal angle which is equal to a right angle without being one[8].

The defence of Geminus against Carpus, who combated his view of theorems and problems, is also probably due to Proclus[9], as well as an observation on I. 38 to the effect that I. 35—38 are really comprehended in VI. 1 as particular cases[10].

Lastly, we can have no hesitation in attributing to Proclus himself (1) the criticism of Ptolemy's attempt to prove the parallel-postulate[11], and (2) the other attempted proof given in the same note[12] (on I. 29) and assuming as an axiom that " if from one point two straight lines forming an angle be produced *ad infinitum* the distance between them when so produced *ad infinitum* exceeds any finite magnitude (i.e. length)," an assumption which purports to be the equivalent of a statement in Aristotle[13]. It is introduced by words in which the writer appears to claim originality for his proof : " To him who desires to see this proved (κατασκευαζόμενον) *let it be said by us* (λεγέσθω παρ᾽ ἡμῶν)" etc.[14] Moreover, Philoponus, in a note on Aristotle's *Anal. post.* I. 10, says that " the geometer (Euclid) assumes this as an axiom, but it wants a great deal of proof, insomuch that both Ptolemy and Proclus wrote a whole book upon it[15]."

---

[1] Proclus, p. 84, 13, p. 432, 14, 15.   [2] Cf. *ibid.* p. 289, 11—15; p. 432, 15—17.
[3] *ibid.* p. 432, 17.   [4] *ibid.* p. 202, 9—25.
[5] See Proclus, p. 270, 5—24 (I. 8); pp. 309, 3—310, 8 (I. 16); pp. 310, 19—311, 23 (I. 17); pp. 316, 14—318, 2 (I. 18); pp. 384, 13—21 (I. 32); pp. 426, 22—427, 8 (I. 47).
[6] Proclus, p. 198, 5—15.   [7] *ibid.* p. 250, 12—19.   [8] *ibid.* p. 190, 9—23.
[9] *ibid.* p. 243, 12—29.   [10] *ibid.* pp. 405, 6—406, 9.
[11] *ibid.* p. 368, 1—23.   [12] *ibid.* pp. 371, 11—373, 2.
[13] Aristotle, *de caelo*, I. 5 (271 b 28—30).   [14] Proclus, p. 371, 10.
[15] Berlin Aristotle, vol. IV. p. 214 a 9—12.

CHAPTER V.

THE TEXT[1].

IT is well known that the title of Simson's edition of Euclid (first brought out in ʾLatin and English in 1756) claims that, in it, "the errors by which Theon, or others, have long ago vitiated these books are corrected, and some of Euclid's demonstrations are restored"; and readers of Simson's notes are familiar with the phrases used, where anything in the text does not seem to him satisfactory, to the effect that the demonstration has been spoiled, or things have been interpolated or omitted, by Theon "or some other unskilful editor." Now most of the MSS. of the Greek text prove by their titles that they proceed from the recension of the *Elements* by Theon; they purport to be either "from the edition of Theon" (ἐκ τῆς Θέωνος ἐκδόσεως) or "from the lectures of Theon" (ἀπὸ συνουσιῶν τοῦ Θέωνος). This was Theon of Alexandria (4th c. A.D.) who also wrote a commentary on Ptolemy, in which there occurs a passage of the greatest importance in this connexion[2]: "But that sectors in equal circles are to one another as the angles on which they stand *has been proved by me in my edition of the Elements at the end of the sixth book.*" Thus Theon himself says that he edited the *Elements* and also that the second part of VI. 33, found in nearly all the MSS., is his addition.

This passage is the key to the whole question of Theon's changes in the text of Euclid; for, when Peyrard found in the Vatican the MS. 190 which contained neither the words from the titles of the other MSS. quoted above nor the interpolated second part of VI. 33, he was justified in concluding, as he did, that in the Vatican MS. we have an edition more ancient than Theon's. It is also clear that the copyist of P, or rather of its archetype, had before him the two recensions and systematically gave the preference to the earlier one; for at XIII. 6 in P the first hand has added a note in the margin: "This theorem is not given in most copies of the *new edition,* but is found in those of the old." Thus we are more fortunate than Simson, since our judgment of Theon's recension can be formed on the basis, not of mere conjecture, but of the documentary evidence afforded by a comparison of the Vatican MS. just mentioned with what we may conveniently call, after Heiberg, the Theonine MSS.

[1] The material for the whole of this chapter is taken from Heiberg's edition of the *Elements,* introduction to vol. v., and from the same scholar's *Litterargeschichtliche Studien über Euklid,* p. 174 sqq. and *Paralipomena zu Euklid* in *Hermes,* XXXVIII., 1903.
[2] I. p. 201 ed. Halma = p. 50 ed. Basel.

The MSS. used for Heiberg's edition of the *Elements* are the following :

(1) P = Vatican MS. numbered 190, 4to, in two volumes (doubt-less one originally) ; 10th c.

This is the MS. which Peyrard was able to use ; it was sent from Rome to Paris for his use and bears the stamp of the Paris Imperial Library on the last page. It is well and carefully written. There are corrections some of which are by the original hand, but generally in paler ink, others, still pretty old, by several different hands, or by one hand with different ink in different places (P m. 2), and others again by the latest hand (P m. rec.). It contains, first, the *Elements* I.—XIII. with scholia, then Marinus' commentary on the *Data* (without the name of the author), followed by the *Data* itself and scholia, then the *Elements* XIV., XV. (so called), and lastly three books and a part of a fourth of a commentary by Theon εἰς τοὺς προχείρους κανόνας Πτολε-μαίου.

The other MSS. are " Theonine."

(2) F = MS. XXVIII, 3, in the Laurentian Library at Florence, 4to ; 10th c.

This MS. is written in a beautiful and scholarly hand and contains the *Elements* I.—XV., the *Optics* and the *Phaenomena*, but is not well preserved. Not only is the original writing renewed in many places, where it had become faint, by a later hand of the 16th c., but the same hand has filled certain smaller lacunae by gumming on to torn pages new pieces of parchment, and has replaced bodily certain portions of the MS., which had doubtless become illegible, by fresh leaves. The larger gaps so made good extend from Eucl. VII. 12 to IX. 15, and from XII. 3 to the end ; so that, besides the conclusion of the *Elements*, the *Optics* and *Phaenomena* are also in the later hand, and we cannot even tell what in addition to the *Elements* I.—XIII. the original MS. contained. Heiberg denotes the later hand by φ and observes that, while in restoring words which had become faint and filling up minor lacunae the writer used no other MS., yet in the two larger restorations he used the Laurentian MS. XXVIII, 6, belonging to the 13th—14th c. The latter MS. (which Heiberg denotes by f) was copied from the Viennese MS. (V) to be described below.

(3) B = Bodleian MS., D'Orville X. 1 inf. 2, 30, 4to ; A.D. 888.

This MS. contains the *Elements* I.—XV. with many scholia. Leaves 15—118 contain I. 14 (from about the middle of the proposition) to the end of Book VI., and leaves 123—387 (wrongly numbered 397) Books VII.—XV. in one and the same elegant hand (9th c.). The leaves preceding leaf 15 seem to have been lost at some time, leaves 6 to 14 (containing *Elem.* I. to the place in I. 14 above referred to) being carelessly written by a later hand on thick and common parch-ment (13th c.). On leaves 2 to 4 and 122 are certain notes in the hand of Arethas, who also wrote a two-line epigram on leaf 5, the greater part of the scholia in uncial letters, a few notes and corrections, and two sentences on the last leaf, the first of which states that the MS. was written by one Stephen *clericus* in the year of the world 6397

(= 888 A.D.), while the second records Arethas' own acquisition of it. Arethas lived from, say, 865 to 939 A.D. He was Archbishop of Caesarea and wrote a commentary on the Apocalypse. The portions of his library which survive are of the greatest interest to palaeography on account of his exact notes of dates, names of copyists, prices of parchment etc. It is to him also that we owe the famous Plato MS. from Patmos (Cod. Clarkianus) which was written for him in November 895[1].

(4)   V = Viennese MS. Philos. Gr. No. 103; probably 12th c.

This MS. contains 292 leaves, Eucl. *Elements* I.—XV. occupying leaves 1 to 254, after which come the *Optics* (to leaf 271), the *Phaenomena* (mutilated at the end) from leaf 272 to leaf 282, and lastly scholia, on leaves 283 to 292, also imperfect at the end. The different material used for different parts and the varieties of handwriting make it necessary for Heiberg to discuss this MS. at some length[2]. The handwriting on leaves 1 to 183 (Book I. to the middle of X. 105) and on leaves 203 to 234 (from XI. 31, towards the end of the proposition, to XIII. 7, a few lines down) is the same; between leaves 184 and 202 there are two varieties of handwriting, that of leaves 184 to 189 and that of leaves 200 (verso) to 202 being the same. Leaf 235 begins in the same handwriting, changes first gradually into that of leaves 184 to 189 and then (verso) into a third more rapid cursive writing which is the same as that of the greater part of the scholia, and also as that of leaves 243 and 282, although, as these leaves are of different material, the look of the writing and of the ink seems altered. There are corrections both by the first and a second hand, and scholia by many hands. On the whole, in spite of the apparent diversity of handwriting in the MS., it is probable that the whole of it was written at about the same time, and it may (allowing for changes of material, ink etc.) even have been written by the same man. It is at least certain that, when the Laurentian MS. XXVIII, 6 was copied from it, the whole MS. was in the condition in which it is now, except as regards the later scholia and leaves 283 to 292 which are not in the Laurentian MS., that MS. coming to an end where the *Phaenomena* breaks off abruptly in V. Hence Heiberg attributes the whole MS. to the 12th c.

But it was apparently in two volumes originally, the first consisting of leaves 1 to 183; and it is certain that it was not all copied at the same time or from one and the same original. For leaves 184 to 202 were evidently copied from two MSS. different both from one another and from that from which the rest was copied. Leaves 184 to the middle of leaf 189 (recto) must have been copied from a MS. similar to P, as is proved by similarity of readings, though not from P itself. The rest, up to leaf 202, were copied from the Bologna MS. (b) to be mentioned below. It seems clear that the content of leaves 184 to 202 was supplied from other MSS. because there was a lacuna in the original from which the rest of V was copied.

---

[1] See Pauly-Wissowa, *Real-Encyclopädie der class. Altertumswissenschaft*, vol. II., 1896, p. 675.

[2] Heiberg, vol. V. pp. xxix—xxxiii.

Heiberg sums up his conclusions thus. The copyist of V first copied leaves 1 to 183 from an original in which two *quaterniones* were missing (covering from the middle of Eucl. X. 105 to near the end of XI. 31). Noticing the lacuna he put aside one *quaternio* of the parchment used up to that point. Then he copied onwards from the end of the lacuna in the original to the end of the *Phaenomena*. After this he looked about him for another MS. from which to fill up the lacuna; finding one, he copied from it as far as the middle of leaf 189 (recto). Then, noticing that the MS. from which he was copying was of a different class, he had recourse to yet another MS. from which he copied up to leaf 202. At the same time, finding that the lacuna was longer than he had reckoned for, he had to use twelve more leaves of a different parchment in addition to the *quaternio* which he had put aside. The whole MS. at first formed two volumes (the first containing leaves 1 to 183 and the second leaves 184 to 282); then, after the last leaf had perished, the two volumes were made into one to which two more *quaterniones* were also added. A few leaves of the latter of these two have since perished.

(5)   b = MS. numbered 18—19 in the Communal Library at Bologna, in two volumes, 4to; 11th c.

This MS. has scholia in the margin written both by the first hand and by two or three later hands; some are written by the latest hand, Theodorus Cabasilas (a descendant apparently of Nicolaus Cabasilas, 14th c.) who owned the MS. at one time. It contains (*a*) in 14 *quaterniones* the definitions and the enunciations (without proofs) of the *Elements* I.—XIII. and of the *Data*, (*b*) in the remainder of the volumes the *Proem to Geometry* (published among the *Variae Collectiones* in Hultsch's edition of Heron, pp. 252, 24 to 274, 14) followed by the *Elements* I.—XIII. (part of XIII. 18 to the end being missing), and then by part of the *Data* (from the last three words of the enunciation of Prop. 38 to the end of the penultimate clause in Prop. 87, ed. Menge). From XI. 36 inclusive to the end of XII. this MS. appears to represent an entirely different recension. Heiberg is compelled to give this portion of b separately in an appendix. He conjectures that it is due to a Byzantine mathematician who thought Euclid's proofs too long and tiresome and consequently contented himself with indicating the course followed[1]. At the same time this Byzantine must have had an excellent MS. before him, probably of the ante-Theonine variety of which the Vatican MS. 190 (P) is the sole representative.

(6)   p = Paris MS. 2466, 4to; 12th c.

This manuscript is written in two hands, the finer hand occupying leaves 1 to 53 (recto), and a more careless hand leaves 53 (verso) to 64, which are of the same parchment as the earlier leaves, and leaves 65 to 239, which are of a thinner and rougher parchment showing traces of writing of the 8th—9th c. (a Greek version of the Old Testament). The MS. contains the *Elements* I.—XIII. and some scholia after Books XI., XII. and XIII.

[1] *Zeitschrift für Math. u. Physik*, XXIX., hist.-litt. Abtheilung, p. 13.

(7) q = Paris MS. 2344, folio; 12th c.

It is written by one hand but includes scholia by many hands. On leaves 1 to 16 (recto) are scholia with the same title as that found by Wachsmuth in a Vatican MS. and relied upon by him to prove that Proclus continued his commentaries beyond Book I.[1] Leaves 17 to 357 contain the *Elements* I.—XIII. (except that there is a lacuna from the middle of VIII. 25 to the ἔκθεσις of IX. 14); before Books VII. and X. there are some leaves filled with scholia only, and leaves 358 to 366 contain nothing but scholia.

(8) Heiberg also used a palimpsest in the British Museum (Add. 17211). Five pages are of the 7th—8th c. and are contained (leaves 49—53) in the second volume of the Syrian MS. Brit. Mus. 687 of the 9th c.; half of leaf 50 has perished. The leaves contain various fragments from Book X. enumerated by Heiberg, Vol. III., p. v, and nearly the whole of XIII. 14.

Since his edition of the *Elements* was published, Heiberg has collected further material bearing on the history of the text[2]. Besides giving the results of further or new examination of MSS., he has collected the fresh evidence contained in an-Nairīzī's commentary, and particularly in the quotations from Heron's commentary given in it (often word for word), which enable us in several cases to trace differences between our text and the text as Heron had it, and to identify some interpolations which actually found their way into the text from Heron's commentary itself; and lastly he has dealt with some valuable fragments of ancient papyri which have recently come to light, and which are especially important in that the evidence drawn from them necessitates some modification in the views expressed in the preface to Vol. V. as to the nature of the changes made in Theon's recension, and in the principles laid down for differentiating between Theon's recension and the original text, on the basis of a comparison between P and the Theonine MSS. alone.

The fragments of ancient papyri referred to are the following.

1. *Papyrus Herculanensis* No. 1061[3].

This fragment quotes Def. 15 of Book I. in Greek, and *omits* the words ἣ καλεῖται περιφέρεια, "which is called the circumference," found in all our MSS., and the further addition πρὸς τὴν τοῦ κύκλου περιφέρειαν also found in practically all the MSS. Thus Heiberg's assumption that both expressions are interpolations is now confirmed by this oldest of all sources.

2. *The Oxyrhynchus Papyri* I. p. 58, No. XXIX. of the 3rd or 4th c.

This fragment contains the enunciation of Eucl. II. 5 (with figure, apparently without letters, immediately following, and not, as usual in our MSS., at the end of the proof) and before it the part of a word περιεχομε belonging to II. 4 (with room for -νῳ ὀρθογωνίῳ· ὅπερ ἔδει

---

[1] [εἰς τ]ὰ τοῦ Εὐκλείδου στοιχεῖα προλαμβανόμενα ἐκ τῶν Πρόκλου σποράδην καὶ κατ᾽ ἐπιτομήν. Cf. p. 32, note 8, above.

[2] Heiberg, *Paralipomena zu Euklid* in *Hermes*, XXXVIII., 1903, pp. 46—74, 161—201, 321—356.

[3] Described by Heiberg in *Oversigt over det kngl. danske Videnskabernes Selskabs Forhandlinger*, 1900, p. 161.

δεῖξαι and a stroke to mark the end), showing that the fragment *had not* the Porism which appears in all the Theonine MSS. and (in a later hand) in P, and thereby confirming Heiberg's assumption that the Porism was due to Theon.

3. A fragment in *Fayum towns and their papyri*, p. 96, No. IX. of 2nd or 3rd c.

This contains I. 39 and I. 41 following one another and almost complete, showing that I. 40 was wanting, whereas it is found in all the MSS. and is recognised by Proclus. Moreover the text of the beginning of I. 39 is better than ours, since it has no double διορισμός but omits the first ("I say that they are also in the same parallels") and has "*and*" instead of "*for* let AD be joined" in the next sentence. It is clear that I. 40 was interpolated by someone who thought there ought to be a proposition following I. 39 and related to it as I. 38 is related to I. 37 and I. 36 to I. 35, although Euclid nowhere uses I. 40, and therefore was not likely to include it. The same interpolator failed to realise that the words "let AD be joined" were part of the ἔκθεσις or *setting-out*, and took them for the κατασκευή or "construction" which generally follows the διορισμός or "particular statement" of the conclusion to be proved, and consequently thought it necessary to insert a διορισμός *before* the words.

The conclusions drawn by Heiberg from a consideration of particular readings in this papyrus along with those of our MSS. will be referred to below.

We now come to the principles which Heiberg followed, when preparing his edition, in differentiating the original text from the Theonine recension by means of a comparison of the readings of P and of the Theonine MSS. The rules which he gives are subject to a certain number of exceptions (mostly in cases where one MS. or the other shows readings due to copyists' errors), but in general they may be relied upon to give conclusive results.

The possible alternatives which the comparison of P with the Theonine MSS. may give in particular passages are as follows:

I. There may be *agreement* in three different degrees.

(1) P and *all* the Theonine MSS. may agree.

In this case the reading common to all, even if it is corrupt or interpolated, is more ancient than Theon, i.e. than the 4th c.

(2) P may agree with *some* (only) of the Theonine MSS.

In this case Heiberg considered that the latter give the true reading of Theon's recension, and the other Theonine MSS. have departed from it.

(3) P and *one* only of the Theonine MSS may agree.

In this case too Heiberg assumed that the *one* Theonine MS. which agrees with P gives the true Theonine reading, and that this rule even supplies a sort of measure of the quality and faithfulness of the Theonine MSS. Now none of them agrees *alone* with P in preserving the true reading so often as F. Hence F must be held to have preserved Theon's recension more faithfully than the other Theonine MSS.; and it would follow that in those portions where F fails us P must carry rather more weight even though it may differ from the Theonine MSS. BVpq. (Heiberg gives many examples in proof of this, as of his main rules generally, for which reference must be made to his *Prolegomena* in Vol. V.) The specially close relation of F and P is also illustrated by passages in which they have the same *errors*; the explanation of these common errors (where not due to accident) is found by Heiberg in the supposition that they existed, but were not noticed by Theon, in the original copy in which he made his changes.

Although however F is by far the best of the Theonine MSS., there are a considerable number of passages where one of the others (B, V, p or q) *alone* with P gives the genuine reading of Theon's recension.

As the result of the discovery of the papyrus fragment containing I. 39, 41, the principles above enunciated under (2) and (3) are found by Heiberg to require some qualification. For there is in some cases a remarkable agreement between the papyrus and the Theonine MSS. (some or all) as against P. This shows that Theon took more trouble to follow older MSS., and made fewer arbitrary changes of his own, than has hitherto been supposed. Next, when the papyrus agrees with some of the Theonine MSS. against P, it must now be held that these MSS. (and not, as formerly supposed, those which agree with P) give the true reading of Theon. If it were otherwise, the agreement between the papyrus and the Theonine MSS. would be accidental: but it happens too often for this. It is clear also that there must have been contamination between the two recensions; otherwise, whence could the Theonine MSS. which agree with P and not with the papyrus have got their readings? The influence of the P class on the Theonine F is especially marked.

II. There may be *disagreement* between P and all the Theonine MSS.

The following possibilities arise.

(1) The Theonine MSS. differ also among themselves.

In this case Heiberg considered that P nearly always has the true reading, and the Theonine MSS. have suffered interpolation in different ways after Theon's time.

(2) The Theonine MSS. all combine against P.

In this case the explanation was assumed by Heiberg to be one or other of the following.

(*a*) The common reading is due to an error which cannot be imputed to Theon (though it may have escaped him when putting together the archetype of his edition); such error may either have arisen accidentally in all alike, or (more frequently) may be referred to a common archetype of all the MSS.

(*β*) There may be an accidental error in P; e.g. something has dropped out of P in a good many places, generally through ὁμοιοτέλευτον

(*γ*) There may be words interpolated in P.

(*δ*) Lastly, *we may have in the Theonine* MSS. *a change made by Theon himself.*

(The discovery of the ancient papyrus showing readings agreeing

with some, or with all, of the Theonine MSS. against P now makes it
necessary to be very cautious in applying these criteria.)

It is of course the last class (δ) of changes which we have to
investigate in order to get a proper idea of Theon's recension.

Heiberg first observes, as regards these, that we shall find that
Theon, in editing the *Elements*, altered hardly anything without some
reason, often inadequate according to our ideas, but still some reason
which seemed to him sufficient. Hence, in cases of very slight differences where both the Theonine MSS. and P have readings good and
probable in themselves, Heiberg is not prepared to put the differences
down to Theon. In those passages where we cannot see the least
reason why Theon, if he had the reading of P before him, should have
altered it, Heiberg would not at once assume the superiority of P
unless there was such a consistency in the differences as would indicate
that they were due not to accident but to design. In the absence of
such indications, he thinks that the ordinary principles of criticism
should be followed and that proper weight should be attached to the
antiquity of the sources. And it cannot be denied that the sources of
the Theonine version are the more ancient. For not only is the
British Museum palimpsest (L), which is intimately connected with
the rest of our MSS., at least two centuries older than P, but the other
Theonine MSS. are so nearly allied that they must be held to have
had a common archetype intermediate between them and the actual
edition of Theon; and, since they themselves are as old as, or older
than P, their archetype must have been much older. Heiberg gives
(pp. xlvi, xlvii) a list of passages where, for this reason, he has
followed the Theonine MSS. in preference to P.

It has been mentioned above that the copyist of P or rather of its
archetype wished to give an ancient recension. Therefore (apart from
clerical errors and interpolations) the first hand in P may be relied
upon as giving a genuine reading even where a correction by the first
hand has been made *at the same time*. But in many places the first
hand has made corrections *afterwards*; on these occasions he must
have used new sources, e.g. when inserting the scholia to the first
Book which P *alone* has, and in a number of passages he has made
additions from Theonine MSS.

We cannot make out any "family tree" for the different Theonine
MSS. Although they all proceeded from a common archetype later
than the edition of Theon itself, they cannot have been copied one
from the other; for, if they had been, how could it have come about
that in one place or other each of them agrees *alone* with P in preserving the genuine reading? Moreover the great variety in their
agreements and disagreements indicates that they have all diverged
to about the same extent from their archetype. As we have seen that
P contains corrections from the Theonine family, so they show corrections from P or other MSS. of the same family. Thus V has part of
the lacuna in the MS. from which it was copied filled up from a MS.
similar to P, and has corrections apparently derived from the same;
the copyist, however, in correcting V, also used another MS. to which

he alludes in the additions to IX. 19 and 30 (and also on X. 23 Por.) : "in the book of the Ephesian (this) is not found." Who this Ephesian of the 12th c. was, we do not know.

We now come to the alterations made by Theon in his edition of the *Elements*. I shall indicate *classes* into which these alterations may be divided but without details (except in cases where they affect the *mathematical content* as distinct from form or language pure and simple)[1].

I. *Alterations made by Theon where he found, or thought he found, mistakes in the original.*

1. Real blots in the original which Theon saw and tried to remove.

(*a*) Euclid has a porism (corollary) to VI. 19, the enunciation of which speaks of similar and similarly described *figures* though the proposition itself refers only to triangles, and therefore the porism should have come after VI. 20. Theon substitutes *triangle* for *figure* and proves the more general porism after VI. 20.

(*b*) In IX. 19 there is a statement which is obviously incorrect. Theon saw this and altered the proof by reducing four alternatives to two, with the result that it fails to correspond to the enunciation even with Theon's substitution of " if " for " when " in the enunciation.

(*c*) Theon omits a porism to IX. 11, although it is necessary for the proof of the succeeding proposition, apparently because, owing to an error in the text (κατὰ τὸν corrected by Heiberg into ἐπὶ τὸ), he could not get out of it the right sense.

(*d*) I should also put into this category a case which Heiberg classifies among those in which Theon merely fancied that he found mistakes, viz. the porism to V. 7 stating that, if four magnitudes are proportional, they are proportional inversely. Theon puts this after V. 4 with a proof, which however has no necessary connexion with V. 4 but is obvious from the definition of proportion.

(*e*) I should also put under this head XI. 1, where Euclid's argument to prove that two straight lines cannot have a common segment is altered.

2. Passages which seemed to Theon to contain blots, and which he therefore set himself to correct, though more careful consideration would have shown that Euclid's words are right or at least may be excused and offer no difficulty to an intelligent reader. Under this head come :

(*a*) an alteration in III. 24.

(*b*) a perfectly unnecessary alteration, in VI. 14, of " equiangular parallelograms " into " parallelograms having one angle equal to one angle," where Theon followed the false analogy of VI. 15.

(*c*) an omission of words in V. 26, owing to his having been misled by a wrong figure.

(*d*) an alteration of the order of XI. Deff. 27, 28.

(*e*) the substitution of " parallelepipedal solid " for " cube " in XI.

---

[1] Exhaustive details under all the different heads are given by Heiberg (Vol. v. pp. lii—lxxv).

38, because Theon observed, correctly enough, that it was true of the parallelepipedal solid in general as well as of the cube, but failed to give weight to the fact that Euclid must have given the particular case of the cube for the simple reason that that was all he wanted for use in XIII. 17.

(*f*) the substitution of the letter Φ for Ω (*V* for *Z* in my figure) because he saw that the perpendicular from K to BΦ would fall on Φ itself, so that Φ, Ω coincide. But, if the substitution is made, it should be *proved* that Φ, Ω coincide. Euclid can hardly have failed to notice the fact, but it may be that he deliberately ignored it as unnecessary for his purpose, because he did not want to lengthen his proposition by giving the proof.

II.   *Emendations intended to improve the form or diction of Euclid.*

Some of these emendations of Theon affect passages of appreciable length. Heiberg notes about ten such passages; the longest is in Eucl. XII. 4 where a whole page of Heiberg's text is affected and Theon's version is put in the Appendix. The kind of alteration may be illustrated by that in IX. 15 where Euclid uses successively the propositions VII. 24, 25, quoting the enunciation of the former but not of the latter; Theon does exactly the reverse. In a few of the cases here quoted by Heiberg, Theon shortened the original somewhat.

But, as a rule, the emendations affect only a few words in each sentence. Sometimes they are considerable enough to alter the conformation of the sentence, sometimes they are trifling alterations "more magistellorum ineptorum" and unworthy of Theon. Generally speaking, they were prompted by a desire to change anything which was out of the common in expression or in form, in order to reduce the language to one and the same standard or norm. Thus Theon changed the order of words, substituted one word for another where the latter was used in a sense unusual with Euclid (e.g. ἐπειδήπερ, "since," for ὅτι in the sense of "because"), or one expression for another in like circumstances (e.g. where, finding "that which was enjoined would be done" in a *theorem*, VII. 31, and deeming the phrase more appropriate to a *problem*, he substituted for it "that which is sought would be manifest"; probably also and for similar reasons he made certain variations between the two expressions usual at the end of propositions ὅπερ ἔδει δεῖξαι and ὅπερ ἔδει ποιῆσαι, *quod erat demonstrandum* and *quod erat faciendum*). Sometimes his alterations show carelessness in the use of technical terms, as when he uses ἅπτεσθαι (to *meet*) for ἐφάπτεσθαι (to *touch*) although the ancients carefully distinguished the two words. The desire of keeping to a standard phraseology also led Theon to omit or add words in a number of cases, and also, sometimes, to change the lettering of figures.

But Theon seems, in editing the *Elements*, to have bestowed the most attention upon

III.   *Additions designed to supplement or explain Euclid.*

First, he did not hesitate to interpolate whole propositions where he thought there was room or use for them. We have already mentioned the addition to VI. 33 of the second part relating to *sectors*, for which Theon himself takes credit in his commentary on Ptolemy. Again, he interpolated the proposition commonly known as VII. 22 (*ex aequo in proportione perturbata* for numbers, corresponding to V. 23), and perhaps also VII. 20, a particular case of VII. 19 as VI. 17 is of VI. 16. He added a second case to VI. 27, a porism to II. 4, a second porism to III. 16, and a lemma after X. 12; perhaps also the porism to V. 19 and the first porism to VI. 20. He also inserted alternative proofs here and there, e.g. in II. 4 (where the alternative differs little from the original) and in VII. 31; perhaps also in X. 1, 6, and 9.

Secondly, he sometimes repeats an argument where Euclid had said "For the same reason," adds specific references to points, straight lines etc. in the figures in order to exclude the possibility of mistake arising from Euclid's reference to them in general terms, or inserts words to make the meaning of Euclid more plain, e.g. *componendo* and *alternately*, where Euclid had left them out. Sometimes he thought to increase by his additions the mathematical precision of Euclid's language in enunciations or elsewhere, sometimes to make smoother and clearer things which Euclid had expressed with unusual brevity and harshness or carelessness, in reliance on the intelligence of his readers.

Thirdly, he supplied intermediate steps where Euclid's argument seemed too rapid and not easy enough to follow. The form of these additions varies; they are sometimes placed as a definite intermediate step with "therefore" or "so that," sometimes they are additions to the statement of premises, sometimes phrases introduced by "since," "for" and the like, after the inference.

Lastly, there is a very large class of additions of a word, or one or two words, for the sake of clearness or consistency. Heiberg gives a number of examples of the addition of such nouns as "triangle," "square," "rectangle," "magnitude," "number," "point," "side," "circle," "straight line," "area" and the like, of adjectives such as "remaining," "right," "whole," "proportional," and of other parts of speech, even down to words like "is" (ἐστί) which is added 600 times, δή, ἄρα, μέν, γάρ, καί and the like.

IV. *Omissions by Theon.*

Heiberg remarks that, Theon's object having been, as above shown, to amplify and explain Euclid, we should not naturally have expected to find him doing much in the contrary process of compression, and it is only owing to the recurrence of a certain sort of omissions so frequently (especially in the first Books) as to exclude the hypothesis of their being all due to chance that we are bound to credit him with alterations making for greater brevity. We have seen, it is true, that he made omissions as well as additions for the purpose of reducing the language to a certain standard form. But there are also a good number of cases where in the enunciation of propositions, and in the *exposition* (the re-statement of them with reference to the figure), he has left out words because, apparently, he regarded Euclid's language as being *too* careful and precise.

Again, he is apparently responsible for the frequent omission of the words ὅπερ ἔδει δεῖξαι (or ποιῆσαι), Q.E.D. (or F.), at the end of propositions. This is often the case at the end of porisms, where, in omitting the words, Theon seems to have deliberately departed from Euclid's practice. The MS. P seems to show clearly that, where Euclid put a porism at the end of a proposition, he omitted the Q.E.D. at the end of the proposition but inserted it at the end of the porism, as if he regarded the latter as being actually a part of the proposition itself. As in the Theonine MSS. the Q.E.D. is generally omitted, the omission would seem to have been due to Theon. Sometimes in these cases the Q.E.D. is interpolated at the end of the proposition.

Heiberg summed up the discussion of Theon's edition by the remark that Theon evidently took no pains to discover and restore from MSS. the actual words which Euclid had written, but aimed much more at removing difficulties that might be felt by learners in studying the book. His edition is therefore not to be compared with the editions of the Alexandrine grammarians, but rather with the work done by Eutocius in editing Apollonius and with an interpolated recension of some of the works of Archimedes by a certain Byzantine, Theon occupying a position midway between these two editors, being superior to the latter in mathematical knowledge but behind Eutocius in industry (these views now require to be somewhat modified, as above stated). But however little Theon's object may be approved by those of us who would rather know the *ipsissima verba* of Euclid, there is no doubt that his work was approved by his pupils at Alexandria for whom it was written ; and his edition was almost exclusively used by later Greeks, with the result that the more ancient text is only preserved to us in one MS.

As the result of the above investigation, we may feel satisfied that, where P and the Theonine MSS. agree, they give us (except in a few accidental instances) Euclid as he was read by the Greeks of the 4th c. But even at that time the text had been passed from hand to hand through more than six centuries, so that it is certain that it had already suffered changes, due partly to the fault of copyists and partly to the interpolations of mathematicians. Some errors of copyists escaped Theon and were corrected in some MSS. by later hands. Others appear in all our MSS. and, as they cannot have arisen accidentally in all, we must put them down to a common source more ancient than Theon. A somewhat serious instance is to be found in III. 8 ; and the use of ἀπτέσθω for ἐφαπτέσθω in the sense of "touch" may also be mentioned, the proper distinction between the words having been ignored as it was by Theon also. But there are a number of imperfections in the ante-Theonine text which it would be unsafe to put down to the errors of copyists, those namely where the good MSS. agree and it is not possible to see any motive that a copyist could have had for altering a correct reading. In these cases it is possible that the imperfections are due to a certain degree of carelessness on the part of Euclid himself; for it is not possible "Euclidem ab omni naevo vindicare," to use the words of Saccheri[1], and consequently Simson is not right in attributing to Theon and other editors all the things in Euclid to which mathematical objection can be taken. Thus, when Euclid speaks of "the ratio compounded of the sides" for "the ratio compounded of the *ratios of the* sides," there is no reason for doubting that Euclid himself is responsible for the more slip-shod expression. Again, in the Books XI.—XIII. relating to solid geometry there are blots neither few nor altogether unimportant which can only be attributed to Euclid himself[2]; and there is the less reason for hesitation in so attributing them because solid geometry was then being treated in a thoroughly systematic manner for the first time. Sometimes the *conclusion* ($\sigma\upsilon\mu\pi\acute{\epsilon}\rho\alpha\sigma\mu\alpha$) of a proposition does not correspond exactly to the enunciation, often it is cut short with the words $\kappa\alpha\grave{\iota}\ \tau\grave{\alpha}\ \acute{\epsilon}\xi\hat{\eta}\varsigma$ "and the rest" (especially from Book X. onwards), and very often in Books VIII., IX. it is omitted. Where all the MSS. agree, there is no ground for hesitating to attribute the abbreviation or omission to Euclid; though, of course, where one or more MSS. have the longer form, it must be retained because this is one of the cases where a copyist has a temptation to abbreviate.

Where the true reading is preserved in one of the Theonine MSS. alone, Heiberg attributes the wrong reading to a mistake which arose before Theon's time, and the right reading of the single MS. to a successful correction.

We now come to the most important question of the *Interpolations introduced before Theon's time.*

I. Alternative proofs or additional cases.

It is not in itself probable that Euclid would have given two proofs of the same proposition; and the doubt as to the genuineness of the alternatives is increased when we consider the character of some of them and the way in which they are introduced. First of all, we have those of VI. 20 and XII. 17 introduced by "we shall prove this otherwise *more readily* ($\pi\rho\omega\chi\epsilon\iota\rho\acute{o}\tau\epsilon\rho\omega\nu$)" or that of X. 90 "it is possible to prove *more shortly* ($\sigma\upsilon\nu\tau\omega\mu\acute{\omega}\tau\epsilon\rho\omega\nu$)." Now it is impossible to suppose that Euclid would have given one proof as that definitely accepted by him and then added another with the express comment that the latter has certain advantages over the former. Had he considered the two proofs and come to this conclusion, he would have inserted the latter in the received text instead of the former. These alternative proofs must therefore have been interpolated. The same argument applies to alternatives introduced with the words "or even thus" ($\mathring{\eta}\ \kappa\alpha\grave{\iota}\ o\mathring{\upsilon}\tau\omega\varsigma$), "or even otherwise" ($\mathring{\eta}\ \kappa\alpha\grave{\iota}\ \mathring{\alpha}\lambda\lambda\omega\varsigma$). Under this head come the alternatives for the last portions of III. 7, 8; and Heiberg also compares the alternatives for parts of III. 31 (that the angle in a semicircle is a right angle) and XIII. 18, and the alternative proof of the lemma after X. 32. The alternatives to X. 105 and 106,

---

[1] *Euclides ab omni naevo vindicatus,* Mediolani, 1733.
[2] Cf. especially the assumption, without proof or definition, of the criterion for *equal* solid angles, and the incomplete proof of XII. 17.

again, are condemned by the place in which they occur, namely after an alternative proof to X. 115. The above alternatives being all admitted to be spurious, suspicion must necessarily attach to the few others which are in themselves unobjectionable. Heiberg instances the alternative proofs to III. 9, III. 10, VI. 30, VI. 31 and XI. 22, observing that it is quite comprehensible that any of these might have occurred to a teacher or editor and seemed to him, rightly or wrongly, to be better than the corresponding proofs in Euclid. Curiously enough, Simson adopted the alternatives to III. 9, 10 in preference to the genuine proofs. Since Heiberg's preface was written, his suspicion has been amply confirmed as regards III. 10 by the commentary of an-Nairīzī (ed. Curtze) which shows not only that this alternative is Heron's, but also that the substantive proposition III. 12 in Euclid is also Heron's, having been given by him to supplement III. 11 which must originally have been enunciated of circles "touching one another" simply, i.e. so as to include the case of external as well as internal contact, though the proof covered the case of internal contact only. "Euclid, in the 11th proposition," says Heron, "supposed two circles touching one another internally and wrote the proposition on this case, proving what it was required to prove in it. *But I will show how it is to be proved if the contact be external*[1]." This additional proposition of Heron's is by way of adding another *case*, which brings us to that class of interpolation. It was the practice of Euclid and the ancients to give only one case (generally the most difficult one) and to leave the others to be investigated by the reader for himself. One interpolation of a second case (VI. 27) is due, as we have seen, to Theon. The two extra cases of XI. 23 were manifestly interpolated before Theon's time, for the preliminary distinction of three cases, "(the centre) will either be within' the triangle *LMN*, or on one of the sides, or outside. First let it be within," is a spurious addition (B and V only). Similarly an unnecessary case is interpolated in III. 11.

II. Lemmas.

Heiberg has unhesitatingly placed in his Appendix to Vol. III. certain lemmas interpolated either by Theon (on X. 13) or later writers (on X. 27, 29, 31, 32, 33, 34, where V only has the lemmas). But we are here concerned with the lemmas found in all the MSS., which however are, for different reasons, necessarily suspected. We will deal with the Book X. lemmas last.

(1) There is an *a priori* ground of objection to those lemmas which come *after* the propositions to which they relate and prove properties used in those propositions; for, if genuine, they would be a sign of faulty arrangement such as would not be likely in a systematic work so carefully ordered as the *Elements*. The lemma to VI. 22 is one of this class, and there is the further objection to it that in VI. 28 Euclid makes an assumption which would equally require a lemma though none is found. The lemma after XII. 4 is open to the further objections that certain altitudes are used but are not drawn in the

---

[1] An-Nairīzī, ed. Curtze, p. 121.

figure (which is not in the manner of Euclid), and that a peculiar expression "parallelepipedal solids *described on* (ἀναγραφόμενα ἀπό) *prisms*" betrays a hand other than Euclid's. There is an objection on the score of language to the lemma after XIII. 2. The lemmas on XI. 23, XIII. 13, XIII. 18, besides coming after the propositions to which they relate, are not very necessary in themselves and, as regards the lemma to XIII. 13, it is to be noticed that the writer of a gloss in the proposition could not have had it, and the words "as will be proved afterwards" in the text are rightly suspected owing to differences between the MS. readings. The lemma to XII. 2 also, to which Simson raised objection, comes *after* the proposition; but, if it is rejected, the words "as was proved before" used in XII. 5 and 18, and referring to this lemma, must be struck out.

(2) Reasons of substance are fatal to the lemma before X. 60, which is really assumed in X. 44 and therefore should have appeared there if anywhere, and to the lemma on X. 20, which tries to prove what is already stated in X. Def. 4.

We now come to the remaining lemmas in Book X., eleven in number, which come *before* the propositions to which they relate and remove difficulties in the way of their demonstration. That before X. 42 introduces a set of propositions with the words "that the said irrational straight lines are uniquely divided ... we will prove after premising the following lemma," and it is not possible to suppose that these words are due to an interpolator; nor are there any objections to the lemmas before X. 14, 17, 22, 33, 54, except perhaps that they are rather easy. The lemma before X. 10 and X. 10 itself should probably be removed from the *Elements*; for X. 10 really uses the following proposition X. 11, which is moreover numbered 10 by the first hand in P, and the words in X. 10 referring to the lemma "for we learnt (how to do this)" betray the interpolator. Heiberg gives reason also for rejecting the lemmas before X. 19 and 24 with the words "in any of the aforesaid ways" (omitted in the Theonine MSS.) in the enunciations of X. 19, 24 and in the *exposition* of X. 20. Lastly, the lemmas before X. 29 may be genuine, though there is an addition to the second of them which is spurious.

Heiberg includes under this heading of interpolated lemmas two which purport to be substantive propositions, XI. 38 and XIII. 6. These must be rejected as spurious for reasons which will be found in detail in my notes on XI. 37 and XIII. 6 respectively. The latter proposition is only quoted once (in XIII. 17); probably the words quoting it (with γραμμή instead of εὐθεῖα) are themselves interpolated, and Euclid thought the fact stated a sufficiently obvious inference from XIII. 1.

III. Porisms (or corollaries).

Most of the porisms in the text are both genuine and necessary; but some are shown by differences in the MSS. not to be so, e.g. those to I. 15 (though Proclus has it), III. 31 and VI. 20 (Por. 2). Sometimes parts of porisms are interpolated. Such are the last few lines in the porisms to IV. 5, VI. 8; the latter addition is proved later by means of VI. 4, 8, so that the writer of these proofs could not have had the addition to VI. 8 Por. before him. Lastly, interpolators have added a sort of proof to some porisms, as though they were not quite obvious enough; but to add a demonstration is inconsistent with the idea of a porism, which, according to Proclus, is a by-product of a proposition appearing without our seeking it.

IV. Scholia.

Several interpolated scholia betray themselves by their wording, e.g. those given by Heiberg in the Appendix to Book X. and containing the words $\kappa\alpha\lambda\epsilon\hat{\iota}$, $\dot{\epsilon}\kappa\dot{\alpha}\lambda\epsilon\sigma\epsilon$ ("he calls" or "called"); these scholia were apparently written as marginal notes before Theon's time, and, being adopted as such by Theon, found their way into the text in P and some of the Theonine MSS. The same thing no doubt accounts for the interpolated analyses and syntheses to XIII. 1—5, as to which see my note on XIII. 1.

V. Interpolations in Book X.

First comes the proposition " *Let it be proposed to us* to show that in square figures the diameter is incommensurable in length with the side," which, with a scholium after it, ends the tenth Book. The form of the enunciation is suspicious enough and the proposition, the proof of which is indicated by Aristotle and perhaps was Pythagorean, is perfectly unnecessary when X. 9 has preceded. The scholium ends with remarks about commensurable and incommensurable *solids*, which are of course out of place before the Books on solids. The scholiast on Book X. alludes to this particular scholium as being due to " Theon and some others." But it is doubtless much more ancient, and may, as Heiberg conjectures have been the beginning of Apollonius' more advanced treatise on incommensurables. Not only is everything in Book X. after X. 115 interpolated, but Heiberg doubts the genuineness even of X. 112—115, on the ground that X. 111 rounds off the theory of incommensurables as we want it in the Books on solid geometry, while X. 112—115 are not really connected with what precedes, nor wanted for the later Books, but seem to form the starting-point of a new and more elaborate theory of irrationals.

VI. Other minor interpolations are found of the same character as those above attributed to Theon. First there are two places (XI. 35 and XI. 26) where, after " similarly we shall prove " and " for the same reason," an actual proof is nevertheless given. Clearly the proofs are interpolated; and there are other similar interpolations. There are also interpolations of intermediate steps in proofs, unnecessary explanations and so on, as to which I need not enter into details.

Lastly, following Heiberg's order, I come to

VII. Interpolated definitions, axioms etc.

Apart from VI. Def. 5 (which may have been interpolated by Theon although it is found written in the margin of P by the first hand), the definition of a segment of a circle in Book I. is interpolated, as is clear from the fact that it occurs in a more appropriate place in Book III. and Proclus omits it. VI. Def. 2 (reciprocal figures) is rightly condemned by Simson—perhaps it was taken from Heron—and

Heiberg would reject VII. Def. 10, as to which see my note on that definition. Lastly the double definition of a solid angle (XI. Def. 11) constitutes a difficulty. The use of the word ἐπιφάνεια suggests that the first definition may have been older than Euclid, and he may have quoted it from older *elements*, especially as his own definition which follows only includes solid angles contained by *planes*, whereas the other includes other sorts (cf. the words γραμμῶν, γραμμαῖς) which are also distinguished by Heron (Def. 22). If the first definition had come last, it could have been rejected without hesitation : but it is not so easy to reject the first part up to and including "otherwise" (ἄλλως). No difficulty need be felt about the definitions of "oblong," "rhombus," and "rhomboid," which are not actually used in the *Elements*; they were no doubt taken from earlier *elements* and given for the sake of completeness.

As regards the axioms or, as they are called in the text, *common notions* (κοιναὶ ἔννοιαι), it is to be observed that Proclus says[1] that Apollonius tried to prove "the axioms," and he gives Apollonius' attempt to prove Axiom 1. This shows at all events that Apollonius had *some* of the axioms now appearing in the text. But how could Apollonius have taken a controversial line against Euclid on the subject of axioms if these axioms had not been Euclid's to his knowledge? And, if they had been interpolated between Euclid's time and his own, how could Apollonius, living so comparatively short a time after Euclid, have been ignorant of the fact? Therefore *some* of the axioms are Euclid's (whether he called them *common notions*, or *axioms*, as is perhaps more likely since Proclus calls them axioms): and we need not hesitate to accept as genuine the first three discussed by Proclus, viz. (1) things equal to the same equal to one another, (2) if equals be added to equals, wholes equal, (3) if equals be subtracted from equals, remainders equal. The other two mentioned by Proclus (whole greater than part, and congruent figures equal) are more doubtful, since they are omitted by Heron, Martianus Capella, and others. The axiom that "two lines cannot enclose a space" is however clearly an interpolation due to the fact that I. 4 appeared to require it. The others about equals added to unequals, doubles of the same thing, and halves of the same thing are also interpolated; they are connected with other interpolations, and Proclus clearly used some source which did not contain them.

Euclid evidently limited his formal axioms to those which seemed to him most essential and of the widest application; for he not unfrequently assumes other things as axiomatic, e.g. in VII. 28 that, if a number measures two numbers, it measures their difference.

The differences of reading appearing in Proclus suggest the question of the comparative purity of the sources used by Proclus, Heron and others, and of our text. The omission of the definition of a segment in Book I. and of the old gloss "which is called the circumference" in I. Def. 15 (also omitted by Heron, Taurus, Sextus

---

[1] Proclus, pp. 194, 10 sqq.

Empiricus and others) indicates that Proclus had better sources than we have ; and Heiberg gives other cases where Proclus omits words which are in all our MSS. and where Proclus' reading should perhaps be preferred. But, except in these instances (where Proclus may have drawn from some ancient source such as one of the older commentaries), Proclus' MS. does not seem to have been among the best. Often it agrees with our worst MSS., sometimes it agrees with F where F alone has a certain reading in the text, so that (e.g. in I. 15 Por.) the common reading of Proclus and F must be rejected, thrice only does it agree with P alone, sometimes it agrees with P and some Theonine MSS., and once it agrees with the Theonine MSS. against P and other sources.

Of the other external sources, those which are older than Theon generally agree with our best MSS., e.g. Heron, allowing for the difference in the plan of his definitions and the somewhat free adaptation to his purpose of the Euclidean definitions in Books X., XI.

Heiberg concludes that the *Elements* were most spoiled by interpolations about the 3rd c., for Sextus Empiricus had a correct text, while Iamblichus had an interpolated one ; but doubtless the purer text continued for a long time in circulation, as we conclude from the fact that our MSS. are free from interpolations already found in Iamblichus' MS.

CHAPTER VI.

THE SCHOLIA.

HEIBERG has collected scholia, to the number of about 1500, in Vol. V. of his edition of Euclid, and has also discussed and classified them in a separate short treatise, in which he added a few others[1].

These scholia cannot be regarded as doing much to facilitate the reading of the *Elements*. As a rule, they contain only such observations as any intelligent reader could make for himself. Among the few exceptions are XI. Nos. 33, 35 (where XI. 22, 23 are extended to solid angles formed by any number of plane angles), XII. No. 85 (where an assumption tacitly made by Euclid in XII. 17 is proved), IX. Nos. 28, 29 (where the scholiast has pointed out the error in the text of IX. 19).

Nor are they very rich in historical information; they cannot be compared in this respect with Proclus' commentary on Book I. or with those of Eutocius on Archimedes and Apollonius. But even under this head they contain some things of interest, e.g. II. No. 11 explaining that the gnomon was invented by geometers for the sake of brevity, and that its name was suggested by an incidental characteristic, namely that "from it the whole is known ($\gamma\nu\omega\rho\iota\zeta\epsilon\tau\alpha\iota$), either of the whole area or of the remainder, when it (the $\gamma\nu\omega\mu\omega\nu$) is either placed round or taken away"; II. No. 13, also on the gnomon; IV. No. 2 stating that Book IV. was the discovery of the Pythagoreans; V. No. 1 attributing the content of Book V. to Eudoxus; X. No. 1 with its allusion to the discovery of incommensurability by the Pythagoreans and to Apollonius' work on irrationals; X. No. 62 definitely attributing X. 9 to Theaetetus; XIII. No. 1 about the "Platonic" figures, which attributes the cube, the pyramid, and the dodecahedron to the Pythagoreans, and the octahedron and icosahedron to Theaetetus.

Sometimes the scholia are useful in connexion with the settlement of the text, (1) directly, e.g. III. No. 16 on the interpolation of the word "within" ($\epsilon\nu\tau\delta s$) in the enunciation of III. 6, and X. No. 1 alluding to the discussion by "Theon and some others" of irrational "surfaces" and "solids," as well as "lines," from which we may

---

[1] Heiberg, *Om Scholierne til Euklids Elementer*, Kjøbenhavn, 1888. The tract is written in Danish, but, fortunately for those who do not read Danish easily, the author has appended (pp. 70—78) a résumé in French.

conclude that the scholium at the end of Book X. is not genuine; (2) indirectly in that they sometimes throw light on the connexion of certain MSS.

Lastly, they have their historical importance as enabling us to judge of the state of mathematical science at the times when they were written.

Before passing to the classification of the scholia, Heiberg remarks that we must separate from them a number of additions in the nature of scholia which are found in the text of our MSS. but which can, in one way or another, be proved to be spurious. As they are found both in P and in the Theonine MSS., they must have been in the MSS. anterior to Theon (4th c.). But they are, in great part, only found in the margin of P and the Theonine MSS.; in V they are half in the text and half in the margin. This can hardly be explained except on the supposition that these additions were originally (in the MSS. before Theon) in the margin, and that Theon kept them there in his edition, but that they afterwards found their way gradually into the text of P as well as of the Theonine MSS., or were omitted altogether, while particular MSS. have in certain places preserved the old arrangement. Of such spurious additions Heiberg enumerates the following: the axiom about equals subtracted from unequals, the last lines of the porism to VI. 8, second porisms to V. 19 and to VI. 20, the porism to III. 31, VI. Def. 5, various additions in Book X., the analyses and syntheses of XIII. 1—5, and the proposition XIII. 6.

The two first classes of scholia distinguished by Heiberg are denoted by the convenient abbreviations "Schol. Vat." and "Schol. Vind."

I. Schol. Vat.

It is first necessary to set out the letters by which Heiberg denotes certain collections of scholia.

P = Scholia in P written by the first hand.

B = Scholia in B by a hand of the same date as the MS. itself, generally that of Arethas.

F = Scholia in F by the first hand.

Vat. = Scholia of the Vatican MS. 204 of the 10th c., which has these scholia on leaves 198—205 (the end is missing) as an independent collection. It does not contain the text of the *Elements*.

V^c = Scholia found on leaves 283—292 of V and written in the same hand as that part of the MS. itself which begins at leaf 235.

Vat. 192 = a Vatican MS. of the 14th c. which contains, after (1) the *Elements* I.—XIII. (without scholia), (2) the *Data* with scholia, (3) Marinus on the *Data*, the Schol. Vat. as an independent collection and in their entirety, beginning with I. No. 88 and ending with XIII. No. 44.

The Schol. Vat., the most ancient and important collection of scholia, comprise those which are found in PBF Vat. and, from VII. 12 to IX. 15, in PB Vat. only, since in that portion of the *Elements* F was restored by a later hand without scholia; they also include I.

No. 88 which only happens to be erased in F, and IX. Nos. 28, 29
which may be left out because F. here has a different text. In F
and Vat. the collection ends with Book X.; but it must also include
Schol. PB of Books XI.—XIII., since these are found along with Schol.
Vat. to Books I.—X. in several MSS. (of which Vat. 192 is one) as a
separate collection. The Schol. Vat. to Books X.—XIII. are also
found in the collection V^c (where, curiously enough, XIII. Nos. 43, 44
are at the beginning). The Schol. Vat. accordingly include Schol.
PBV^c Vat. 192, and doubtless also those which are found in two of
these sources. The total number of scholia classified by Heiberg as
Schol. Vat. is 138.

As regards the contents of Schol. Vat. Heiberg has the following
observations. The thirteen scholia to Book I. are extracts made
from Proclus by a writer thoroughly conversant with the subject,
and cleverly recast (with some additions). Their author does not
seem to have had the two lacunae which our text of Proclus has
(at the end of the note on I. 36 and the beginning of the next note,
and at the beginning of the note on I. 43), for the scholia I. Nos. 125
and 137 seem to fill the gaps appropriately, at least in part. In
some passages he had better readings than our MSS. have. The rest
of Schol. Vat. (on Books II.—XIII.) are essentially of the same
character as those on Book I., containing prolegomena, remarks on
the object of the propositions, critical remarks on the text, converses,
lemmas; they are, in general, exact and true to tradition. The
reason of the resemblance between them and Proclus appears to be
due to the fact that they have their origin in the commentary of
Pappus, of which we know that Proclus also made use. In support
of the view that Pappus is the source, Heiberg places some of the
Schol. Vat. to Book X. side by side with passages from the commentary of Pappus in the Arabic translation discovered by Woepcke[1];
he also refers to the striking confirmation afforded by the fact that
XII. No. 2 contains the solution of the problem of inscribing in a
given circle a polygon similar to a polygon inscribed in another circle,
which problem Eutocius says[2] that Pappus gave in his commentary
on the *Elements*.

But, on the other hand, Schol. Vat. contain some things which
cannot have come from Pappus, e.g. the allusion in X. No. 1 to Theon
and irrational surfaces and solids, Theon being later than Pappus;
III. No. 10 about *porisms* is more like Proclus' treatment of the
subject than Pappus', though one expression recalls that of Pappus
about *forming* ($\sigma\chi\eta\mu\alpha\tau i\zeta\epsilon\sigma\theta\alpha\iota$) the enunciations of porisms like those
of either theorems or problems.

The Schol. Vat. give us important indications as regards the
text of the *Elements* as Pappus had it. In particular, they show that
he could not have had in his text certain of the lemmas in Book X.
For example, three of these are identical with what we find in Schol.

---

[1] *Om Scholierne til Euklids Elementer*, pp. 11, 12 : cf. *Euklid-Studien*, pp. 170, 171 ;
Woepcke, *Mémoires présent. à l'Acad. des Sciences*, 1856, XIV. p. 658 sqq.

[2] Archimedes, ed. Heiberg, III. p. 28, 19—22.

Vat. (the lemma to X. 17 = Schol. X. No. 106, and the lemmas to
X. 54, 60 come in Schol. X. No. 328); and it is not possible to suppose
that these lemmas, if they were already in the text, would also be
given as scholia. Of these three lemmas, that before X. 60 has
already been condemned for other reasons; the other two, unobjectionable in themselves, must be rejected on the ground now
stated. There were four others against which Heiberg found nothing
to urge when writing his prolegomena to Vol. V., viz. the lemmas
before X. 42, X. 14, X. 22 and X. 33. Of these, the lemma to X. 22
is not reconcilable with Schol. X. No. 161, which takes up the
assumption in the text of Eucl. X. 22 as if no lemma had gone before.
The lemma to X. 42, which, on account of the words introducing it
(see p. 60 above), Heiberg at first hesitated to regard as an interpolation, is identical with Schol. X. No. 270. It is true that in
Schol. X. No. 269 we find the words "this lemma has been proved
before (ἐν τοῖς ἔμπροσθεν), but it shall also be proved now for
convenience' sake (τοῦ ἑτοίμου ἔνεκα)," and it is possible to suppose
that "before" may mean in Euclid's text *before* X. 42; but a proof
in that place would surely have been as "convenient" as could be
desired, and it is therefore more probable that the proof had been
given by *Pappus* in some earlier place. (It may be added that the
lemma to X. 14, which is identical with the lemma to XI. 23, condemned on other grounds, is for that reason open to suspicion.)

Heiberg's conclusion is that *all* the lemmas are spurious, and that
most or all of them have found their way into the text from Pappus'
commentary, though at a time anterior to Theon's edition, since
they are found in all our MSS. This enables us to fix a date for these
interpolations, namely the first half of the 4th c.

Of course Pappus had not in his text the interpolations which,
from the fact of their appearing only in some of our MSS., are seen to
be later than those above-mentioned. Such are the lemmas which
are found in the text of V only after X. 29 and X. 31 respectively and
are given in Heiberg's Appendix to Book X. (numbered 10 and 11).
On the other hand it appears from Woepcke's tract[1] that Pappus
already had X. 115 in his text: though it does not follow from this
that the proposition is genuine but only that interpolations began
very early.

Theon interpolated a proposition (or lemma) between X. 12 and
X. 13 (No. 5 in Heiberg's Appendix). Schol. Vat. has the same
thing (X. No. 125). The writer of the scholia therefore did not find
this lemma in the text. Schol. Vat. IX. Nos. 28, 29 show that neither
did he find in his text the alterations which Theon made in Eucl. IX.
19; the scholia in fact only agree with the text of P, not with Theon's.
This suggests that Schol. Vat. were written for use with a MS. of the
ante-Theonine recension such as P is. This probability is further
confirmed by a certain independence which P shows in several places
when compared with the Theonine MSS. Not only has P better
readings in some passages, but more substantial divergences; and,

[1] Woepcke, *op. cit.* p. 702.

in particular, the absence in P of three notes of a historical character which are added, wholly or partly from Proclus, in the Theonine MSS. attests an independent and more primitive point of view in P.

In view of the distinctive character of P, it is possible that some of the scholia found in it in the first hand, but not in the other sources of Schol. Vat., also belong to that collection; and several circumstances confirm this. Schol. XIII. No. 45, found in P only, which relates to a passage in Eucl. XIII. 13, shows that certain words in the text, though older than Theon, are interpolated; and, as the scholium is itself older than Theon, is headed "*third* lemma," and follows a "second lemma" relating to a passage in the text immediately preceding, which "second lemma" belongs to Schol. Vat. and is taken from Pappus, the "third" in all probability came from Pappus also. The same is true of Schol. XII. No. 72 and XIII. No. 69, which are respectively identical with the propositions *vulgo* XI. 38 (Heiberg, App. to Book XI., No. 3) and XIII. 6; for both of these interpolations are older than Theon. Moreover most of the scholia which P in the first hand alone has are of the same character as Schol. Vat. Thus VII. No. 7 and XIII. No. 1 introducing Books VII. and XIII. respectively are of the same historical character as several of Schol. Vat.; that VII. No. 7 appears in the *text* of P at the beginning of Book VII. constitutes no difficulty. There are a number of *converses*, remarks on the relation of propositions to one another, explanations such as XII. No. 89 in which it is remarked that Φ, Ω in Euclid's figure to XII. 17 (Z, V in my figure) are really the same point but that this makes no difference in the proof. Two other Schol. P on XII. 17 are connected by their headings with XII. No. 72 mentioned above. XI. No. 10 (P) is only another form of XI. No. 11 (B); and B often, alone with P, has preserved Schol. Vat. On the whole Heiberg considers some 40 scholia found in P alone to belong to Schol. Vat.

The history of Schol. Vat. appears to have been, in its main outlines, the following. They were put together after 500 A.D., since they contain extracts from Proclus, to which we ought not to assign a date too near to that of Proclus' work itself; and they must at least be earlier than the latter half of the 9th c., in which B was written. As there must evidently have been several intermediate links between the archetype and B, we must assign them 'rather to the first half of the period between the two dates, and it is not improbable that they were a new product of the great development of mathematical studies at the end of the 6th c. (Isidorus of Miletus). The author extracted what he found of interest in the commentary of Proclus on Book I. and in that of Pappus on the rest of the work, and put these extracts in the margin of a MS. of the class of P. As there are no scholia to I. 1—22, the first leaves of the archetype or of one of the earliest copies must have been lost at an early date, and it was from that mutilated copy that partly P and partly a MS. of the Theonine class were taken, the scholia being put in the margin in both. Then the collection spread through the Theonine MSS., gradually losing some

scholia which could not be read or understood, or which were
accidentally or deliberately omitted. Next it was extracted from
one of these MSS. and made into a separate work which has been
preserved, in part, in its entirety (Vat. 192 etc.) and, in part, divided
into sections, so that the scholia to Books X.—XIII. were detached
(V^c). It had the same fate in the MSS. which kept the original
arrangement (in the margin), and in consequence there are some MSS.
where the scholia to the stereometric Books are missing, those Books
having come to be less read in the period of decadence. It is from
one of these MSS. that the collection was extracted as a separate work
such as we find it in Vat. (10th c.).

II.   The second great division of the scholia is **Schol. Vind.**

This title is taken from the Viennese MS. (V), and the letters used
by Heiberg to indicate the sources here in question are as follows.

V^a = scholia in V written by the same hand that copied the MS.
itself from fol. 235 onward.

q = scholia of the Paris MS. 2344 (q) written by the first hand.

l = scholia of the Florence MS. Laurent. XXVIII, 2 written in the
13th—14th c., mostly in the first hand, but partly in two later
hands.

V^b = scholia in V written by the same hand as the first part
(leaves 1—183) of the MS. itself; V^b wrote his scholia after V^a.

q^1 = scholia of the Paris MS. (q) found here and there in another
hand of early date.

Schol. Vind. include scholia found in V^{a}q. l is nearly related to
q ; and in fact the three MSS. which, so far as Euclid's text is concerned, show no direct interdependence, are, as regards their scholia,
derived from one original. Heiberg proves this by reference to the
readings of the three in two passages (found in Schol. I. No. 109 and
X. No. 39 respectively). The common source must have contained,
besides the scholia found in the three MSS. V^{a}ql, those also which
are contained in two of them, for it is more unlikely that two of the
three should contain common interpolations than that a particular
scholium should drop out of one of them. Besides V^a and q, the
scholia V^b and q^1 must equally be referred to Schol. Vind., since the
greater part of their scholia are found in l. There is a lacuna in q
from Eucl. VIII. 25 to IX. 14, so that for this portion of the *Elements*
Schol. Vind. are represented by Vl only. Heiberg gives about 450
numbers in all as belonging to this collection.

Schol. Vind. did not all come from one source; this is shown by
differences of substance, e.g. between X. Nos. 36 and 39, and by
differences of time of writing: e.g. VI. No. 52 refers at the beginning
to No. 55 with the words "as the scholium has it" and is therefore
later than that scholium ; X. No. 247 is also later than X. No. 246.

The scholia to Book I. are here also extracts from Proclus, but
more copious and more verbatim than in Schol. Vat. The author
has not always understood Proclus; and he had a text as bad as
that of our MSS., with the same lacunae. The scholia to the other

Books are partly drawn (1) from Schol. Vat., the MSS. representing
Schol. Vind. and Schol. Vat. in these cases showing nearly all possible
combinations; but there is no certain trace in Schol. Vind. of the
scholia peculiar to P.   The author used a copy of Schol. Vat. in the
form in which they were attached to the Theonine text; thus Schol.
Vind. correspond to BF Vat., where these diverge from P, and
especially closely to B.   Besides Schol. Vat., the editors of Schol.
Vind. used (2) other old collections of scholia of which we find traces
in B and F; Schol. Vind. have also some scholia common with b.
The scholia which Schol. Vind. have in common with BF come from
two different sources, and were apparently afterwards introduced
into the other MSS.; one result of this is that several scholia are
reproduced twice.

But, besides the scholia derived from these sources, Schol. Vind.
contain a large number of others of late date, characterised by incorrect language or by triviality of content (there are many examples
in numbers, citations of propositions used, absurd ἀπορίαι, and the
like).   Unlike Schol. Vat., these scholia often quote words from Euclid
as a heading (in one case a heading is inserted in Schol. Vind. where
a scholium without the heading is quoted from Schol. Vat., see V.
No. 14).   The explanations given often presuppose very little knowledge on the part of the reader and frequently contain obscurities
and gross errors.

Schol. Vind. were collected for use with a MS. of the Theonine
class; this follows from the fact that they contain a note on the
proposition *vulgo* VII. 22 interpolated by Theon (given in Heiberg's
App. to Vol. II. p. 430).   Since the scholium to VII. 39 given in V and
p in the text after the title of Book VIII. quotes the proposition as
VII. 39, it follows that this scholium must have been written before
the interpolation of the two propositions *vulgo* VII. 20, 22; Schol.
Vind. contain (VII. No. 80) the first sentence of it, but without the
heading referring to VII. 39.   Schol. VII. No. 97 quotes VII. 33 as
VII. 34, so that the proposition *vulgo* VII. 22 may have stood in the
scholiast's text but not the later interpolation *vulgo* VII. 20 (later
because only found in B in the margin by the first hand).   Of course
the scholiast had also the interpolations earlier than Theon.

For the date of the collection we have a lower limit in the date
(12th c.) of MSS. in which the scholia appear.   That it was not much
earlier than the 12th c. is indicated (1) by the poverty of its contents,
(2) by the quality of the MS. of Proclus which was used in the
compilation of it (the Munich MS. used by Friedlein with which the
scholiast's excerpts are essentially in agreement belongs to the 11th—
12th c.), (3) by the fact that Schol. Vind. appear only in MSS. of the
12th c. and no trace of them is found in our MSS. belonging to
the 9th—10th c. in which Schol. Vat. are found.   The collection may
therefore probably be assigned to the 11th c.   Perhaps it may be in
part due to Psellus who lived towards the end of that century: for in
a Florence MS. (Magliabecch. XI, 53 of the 15th c.) containing a
mathematical compendium intended for use in the reading of Aristotle

the scholia I. Nos. 40 and 49 appear with the name of Psellus attached.

Schol. Vind. are not found without the admixture of foreign elements in any of our three sources. In l there are only very few such in the first hand. In q there are several new scholia in the first hand, for the most part due to the copyist himself. The collection of scholia on Book X. in q (Heiberg's q^c) is also in the first hand ; it is not original, and it may perhaps be due to Psellus (Maglb. has some definitions of Book X. with a heading "scholia of ... Michael Psellus on the definitions of Euclid's 10th *Element*" and Schol. X. No. 9), whose name must have been attached to it in the common source of Maglb. and q ; to a great extent it consists of extracts from Schol. Vind. taken from the same source as Vl. The scholia q^l (in an ancient hand in q), confined to Book II., partly belong to Schol. Vind. and partly correspond to b^l (Bologna MS.). q^a and q^b are in one hand (Theodorus Antiochita), the nearest to the first hand of q ; they are doubtless due to an early possessor of the MS. of whom we know nothing more.

V^a has, besides Schol. Vind., a number of scholia which also appear in other MSS., one in BFb, some others in P, and some in v (Codex Vat. 1038, 13th c.) ; these scholia were taken from a source in which many abbreviations were used, as they were often misunderstood by V^a. Other scholia in V^a which are not found in the older sources—some appearing in V^a alone—are also not original, as is proved by mistakes or corruptions which they contain ; some others may be due to the copyist himself.

V^b seldom has scholia common with the other older sources ; for the most part they either appear in V^b alone or only in the later sources as v or F^2 (later scholia in F), some being original, others not. In Book X. V^b has three series of numerical examples, (1) with Greek numerals, (2) alternatives added later, also mostly with Greek numerals, (3) with Arabic numerals. The last class were probably the work of the copyist himself. These examples (cf. p. 74 below) show the facility with which the Byzantines made calculations at the date of the MS. (12th c.). They prove also that the use of the Arabic numerals (in the East-Arabian form) was thoroughly established in the 12th c.; they were actually known to the Byzantines a century earlier, since they appear, in the first hand, in an Escurial MS. of the 11th c.

Of collections in other hands in V distinguished by Heiberg (see preface to Vol. v.), V^1 has very few scholia which are found in other sources, the greater part being original ; V^2, V^3 are the work of the copyist himself ; V^4 are so in part only, and contain several scholia from Schol. Vat. and other sources. V^3 and V^4 are later than 13th —14th c., since they are not found in f (cod. Laurent. XXVIII, 6) which was copied from V and contains, besides V^a V^b, the greater part of V^1 and VI. No. 20 of V^2 (in the text).

In P there are, besides P^3 (a quite late hand, probably one of the old Scriptores Graeci at the Vatican), two late hands (P^2), one of which has some new and independent scholia, while the other has added the greater part of Schol. Vind., partly in the margin and partly on pieces of leaves stitched on.

Our sources for Schol. Vat. also contain other elements. In P there were introduced a certain number of extracts from Proclus, to supplement Schol. Vat. to Book I.; they are all written with a different ink from that used for the oldest part of the MS., and the text is inferior. There are additions in the other sources of Schol. Vat. (F and B) which point to a common source for FB and which are nearly all found in other MSS., and, in particular, in Schol. Vind., which also used the same source; that they are not assignable to Schol. Vat. results only from their not being found in Vat. Of other additions in F, some are peculiar to F and some common to it and b; but they are not original. F² (scholia in a later hand in F) contains three original scholia; the rest come from V. B contains, besides scholia common to it and F, b or other sources, several scholia which seem to have been put together by Arethas, who wrote at least a part of them with his own hand.

Heiberg has satisfied himself, by a closer study of b, that the scholia which he denotes by b, β and b¹ are by one hand; they are mostly to be found in other sources as well, though some are original. By the same hand (Theodorus Cabasilas, 15th c.) are also the scholia denoted by b², B², b³ and B³. These scholia come in great part from Schol. Vind., and in making these extracts Theodorus probably used one of our sources, l, mistakes in which often correspond to those of Theodorus. To one scholium is attached the name of Demetrius (who must be Demetrius Cydonius, a friend of Nicolaus Cabasilas, 14th c.); but it could not have been written by him, since it appears in B and Schol. Vind. Nor are all the scholia which bear the name of Theodorus due to Theodorus himself, though some are so.

As B³ (a late hand in B) contains several of the original scholia of b², B³ must have used b itself as his source, and, as all the scholia in B³ are in b, the latter is also the source of the scholia in B³ which are found in other MSS. B and b were therefore, in the 15th c., in the hands of the same person; this explains, too, the fact that b in a late hand has some scholia which can only come from B. We arrive then at the conclusion that Theodorus Cabasilas, in the 15th c., owned both the MSS. B and b, and that he transferred to B scholia which he had before written in b, either independently or after other sources, and inversely transferred some scholia from B to b. Further, B² are earlier than Theodorus Cabasilas, who certainly himself wrote B³ as well as b² and b³.

An author's name is also attached to the scholia VI. No. 6 and X. No. 223, which are attributed to Maximus Planudes (end of 13th c.) along with scholia on I. 31, X. 14 and X. 18 found in l in a quite late hand and published on pp. 46, 47 of Heiberg's dissertation. These seem to have been taken from lectures of Planudes on the *Elements* by a pupil who used l as his copy.

There are also in l two other Byzantine scholia, written by a late hand, and bearing the names Ioannes and Pediasimus respectively;

these must in like manner have been written by a pupil after lectures of Ioannes Pediasimus (first half of 14th c.), and this pupil must also have used l.

Before these scholia were edited by Heiberg, very few of them had been published in the original Greek. The Basel *editio princeps* has a few (V. No. I, VI. Nos. 3, 4 and some in Book X.) which are taken, some from the Paris MS. (Paris. Gr. 2343) used by Grynaeus, others probably from the Venice MS. (Marc. 301) also used by him; one published by Heiberg, not in his edition of Euclid but in his paper on the scholia, may also be from Venet. 301, but appears also in Paris. Gr. 2342. The scholia in the Basel edition passed into the Oxford edition in the text, and were also given by August in the Appendix to his Vol. II.

Several specimens of the two series of scholia (Vat. and Vind.) were published by C. Wachsmuth (*Rhein. Mus.* XVIII. p. 132 sqq.) and by Knoche (*Untersuchungen über die neu aufgefundenen Scholien des Proklus*, Herford, 1865).

The scholia published in Latin were much more numerous. G. Valla (*De expetendis et fugiendis rebus*, 1501) reproduced apparently some 200 of the scholia included in Heiberg's edition. Several of these he obtained from two Modena MSS. which at one time were in his possession (Mutin. III B, 4 and II E, 9, both of the 15th c.); but he must have used another source as well, containing extracts from other series of scholia, notably Schol. Vind. with which he has some 87 scholia in common. He has also several that are new.

Commandinus included in his translation under the title " Scholia antiqua" the greater part of the Schol. Vat. which he certainly obtained from a MS. of the class of Vat. 192; on the whole he adhered closely to the Greek text. Besides these scholia Commandinus has the scholia and lemmas which he found in the Basel *editio princeps*, and also three other scholia not belonging to Schol. Vat., as well as one new scholium (to XII. 13) not included in Heiberg's edition, which are distinguished by different type and were doubtless taken from the Greek MS. used by him along with the Basel edition.

In Conrad Dasypodius' *Lexicon mathematicum* published in 1573 there is (on fol. 42—44) "Graecum scholion in definitiones Euclidis libri quinti elementorum appendicis loco propter pagellas vacantes annexum." This contains four scholia, and part of two others, published in Heiberg's edition, with some variations of readings, and with some new matter added (for which see pp. 64—6 of Heiberg's pamphlet). The source of these scholia is revealed to us by another work of Dasypodius, *Isaaci Monachi Scholia in Euclidis elementorum geometriae sex priores libros per C. Dasypodium in latinum sermonem translata et in lucem edita* (1579). This work contains, besides excerpts from Proclus on Book I. (in part closely related to Schol. Vind.), some 30 scholia included in Heiberg's edition, several new scholia, and the above-mentioned scholia to the definitions of Book V. published in Greek in 1573. After the scholia follow "Isaaci Monachi prolegomena in Euclidis Elementorum geometriae libros" (two definitions of geometry) and "Varia miscellanea ad geometriae cognitionem necessaria ab Isaaco Monacho collecta" (mostly the same as pp. 252, 24—272, 27 in the *Variae Collectiones* included in Hultsch's Heron); lastly, a note of Dasypodius to the reader says that these scholia were taken "ex clarissimi viri Joannis Sambuci antiquo codice manu propria Isaaci Monachi scripto." Isaak Monachus is doubtless Isaak Argyrus, 14th c. ; and Dasypodius used a MS. in whioh, besides the passage in Hultsch's *Variae Collectiones*, there were a number of scholia marked in the margin with the name of Isaak (cf. those in b under the name of Theodorus Cabasilas). Whether the new scholia are original cannot be decided until they are published in Greek ; but it is not improbable that they are at all events independent arrangements of older scholia. All but five of the others, and all but one of the Greek scholia to Book V., are taken from Schol. Vat. ; three of the excepted ones are from Schol. Vind., and the other three seem to come from F (where some words of them are illegible, but can be supplied by means of Mut. III B, 4, which has these three scholia and generally shows a certain likeness to Isaak's scholia).

Dasypodius also published in 1564 the arithmetical commentary of Barlaam the monk (14th c.) on Eucl. Book II., which finds a place in Appendix IV. to the Scholia in Heiberg's edition.

Hultsch has some remarks on the origin of the scholia[1]. He observes that the scholia to Book I. contain a considerable portion of Geminus' commentary on the definitions and are specially valuable because they contain extracts from Geminus *only*, whereas Proclus, though drawing mainly upon him, quotes from others as well. On the postulates and axioms the scholia give more than is found in Proclus. Hultsch conjectures that the scholium on Book V., No. 3, attributing the discovery of the theorems to Eudoxus but their arrangement to Euclid, represents the tradition going back to Geminus, and that the scholium XIII., No. 1, has the same origin.

A word should be added about the numerical illustrations of Euclid's propositions in the scholia to Book X. They contain a large number of calculations with sexagesimal fractions[2]; the fractions go as far as *fourth-sixtieths* ($1/60^4$). Numbers expressed in these fractions are handled with skill and include some results of surprising accuracy[3]

---

[1] Art. "Eukleides" in Pauly-Wissowa's *Real-Encyclopädie.*

[2] Hultsch has written upon these in *Bibliotheca Mathematica*, v_3, 1904, pp. 225—233.

[3] Thus $\sqrt{(27)}$ is given (allowing for a slight correction by means of the context) as 5 11' 46" 10''', which gives for $\sqrt{3}$ the value 1 43' 55" 23''', being the same value as that given by Hipparchus in his Table of Chords, and correct to the seventh decimal place. Similarly $\sqrt{8}$ is given as 2 49' 42" 20''' 10'''', which is equivalent to $\sqrt{2} = 1·41421335$. Hultsch gives instances of the various operations, addition, subtraction, etc., carried out in these fractions, and shows how the extraction of the square root was effected. Cf. T. L. Heath, *History of Greek Mathematics*, I., pp. 59—63.

CHAPTER VII.

EUCLID IN ARABIA.

WE are told by Hājī Khalfa[1] that the Caliph al-Mansūr (754–775) sent a mission to the Byzantine Emperor as the result of which he obtained from him a copy of Euclid among other Greek books, and again that the Caliph al-Ma'mūn (813–833) obtained manuscripts of Euclid, among others, from the Byzantines. The version of the *Elements* by al-Ḥajjāj b. Yūsuf b. Matar is, if not the very first, at least one of the first books translated from the Greek into Arabic[2]. According to the *Fihrist*[3] it was translated by al-Ḥajjāj twice; the first translation was known as "Hārūnī" ("for Hārūn"), the second bore the name "Ma'mūnī" ("for al-Ma'mūn") and was the more trustworthy. Six Books of the second of these versions survive in a Leiden MS. (Codex Leidensis 399, 1) now in part published by Besthorn and Heiberg[4]. In the preface to this MS. it is stated that, in the reign of Hārūn ar-Rashīd (786–809), al-Ḥajjāj was commanded by Yahyā b. Khālid b. Barmak to translate the book into Arabic. Then, when al-Ma'mūn became Caliph, as he was devoted to learning, al-Ḥajjāj saw that he would secure the favour of al-Ma'mūn "if he illustrated and expounded this book and reduced it to smaller dimensions. He accordingly left out the superfluities, filled up the gaps, corrected or removed the errors, until he had gone through the book and reduced it, when corrected and explained, to smaller dimensions, as in this copy, but without altering the substance, for the use of men endowed with ability and devoted to learning, the earlier edition being left in the hands of readers."

The *Fihrist* goes on to say that the work was next translated by Isḥāq b. Ḥunain, and that this translation was improved by Thābit b. Qurra. This Abū Ya'qūb Isḥāq b. Ḥunain b. Isḥāq al-'Ibādī (d. 910) was the son of the most famous of Arabic translators, Ḥunain b. Isḥāq al-'Ibādī (809–873), a Christian and physician to the Caliph al-Mutawakkil (847–861). There seems to be no doubt that Isḥāq, who

---

[1] *Lexicon bibliogr. et encyclop.* ed. Flügel, III. pp. 91, 92.
[2] Klamroth, *Zeitschrift der Deutschen Morgenländischen Gesellschaft*, XXXV. p. 303.
[3] *Fihrist* (tr. Suter), p. 16.
[4] *Codex Leidensis* 399, I. *Euclidis Elementa ex interpretatione al-Hadschdschadschii cum commentariis al-Narizii*, Hauniae, part I. i. 1893, part I. ii. 1897, part II. i. 1900, part II. ii. 1905, part III. i. 1910.

must have known Greek as well as his father, made his translation direct from the Greek. The revision must apparently have been the subject of an arrangement between Isḥāq and Thābit, as the latter died in 901 or nine years before Isḥāq. Thābit undoubtedly consulted Greek MSS. for the purposes of his revision. This is expressly stated in a marginal note to a Hebrew version of the *Elements*, made from Isḥāq's, attributed to one of two scholars belonging to the same family, viz. either to Moses b. Tibbon (about 1244–1274) or to Jakob b. Machir (who died soon after 1306)[1]. Moreover Thābit observes, on the proposition which he gives as IX. 31, that he had not found this proposition and the one before it in the Greek but only in the Arabic; from which statement Klamroth draws two conclusions, (1) that the Arabs had already begun to interest themselves in the authenticity of the text and (2) that Thābit did not alter the numbers of the propositions in Isḥāq's translation[2]. The *Fihrist* also says that Yuḥannā al-Qass (i.e. "the Priest") had seen in the Greek copy in his possession the proposition in Book I. which Thābit took credit for, and that this was confirmed by Nazīf, the physician, to whom Yuḥannā had shown it. This proposition may have been wanting in Isḥāq, and Thābit may have added it, but without claiming it as his own discovery[3]. As a fact, I. 45 is missing in the translation by al-Ḥajjāj.

The original version of Isḥāq *without* the improvements by Thābit has probably not survived any more than the first of the two versions by al-Ḥajjāj; the divergences between the MSS. are apparently due to the voluntary or involuntary changes of copyists, the former class varying according to the degree of mathematical knowledge possessed by the copyists and the extent to which they were influenced by considerations of practical utility for teaching purposes[4]. Two MSS. of the Isḥāq-Thābit version exist in the Bodleian Library (No. 279 belonging to the year 1238, and No. 280 written in 1260–1)[5]; Books I.—XIII. are in the Isḥāq-Thābit version, the non-Euclidean Books XIV., XV. in the translation of Qusṭā b. Lūqā al-Baʿlabakkī (d. about 912). The first of these MSS. (No. 279) is that (O) used by Klamroth for the purpose of his paper on the Arabian Euclid. The other MS. used by Klamroth is (K) Kjøbenhavn LXXXI, undated but probably of the 13th c., containing Books V.—XV., Books V.—X. being in the Isḥāq-Thābit version, Books XI.—XIII. purporting to be in al-Ḥajjāj's translation, and Books XIV., XV. in the version of Qusṭā b. Lūqā. In not a few propositions K and O show not the slightest difference, and, even where the proofs show considerable differences, they are generally such that, by a careful comparison, it is possible to reconstruct the common archetype, so that it is fairly clear that we have in these cases, not two recensions of one translation, but arbitrarily altered and

---

[1] Steinschneider, *Zeitschrift für Math. u. Physik*, XXXI., hist.-litt. Abtheilung, pp. 85, 86, 99.
[2] Klamroth, p. 279.   [3] Steinschneider, p. 88.
[4] Klamroth, p. 306.
[5] These MSS. are described by Nicoll and Pusey, *Catalogus cod. mss. orient. bibl. Bodleianae*, pt. II. 1835 (pp. 257—262).

shortened copies of one and the same recension[1]. The Bodleian MS. No. 280 contains a preface, translated by Nicoll, which cannot be by Thābit himself because it mentions Avicenna (980–1037) and other later authors. The MS. was written at Marāġa in the year 1260–1 and has in the margin readings and emendations from the edition of Naṣīraddīn aṭ-Ṭūsī (shortly to be mentioned) who was living at Marāġa at the time. Is it possible that aṭ-Ṭūsī himself is the author of the preface[2]? Be this as it may, the preface is interesting because it throws light on the liberties which the Arabians allowed themselves to take with the text. After the observation that the book (in spite of the labours of many editors) is not free from errors, obscurities, redundancies, omissions etc., and is without certain definitions necessary for the proofs, it goes on to say that the man has not yet been found who could make it perfect, and next proceeds to explain (1) that Avicenna "cut out postulates and many definitions" and attempted to clear up difficult and obscure passages, (2) that Abū'l Wafā al-Būzjānī (939–997) "introduced unnecessary additions and left out many things of great importance and entirely necessary," inasmuch as he was too long in various places in Book VI. and too short in Book X. where he left out entirely the proofs of the *apotomae*, while he made an unsuccessful attempt to emend XII. 14, (3) that Abū Jaʿfar al-Khāzin (d. between 961 and 971) arranged the postulates excellently but "disturbed the number and order of the propositions, reduced several propositions to one" etc. Next the preface describes the editor's own claims[3] and then ends with the sentences, "But we have kept to the order of the books and propositions in the work itself (i.e. Euclid's) except in the twelfth and thirteenth books. For we have dealt in Book XIII. with the (solid) bodies and in Book XII. with the surfaces by themselves."

After Thābit the *Fihrist* mentions Abū ʿUthmān ad-Dimashqī as having translated some Books of the *Elements* including Book X. (It is Abū ʿUthmān's translation of Pappus' commentary on Book X. which Woepcke discovered at Paris.) The *Fihrist* adds also that "Nazīf the physician told me that he had seen the tenth Book of Euclid in Greek, that it had 40 propositions more than the version in common circulation which had 109 propositions, and that he had determined to translate it into Arabic."

But the third form of the Arabian Euclid actually accessible to us is the edition of Abū Jaʿfar Muḥ. b. Muḥ. b. al-Ḥasan Naṣīraddīn aṭ-Ṭūsī (whom we shall call aṭ-Ṭūsī for short), born at Ṭūs (in Khurāsān) in 1201 (d. 1274). This edition appeared in two forms, a larger and a smaller. The larger is said to survive in Florence only (Pal. 272 and 313, the latter MS. containing only six Books); this was published at Rome in 1594, and, remarkably enough, some copies of

---

[1] Klamroth, pp. 306—8.
[2] Steinschneider, p. 98. Heiberg has quoted the whole of this preface in the *Zeitschrift für Math. u. Physik*, XXIX., hist.-litt. Abth. p. 16.
[3] This seems to include a rearrangement of the contents of Books XIV., XV. added to the *Elements*.

this edition are to be found with 12 and some with 13 Books, some with a Latin title and some without[1]. But the book was printed in Arabic, so that Kästner remarks that he will say as much about it as can be said about a book which one cannot read[2]. The shorter form, which however, in most MSS., is in 15 Books, survives at Berlin, Munich, Oxford, British Museum (974, 1334[3], 1335), Paris (2465, 2466), India Office, and Constantinople; it was printed at Constantinople in 1801, and the first six Books at Calcutta in 1824[4].

At-Ṭūsī's work is however not a *translation* of Euclid's text, but a re-written Euclid based on the older Arabic translations. In this respect it seems to be like the Latin version of the *Elements* by Campanus (Campano), which was first published by Erhard Ratdolt at Venice in 1482 (the first printed edition of Euclid[5]). Campanus (13th c.) was a mathematician, and it is likely enough that he allowed himself the same liberty as at-Ṭūsī in reproducing Euclid. Whatever may be the relation between Campanus' version and that of Athelhard of Bath (about 1120), and whether, as Curtze thinks[6], they both used one and the same Latin version of 10th—11th c., or whether Campanus used Athelhard's version in the same way as at-Ṭūsī used those of his predecessors[7], it is certain that both versions came from an Arabian source, as is evident from the occurrence of Arabic words in them[8]. Campanus' version is not of much service for the purpose of forming a judgment on the relative authenticity of the Greek and Arabian tradition; but it sometimes preserves traces of the purer source, as when it omits Theon's addition to VI. 33[9]. A curious circumstance is that, while Campanus' version agrees with at-Ṭūsī's in the number of the propositions in all the genuine Euclidean Books except V. and IX., it agrees with Athelhard's in having 34 propositions in Book V. (as against 25 in other versions), which confirms the view that the two are not independent, and also leads, as Klamroth says, to this dilemma: either the additions to Book V. are Athelhard's own, or he used an Arabian Euclid which is not known to us[10]. Heiberg also notes that Campanus' Books XIV., XV. show a certain agreement with the preface to the Thābit-Isḥāq version, in which the author claims to have (1) given a method of inscribing spheres in the five regular solids, (2) carried further the solution of the problem how

---

[1] Suter, *Die Mathematiker und Astronomen der Araber*, p. 151. The Latin title is *Euclidis elementorum geometricorum libri tredecim. Ex traditione doctissimi Nasiridini Tusini nunc primum arabice impressi.* Romae in typographia Medicea MDXCIV. Cum licentia superiorum.
[2] Kästner, *Geschichte der Mathematik*, I. p. 367.
[3] Suter has a note that this MS. is very old, having been copied from the original in the author's lifetime.
[4] Suter, p. 151.
[5] Described by Kästner, *Geschichte der Mathematik*, I. pp. 289—299, and by Weissenborn, *Die Uebersetzungen des Euklid durch Campano und Zamberti*, Halle a. S., 1882, pp. 1—7. See also *infra*, Chapter VIII, p. 97.
[6] Sonderabdruck des *Jahresberichtes über die Fortschritte der klassischen Alterthumswissenschaft vom Okt.* 1879—1882, Berlin, 1884.
[7] Klamroth, p. 271.
[8] Curtze, *op. cit.* p. 20; Heiberg, *Euklid-Studien*, p. 178.
[9] Heiberg's Euclid, vol. v. p. ci.　　　[10] Klamroth, pp. 273—4.

to inscribe any one of the solids in any other and (3) noted the cases where this could not be done[1].

With a view to arriving at what may be called a common measure of the Arabian tradition, it is necessary to compare, in the first place, the numbers of propositions in the various Books. Ḥājī Khalfa says that al-Ḥajjāj's translation contained 468 propositions, and Thābit's 478; this is stated on the authority of aṭ-Ṭūsī, whose own edition contained 468[2]. The fact that Thābit's version had 478 propositions is confirmed by an index in the Bodleian MS. 279 (called O by Klamroth). A register at the beginning of the Codex Leidensis 399, 1 which gives Isḥaq's numbers (although the translation is that of al-Ḥajjāj) apparently makes the total 479 propositions (the number in Book XIV. being apparently 11, instead of the 10 of O[3]). I subjoin a table of relative numbers taken from Klamroth, to which I have added the corresponding numbers in August's and Heiberg's editions of the Greek text.

	The Arabian Euclid			The Greek Euclid		
Books	Isḥāq	aṭ-Ṭūsī	Campanus	Gregory	August	Heiberg
I	48	48	48	48	48	48
II	14	14	14	14	14	14
III	36	36	36	37	37	37
IV	16	16	16	16	16	16
V	25	25	34	25	25	25
VI	33	32	32	33	33	33
VII	39	39	39	41	41	39
VIII	27	25	25	27	27	27
IX	38	36	39	36	36	36
X	109	107	107	117	116	115
XI	41	41	41	40	40	39
XII	15	15	15	18	18	18
XIII	21	18	18	18	18	18
	462	452	464	470	469	465
[XIV	10	10	18	7		?
XV	6	6	13	10		
	478	468	495	487		?]

The numbers in the case of Heiberg include all propositions which he has printed in the text; they include therefore XIII. 6 and III. 12 now to be regarded as spurious, and X. 112—115 which he brackets as doubtful. He does not number the propositions in Books XIV., XV., but I conclude that the numbers in P reach at least 9 in XIV., and 9 in XV.

[1] Heiberg, *Zeitschrift für Math. u. Physik*, XXIX., hist.-litt. Abtheilung, p. 21.
[2] Klamroth, p. 274; Steinschneider, *Zeitschrift für Math. u. Physik*, XXXI., hist.-litt. Abth. p. 98.
[3] Besthorn-Heiberg read "11?" as the number, Klamroth had read it as 21 (p. 273).

The *Fihrist* confirms the number 109 for Book X., from which Klamroth concludes that Ishāq's version was considered as by far the most authoritative.

In the text of O, Book IV. consists of 17 propositions and Book XIV. of 12, differing in this respect from its own table of contents; IV. 15, 16 in O are really two proofs of the same proposition.

In al-Ḥajjāj's version Book I. consists of 47 propositions only, I. 45 being omitted. It has also one proposition fewer in Book III., the Heronic proposition III. 12 being no doubt omitted.

In speaking of particular propositions, I shall use Heiberg's numbering, except where otherwise stated.

The difference of 10 propositions between Thābit-Ishāq and aṭ-Ṭūsī is accounted for thus:

(1) The three propositions VI. 12 and X. 28, 29 which both Ishāq and the Greek text have are omitted in aṭ-Ṭūsī.

(2) Ishāq divides each of the propositions XIII. 1—3 into two, making six instead of three in aṭ-Ṭūsī and in the Greek.

(3) Ishāq has four propositions (numbered by him VIII. 24, 25, IX. 30, 31) which are neither in the Greek Euclid nor in aṭ-Ṭūsī.

Apart from the above differences al-Ḥajjāj (so far as we know), Ishāq and aṭ-Ṭūsī agree; but their Euclid shows many differences from our Greek text. These differences we will classify as follows[1].

1. *Propositions.*

The Arabian Euclid omits VII. 20, 22 of Gregory's and August's editions (Heiberg, App. to Vol. II. pp. 428–32); VIII. 16, 17; X. 7, 8, 13, 16, 24, 112, 113, 114, besides a lemma *vulgo* X. 13, the proposition X. 117 of Gregory's edition, and the scholium at the end of the Book (see for these Heiberg's Appendix to Vol. III. pp. 382, 408—416); XI. 38 in Gregory and August (Heiberg, App. to Vol. IV. p. 354); XII. 6, 13, 14; (also all but the first third of Book XV.).

The Arabian Euclid makes III. 11, 12 into one proposition, and divides some propositions (X. 31, 32; XI. 31, 34; XIII. 1—3) into two each.

The order is also changed in the Arabic to the following extent. V. 12, 13 are interchanged and the order in Books VI., VII., IX.—XIII. is:

VI. 1—8, 13, 11, 12, 9, 10, 14—17, 19, 20, 18, 21, 22, 24, 26, 23, 25, 27—30, 32, 31, 33.

VII. 1—20, 22, 21, 23—28, 31, 32, 29, 30, 33—39.

IX. 1—13, 20, 14—19, 21—25, 27, 26, 28—36, with two new propositions coming before prop. 30.

X. 1—6, 9—12, 15, 14, 17—23, 26—28, 25, 29—30, 31, 32, 33—111, 115.

XI. 1—30, 31, 32, 34, 33, 35—39.

XII. 1—5, 7, 9, 8, 10, 12, 11, 15, 16—18.

XIII. 1—3, 5, 4, 6, 7, 12, 9, 10, 8, 11, 13, 15, 14, 16—18.

[1] See Klamroth, pp. 275—6, 280, 282—4, 314—15, 326; Heiberg, vol. v. pp. xcvi, xcvii.

2. *Definitions.*

The Arabic omits the following definitions: IV. Deff. 3—7, VII. Def. 9 (or 10), XI. Deff. 5—7, 15, 17. 23, 25—28; but it has the spurious definitions VI. Deff. 2, 5, and those of *proportion* and *ordered proportion* in Book V. (Deff. 8, 19 August), and wrongly interchanges V. Deff. 11, 12 and also VI. Deff. 3, 4.

The order of the definitions is also different in Book VII. where, after Def. 11, the order is 12, 14, 13, 15, 16, 19, 20, 17, 18, 21, 22, 23, and in Book XI. where the order is 1, 2, 3, 4, 8, 10, 9, 13, 14, 16, 12, 21, 22, 18, 19, 20, 11, 24.

3. *Lemmas and porisms.*

All are omitted in the Arabic except the porisms to VI. 8, VIII. 2, X. 3; but there are slight additions here and there, not found in the Greek, e.g. in VIII. 14, 15 (in K).

4. *Alternative proofs.*

These are all omitted in the Arabic, except that in X. 105, 106 they are substituted for the genuine proofs; but one or two alternative proofs are peculiar to the Arabic (VI. 32 and VIII. 4, 6).

The analyses and syntheses to XIII. 1—5 are also omitted in the Arabic.

Klamroth is inclined, on a consideration of all these differences, to give preference to the Arabian tradition over the Greek (1) "on historical grounds," subject to the proviso that no Greek MS. as ancient as the 8th c. is found to contradict his conclusions, which are based generally (2) on the improbability that the Arabs would have *omitted* so much if they had found it in their Greek MSS., it being clear from the *Fihrist* that the Arabs had already shown an anxiety for a pure text, and that the old translators were subjected in this matter to the check of public criticism. Against the "historical grounds," Heiberg is able to bring a considerable amount of evidence[1]. First of all there is the British Museum palimpsest (L) of the 7th or the beginning of the 8th c. This has fragments of propositions in Book X. which are omitted in the Arabic; the numbering of one proposition, which agrees with the numbering in other Greek MS., is not comprehensible on the assumption that eight preceding propositions were omitted in it, as they are in the Arabic; and lastly, the readings in L are tolerably like those of our MSS., and surprisingly like those of B. It is also to be noted that, although P dates from the 10th c. only, it contains, according to all appearance, an ante-Theonine recension.

Moreover there is positive evidence against certain omissions by the Arabians. At-Ṭūsī omits VI. 12, but it is scarcely possible that, if Eutocius had not had it, he would have quoted VI. 23 by that number[2]. This quotation of VI. 23 by Eutocius also tells against Isḥāq who has the proposition as VI. 25. Again, Simplicius quotes VI. 10 by that number, whereas it is VI. 13 in Isḥāq; and Pappus quotes, by number, XIII. 2 (Isḥāq 3, 4), XIII. 4 (Isḥāq 8), XIII. 16 (Isḥāq 19).

---

[1] Heiberg in *Zeitschrift für Math. u. Physik*, XXIX., hist.-litt. Abth. p. 3 sqq.
[2] Apollonius, ed. Heiberg, vol. II. p. 218, 3—5.

On the other hand the contraction of III. 11, 12 into one proposition in the Arabic tells in favour of the Arabic.

Further, the omission of certain porisms in the Arabic cannot be supported; for Pappus quotes the porism to XIII. 17[1], Proclus those to II. 4, III. 1, VII. 2[2], and Simplicius that to IV. 15.

Lastly, some propositions omitted in the Arabic are required in later propositions. Thus X. 13 is used in X. 18, 22, 23, 26 etc.; X. 17 is wanted in X. 18, 26, 36; XII. 6, 13 are required for XII. 11 and XII. 15 respectively.

It must also be remembered that some of the things which were properly omitted by the Arabians are omitted or marked as doubtful in Greek MSS. also, especially in P, and others are rightly suspected for other reasons (e.g. a number of alternative proofs, lemmas, and porisms, as well as the analyses and syntheses of XIII. 1—5). On the other hand, the Arabic has certain interpolations peculiar to our inferior MSS. (cf. the definition VI. Def. 2 and those of *proportion* and *ordered proportion*).

Heiberg comes to the general conclusion that, not only is the Arabic tradition not to be preferred offhand to that of the Greek MSS., but it must be regarded as inferior in authority. It is a question how far the differences shown in the Arabic are due to the use of Greek MSS. differing from those which have been most used as the basis of our text, and how far to the arbitrary changes made by the Arabians themselves. Changes of order and arbitrary omissions could not surprise us, in view of the preface above quoted from the Oxford MS. of Thābit-Ishāq, with its allusion to the many important and necessary things left out by Abū 'l Wafā and to the author's own rearrangement of Books XII., XIII. But there is evidence of differences due to the use by the Arabs of other Greek MSS. Heiberg[3] is able to show considerable resemblances between the Arabic text and the Bologna MS. b in that part of the MS. where it diverges so remarkably from our other MSS. (see the short description of it above, p. 49); in illustration he gives a comparison of the proofs of XII. 7 in b and in the Arabic respectively, and points to the omission in both of the proposition given in Gregory's edition as XI. 38, and to a remarkable agreement between them as regards the order of the propositions of Book XII. As above stated, the remarkable divergence of b only affects Books XI. (at end) and XII.; and Book XIII. in b shows none of the transpositions and other peculiarities of the Arabic. There are many differences between b and the Arabic, especially in the definitions of Book XI., as well as in Book XIII. It is therefore a question whether the Arabians made arbitrary changes, or the Arabic form is the more ancient, and b has been altered through contact with other MSS. Heiberg points out that the Arabians must be alone responsible for their definition of a prism, which only covers a prism with a triangular base. This could not have been Euclid's own, for the word *prism* already has the wider meaning in Archimedes, and

<hr />

[1] Pappus, v. p. 436, 5.   [2] Proclus, pp. 303—4.
[3] *Zeitschrift für Math. u. Physik*, XXIX., hist.-litt. Abth. p. 6 sqq.

Euclid himself speaks of prisms with parallelograms and polygons as bases (XI. 39; XII. 10). Moreover, a Greek would not have been likely to leave out the definitions of the " Platonic" regular solids.

Heiberg considers that the Arabian translator had before him a MS. which was related to b, but diverged still further from the rest of our MSS. He does not think that there is evidence of the existence of a redaction of Books I.—X. similar to that of Books XI., XII. in b; for Klamroth observes that it is the Books on solid geometry (XI.—XIII.) which are more remarkable than the others for omissions and shorter proofs, and it is a noteworthy coincidence that it is just in these Books that we have a divergent text in b.

An advantage in the Arabic version is the omission of VII. Def. 10, although, as Iamblichus had it, it may have been deliberately omitted by the Arabic translator. Another advantage is the omission of the analyses and syntheses of XIII. 1—5; but again these may have been omitted purposely, as were evidently a number of porisms which are really necessary.

One or two remarks may be added about the Arabic versions as compared with one another. Al-Ḥajjāj's object seems to have been less to give a faithful reflection of the original than to write a useful and convenient mathematical text-book. One characteristic of it is the careful references to earlier propositions when their results are used. Such specific quotations of earlier propositions are rare in Euclid; but in al-Ḥajjāj we find not only such phrases as " by prop. so and so," " which was proved" or "which we showed how to do in prop. so and so," but also still longer phrases. Sometimes he repeats a construction, as in I. 44 where, instead of constructing "the parallelogram $BEFG$ equal to the triangle C in the angle EBG which is equal to the angle D" and placing it in a certain position, he produces AB to G, making BG equal to half DE (the base of the triangle CDE in his figure), and on GB so constructs the parallelogram $BHKG$ by I. 42 that it is equal to the triangle CDE, and its angle GBH is equal to the given angle.

Secondly, al-Ḥajjāj, in the arithmetical books, in the theory of proportion, in the applications of the Pythagorean I. 47, and generally where possible, illustrates the proofs by numerical examples. It is true, observes Klamroth, that these examples are not apparently separated from the commentary of an-Nairīzī, and might not therefore have been due to al-Ḥajjāj himself; but the marginal notes to the Hebrew translation in Municn MS. 36 show that these additions were in the copy of al-Ḥajjāj used by the translator, for they expressly give these proofs in numbers as variants taken from al-Ḥajjāj[1].

These characteristics, together with al-Ḥajjāj's freer formulation of the propositions and expansion of the proofs, constitute an intelligible reason why Isḥāq should have undertaken a fresh translation from the Greek. Klamroth calls Isḥāq's version a model of a good translation of a mathematical text; the introductory and transitional

---

[1] Klamroth, p. 310; Steinschneider, pp. 85—6.

phrases are stereotyped and few in number, the technical terms are simply and consistently rendered, and the less formal expressions connect themselves as closely with the Greek as is consistent with intelligibility and the character of the Arabic language. Only in isolated cases does the formulation of definitions and enunciations differ to any considerable extent from the original. In general, his object seems to have been to get rid of difficulties and unevennesses in the Greek text by neat devices, while at the same time giving a faithful reproduction of it[1].

There are curious points of contact between the versions of al-Ḥajjāj and Thābit-Isḥāq. For example, the definitions and enunciations of propositions are often word for word the same. Presumably this is owing to the fact that Isḥāq found these definitions and enunciations already established in the schools in his time, where they would no doubt be learnt by heart, and refrained from translating them afresh, merely adopting the older version with some changes[2]. Secondly, there is remarkable agreement between the Arabic versions as regards the figures, which show considerable variations from the figures of the Greek text, especially as regards the letters; this is also probably to be explained in the same way, all the later translators having most likely borrowed al-Ḥajjāj's adaptation of the Greek figures[3]. Lastly, it is remarkable that the version of Books XI.—XIII. in the Kjøbenhavn MS. (K), purporting to be by al-Ḥajjāj, is almost exactly the same as the Thābit-Isḥāq version of the same Books in O. Klamroth conjectures that Isḥāq may not have translated the Books on solid geometry at all, and that Thābit took them from al-Ḥajjāj, only making some changes in order to fit them to the translation of Isḥāq[4].

From the facts (1) that aṭ-Ṭūsī's edition had the same number of propositions (468) as al-Ḥajjāj's version, while Thābit-Isḥāq's had 478, and (2) that aṭ-Ṭūsī has the same careful references to earlier propositions, Klamroth concludes that aṭ-Ṭūsī deliberately preferred al-Ḥajjāj's version to that of Isḥāq[5]. Heiberg, however, points out (1) that aṭ-Ṭūsī left out VI. 12 which, if we may judge by Klamroth's silence, al-Ḥajjāj had, and (2) al-Ḥajjāj's version had one proposition less in Books I. and III. than aṭ-Ṭūsī has. Besides, in a passage quoted by Ḥājī Khalfa[6] from aṭ-Ṭūsī, the latter says that "he separated the things which, in the approved editions, were taken from the archetype from the things which had been added thereto," indicating that he had compiled his edition from *both* the earlier translations[7].

There were a large number of Arabian commentaries on, or reproductions of, the *Elements* or portions thereof, which will be

---

[1] Klamroth, p. 290, illustrates Isḥāq's method by his way of distinguishing ἐφαρμόζειν (to be congruent with) and ἐφαρμόζεσθαι (to be applied to), the confusion of which by translators was animadverted on by Savile. Isḥāq avoided the confusion by using two entirely different words.

[2] Klamroth, pp. 310—1.   [3] *ibid.* p. 287.
[4] *ibid.* pp. 304—5.   [5] *ibid.* p. 274.
[6] Ḥājī Khalfa, I. p. 383.
[7] Heiberg, *Zeitschrift für Math. u. Physik*, XXIX., hist.-litt. Abth. pp. 2, 3.

found fully noticed by Steinschneider[1]. I shall mention here the commentators etc. referred to in the *Fihrist*, with a few others.

1. Abū 'l 'Abbās al-Faḍl b. Ḥātim an-Nairīzī (born at Nairīz, died about 922) has already been mentioned[2]. His commentary survives, as regards Books I.—VI., in the Codex Leidensis 399, 1, now edited, as to four Books, by Besthorn and Heiberg, and as regards Books I.—X. in the Latin translation made by Gherard of Cremona in the 12th c. and now published by Curtze from a Cracow MS.[3] Its importance lies mainly in the quotations from Heron and Simplicius.

2. Aḥmad b. 'Umar al-Karābīsī (date uncertain, probably 9th—10th c.), "who was among the most distinguished geometers and arithmeticians[4]."

3. Al-'Abbās b. Sa'īd al-Jauharī (fl. 830) was one of the astronomical observers under al-Ma'mūn, but devoted himself mostly to geometry. He wrote a commentary to the whole of the *Elements*, from the beginning to the end; also the "Book of the propositions which he added to the first book of Euclid[5]."

4. Muḥ. b. 'Īsā Abū 'Abdallāh al-Māhānī (d. between 874 and 884) wrote, according to the *Fihrist*, (1) a commentary on Eucl. Book V., (2) "On proportion," (3) "On the 26 propositions of the first Book of Euclid which are proved without *reductio ad absurdum*[6]." The work "On proportion" survives and is probably identical with, or part of, the commentary on Book V.[7] He also wrote, what is not mentioned by the *Fihrist*, a commentary on Eucl. Book X., a fragment of which survives in a Paris MS.[8]

5. Abū Ja'far al-Khāzin (i.e. "the treasurer" or "librarian"), one of the first mathematicians and astronomers of his time, was born in Khurāsān and died between the years 961 and 971. The *Fihrist* speaks of him as having written a commentary on the whole of the *Elements*[9], but only the commentary on the beginning of Book X. survives (in Leiden, Berlin and Paris); therefore either the notes on the rest of the Books have perished, or the *Fihrist* is in error[10]. The latter would seem more probable, for, at the end of his commentary, al-Khāzin remarks that the rest had already been commented on by Sulaimān b. 'Uṣma (Leiden MS.)[11] or 'Oqba (Suter), to be mentioned below. Al-Khāzin's method is criticised unfavourably in the preface to the Oxford MS. quoted by Nicoll (see p. 77 above).

6. Abū 'l Wafā al-Būzjānī (940–997), one of the greatest Arabian mathematicians, wrote a commentary on the *Elements*, but

---

[1] Steinschneider, *Zeitschrift für Math. u. Physik*, XXXI., hist.-litt. Abth. pp. 86 sqq.
[2] Steinschneider, p. 86, *Fihrist* (tr. Suter), pp. 16, 67; Suter, *Die Mathematiker und Astronomen der Araber* (1900), p. 45.
[3] *Supplementum ad Euclidis opera omnia*, ed. Heiberg and Menge, Leipzig, 1899.
[4] *Fihrist*, pp. 16, 38; Steinschneider, p. 87; Suter, p. 65.
[5] *Fihrist*, pp. 16, 25; Steinschneider, p. 88; Suter, p. 12.
[6] *Fihrist*, pp. 16, 25, 58.
[7] Suter, p. 26, note, quotes the Paris MS. 2467, 16° containing the work "on proportion" as the authority for this conjecture.
[8] MS. 2457, 39° (cf. Woepcke in *Mém. prés. à l'acad. des sciences*, XIV., 1856, p. 669).
[9] *Fihrist*, p. 17.　　　[10] Suter, p. 58, note b.　　　[11] Steinschneider, p. 89.

did not complete it[1]. His method is also unfavourably regarded in the same preface to the Oxford MS. 280. According to Ḥājī Khalfa, he also wrote a book on geometrical constructions, in thirteen chapters. Apparently a book answering to this description was compiled by a gifted pupil from lectures by Abū 'l Wafā, and a Paris MS. (Anc. fonds 169) contains a Persian translation of this work, not that of Abū 'l Wafā himself. An analysis of the work was given by Woepcke[2], and some particulars will be found in Cantor[3]. Abū 'l Wafā also wrote a commentary on Diophantus, as well as a separate "book of proofs to the propositions which Diophantus used in his book and to what he (Abū 'l Wafā) employed in his commentary[4]."

7. **Ibn Rāhawaihi al-Arjānī** also commented on Eucl. Book X.[5]

8. **ʿAlī b. Ahmad Abū 'l-Qāsim al-Anṭākī** (d. 987) wrote a commentary on the whole book[6]; part of it seems to survive (from the 5th Book onwards) at Oxford (Catal. MSS. orient. II. 281)[7].

9. **Sind b. ʿAlī Abū 'ṭ-Ṭaiyib** was a Jew who went over to Islam in the time of al-Maʾmūn, and was received among his astronomical observers, whose head he became[8] (about 830); he died after 864. He wrote a commentary on the whole of the *Elements*; "Abū ʿAlī saw nine books of it, and a part of the tenth[9]." His book "On the Apotomae and the Medials," mentioned by the *Fihrist*, may be the same as, or part of, his commentary on Book X.

10. **Abū Yūsuf Yaʿqūb b. Muḥ. ar-Rāzī** "wrote a commentary on Book X., and that an excellent one, at the instance of Ibn al-ʿAmīd[10]."

11. The *Fihrist* next mentions **al-Kindī** (Abū Yūsuf Yaʿqūb b. Ishāq b. as-Ṣabbāḥ al-Kindī, d. about 873), as the author (1) of a work "on the objects of Euclid's book," in which occurs the statement that the *Elements* were originally written by Apollonius, the carpenter (see above, p. 5 and note), (2) of a book "on the improvement of Euclid's work," and (3) of another "on the improvement of the 14th and 15th Books of Euclid." "He was the most distinguished man of his time, and stood alone in the knowledge of the old sciences collectively; he was called 'the philosopher of the Arabians'; his writings treat of the most different branches of knowledge, as logic, philosophy, geometry, calculation, arithmetic, music, astronomy and others[11]." Among the other geometrical works of al-Kindī mentioned by the *Fihrist*[12] are treatises on the closer investigation of the results of Archimedes concerning the measure of the diameter of a circle in terms of its circumference, on the construction of the figure of the two mean proportionals, on the approximate determination of the chords

---

[1] *Fihrist*, p. 17.
[2] Woepcke, *Journal Asiatique*, Sér. v. T. v. pp. 218—256 and 309—359.
[3] *Gesch. d. Math.* vol. I_3, pp. 743—6.
[4] *Fihrist*, p. 39; Suter, p. 71.
[5] *Fihrist*, p. 17; Suter, p. 17.
[6] *Fihrist*, p. 17.
[7] Suter, p. 64.
[8] *Fihrist*, p. 17, 29; Suter, pp. 13, 14.
[9] *Fihrist*, p. 17.
[10] *Fihrist*, p. 17; Suter, p. 66.
[11] *Fihrist*, p. 17, 10—15.
[12] The mere catalogue of al-Kindī's works on the various branches of science takes up four octavo pages (11—15) of Suter's translation of the *Fihrist*.

of the circle, on the approximate determination of the chord (side) of the nonagon, on the division of triangles and quadrilaterals and constructions for that purpose, on the manner of construction of a circle which is equal to the surface of a given cylinder, on the division of the circle, in three chapters etc.

12. The physician Naẓīf b. Yumn (or Yaman) al-Qass ("the priest") is mentioned by the *Fihrist* as having seen a Greek copy of Eucl. Book X. which had 40 more propositions than that which was in general circulation (containing 109), and having determined to translate it into Arabic[1]. Fragments of such a translation exist at Paris, Nos. 18 and 34 of the MS. 2457 (952, 2 Suppl. Arab. in Woepcke's tract); No. 18 contains "additions to some propositions of the 10th Book, existing in the Greek language[2]." Naẓīf must have died about 990[3].

13. Yūḥannā b. Yūsuf b. al-Ḥārith b. al-Biṭrīq al-Qass (d. about 980) lectured on the *Elements* and other geometrical books, made translations from the Greek, and wrote a tract on the "proof" of the case of two straight lines both meeting a third and making with it, on one side, two angles together less than two right angles[4]. Nothing of his appears to survive, except that a tract "on rational and irrational magnitudes," No. 48 in the Paris MS. just mentioned, is attributed to him.

14. Abū Muḥ. al-Ḥasan b. ʿUbaidallāh b. Sulaimān b. Wahb (d. 901) was a geometer of distinction, who wrote works under the two distinct titles "A commentary on the difficult parts of the work of Euclid" and "The Book on Proportion[5]." Suter thinks that another reading is possible in the case of the second title, and that it may refer to the Euclidean work "on the divisions (of figures)[6]."

15. Qusṭā b. Lūqā al-Baʿlabakkī (d. about 912), a physician, philosopher, astronomer, mathematician and translator, wrote "on the difficult passages of Euclid's book" and "on the solution of arithmetical problems from the third book of Euclid[7]"; also an "introduction to geometry," in the form of question and answer[8].

16. Thābit b. Qurra (826–901), besides translating some parts of Archimedes and Books V.—VII. of the *Conics* of Apollonius, and revising Isḥāq's translation of Euclid's *Elements*, also revised the translation of the *Data* by the same Isḥāq and the book *On divisions of figures* translated by an anonymous writer. We are told also that he wrote the following works: (1) On the Premisses (Axioms, Postulates etc.) of Euclid, (2) On the Propositions of Euclid, (3) On the propositions and questions which arise when two straight lines are cut by a third (or on the "proof" of Euclid's famous postulate). The last tract is extant in the MS. discovered by Woepcke (Paris 2457, 32°). He is also credited with "an excellent work" in the shape of an "Introduction to the Book of Euclid," a treatise on

[1] *Fihrist*, pp. 16, 17.
[2] Woepcke, *Mém. prés. à l'acad. des sciences.* XIV. pp. 666, 668.
[3] Suter, p. 68.   [4] *Fihrist*, p. 38; Suter, p. 60.
[5] *Fihrist*, p. 26, and Suter's note, p. 60.   [6] Suter, p. 211, note 23.
[7] *Fihrist*, p. 43.   [8] *Fihrist*, p. 43; Suter, p. 41.

Geometry dedicated to Ismāʿil b. Bulbul, a Compendium of Geometry, and a large number of other works for the titles of which reference may be made to Suter, who also gives particulars as to which are extant[1].

17. Abū Saʿīd **Sinān** b. Thābit b. Qurra, the son of the translator of Euclid, followed in his father's footsteps as geometer, astronomer and physician. He wrote an "improvement of the book of ...... on the Elements of Geometry, in which he made various additions to the original." It is natural to conjecture that *Euclid* is the name missing in this description (by Ibn abī Usaibiʿa); Casiri has the name Aqāton[2]. The latest editor of the *Taʾrīkh al-Ḥukamā*, however, makes the name to be Iflāton (= Plato), and he refers to the statement by the *Fihrist* and Ibn al-Qiftī attributing to Plato a work on the Elements of Geometry translated by Qustā. It is just possible, therefore, that at the time of Qustā the Arabs were acquainted with a book on the Elements of Geometry translated from the Greek, which they attributed to Plato[3]. Sinān died in 943.

18. Abū Sahl Wījan (or Waijan) b. Rustam **al-Kūhī** (fl. 988), born at Kūh in Ṭabaristān, a distinguished geometer and astronomer, wrote, according to the *Fihrist*, a "Book of the Elements" after that of Euclid[4]; the 1st and 2nd Books survive at Cairo, and a part of the 3rd Book at Berlin (5922)[5]. He wrote also a number of other geometrical works: Additions to the 2nd Book of Archimedes on the Sphere and Cylinder (extant at Paris, at Leiden, and in the India Office), On the finding of the side of a heptagon in a circle (India Office and Cairo), On two mean proportionals (India Office), which last may be only a part of the Additions to Archimedes' On the Sphere and Cylinder, etc.

19. Abū Naṣr Muḥ. b. Muḥ. b. Ṭarkhān b. Uzlaġ **al-Fārābī** (870–950) wrote a commentary on the difficulties of the introductory matter to Books I. and V.[6] This appears to survive in the Hebrew translation which is, with probability, attributed to Moses b. Tibbon[7].

20. Abū ʿAlī al-Ḥasan b. al-Hasan b. **al-Haitham** (about 965–1039), known by the name Ibn al-Haitham or Abū ʿAlī-al-Baṣrī, was a man of great powers and knowledge, and no one of his time approached him in the field of mathematical science. He wrote several works on Euclid the titles of which, as translated by Woepcke from Usaibiʿa, are as follows[8]:

1. Commentary and abridgment of the *Elements*.
2. Collection of the Elements of Geometry and Arithmetic, drawn from the treatises of Euclid and Apollonius.
3. Collection of the Elements of the Calculus deduced from the principles laid down by Euclid in his *Elements*.

---

[1] Suter, pp. 34—8.
[2] *Fihrist* (ed. Suter), p. 59, note 132 ; Suter, p. 52, note b.
[3] See Suter in *Bibliotheca Mathematica*, IV₃, 1903-4, pp. 296—7, review of Julius Lippert's *Ibn al-Qiftī. Taʾrīch al-ḥukamā*, Leipzig, 1903.
[4] *Fihrist*, p. 40.   [5] Suter, p. 75.
[6] Suter, p. 55.   [7] Steinschneider, p. 92.
[8] Steinschneider, pp. 92—3.

4. Treatise on "measure" after the manner of Euclid's *Elements*.

5. Memoir on the solution of the difficulties in Book I.

6. Memoir for the solution of a doubt about Euclid, relative to Book V.

7. Memoir on the solution of a doubt about the stereometric portion.

8. Memoir on the solution of a doubt about Book XII.

9. Memoir on the division of the two magnitudes mentioned in X. 1 (the theorem of exhaustion).

10. Commentary on the definitions in the work of Euclid (where Steinschneider thinks that some more general expression should be substituted for "definitions").

The last-named work (which Suter calls a commentary on the *Postulates* of Euclid) survives in an Oxford MS. (Catal. MSS. orient. I. 908) and in Algiers (1446, 1°).

A Leiden MS. (966) contains his Commentary "on the difficult places" up to Book V. We do not know whether in this commentary, which the author intended to form, with the commentary on the Musādarāt, a sort of complete commentary, he had collected the separate memoirs on certain doubts and difficult passages mentioned in the above list.

A commentary on Book V. and following Books found in a Bodleian MS. (Catal. II. p. 262) with the title "Commentary on Euclid and solution of his difficulties" is attributed to b. Haitham; this might be a continuation of the Leiden MS.

The memoir on X. 1 appears to survive at St Petersburg, MS. de l'Institut des langues orient. 192, 5° (Rosen, Catal. p. 125).

21. **Ibn Sīnā**, known as **Avicenna** (980–1037), wrote a Compendium of Euclid, preserved in a Leiden MS. No. 1445, and forming the geometrical portion of an encyclopaedic work embracing Logic, Mathematics, Physics and Metaphysics[1].

22. Aḥmad b. al-Ḥusain **al-Ahwāzī** al-Kātib wrote a commentary on Book X., a fragment of which (some 10 pages) is to be found at Leiden (970), Berlin (5923) and Paris (2467, 18°)[2].

23. Naṣīraddīn aṭ-Ṭūsī (1201–1274) who, as we have seen, brought out a Euclid in two forms, wrote:

1. A treatise on the postulates of Euclid (Paris, 2467, 5°).

2. A treatise on the 5th postulate, perhaps only a part of the foregoing (Berlin, 5942, Paris, 2467, 6°).

3. Principles of Geometry taken from Euclid, perhaps identical with No. 1 above (Florence, Pal. 298).

4. 105 problems out of the *Elements* (Cairo). He also edited the *Data* (Berlin, Florence, Oxford etc.)[3].

24. Muḥ. b. Ashraf Shamsaddīn **as-Samarqandī** (fl. 1276) wrote "Fundamental Propositions, being elucidations of 35 selected proposi-

---

[1] Steinschneider, p. 92; Suter, p. 89.   [2] Suter, p. 57.
[3] Suter, pp. 150—1.

tions of the first Books of Euclid," which are extant at Gotha (1496 and 1497), Oxford (Catal. I. 967, 2°), and Brit. Mus.[1].

25. Mūsā b. Muḥ. b. Maḥmūd, known as **Qāḍīzāde** ar-Rūmī (i.e. the son of the judge from Asia Minor), who died between 1436 and 1446, wrote a commentary on the "Fundamental Propositions" just mentioned, of which many MSS. are extant[2]. It contained biographical statements about Euclid alluded to above (p. 5, note).

26. Abū Dā'ūd Sulaimān b. ʿUqba, a contemporary of al-Khāzin (see above, No. 5), wrote a commentary on the second half of Book X., which is, at least partly, extant at Leiden (974) under the title "On the binomials and apotomae found in the 10th Book of Euclid[3]."

27. The Codex Leidensis 399, 1 containing al-Ḥajjāj's translation of Books I.—VI. is said to contain glosses to it by Saʿīd b. Masʿūd b. al-Qass, apparently identical with Abū Naṣr Gars al-Naʿma, son of the physician Masʿūd b. al-Qass al-Baġdādī, who lived in the time of the last Caliph al-Mustaʿṣim (d. 1258)[4].

28. Abū **Muḥammad b. Abdalbāqī** al-Baġdādī al-Faraḍī (d. 1141, at the age of over 70 years) is stated in the *Taʾrīkh al-Ḥukamā* to have written an excellent commentary on Book X. of the *Elements*, in which he gave numerical examples of the propositions[5]. This is published in Curtze's edition of an-Nairīzī where it occupies pages 252—386[6].

29. Yaḥyā b. Muḥ. b. ʿAbdān b. ʿAbdalwāḥid, known by the name of Ibn al-Lubūdī (1210–1268), wrote a Compendium of Euclid, and a short presentation of the postulates[7].

30. Abū ʿAbdallāh Muḥ. b. Muʿādh al-Jayyānī wrote a commentary on Eucl. Book V. which survives at Algiers (1446, 3°)[8].

31. Abū Naṣr Mansūr b. ʿAlī b. ʿIrāq wrote, at the instance of Muḥ. b. Aḥmad Abū 'r-Raiḥān al-Bīrūnī (973–1048), a tract "on a doubtful (difficult) passage in Eucl. Book XIII." (Berlin, 5925)[9].

---

[1] Suter, p. 157.   [2] *ibid.* p. 175.   [3] *ibid.* p. 56.

[4] *ibid.* pp. 153—4, 227.

[5] Gartz, p. 14; Steinschneider, pp. 94—5.

[6] Suter in *Bibliotheca Mathematica*, IV$_3$, 1903, pp. 25, 295; Suter has also an article on its contents, *Bibliotheca Mathematica*, VII$_3$, 1906–7, pp. 234—251.

[7] Steinschneider, p. 94; Suter, p. 146.

[8] Suter, *Nachträge und Berichtigungen*, in *Abhandlungen zur Gesch. der math. Wissenschaften*, XIV., 1902, p. 170.

[9] Suter, p. 81, and *Nachträge*, p. 172.

CHAPTER VIII.

PRINCIPAL TRANSLATIONS AND EDITIONS OF THE ELEMENTS.

CICERO is the first Latin author to mention Euclid[1]; but it is not likely that in Cicero's time Euclid had been translated into Latin or was studied to any considerable extent by the Romans; for, as Cicero says in another place[2], while geometry was held in high honour among the Greeks, so that nothing was more brilliant than their mathematicians, the Romans limited its scope by having regard only to its utility for measurements and calculations. How very little theoretical geometry satisfied the Roman *agrimensores* is evidenced by the work of Balbus *de mensuris*[3], where some of the definitions of Eucl. Book I. are given. Again, the extracts from the *Elements* found in the fragment attributed to Censorinus (fl. 238 A.D)[4] are confined to the definitions, postulates, and common notions. But by degrees the *Elements* passed even among the Romans into the curriculum of a liberal education; for Martianus Capella speaks of the effect of the enunciation of the proposition "how to construct an equilateral triangle on a given straight line" among a company of philosophers, who, recognising the first proposition of the *Elements*, straightway break out into encomiums on Euclid[5]. But the *Elements* were then (*c.* 470 A.D.) doubtless read in Greek; for what Martianus Capella gives[6] was drawn from a Greek source, as is shown by the occurrence of Greek words and by the wrong translation of I. def. 1 ("punctum vero est cuius pars *nihil* est"). Martianus may, it is true, have quoted, not from Euclid himself, but from Heron or some other ancient source.

But it is clear from a certain palimpsest at Verona that some scholar had already attempted to translate the *Elements* into Latin. This palimpsest[7] has part of the "Moral reflections on the Book of Job" by Pope Gregory the Great written in a hand of the 9th c. above certain fragments which in the opinion of the best judges date from the 4th c. Among these are fragments of Vergil and of Livy, as well as a geometrical fragment which purports to be taken from the 14tn and 15th Books of Euclid. As a matter of fact it is from Books XII. and XIII. and is of the nature of a free rendering, or rather a new

---

[1] *De oratore* III. 132.   [2] *Tusc.* I. 5.
[3] *Gromatici veteres*, I. 97 sq. (ed. F. Blume, K. Lachmann and A. Rudorff. Berlin, 1848, 1852). ·
[4] Censorinus, ed. Hultsch, pp. 60—3.
[5] Martianus Capella, VI. 724.   [6] *ibid.* VI. 708 sq.
[7] Cf. Cantor, I₃, p. 565.

arrangement, of Euclid with the propositions in different order[1]. The MS. was evidently the translator's own copy, because some words are struck out and replaced by synonyms. We do not know whether the translator completed the translation of the whole, or in what relation his version stood to our other sources.

Magnus Aurelius Cassiodorus (b. about 475 A.D.) in the geometrical part of his encyclopaedia *De artibus ac disciplinis liberalium literarum* says that geometry was represented among the Greeks by Euclid, Apollonius, Archimedes, and others, " of whom Euclid was given us translated into the Latin language by the same great man Boethius.'' ; also in his collection of letters[2] is a letter from Theodoric to Boethius containing the words, "for in your translations... Nicomachus the arithmetician, and Euclid the geometer, are heard in the Ausonian tongue." The so-called Geometry of Boethius which has come down to us by no means constitutes a translation of Euclid. The MSS. variously give five, four, three or two Books, but they represent only two distinct compilations, one normally in five Books and the other in two. Even the latter, which was edited by Friedlein, is not genuine[3], but appears to have been put together in the 11th c., from various sources. It begins with the definitions of Eucl. I., and in these are traces of perfectly correct readings which are not found even in the MSS. of the 10th c., but which can be traced in Proclus and other ancient sources; then come the Postulates (five only), the Axioms (three only), and after these some definitions of Eucl. II., III., IV. Next come the enunciations of Eucl. I., of ten propositions of Book II., and of some from Books III., IV., but always without proofs ; there follows an extraordinary passage which indicates that the author will now give something of his own in elucidation of Euclid, though what follows is a literal translation of the proofs of Eucl. I. 1—3. This latter passage, although it affords a strong argument against the genuineness of this part of the work, shows that the Pseudoboethius had a Latin translation of Euclid from which he extracted the three propositions.

Curtze has reproduced, in the preface to his edition of the trans-lation by Gherard of Cremona of an-Nairīzī's Arabic commentary on Euclid, some interesting fragments of a translation of Euclid taken from a Munich MS. of the 10th c. They are on two leaves used for the cover of the MS. (Bibliothecae Regiae Universitatis Monacensis 2° 757) and consist of portions of Eucl. I. 37, 38 and II. 8, translated literally word for word from the Greek text. The translator seems to have been an Italian (cf. the words "capitolo nono" used for the ninth prop. of Book II.) who knew very little Greek and had moreover little mathematical knowledge. For example, he translates the capital letters denoting points in figures as if they were numerals: thus τὰ ΑΒΓ,

---

[1] The fragment was deciphered by W. Studemund, who communicated his results to Cantor.

[2] Cassiodorus, *Variae*, I. 45, p. 40, 12 ed. Mommsen.

[3] See especially Weissenborn in *Abhandlungen zur Gesch. d. Math.* II. p. 185 sq.; Heiberg in *Philologus*, XLIII. p. 507 sq. ; Cantor, I₃, p. 580 sq.

ΔEZ is translated "que primo secundo et tertio quarto quinto et septimo," T becomes "tricentissimo" and so on. The Greek MS. which he used was evidently written in uncials, for ΔEZΘ becomes in one place "quod autem septimo nono," showing that he mistook ΔE for the particle δέ, and καὶ ὁ ΣTU is rendered "sicut tricentissimo et quadringentissimo," showing that the letters must have been written ΚΑΙΟCΤU.

The date of the Englishman Athelhard (Æthelhard) is approximately fixed by some remarks in his work *Perdifficiles Quaestiones Naturales* which, on the ground of the personal allusions they contain, must be assigned to the first thirty years of the 12th c.[1] He wrote a number of philosophical works. Little is known about his life. He is said to have studied at Tours and Laon, and to have lectured at the latter school. He travelled to Spain, Greece, Asia Minor and Egypt, and acquired a knowledge of Arabic, which enabled him to translate from the Arabic into Latin, among other works, the *Elements* of Euclid. The date of this translation must be put at about 1120. MSS. purporting to contain Athelhard's version are extant in the British Museum (Harleian No. 5404 and others), Oxford (Trin. Coll. 47 and Ball. Coll. 257 of 12th c.), Nürnberg (Johannes Regiomontanus' copy) and Erfurt.

Among the very numerous works of Gherard of Cremona (1114—1187) are mentioned translations of "15 Books of Euclid" and of the *Data*[2]. Till recently this translation of the *Elements* was supposed to be lost; but Axel Anthon Björnbo has succeeded (1904) in discovering a translation from the Arabic which is different from the two others known to us (those by Athelhard and Campanus respectively), and which he, on grounds apparently convincing, holds to be Gherard's. Already in 1901 Björnbo had found Books X.—XV. of this translation in a MS. at Rome (Codex Reginensis lat. 1268 of 14th c.)[3]; but three years later he had traced three MSS. containing the whole of the same translation at Paris (Cod. Paris. 7216, 15th c.), Boulogne-sur-Mer (Cod. Bononiens. 196, 14th c.), and Bruges (Cod. Brugens. 521, 14th c.), and another at Oxford (Cod. Digby 174, end of 12th c.) containing a fragment, XI. 2 to XIV. The occurrence of Greek words in this translation such as *rombus*, *romboides* (where Athelhard keeps the Arabic terms), *ambligonius*, *orthogonius*, *gnomo*, *pyramis* etc., show that the translation is independent of Athelhard's. Gherard appears to have had before him an old translation of Euclid from the Greek which Athelhard also often followed, especially in his terminology, using it however in a very different manner. Again, there are some Arabic terms, e.g. *meguar* for *axis of rotation*, which Athelhard did not use, but which is found in almost all the translations that are with certainty attributed to Gherard of Cremona; there occurs also the

---

[1] Cantor, *Gesch. d. Math.* I₃, p. 906.
[2] Boncompagni, *Della vita e delle opere di Gherardo Cremonese*, Rome, 1851, p. 5.
[3] Described in an appendix to *Studien über Menelaos' Sphärik* (*Abhandlungen zur Geschichte der mathematischen Wissenschaften*, XIV., 1902).
[4] See *Bibliotheca Mathematica*, VI₃, 1905–6, pp. 242—8.

expression "superficies equidistantium laterum et rectorum angulorum," found also in Gherard's translation of an-Nairīzī, where Athelhard says "parallelogrammum rectangulum." The translation is much clearer than Athelhard's: it is neither abbreviated nor "edited" as Athelhard's appears to have been ; it is a word-for-word translation of an Arabic MS. containing a revised and critical edition of Thābit's version. It contains several notes quoted from Thābit himself (*Thebit dixit*), e.g. about alternative proofs etc. which Thābit found "in another Greek MS.," and is therefore a further testimony to Thābit's critical treatment of the text after Greek MSS. The new editor also added critical remarks of his own, e.g. on other proofs which he found in other Arabic versions, but not in the Greek: whence it is clear that he compared the Thābit version before him with other versions as carefully as Thābit collated the Greek MSS. Lastly, the new editor speaks of "Thebit qui transtulit hunc librum in arabicam linguam" and of "translatio Thebit," which may tend to confirm the statement of al-Qiftī who credited Thābit with an independent translation, and not (as the *Fihrist* does) with a mere improvement of the version of Ishāq b. Hunain.

Gherard's translation of the Arabic commentary of an-Nairīzī on the first ten Books of the *Elements* was discovered by Maximilian Curtze in a MS. at Cracow and published as a supplementary volume to Heiberg and Menge's Euclid[1]: it will often be referred to in this work.

Next in chronological order comes Johannes Campanus (Campano) of Novara. He is mentioned by Roger Bacon (1214–1294) as a prominent mathematician of his time[2], and this indication of his date is confirmed by the fact that he was chaplain to Pope Urban IV, who was Pope from 1261 to 1281[3]. His most important achievement was his edition of the *Elements* including the two Books XIV. and XV. which are not Euclid's. The sources of Athelhard's and Campanus' translations, and the relation between them, have been the subject of much discussion, which does not seem to have led as yet to any definite conclusion. Cantor (II₁, p. 91) gives references[4] and some particulars. It appears that there is a MS. at Munich (Cod. lat. Mon. 13021) written by Sigboto in the 12th c. at Prüfning near Regensburg, and denoted by Curtze by the letter R, which contains the enunciations of part of Euclid. The Munich MSS. of Athelhard and Campanus' translations have many enunciations textually identical with those in R, so that the source of all three must, for these enunciations, have

---

[1] *Anaritii in decem libros priores Elementorum Euclidis Commentarii ex interpretatione Gherardi Cremonensis in codice Cracoviensi 569 servata* edidit Maximilianus Curtze, Leipzig (Teubner), 1899.

[2] Cantor, II₁, p. 88.

[3] Tiraboschi, *Storia della letteratura italiana*, IV. 145—160.

[4] H. Weissenborn in *Zeitschrift für Math. u. Physik*, XXV., Supplement, pp. 143—166, and in his monograph, *Die Übersetzungen des Euklid durch Campano und Zamberti* (1882); Max. Curtze in *Philologische Rundschau* (1881), I. pp. 943—950, and in *Jahresbericht über die Fortschritte der classischen Alterthumswissenschaft*, XL. (1884, III.) pp. 19—22 ; Heiberg in *Zeitschrift für Math. u. Physik*, XXXV., hist.-litt. Abth., pp. 48—58 and pp. 81—6.

been the same; in others Athelhard and Campanus diverge completely from R, which in these places follows the Greek text and is
therefore genuine and authoritative. In the 32nd definition occurs the
word "elinuam," the Arabic term for "rhombus," and throughout the
translation are a number of Arabic figures. But R was not translated
from the Arabic, as is shown by (among other things) its close
resemblance to the translation from Euclid given on pp. 377 sqq. of
the *Gromatici Veteres* and to the so-called geometry of Boethius. The
explanation of the Arabic figures and the word "elinuam" in Def. 32
appears to be that R was a late copy of an earlier original with
corruptions introduced in many places ; thus in Def. 32 a part of the
text was completely lost and was supplied by some intelligent copyist
who inserted the word "elinuam," which was known to him, and also
the Arabic figures. Thus Athelhard certainly was not the first to
translate Euclid into Latin ; there must have been in existence before
the 11th c. a Latin translation which was the common source of R,
the passage in the *Gromatici*, and "Boethius." As in the two latter
there occur the *proofs* as well as the enunciations of I. 1—3, it is
possible that this translation originally contained the proofs also.
Athelhard must have had before him this translation of the
enunciations, as well as the Arabic source from which he obtained his
proofs. That some sort of translation, or at least fragments of one,
were available before Athelhard's time even in England is indicated
by some old English verses[1]:

> "The clerk Euclide on this wyse hit fonde
> Thys craft of gemetry yn Egypte londe
> Yn Egypte he tawghte hyt ful wyde,
> In dyvers londe on every syde.
> Mony erys afterwarde y understonde
> Yer that the craft com ynto thys londe.
> Thys craft com into England, as y yow say,
> Yn tyme of good kyng Adelstone's day,"

which would put the introduction of Euclid into England as far back
as 924–940 A.D.

We now come to the relation between Athelhard and Campanus.
That their translations were not independent, as Weissenborn would
have us believe, is clear from the fact that in all MSS. and editions,
apart from orthographical differences and such small differences as
are bound to arise when MSS. are copied by persons with some
knowledge of the subject-matter, the definitions, postulates, axioms,
and the 364 enunciations are word for word identical in Athelhard
and Campanus ; and this is the case not only where both have the
same text as R but where they diverge from it. Hence it would seem
that Campanus used Athelhard's translation and only developed the
proofs by means of another redaction of the Arabian Euclid. It is
true that the difference between the proofs of the propositions in the
two translations is considerable ; Athelhard's are short and com-

---

[1] Quoted by Halliwell in *Rara Mathematica* (p. 56 note) from MS. Bib. Reg. Mus. Brit.
17 A. 1. f. 2ᵇ–3.

pressed, Campanus' clearer and more complete, following the Greek text more closely, though still at some distance. Further, the arrangement in the two is different; in Athelhard the proofs regularly precede the enunciations, Campanus follows the usual order. It is a question how far the differences in the proofs, and certain additions in each, are due to the two translators themselves or go back to Arabic originals. The latter supposition seems to Curtze and Cantor the more probable one. Curtze's general view of the relation of Campanus to Athelhard is to the effect that Athelhard's translation was gradually altered, from the form in which it appears in the two Erfurt MSS. described by Weissenborn, by successive copyists and commentators *who had Arabic originals before them,* until it took the form which Campanus gave it and in which it was published. In support of this view Curtze refers to Regiomontanus' copy of the Athelhard-Campanus translation. In Regiomontanus' own preface the title is given, and this attributes the translation to Athelhard; but, while this copy agrees almost exactly with Athelhard in Book I., yet, in places where Campanus is more lengthy, it has similar additions, and in the later Books, especially from Book III. onwards, agrees absolutely with Campanus; Regiomontanus, too, himself implies that, though the translation was Athelhard's, Campanus had revised it; for he has notes in the margin such as the following, "Campani est hec," "dubito an demonstret hic Campanus" etc.

We come now to the printed editions of the whole or of portions of the *Elements*. This is not the place for a complete bibliography, such as Riccardi has attempted in his valuable memoir issued in five parts between 1887 and 1893, which makes a large book in itself[1]. I shall confine myself to saying something of the most noteworthy translations and editions. It will be convenient to give first the Latin translations which preceded the publication of the *editio princeps* of the Greek text in 1533, next the most important editions of the Greek text itself, and after them the most important translations arranged according to date of first appearance and languages, first the Latin translations after 1533, then the Italian, German, French and English translations in order.

It may be added here that the first allusion, in the West, to the Greek text as still extant is found in Boccaccio's commentary on the *Divina Commedia* of Dante[2]. Next Johannes Regiomontanus, who intended to publish the *Elements* after the version of Campanus, but with the latter's mistakes corrected, saw in Italy (doubtless when staying with his friend Bessarion) some Greek MSS. and noticed how far they differed from the Latin version (see a letter of his written in the year 1471 to Christian Roder of Hamburg)[3].

---

[1] *Saggio di una Bibliografia Euclidea,* memoria del Prof. Pietro Riccardi (Bologna, 1887, 1888, 1890, 1893).

[2] I. p. 404.

[3] Published in C. T. de Murr's *Memorabilia Bibliothecarum Norimbergensium,* Part I. p. 190 sqq.

I. LATIN TRANSLATIONS PRIOR TO 1533.

1482. In this year appeared the first printed edition of Euclid, which was also the first printed mathematical book of any importance. This was printed at Venice by Erhard Ratdolt and contained Campanus' translation[1]. Ratdolt belonged to a family of artists at Augsburg, where he was born about 1443. Having learnt the trade of printing at home, he went in 1475 to Venice, and founded there a famous printing house which he managed for 11 years, after which he returned to Augsburg and continued to print important books until 1516. He is said to have died in 1528. Kästner[2] gives a short description of this first edition of Euclid and quotes the dedication to Prince Mocenigo of Venice which occupies the page opposite to the first page of text. The book has a margin of 2½ inches, and in this margin are placed the figures of the propositions. Ratdolt says in his dedication that at that time, although books by ancient and modern authors were printed every day in Venice, little or nothing mathematical had appeared : a fact which he puts down to the difficulty involved by the figures, which no one had up to that time succeeded in printing. He adds that after much labour he had discovered a method by which figures could be produced as easily as letters[3]. Experts are in doubt as to the nature of Ratdolt's discovery. Was it a method of making figures up out of separate parts of figures, straight or curved lines, put together as letters are put together to make words ? In a life of Joh. Gottlob Immanuel Breitkopf, a contemporary of Kästner's own, this member of the great house of Breitkopf is credited with this particular discovery. Experts in that same house expressed the opinion that Ratdolt's figures were woodcuts, while the letters denoting points in the figures were like the other letters in the text ; yet it was with carved wooden blocks that printing began. If Ratdolt was the first to print geometrical figures, it was not long before an emulator arose ; for in the very same year Mattheus Cordonis of Windischgrätz employed woodcut mathematical figures in printing Oresme's *De latitudinibus*[4]. How eagerly the opportunity of spreading geometrical knowledge was seized upon is proved by the number of editions which followed in the next few years. Even the year 1482 saw two forms of the book, though they only differ in the first sheet. Another edition came out in 1486 (*Ulmae, apud Io. Regerum*) and another in 1491 (*Vincentiae per*

---

[1] Curtze (An-Nairīzī, p. xiii) reproduces the heading of the first page of the text as follows (there is no title-page): Preclariſſimũ opus elemento꜓ Euclidis megarẽſis vna cũ cõmentis Campani pſpicaciſſimi in artē geometriã incipit felicit', after which the definitions begin at once. Other copies have the shorter heading : Preclarissimus liber elementorum Euclidis perspicacissimi : in artem Geometrie incipit quam foelicissime. At the end stands the following : ⊄ Opus elementorũ euclidis megarenſis in geometriã artē Jn id quoꝗ Campani pſpicaciſſimi Cõmentationes finiũt. Erhardus ratdolt Augustensis impreſſor ſolertiſſimus . venetijs impreſſit . Anno ſalutis . M.cccc.lxxxij . Octauis . Caleñ . Juñ . Lector . Vale.

[2] Kästner, *Geschichte der Mathematik*, I. p. 289 sqq. See also Weissenborn, *Die Übersetzungen des Euklid durch Campano und Zamberti*, pp. 1—7.

[3] "Mea industria non sine maximo labore effeci vt qua facilitate litterarum elementa imprimuntur ea etiam geometrice figure conficerentur."

[4] Curtze in *Zeitschrift für Math. u. Physik*, XX., hist.-litt. Abth. p. 58.

Leonardum de Basilea et Gulielmum de Papia), but without the dedication to Mocenigo who had died in the meantime (1485). If Campanus added anything of his own, his additions are at all events not distinguished by any difference of type or otherwise; the enunciations are in large type, and the rest is printed continuously in smaller type. There are no superscriptions to particular passages such as *Euclides ex Campano, Campanus, Campani additio,* or *Campani annotatio,* which are found for the first time in the Paris edition of 1516 giving both Campanus' version and that of Zamberti (presently to be mentioned).

1501. G. Valla included in his encyclopaedic work *De expetendis et fugiendis rebus* published in this year at Venice (*in aedibus Aldi Romani*) a number of propositions with proofs and scholia translated from a Greek MS. which was once in his possession (cod. Mutin. III B, 4 of the 15th c.).

1505. In this year Bartolomeo Zamberti (Zambertus) brought out at Venice the first translation, from the Greek text, of the whole of the *Elements*. From the title[1], as well as from his prefaces to the *Catoptrica* and *Data*, with their allusions to previous translators " who take some things out of authors, omit some, and change some," or " to that most barbarous translator" who filled a volume purporting to be Euclid's "with extraordinary scarecrows, nightmares and phantasies," one object of Zamberti's translation is clear. His animus against Campanus appears also in a number of notes, e.g. when he condemns the terms "helmuain" and "helmuariphe" used by Campanus as barbarous, un-Latin etc., and when he is roused to wrath by Campanus' unfortunate mistranslation of V. Def. 5. He does not seem to have had the penetration to see that Campanus was translating from an Arabic, and not from a Greek, text. Zamberti tells us that he spent seven years over his translation of the thirteen Books of the *Elements*. As he seems to have been born in 1473, and the *Elements* were printed as early as 1500, though the complete work (including the *Phaenomena, Optica, Catoptrica, Data* etc.) has the date 1505 at the end, he must have translated Euclid before the age of 30. Heiberg has not been able to identify the MS. of the *Elements* which Zamberti used; but it is clear that it belonged to the worse class of MSS., since it contains most of the interpolations of the Theonine variety. Zamberti, as his title shows, attributed the *proofs* to Theon.

1509. As a counterblast to Zamberti, Luca Paciuolo brought out an edition of Euclid, apparently at the expense of Ratdolt, at Venice (*per Paganinum de Paganinis*), in which he set himself to vindicate Campanus. The title-page of this now very rare edition[2] begins thus: "The works of Euclid of Megara, a most acute philosopher and without

---

[1] The title begins thus: "Euclidis megaresis philosophi platonicj mathematicarum disciplinarum Janitoris: Habent in hoc volumine quicunque ad mathematicam substantiam aspirant: elementorum libros xiij cum expositione Theonis insignis mathematici. quibus multa quae deerant ex lectione graeca sumpta addita sunt nec non plurima peruersa et praepostere: voluta in Campani interpretatione: ordinata digesta et castigata sunt etc." For a description of the book see Weissenborn, p. 12 sqq.

[2] See Weissenborn, p. 30 sqq.

question the chief of all mathematicians, translated by Campanus their *most faithful interpreter*." It proceeds to say that the translation had been, through the fault of copyists, so spoiled and deformed that it could scarcely be recognised as Euclid. Luca Paciuolo accordingly has polished and emended it with the most critical judgment, has corrected 129 figures wrongly drawn and added others, besides supplying short explanations of difficult passages. It is added that Scipio Vegius of Milan, distinguished for his knowledge "*of both languages*" (i.e. of course Latin and Greek), as well as in medicine and the more sublime studies, had helped to make the edition more perfect. Though Zamberti is not once mentioned, this latter remark must have reference to Zamberti's statement that his translation was from the Greek text; and no doubt Zamberti is aimed at in the wish of Paciuolo's "that others too would seek to acquire knowledge instead of merely showing off, or that they would not try to make a market of the things of which they are ignorant, as it were (selling) smoke[1]." Weissenborn observes that, while there are many trivialities in Paciuolo's notes, they contain some useful and practical hints and explanations of terms, besides some new proofs which of course are not difficult if one takes the liberty, as Paciuolo does, of diverging from Euclid's order and assuming for the proof of a proposition results not arrived at till later. Two not inapt terms are used in this edition to describe the figures of III. 7, 8, the former of which is called the *goose's foot* (*pes anseris*), the second the *peacock's tail* (*cauda pavonis*). Paciuolo as the *castigator* of Campanus' translation, as he calls himself, failed to correct the mistranslation of V. Def. 5[2]. Before the fifth Book he inserted a discourse which he gave at Venice on the 15th August, 1508, in S. Bartholomew's Church, before a select audience of 500, as an introduction to his elucidation of that Book.

1516. The first of the editions giving Campanus' and Zamberti's translations in conjunction was brought out at Paris (*in officina Henrici Stephani e regione scholae Decretorum*). The idea that only the enunciations were Euclid's, and that Campanus was the author of the proofs in his translation, while Theon was the author of the proofs in the Greek text, reappears in the title of this edition; and the enunciations of the added Books XIV., XV. are also attributed to Euclid, Hypsicles being credited with the proofs[3]. The date is not on the title-page nor at the

---

[1] "Atque utinam et alii cognoscere vellent non ostentare aut ea quae nesciunt veluti fumum venditare non conarentur."

[2] Campanus' translation in Ratdolt's edition is as follows: "Quantitates quae dicuntur continuam habere proportionalitatem, sunt, quarum equè multiplicia aut equa sunt aut equè sibi sine interruptione addunt aut minuunt" (!), to which Campanus adds the note: "Continuè proportionalia sunt quorum omnia multiplicia equalia sunt continuè proportionalia. Sed noluit ipsam diffinitionem proponere sub hac forma, quia tunc diffiniret idem per idem, aperte (? a parte) tamen rei est istud cum sua diffinitione convertibile."

[3] "Euclidis Megarensis Geometricorum Elementorum Libri XV. Campani Galli transalpini in eosdem commentariorum libri XV. Theonis Alexandrini Bartholomaeo Zamberto Veneto interprete, in tredecim priores, commentationum libri XIII. Hypsiclis Alexandrini in duos posteriores, eodem Bartholomaeo Zamberto Veneto interprete, commentariorum libri II." On the last page (261) is a similar statement of content, but with the difference that the expression "ex Campani...deinde Theonis...et Hypsiclis...*traditionibus*." For description see Weissenborn, p. 56 sqq.

end, but the letter of dedication to François Briconnet by Jacques
Lefèvre is dated the day after the Epiphany, 1516. The figures are
in the margin. The arrangement of the propositions is as follows:
first the enunciation with the heading *Euclides ex Campano*, then the
proof with the note *Campanus*, and after that, as *Campani additio*, any
passage found in the edition of Campanus' translation but not in the
Greek text; then follows the text of the enunciation translated from
the Greek with the heading *Euclides ex Zamberto*, and lastly the proof
headed *Theo ex Zamberto*. There are separate figures for the two proofs.
This edition was reissued with few changes in 1537 and 1546 at Basel
(*apud Iohannem Hervagium*), but with the addition of the *Phaenomena*,
Optica, Catoptrica etc. For the edition of 1537 the Paris edition of
1516 was collated with "a Greek copy" (as the preface says) by
Christian Herlin, professor of mathematical studies at Strassburg,
who however seems to have done no more than correct one or two
passages by the help of the Basel *editio princeps* (1533), and add the
Greek word in cases where Zamberti's translation of it seemed unsuitable or inaccurate.

We now come to

II. Editions of the Greek text.

1533 is the date of the *editio princeps*, the title-page of which reads
as follows:

ΕΤΚΛΕΙΔΟΤ ΣΤΟΙΧΕΙΩΝ ΒΙΒΛ ΙΕ
ΕΚ ΤΩΝ ΘΕΩΝΟΣ ΣΥΝΟΥΣΙΩΝ.
Εἰς τοῦ αὐτοῦ τὸ πρῶτον, ἐξηγημάτων Πρόκλου βιβλ. δ.
Adiecta praefatiuncula in qua de disciplinis
Mathematicis nonnihil.
BASILEAE APVD IOAN. HERVAGIVM ANNO
M.D.XXXIII. MENSE SEPTEMBRI.

The editor was Simon Grynaeus the elder (d. 1541), who, after
working at Vienna and Ofen, Heidelberg and Tübingen, taught last
of all at Basel, where theology was his main subject. His "praefatiuncula" is addressed to an Englishman, Cuthbert Tonstall (1474–
1559), who, having studied first at Oxford, then at Cambridge, where
he became Doctor of Laws, and afterwards at Padua, where in addition he learnt mathematics—mostly from the works of Regiomontanus
and Paciuolo—wrote a book on arithmetic[1] as "a farewell to the
sciences," and then, entering politics, became Bishop of London and
member of the Privy Council, and afterwards (1530) Bishop of Durham.
Grynaeus tells us that he used two MSS. of the text of the *Elements*,
entrusted to friends of his, one at Venice by "Lazarus Bayfius"
(Lazare de Baïf, then the ambassador of the King of France at Venice),
the other at Paris by "Ioann. Rvellius" (Jean Ruel, a French doctor
and a Greek scholar), while the commentaries of Proclus were put at

[1] *De arte supputandi libri quatuor.*

the disposal of Grynaeus himself by "Ioann. Claymundus" at Oxford. Heiberg has been able to identify the two MSS. used for the text; they are (1) cod. Venetus Marcianus 301 and (2) cod. Paris. gr. 2343 of the 16th c., containing Books I.—XV., with some scholia which are embodied in the text. When Grynaeus notes in the margin the readings from "the other copy," this "other copy" is as a rule the Paris MS., though sometimes the reading of the Paris MS. is taken into the text and the "other copy" of the margin is the Venice MS. Besides these two MSS. Grynaeus consulted Zamberti, as is shown by a number of marginal notes referring to "Zampertus" or to "latinum exemplar" in certain propositions of Books IX.—XI. When it is considered that the two MSS. used by Grynaeus are among the worst, it is obvious how entirely unauthoritative is the text of the *editio princeps*. Yet it remained the source and foundation of later editions of the Greek text for a long period, the editions which followed being designed, not for the purpose of giving, from other MSS., a text more nearly representing what Euclid himself wrote, but of supplying a handy compendium to students at a moderate price.

1536. Orontius Finaeus (Oronce Fine) published at Paris (*apud Simonem Colinaeum*) "demonstrations on the first six books of Euclid's elements of geometry," "in which the Greek text of Euclid himself is inserted in its proper places, with the Latin translation of Barth. Zamberti of Venice," which seems to imply that only the enunciations were given in Greek. The preface, from which Kästner quotes[1], says that the University of Paris at that time required, from all who aspired to the laurels of philosophy, a most solemn oath that they had attended lectures on the said first six Books. Other editions of Fine's work followed in 1544 and 1551.

1545. The *enunciations* of the fifteen Books were published in Greek, with an Italian translation by Angelo Caiani, at Rome (*apud Antonium Bladum Asulanum*). The translator claims to have corrected the books and "purged them of six hundred things which did not seem to savour of the almost divine genius and the perspicuity of Euclid[2]"

1549. Joachim Camerarius published the enunciations of the first six Books in Greek and Latin (Leipzig). The book, with preface, purports to be brought out by Rhaeticus (1514–1576), a pupil of Copernicus. Another edition with proofs of the propositions of the first three Books was published by Moritz Steinmetz in 1577 (Leipzig); a note by the printer attributes the preface to Camerarius himself.

1550. Ioan. Scheubel published at Basel (also *per Ioan. Hervagium*) the first six Books in Greek and Latin "together with true and appropriate proofs of the propositions, without the use of letters" (i.e. letters denoting points in the figures), the various straight lines and angles being described in words[3].

1557 (also 1558). Stephanus Gracilis published another edition (repeated 1573, 1578, 1598) of the enunciations (alone) of Books I.—XV.

---

[1] Kästner, I. p. 260.   [2] Heiberg, vol. V. p. cvii.   [3] Kästner, I. p. 359.

in Greek and Latin at Paris (*apud Gulielmum Cavellat*). He remarks in the preface that for want of time he had changed scarcely anything in Books I.—VI., but 'n the remaining Books he had emended what seemed obscure or inelegant in the Latin translation, while he had adopted in its entirety the translation of Book X. by Pierre Mondoré (Petrus Montaureus), published separately at Paris in 1551. Gracilis also added a few "scholia."

1564.   In this year Conrad Dasypodius (Rauchfuss), the inventor and maker of the clock in Strassburg cathedral, similar to the present one, which did duty from 1571 to 1789, edited (Strassburg, Chr. Mylius) (1) Book I. of the *Elements* in Greek and Latin with scholia, (2) Book II. in Greek and Latin with Barlaam's arithmetical version of Book II., and (3) the *enunciations* of the remaining Books III.—XIII. Book I. was reissued with "vocabula quaedam geometrica" of Heron, the enunciations of all the Books of the *Elements*, and the other works of Euclid, all in Greek and Latin. In the preface to (1) he says that it had been for twenty-six years the rule of his school that all who were promoted from the classes to public lectures should learn the first Book, and that he brought it out, because there were then no longer any copies to be had, and in order to prevent a good and fruitful regulation of his school from falling through. In the preface to the edition of 1571 he says that the first Book was generally taught in all gymnasia and that it was prescribed in his school for the first class. In the preface to (3) he tells us that he published the enunciations of Books III.—XIII. in order not to leave his work unfinished, but that, as it would be irksome to carry about the whole work of Euclid in extenso, he thought it would be more convenient to students of geometry to learn the *Elements* if they were compressed into a smaller book.

1620.   Henry Briggs (of Briggs' logarithms) published the first six Books in Greek with a Latin translation after Commandinus, "corrected in many places" (London, G. Jones).

1703 is the date of the Oxford edition by David Gregory which, until the issue of Heiberg and Menge's edition, was still the only edition of the complete works of Euclid[1]. In the Latin translation attached to the Greek text Gregory says that he followed Commandinus in the main, but corrected numberless passages in it by means of the books in the Bodleian Library which belonged to Edward Bernard (1638–1696), formerly Savilian Professor of Astronomy, who had conceived the plan of publishing the complete works of the ancient mathematicians in fourteen volumes, of which the first was to contain Euclid's *Elements* I.—XV. As regards the Greek text, Gregory tells us that he consulted, as far as was necessary, not a few MSS. of the better sort, bequeathed by the great Savile to the University, as well as the corrections made by Savile in his own hand in the margin of the Basel edition. He had the help of John Hudson, Bodley's Librarian, who

---

[1] ΕΥΚΛΕΙΔΟΥ ΤΑ ΣΩΖΟΜΕΝΑ. Euclidis qúae supersunt omnia. Ex recensione Davidis Gregorii M.D. Astronomiae Professoris Saviliani et R.S.S. Oxoniae, e Theatro Sheldoniano, An. Dom. MDCCIII.

punctuated the Basel text before it went to the printer, compared the Latin version with the Greek throughout, especially in the *Elements* and *Data*, and, *where they differed* or *where he suspected the Greek text*, consulted the Greek MSS. and put their readings in the margin if they agreed with the Latin and, if they did not agree, affixed an asterisk in order that Gregory might judge which reading was geometrically preferable. Hence it is clear that no Greek MS., but the Basel edition, was the foundation of Gregory's text, and that Greek MSS. were only referred to in the special passages to which Hudson called attention.

1814–1818. A most important step towards a good Greek text was taken by F. Peyrard, who published at Paris, between these years, in three volumes, the *Elements* and *Data* in Greek, Latin and French[1]. At the time (1808) when Napoleon was having valuable MSS. selected from Italian libraries and sent to Paris, Peyrard managed to get two ancient Vatican MSS. (190 and 1038) sent to Paris for his use (Vat. 204 was also at Paris at the time, but all three were restored to their owners in 1814). Peyrard noticed the excellence of Cod. Vat. 190, adopted many of its readings, and gave in an appendix a conspectus of these readings and those of Gregory's edition ; he also noted here and there readings from Vat. 1038 and various Paris MSS. He therefore pointed the way towards a better text, but committed the error of correcting the Basel text instead of rejecting it altogether and starting afresh.

1824–1825. A most valuable edition of Books I.—VI. is that of J. G. Camerer (and C. F. Hauber) in two volumes published at Berlin[2]. The Greek text is based on Peyrard, although the Basel and Oxford editions were also used. There is a Latin translation and a collection of notes far more complete than any other I have seen and well nigh inexhaustible. There is no editor or commentator of any mark who is not quoted from ; to show the variety of important authorities drawn upon by Camerer, I need only mention the following names : Proclus, Pappus, Tartaglia, Commandinus, Clavius, Peletier, Barrow, Borelli, Wallis, Tacquet, Austin, Simson, Playfair. No words of praise would be too warm for this veritable encyclopaedia of information.

1825. J. G. C. Neide edited, from Peyrard, the text of Books I.—VI., XI. and XII. (*Halis Saxoniae*).

1826–9. The last edition of the Greek text before Heiberg's is that of E. F. August, who followed the Vatican MS. more closely than Peyrard did, and consulted at all events the Viennese MS. Gr. 103 (Heiberg's V). August's edition (Berlin, 1826–9) contains Books I.—XIII.

---

[1] *Euclidis quae supersunt. Les Œuvres d'Euclide, en Grec, en Latin et en Français d'après un manuscrit très-ancien, qui était resté inconnu jusqu'à nos jours.* Par F. Peyrard. Ouvrage approuvé par l'Institut de France (Paris, chez M. Patris).
[2] *Euclidis elementorum libri sex priores graece et latine commentario e scriptis veterum ac recentiorum mathematicorum et Pfleidereri maxime illustrati* (Berolini, sumptibus G. Reimeri). Tom. I. 1824 ; tom. II. 1825.

III. Latin Versions or Commentaries after 1533.

1545. Petrus Ramus (Pierre de la Ramée, 1515–1572) is credited with a translation of Euclid which appeared in 1545 and again in 1549 at Paris[1]. Ramus, who was more rhetorician and logician than geometer, also published in his *Scholae mathematicae* (1559, Frankfurt; 1569, Basel) what amounts to a series of lectures on Euclid's *Elements*, in which he criticises Euclid's arrangement of his propositions, the definitions, postulates and axioms, all from the point of view of logic.

1557. Demonstrations to the geometrical Elements of Euclid, six Books, by Peletarius (Jacques Peletier). The second edition (1610) contained the same with the addition of the "Greek text of Euclid"; but only the *enunciations* of the propositions, as well as the definitions etc., are given in Greek (with a Latin translation), the rest is in Latin only. He has some acute observations, for instance about the "angle" of contact.

1559. Johannes Buteo, or Borrel (1492–1572), published in an appendix to his book *De quadratura circuli* some notes "on the errors of Campanus, Zambertus, Orontius, Peletarius, Pena, interpreters of Euclid." Buteo in these notes proved, by reasoned argument based on original authorities, that Euclid himself and not Theon was the author of the proofs of the propositions.

1566. Franciscus Flussates Candalla (François de Foix, Comte de Candale, 1502–1594) "restored" the fifteen Books, following, as he says, the terminology of Zamberti's translation from the Greek, but drawing, for his proofs, on both Campanus and Theon (i.e. Zamberti) except where mistakes in them made emendation necessary. Other editions followed in 1578, 1602, 1695 (in Dutch).

1572. The most important Latin translation is that of Commandinus (1509–1575) of Urbino, since it was the foundation of most translations which followed it up to the time of Peyrard, including that of Simson and therefore of those editions, numerous in England, which give Euclid "chiefly after the text of Simson." Simson's first (Latin) edition (1756) has "ex versione Latina Federici Commandini" on the title-page. Commandinus not only followed the original Greek more closely than his predecessors but added to his translation some ancient scholia as well as good notes of his own. The title of his work is

> *Euclidis elementorum libri* XV, *una cum scholiis antiquis. A Federico Commandino Urbinate nuper in latinum conversi, commentariisque quibusdam illustrati* (Pisauri, apud Camillum Francischinum).

He remarks in his preface that Orontius Finaeus had only edited six Books without reference to any Greek MS., that Peletarius had followed Campanus' version from the Arabic rather than the Greek text, and that Candalla had diverged too far from Euclid, having rejected as inelegant the proofs given in the Greek text and substituted faulty proofs of his own. Commandinus appears to have

---

[1] Described by Boncompagni, *Bullettino*, II. p. 389.

used, in addition to the Basel *editio princeps*, some Greek MS., so far not identified ; he also extracted his " scholia antiqua " from a MS. of the class of Vat. 192 containing the scholia distinguished by Heiberg as " Schol. Vat." New editions of Commandinus' translation followed in 1575 (in Italian), 1619, 1749 (in English, by Keill and Stone), 1756 (Books I.—VI., XI., XII. in Latin and English, by Simson), 1763 (Keill). Besides these there were many editions of parts of the whole work, e.g. the first six Books.

1574. The first edition of the Latin version by Clavius[1] (Christoph Klau [?], born at Bamberg 1537, died 1612) appeared in 1574, and new editions of it in 1589, 1591, 1603, 1607, 1612. It is not a translation, as Clavius himself states in the preface, but it contains a vast amount of notes collected from previous commentators and editors, as well as some good criticisms and elucidations of his own. Among other things, Clavius finally disposed of the error by which Euclid had been identified with Euclid of Megara. He speaks of the differences between Campanus who followed the Arabic tradition and the " commentaries of Theon," by which he appears to mean the Euclidean proofs as handed down by Theon ; he complains of predecessors who have either only given the first six Books, or have rejected the ancient proofs and substituted worse proofs of their own, but makes an exception as regards Commandinus, " a geometer not of the common sort, who has lately restored Euclid, in a Latin translation, to his original brilliancy." Clavius, as already stated, did not give a translation of the *Elements* but rewrote the proofs, compressing them or adding to them, where he thought that he could make them clearer. Altogether his book is a most useful work.

1621. Henry Savile's lectures (*Praelectiones tresdecim in principium Elementorum Euclidis Oxoniae habitae* MDC.XX., Oxonii 1621), though they do not extend beyond I. 8, are valuable because they grapple with the difficulties connected with the preliminary matter, the definitions etc., and the tacit assumptions contained in the first propositions.

1654. André Tacquet's *Elementa geometriae planae et solidae* containing apparently the eight geometrical Books arranged for general use in schools. It came out in a large number of editions up to the end of the eighteenth century.

1655. Barrow's *Euclidis Elementorum Libri* XV *breviter demonstrati* is a book of the same kind. In the preface (to the edition of 1659) he says that he would not have written it but for the fact that Tacquet gave only eight Books of Euclid. He compressed the work into a very small compass (less than 400 small pages, in the edition of 1659, for the whole of the fifteen Books and the *Data*) by abbreviating the proofs and using a large quantity of symbols (which, he says, are generally Oughtred's). There were several editions up to 1732 (those of 1660 and 1732 and one or two others are in English).

---

[1] *Euclidis elementorum libri* XV. *Accessit* XVI. *de solidorum regularium comparatione. Omnes perspicuis demonstrationibus, accuratisque scholiis illustrati. Auctore Christophoro Clavio* (Romae, apud Vincentium Accoltum), 2 vols.

1658. Giovanni Alfonso Borelli (1608–1679) published *Euclides restitutus*, on apparently similar lines, which went through three more editions (one in Italian, 1663).

1660. Claude François Milliet Dechales' eight geometrical Books of Euclid's *Elements* made easy. Dechales' versions of the *Elements* had great vogue, appearing in French, Italian and English as well as Latin. Riccardi enumerates over twenty editions.

1733. Saccheri's *Euclides ab omni naevo vindicatus sive conatus geometricus quo stabiliuntur prima ipsa geometriae principia* is important for his elaborate attempt to prove the parallel-postulate, forming an important stage in the history of the development of non-Euclidean geometry.

1756. Simson's first edition, in Latin and in English. The Latin title is

> *Euclidis elementorum libri priores sex, item undecimus et duo-decimus, ex versione latina Federici Commandini; sublatis iis quibus olim libri hi a Theone, aliisve, vitiati sunt, et quibusdam Euclidis demonstrationibus restitutis. A Roberto Simson M.D.* Glasguae, in aedibus Academicis excudebant Robertus et Andreas Foulis, Academiae typographi.

1802. *Euclidis elementorum libri priores* XII *ex Commandini et Gregorii versionibus latinis. In usum juventutis Academicae*...by Samuel Horsley, Bishop of Rochester. (Oxford, Clarendon Press.)

IV. ITALIAN VERSIONS OR COMMENTARIES.

1543. Tartaglia's version, a second edition of which was published in 1565[1], and a third in 1585. It does not appear that he used any Greek text, for in the edition of 1565 he mentions as available only "the first translation by Campano," "the second made by Bartolomeo Zamberto Veneto who is still alive," "the editions of Paris or Germany in which they have included both the aforesaid translations," and "our own translation into the vulgar (tongue)."

1575. Commandinus' translation turned into Italian and revised by him.

1613. The first six Books "reduced to practice" by Pietro Antonio Cataldi, re-issued in 1620, and followed by Books VII.—IX. (1621) and Book X. (1625).

1663. Borelli's Latin translation turned into Italian by Domenico Magni.

1680. *Euclide restituto* by Vitale Giordano.

1690. Vincenzo Viviani's *Elementi piani e solidi di Euclide* (Book V. in 1674).

---

[1] The title-page of the edition of 1565 is as follows : *Euclide Megarense philosopho, solo introduttore delle scientie mathematice, diligentemente rassettato, et alla integrità ridotto, per il degno professore di tal scientie Nicolo Tartalea Brisciano. secondo le due tradottioni. con una ampla espositione dello istesso tradottore di nuouo aggiunta. talmente chiara, che ogni mediocre ingegno, senza la notitia, ouer suffragio di alcun' altra scientia con facilità serà capace a poterlo intendere.* In Venetia, Appresso Curtio Troiano, 1565.

1731. *Elementi geometrici piani e solidi di Euclide* by Guido Grandi. No translation, but an abbreviated version, of which new editions followed one another up to 1806.

1749. Italian translation of Dechales with Ozanam's corrections and additions, re-issued 1785, 1797.

1752. Leonardo Ximenes (the first six Books). Fifth edition, 1819.

1818. Vincenzo Flauti's *Corso di geometria elementare e sublime* (4 vols.) contains (Vol. I.) the first six Books, with additions and a dissertation on Postulate 5, and (Vol. II.) Books XI., XII. Flauti also published the first six Books in 1827 and the *Elements of geometry of Euclid* in 1843 and 1854.

V. GERMAN.

1558. The arithmetical Books VII.—IX. by Scheubel[1] (cf. the edition of the first six Books, with enunciations in Greek and Latin, mentioned above, under date 1550).

1562. The version of the first six Books by Wilhelm Holtzmann (Xylander)[2]. This work has its interest as the first edition in German, but otherwise it is not of importance. Xylander tells us that it was written for practical people such as artists, goldsmiths, builders etc., and that, as the simple amateur is of course content to know facts, without knowing how to prove them, he has often left out the proofs altogether. He has indeed taken the greatest possible liberties with Euclid, and has not grappled with any of the theoretical difficulties, such as that of the theory of parallels.

1651. Heinrich Hoffmann's *Teutscher Euclides* (2nd edition 1653), not a translation.

1694. Ant. Ernst Burkh. v. Pirckenstein's *Teutsch Redender Euclides* (eight geometrical Books), "for generals, engineers etc." "proved in a new and quite easy manner." Other editions 1699, 1744.

1697. Samuel Reyher's *In teutscher Sprache vorgestellter Euclides* (six Books), "made easy, with symbols algebraical or derived from the newest art of solution."

1714. *Euclidis* XV *Bücher teutsch*, "treated in a special and brief manner, yet completely," by Chr. Schessler (another edition in 1729).

1773. The first six Books translated from the Greek for the use of schools by J. F. Lorenz. The first attempt to reproduce Euclid in German word for word.

1781. Books XI., XII. by Lorenz (supplementary to the preceding). Also *Euklid's Elemente fünfzehn Bücher* translated from

---

[1] *Das sibend acht und neunt buch des hochberümbten Mathematici Euclidis Megarensis... durch Magistrum Johann Scheybl, der löblichen universitet zu Tübingen, des Euclidis und Arithmetic Ordinarien, auss dem latein ins teutsch gebracht....*

[2] *Die sechs erste Bücher Euclidis vom anfang oder grund der Geometrj...Auss Griechischer sprach in die Teütsch gebracht aigentlich erklärt...Demassen vormals in Teütscher sprach nie gesehen worden...Durch Wilhelm Holtzman genant Xylander von Augspurg.* Getruckht zu Basel.

the Greek by Lorenz (second edition 1798; editions of 1809, 1818, 1824 by Mollweide, of 1840 by Dippe). The edition of 1824, and I presume those before it, are shortened by the use of symbols and the compression of the enunciation and "setting-out" into one.

1807. Books I.—VI., XI., XII. "newly translated from the Greek," by J. K. F. Hauff.

1828. The same Books by Joh. Jos. Ign. Hoffmann "as guide to instruction in elementary geometry," followed in 1832 by observations on the text by the same editor.

1833. *Die Geometrie des Euklid und das Wesen derselben* by E. S. Unger; also 1838, 1851.

1901. Max Simon, *Euclid und die sechs planimetrischen Bücher.*

VI. FRENCH.

1564–1566. Nine Books translated by Pierre Forcadel, a pupil and friend of P. de la Ramée.

1604. The first nine Books translated and annotated by Jean Errard de Bar-le-Duc; second edition, 1605.

1615. Denis Henrion's translation of the 15 Books (seven editions up to 1676).

1639. The first six Books "demonstrated by symbols, by a method very brief and intelligible," by Pierre Hérigone, mentioned by Barrow as the only editor who, before him, had used symbols for the exposition of Euclid.

1672. Eight Books "rendus plus faciles" by Claude François Milliet Dechales, who also brought out *Les élémens d'Euclide expliqués d'une manière nouvelle et très facile*, which appeared in many editions, 1672, 1677, 1683 etc. (from 1709 onwards revised by Ozanam), and was translated into Italian (1749 etc.) and English (by William Halifax, 1685).

1804. In this year, and therefore before his edition of the Greek text, F. Peyrard published the *Elements* literally translated into French. A second edition appeared in 1809 with the addition of the fifth Book. As this second edition contains Books I.—VI. XI., XII. and X. I, it would appear that the first edition contained Books I.—IV., VI., XI., XII. Peyrard used for this translation the Oxford Greek text and Simson.

VII. DUTCH.

1606. Jan Pieterszoon Dou (six Books). There were many later editions. Kästner, in mentioning one of 1702, says that Dou explains in his preface that he used Xylander's translation, but, having afterwards obtained the French translation of the six Books by Errard de Bar-le-Duc (see above), the proofs in which sometimes pleased him more than those of the German edition, he made his Dutch version by the help of both.

1617. Frans van Schooten, "The Propositions of the Books of Euclid's Elements"; the fifteen Books in this version "enlarged" by Jakob van Leest in 1662.

1695. C. J. Vooght, fifteen Books complete, with Candalla's "16th."

1702. Hendrik Coets, six Books (also in Latin, 1692); several editions up to 1752. Apparently not a translation, but an edition for school use.

1763. Pybo Steenstra, Books I.—VI., XI., XII., likewise an abbreviated version, several times reissued until 1825.

VIII. ENGLISH.

1570 saw the first and the most important translation, that of Sir Henry Billingsley. The title-page is as follows:

THE ELEMENTS
OF GEOMETRIE
of the most auncient Philosopher
EVCLIDE
of Megara

Faithfully (now first) translated into the Englishe toung, by H. Billingsley, *Citizen of London. Whereunto are annexed certaine Scholies, Annotations, and Inuentions, of the best Mathematiciens, both of time past, and in this our age.*

With a very fruitfull Preface by M. I. Dee, *specifying the chiefe Mathematicall Sciĕces, what they are, and whereunto commodious: where, also, are disclosed certaine new Secrets Mathematicall and Mechanicall, vntill these our daies, greatly missed.*

Imprinted at London by *John Daye.*

The Preface by the translator, after a sentence observing that without the diligent study of Euclides Elementes it is impossible to attain unto the perfect knowledge of Geometry, proceeds thus. " Wherefore considering the want and lacke of such good authors hitherto in our Englishe tounge, lamenting also the negligence, and lacke of zeale to their countrey in those of our nation, to whom God hath geuen both knowledge and also abilitie to translate into our tounge, and to publishe abroad such good authors and bookes (the chiefe instrumentes of all learninges): seing moreouer that many good wittes both of gentlemen and of others of all degrees, much desirous and studious of these artes, and seeking for them as much as they can, sparing no paines, and yet frustrate of their intent, by no meanes attaining to that which they seeke: I haue for their sakes, with some charge and great trauaile, faithfully translated into our vulgare toŭge, and set abroad in Print, this booke of Euclide. Whereunto I haue added easie and plaine declarations and examples by figures, of the definitions. In which booke also ye shall in due place finde manifolde additions, Scholies, Annotations, and Inuentions: which I haue gathered out of many of the most famous and chiefe Mathematiciĕs, both of old time, and in our age: as by diligent reading it in course, ye shall well perceaue...."

It is truly a monumental work, consisting of 464 leaves, and therefore 928 pages, of folio size, excluding the lengthy preface by Dee. The notes certainly include all the most important that had ever been written, from those of the Greek commentators, Proclus and the others whom he quotes, down to those of Dee himself on the last books. Besides the fifteen Books, Billingsley included the "sixteenth" added by Candalla. The print and appearance of the book are worthy of its contents; and, in order that it may be understood how no pains were spared to represent everything in the clearest and most perfect form, I need only mention that the figures of the propositions in Book XI. are nearly all duplicated, one being the figure of Euclid, the other an arrangement of pieces of paper (triangular, rectangular etc.) pasted at the edges on to the page of the book so that the pieces can be turned up and made to show the real form of the solid figures represented.

Billingsley was admitted Lady Margaret Scholar of St John's College, Cambridge, in 1551, and he is also said to have studied at Oxford, but he did not take a degree at either University. He was afterwards apprenticed to a London haberdasher and rapidly became a wealthy merchant. Sheriff of London in 1584, he was elected Lord Mayor on 31st December, 1596, on the death, during his year of office, of Sir Thomas Skinner. From 1589 he was one of the Queen's four "customers," or farmers of customs, at the port of London. In 1591 he founded three scholarships at St John's College for poor students, and gave to the College for their maintenance two messuages and tenements in Tower Street and in Mark Lane, Allhallows, Barking. He died in 1606.

1651. *Elements of Geometry. The first* VI *Boocks: In a compendious form contracted and demonstrated* by Captain Thomas Rudd, with the mathematicall preface of John Dee (London).

1660. The first English edition of Barrow's Euclid (published in Latin in 1655), appeared in London. It contained "the whole fifteen books compendiously demonstrated"; several editions followed, in 1705, 1722, 1732, 1751.

1661. *Euclid's Elements of Geometry, with a supplement of divers Propositions and Corollaries. To which is added a Treatise of regular Solids by Campane and Flussat; likewise Euclid's Data and Marinus his Preface. Also a Treatise of the Divisions of Superficies, ascribed to Machomet Bagdedine, but published by Commandine at the request of J. Dee of London.* Published by care and industry of John Leeke and Geo. Serle, students in the Math. (London). According to Potts this was a second edition of Billingsley's translation.

1685. William Halifax's version of Dechales' "Elements of Euclid explained in a new but most easy method" (London and Oxford).

1705. *The English Euclide; being the first six Elements of Geometry, translated out of the Greek, with annotations and usefull supplements by* Edmund Scarburgh (Oxford). A noteworthy and useful edition.

1708. Books I.—VI., XI., XII., translated from Commandinus' Latin version by Dr John Keill, Savilian Professor of Astronomy at Oxford.

Keill complains in his preface of the omissions by such editors as Tacquet and Dechales of many necessary propositions (e.g. VI. 27—29), and of their substitution of proofs of their own for Euclid's. He praises Barrow's version on the whole, though objecting to the "algebraical" form of proof adopted in Book II., and to the excessive use of notes and symbols, which (he considers) make the proofs *too* short and thereby obscure; his edition was therefore intended to hit a proper mean between Barrow's excessive brevity and Clavius' prolixity.

Keill's translation was revised by Samuel Cunn and several times reissued. 1749 saw the eighth edition, 1772 the eleventh, and 1782 the twelfth.

1714. W. Whiston's English version (abridged) of *The Elements of Euclid with select theorems out of Archimedes by the learned Andr. Tacquet.*

1756. Simson's first English edition appeared in the same year as his Latin version under the title:

> *The Elements of Euclid, viz. the first six Books together with the eleventh and twelfth. In this Edition the Errors by which Theon or others have long ago vitiated these Books are corrected and some of Euclid's Demonstrations are restored.* By Robert Simson (Glasgow).

As above stated, the Latin edition, by its title, purports to be " ex versione latina Federici Commandini," but to the Latin edition, as well as to the English editions, are appended

> *Notes Critical and Geometrical; containing an Account of those things in which this Edition differs from the Greek text; and the Reasons of the Alterations which have been made. As also Observations on some of the Propositions.*

Simson says in the Preface to some editions (e.g. the tenth, of 1799) that "the translation is much amended by the friendly assistance of a learned gentleman."

Simson's version and his notes are so well known as not to need any further description. The book went through some thirty successive editions. The first five appear to have been dated 1756, 1762, 1767, 1772 and 1775 respectively; the tenth 1799, the thirteenth 1806, the twenty-third 1830, the twenty-fourth 1834, the twenty-sixth 1844. The *Data* "in like manner corrected" was added for the first time in the edition of 1762 (the first octavo edition).

1781, 1788. In these years respectively appeared the two volumes containing the complete translation of the whole thirteen Books by James Williamson, the last English translation which reproduced Euclid word for word. The title is

> *The Elements of Euclid, with Dissertations intended to assist and encourage a critical examination of these Elements, as the most effectual means of establishing a juster taste upon mathematical subjects than that which at present prevails.* By James Williamson.

In the first volume (Oxford, 1781) he is described as "M.A. Fellow of Hertford College," and in the second (London, printed by T. Spilsbury, 1788) as "B.D." simply. Books V., VI. with the Conclusion in the first volume are paged separately from the rest.

1781. *An examination of the first six Books of Euclid's Elements*, by William Austin (London).

1795. John Playfair's first edition, containing "the first six Books of Euclid with two Books on the Geometry of Solids." The book reached a fifth edition in 1819, an eighth in 1831, a ninth in 1836, and a tenth in 1846.

1826. Riccardi notes under this date *Euclid's Elements of Geometry containing the whole twelve Books translated into English, from the edition of Peyrard*, by George Phillips. The editor, who was President of Queens' College, Cambridge, 1857–1892, was born in 1804 and matriculated at Queens' in 1826, so that he must have published the book as an undergraduate.

1828. A very valuable edition of the first six Books is that of Dionysius Lardner, with commentary and geometrical exercises, to which he added, in place of Books XI., XII., a Treatise on Solid Geometry mostly based on Legendre. Lardner compresses the propositions by combining the enunciation and the setting-out, and he gives a vast number of riders and additional propositions in smaller print. The book had reached a ninth edition by 1846, and an eleventh by 1855. Among other things, Lardner gives an Appendix " on the theory of parallel lines," in which he gives a short history of the attempts to get over the difficulty of the parallel-postulate, down to that of Legendre.

1833. T. Perronet Thompson's *Geometry without axioms, or the first Book of Euclid's Elements with alterations and notes ; and an intercalary book in which the straight line and plane are derived from properties of the sphere, with an appendix containing notices of methods proposed for getting over the difficulty in the twelfth axiom of Euclid.*

Thompson (1783–1869) was 7th wrangler 1802, midshipman 1803, Fellow of Queens' College, Cambridge, 1804, and afterwards general and politician. The book went through several editions, but, having been well translated into French by Van Tenac, is said to have received more recognition in France than at home.

1845. Robert Potts' first edition (and one of the best) entitled :
Euclid's Elements of Geometry chiefly from the text of Dr Simson with explanatory notes...to which is prefixed an introduction containing a brief outline of the History of Geometry. Designed for the use of the higher forms in Public Schools and students in the Universities (Cambridge University Press, and London, John W. Parker), to which was added (1847) *An Appendix to the larger edition of Euclid's Elements of Geometry, containing additional notes on the Elements, a short tract on transversals, and hints for the solution of the problems etc.*

1862. Todhunter's edition.

The later English editions I will not attempt to enumerate ; their name is legion and their object mostly that of adapting Euclid for school use, with all possible gradations of departure from his text and order.

IX. SPANISH.

1576. The first six Books translated into Spanish by Rodrigo Çamorano.

1637. The first six Books translated, with notes, by L. Carduchi.

1689. Books I.—VI., XI., XII., translated and explained by Jacob Knesa.

X. Russian.

1739.　Ivan Astaroff (translation from Latin).
1789.　Pr. Suvoroff and Yos. Nikitin (translation from Greek).
1880.　Vachtchenko-Zakhartchenko.
(1817.　A translation into Polish by Jo. Czecha.)

XI. Swedish.

1744.　Mårten Strömer, the first six Books ; second edition 1748.
The third edition (1753) contained Books XI.—XII. as well ; new
editions continued to appear till 1884.
1836.　H. Falk, the first six Books.
1844, 1845, 1859.　P. R. Bråkenhjelm, Books I.—VI., XI., XII.
1850.　F. A. A. Lundgren.
1850.　H. A. Witt and M. E. Areskong, Books I.—VI., XI., XII.

XII. Danish.

1745.　Ernest Gottlieb Ziegenbalg.
1803.　H. C. Linderup, Books I.—VI.

XIII. Modern Greek.

1820.　Benjamin of Lesbos.

I should add a reference to certain editions which have appeared
in recent years.

A Danish translation (*Euklid's Elementer* oversat af Thyra Eibe)
was completed in 1912 ; Books I.—II. were published (with an Introduction by Zeuthen) in 1897, Books III.—IV. in 1900, Books V.—VI.
in 1904, Books VII.—XIII. in 1912.

The Italians, whose great services to elementary geometry are
more than once emphasised in this work, have lately shown a noteworthy disposition to make the *ipsissima verba* of Euclid once more
the object of study. Giovanni Vacca has edited the text of Book I.
(*Il primo libro degli Elementi*. Testo greco, versione italiana, intro-
duzione e note, Firenze 1916.) Federigo Enriques has begun the
publication of a complete Italian translation (*Gli Elementi d' Euclide
e la critica antica e moderna*) ; Books I.—IV. appeared in 1925 (Alberto
Stock, Roma).

An edition of Book I. by the present writer was published in 1918
(*Euclid in Greek, Book I., with Introduction and Notes*, Camb. Univ.
Press).

CHAPTER IX.

§ 1. ON THE NATURE OF *ELEMENTS*.

It would not be easy to find a more lucid explanation of the terms *element* and *elementary*, and of the distinction between them, than is found in Proclus[1], who is doubtless, here as so often, quoting from Geminus. There are, says Proclus, in the whole of geometry certain leading theorems, bearing to those which follow the relation of a principle, all-pervading, and furnishing proofs of many properties. Such theorems are called by the name of *elements*; and their function may be compared to that of the letters of the alphabet in relation to language, letters being indeed called by the same name in Greek (στοιχεῖα).

The term *elementary*, on the other hand, has a wider application: it is applicable to things "which extend to greater multiplicity, and, though possessing simplicity and elegance, have no longer the same dignity as the *elements*, because their investigation is not of general use in the whole of the science, e.g. the proposition that in triangles the perpendiculars from the angles to the transverse sides meet in a point."

"Again, the term *element* is used in two senses, as Menaechmus says. For that which is the means of obtaining is an element of that which is obtained, as the first proposition in Euclid is of the second, and the fourth of the fifth. In this sense many things may even be said to be elements of each other, for they are obtained from one another. Thus from the fact that the exterior angles of rectilineal figures are (together) equal to four right angles we deduce the number of right angles equal to the internal angles (taken together)[2], and *vice versa*. Such an element is like a *lemma*. But the term *element* is otherwise used of that into which, being more simple, the composite is divided; and in this sense we can no longer say that everything is an element of everything, but only that things which are more of the nature of principles are elements of those which stand to them in the relation of results, as postulates are elements of theorems. It is

---

[1] Proclus, *Comm. on Eucl.* I., ed. Friedlein, pp. 72 sqq.
[2] τὸ πλῆθος τῶν ἐντὸς ὀρθαῖς ἴσων. If the text is right, we must apparently take it as "the number of the angles equal to right angles that there are inside," i.e. that are made up by the internal angles.

according to this signification of the term *element* that the elements found in Euclid were compiled, being partly those of plane geometry, and partly those of stereometry. In like manner many writers have drawn up elementary treatises in arithmetic and astronomy.

"Now it is difficult, in each science, both to select and arrange in due order the elements from which all the rest proceeds, and into which all the rest is resolved. And of those who have made the attempt some were able to put together more and some less; some used shorter proofs, some extended their investigation to an indefinite length; some avoided the method of *reductio ad absurdum*, some avoided *proportion*; some contrived preliminary steps directed against those who reject the principles; and, in a word, many different methods have been invented by various writers of elements.

"It is essential that such a treatise should be rid of everything superfluous (for this is an obstacle to the acquisition of knowledge); it should select everything that embraces the subject and brings it to a point (for this is of supreme service to science); it must have great regard at once to clearness and conciseness (for their opposites trouble our understanding); it must aim at the embracing of theorems in general terms (for the piecemeal division of instruction into the more partial makes knowledge difficult to grasp). In all these ways Euclid's system of elements will be found to be superior to the rest; for its utility avails towards the investigation of the primordial figures[1], its clearness and organic perfection are secured by the progression from the more simple to the more complex and by the foundation of the investigation upon common notions, while generality of demonstration is secured by the progression through the theorems which are primary and of the nature of principles to the things sought. As for the things which seem to be wanting, they are partly to be discovered by the same methods, like the construction of the scalene and isosceles (triangle), partly alien to the character of a selection of elements as introducing hopeless and boundless complexity, like the subject of *unordered irrationals* which Apollonius worked out at length[2], and partly developed from things handed down (in the elements) as causes, like the many species of angles and of lines. These things then have been omitted in Euclid, though they have received full discussion in other works; but the knowledge of them is derived from the simple (elements)."

Proclus, speaking apparently on his own behalf, in another place distinguishes two objects aimed at in Euclid's *Elements*. The first has reference to the *matter* of the investigation, and here, like a good Platonist, he takes the whole subject of geometry to be concerned with the "cosmic figures," the five regular solids, which in Book XIII.

---

[1] τῶν ἀρχικῶν σχημάτων, by which Proclus probably means the regular polyhedra (Tannery, p. 143 *n*.).

[2] We have no more than the most obscure indications of the character of this work in an Arabic MS. analysed by Woepcke, *Essai d'une restitution de travaux perdus d'Apollonius sur les quantités irrationelles d'après des indications tirées d'un manuscrit arabe* in *Mémoires présentés à l'académie des sciences*, XIV. 658—720, Paris, 1856. Cf. Cantor, *Gesch. d. Math.* I₃, pp. 348—9: details are also given in my notes to Book X.

are constructed, inscribed in a sphere and compared with one another. The second object is relative to the learner; and, from this standpoint, the elements may be described as "a means of perfecting the learner's understanding with reference to the whole of geometry. For, starting from these (elements), we shall be able to acquire knowledge of the other parts of this science as well, while without them it is impossible for us to get a grasp of so complex a subject, and knowledge of the rest is unattainable. As it is, the theorems which are most of the nature of principles, most simple, and most akin to the first hypotheses are here collected, in their appropriate order; and the proofs of all other propositions use these theorems as thoroughly well known, and start from them. Thus Archimedes in the books on the sphere and cylinder, Apollonius, and all other geometers, clearly use the theorems proved in this very treatise as constituting admitted principles[1]."

Aristotle too speaks of *elements* of geometry in the same sense. Thus: "in geometry it is well to be thoroughly versed in the elements[2]"; "in general the first of the elements are, given the definitions, e.g. of a straight line and of a circle, most easy to prove, although of course there are not many data that can be used to establish each of them because there are not many middle terms[3]"; "among geometrical propositions we call those 'elements' the proofs of which are contained in the proofs of all or most of such propositions[4]"; "(as in the case of bodies), so in like manner we speak of the elements of geometrical propositions and, generally, of demonstrations; for the demonstrations which come first and are contained in a variety of other demonstrations are called elements of those demonstrations... the term element is applied by analogy to that which, being one and small, is useful for many purposes[5]."

§ 2. *ELEMENTS* ANTERIOR TO EUCLID'S.

The early part of the famous summary of Proclus was no doubt drawn, at least indirectly, from the history of geometry by Eudemus; this is generally inferred from the remark, made just after the mention of Philippus of Medma, a disciple of Plato, that "those who have written histories bring the development of this science up to this point." We have therefore the best authority for the list of writers of *elements* given in the summary. Hippocrates of Chios (fl. in second half of 5th c.) is the first; then Leon, who also discovered *diorismi*, put together a more careful collection, the propositions proved in it being more numerous as well as more serviceable[6]. Leon was a little older than Eudoxus (about 408–355 B.C.) and a little younger than Plato (428/7–347/6 B.C.), but did not belong to the latter's school. The

---

[1] Proclus, pp. 70, 19—71, 21.
[2] *Topics* VIII. 14, 163 b 23.   [3] *Topics* VIII. 3, 158 b 35.   [4] *Metaph.* 998 a 25.
[5] *Metaph.* 1014 a 35—b 5.
[6] Proclus, p. 66, 20 ὥστε τὸν Λέοντα καὶ τὰ στοιχεῖα συνθεῖναι τῷ τε πλήθει καὶ τῇ χρείᾳ τῶν δεικνυμένων ἐπιμελέστερον.

geometrical text-book of the Academy was written by Theudius of Magnesia, who, with Amyclas of Heraclea, Menaechmus the pupil of Eudoxus, Menaechmus' brother Dinostratus and Athenaeus of Cyzicus consorted together in the Academy and carried on their investigations in common. Theudius " put together the elements admirably, making many partial (or limited) propositions more general[1]." Eudemus mentions no text-book after that of Theudius, only adding that Hermotimus of Colophon "discovered many of the elements[2]." Theudius then must be taken to be the immediate precursor of Euclid, and no doubt Euclid made full use of Theudius as well as of the discoveries of Hermotimus and all other available material. Naturally it is not in Euclid's *Elements* that we can find much light upon the state of the subject when he took it up ; but we have another source of information in Aristotle. Fortunately for the historian of mathematics, Aristotle was fond of mathematical illustrations ; he refers to a considerable number of geometrical propositions, definitions etc., in a way which shows that his pupils must have had at hand some textbook where they could find the things he mentions; and this text-book must have been that of Theudius. Heiberg has made a most valuable collection of mathematical extracts from Aristotle[3], from which much is to be gathered as to the changes which Euclid made in the methods of his predecessors ; and these passages, as well as others not included in Heiberg's selection, will often be referred to in the sequel.

§ 3. FIRST PRINCIPLES: DEFINITIONS, POSTULATES, AND AXIOMS.

On no part of the subject does Aristotle give more valuable information than on that of the first principles as, doubtless, generally accepted at the time when he wrote. One long passage in the *Posterior Analytics* is particularly full and lucid, and is worth quoting *in extenso*. After laying it down that every demonstrative science starts from necessary principles[4], he proceeds[5]:

" By first principles in each genus I mean those the truth of which it is not possible to prove. What is *denoted* by the first (terms) and those derived from them is assumed ; but, as regards their *existence*, this must be assumed for the principles but proved for the rest. Thus what a unit is, what the straight (line) is, or what a triangle is (must be assumed); and the existence of the unit and of magnitude must also be assumed, but the rest must be proved. Now of the premises used in demonstrative sciences some are peculiar to each science and others common (to all), the latter being common by analogy, for of course they are actually useful in so far as they are applied to the subject-matter included under the particular science. Instances of first

---

[1] Proclus, p. 67, 14 καὶ γὰρ τὰ στοιχεῖα καλῶς συνέταξεν καὶ πολλὰ τῶν μερικῶν [ὁρικῶν (?) Friedlein] καθολικώτερα ἐποίησεν.

[2] Proclus, p. 67, 22 τῶν στοιχείων πολλὰ ἀνεῦρε.

[3] *Mathematisches zu Aristoteles* in *Abhandlungen zur Gesch. d. math. Wissenschaften*, XVIII. Heft (1904), pp. 1—49.

[4] *Anal. post.* I. 6, 74 b 5.   [5] *ibid.* I. 10, 76 a 31—77 a 4.

principles peculiar to a science are the assumptions that a line is of
such and such a character, and similarly for the straight (line); whereas
it is a common principle, for instance, that, if equals be subtracted
from equals, the remainders are equal.   But it is enough that each of
the common principles is true so far as regards the particular genus
(subject-matter); for (in geometry) the effect will be the same even if
the common principle be assumed to be true, not of everything, but
only of magnitudes, and, in arithmetic, of numbers.

"Now the things peculiar to the science, the existence of which
must be assumed, are the things with reference to which the science
investigates the essential attributes, e.g. arithmetic with reference to
units, and geometry with reference to points and lines.   With these
things it is assumed that they exist and that they are of such and
such a nature.   But, with regard to their essential properties, what is
assumed is only the meaning of each term employed: thus arithmetic
assumes the answer to the question what is (meant by) 'odd' or
'even,' 'a square' or 'a cube,' and geometry to the question
what is (meant by) 'the irrational' or 'deflection' or (the so-called)
'verging' (to a point); but that there are such things is proved by
means of the common principles and of what has already been
demonstrated.   Similarly with astronomy.   For every demonstrative
science has to do with three things, (1) the things which are assumed
to exist, namely the genus (subject-matter) in each case, the essential
properties of which the science investigates, (2) the common axioms
so-called, which are the primary source of demonstration, and (3) the
properties with regard to which all that is assumed is the meaning of
the respective terms used.   There is, however, no reason why some
sciences should not omit to speak of one or other of these things.
Thus there need not be any supposition as to the existence of the
genus, if it is manifest that it exists (for it is not equally clear that
number exists and that cold and hot exist); and, with regard to the
properties, there need be no assumption as to the meaning of terms if
it is clear: just as in the common (axioms) there is no assumption as
to what is the meaning of subtracting equals from equals, because it is
well known.   But none the less is it true that there are three things
naturally distinct, the subject-matter of the proof, the things proved,
and the (axioms) from which (the proof starts).

"Now that which is *per se* necessarily true, and must necessarily be
thought so, is not a hypothesis nor yet a postulate.   For demonstration has not to do with reasoning from outside but with the
reason dwelling in the soul, just as is the case with the syllogism.
It is always possible to raise objection to reasoning from outside,
but to contradict the reason within us is not always possible.   Now
anything that the teacher assumes, though it is matter of proof,
without proving it himself, is a hypothesis if the thing assumed is
believed by the learner, and it is moreover a hypothesis, not absolutely, but relatively to the particular pupil; but, if the same thing
is assumed when the learner either has no opinion on the subject
or is of a contrary opinion, it is a postulate.   This is the difference

between a hypothesis and a postulate; for a postulate is that which is rather contrary than otherwise to the opinion of the learner, or whatever is assumed and used without being proved, although matter for demonstration. Now definitions are not hypotheses, for they do not assert the existence or non-existence of anything, while hypotheses are among propositions. Definitions only require to be understood: a definition is therefore not a hypothesis, unless indeed it be asserted that any audible speech is a hypothesis. A hypothesis is that from the truth of which, if assumed, a conclusion can be established. Nor are the geometer's hypotheses false, as some have said: I mean those who say that 'you should not make use of what is false, and yet the geometer falsely calls the line which he has drawn a foot long when it is not, or straight when it is not straight.' The geometer bases no conclusion on the particular line which he has drawn being that which he has described, but (he refers to) what is *illustrated* by the figures. Further, the postulate and every hypothesis are either universal or particular statements; definitions are neither" (because the subject is of equal extent with what is predicated of it).

Every demonstrative science, says Aristotle, must start from indemonstrable principles: otherwise, the steps of demonstration would be endless. Of these indemonstrable principles some are (*a*) common to all sciences, others are (*b*) particular, or peculiar to the particular science; (*a*) the common principles are the *axioms*, most commonly illustrated by the axiom that, if equals be subtracted from equals, the remainders are equal. Coming now to (*b*) the principles peculiar to the particular science which must be assumed, we have first the *genus* or subject-matter, the *existence* of which must be assumed, viz. magnitude in the case of geometry, the unit in the case of arithmetic. Under this we must assume *definitions* of manifestations or attributes of the genus, e.g. straight lines, triangles, deflection etc. The definition in itself says nothing as to the existence of the thing defined: it only requires to be understood. But in geometry, in addition to the *genus* and the *definitions*, we have to assume the *existence* of a few *primary* things which are defined, viz. points and lines only: the existence of everything else, e.g. the various figures made up of these, as triangles, squares, tangents, and their properties, e.g. incommensurability etc., has to be proved (as it is proved by construction and demonstration). In arithmetic we assume the *existence* of the *unit*: but, as regards the rest, only the *definitions*, e.g. those of odd, even, square, cube, are assumed, and *existence* has to be *proved*. We have then clearly distinguished, among the indemonstrable principles, *axioms* and *definitions*. A *postulate* is also distinguished from a *hypothesis*, the latter being made with the assent of the learner, the former without such assent or even in opposition to his opinion (though, strangely enough, immediately after saying this, Aristotle gives a wider meaning to "postulate" which would cover "hypothesis" as well, namely whatever is assumed, though it is matter for proof, and used without being proved). Heiberg remarks that there is no trace in Aristotle of Euclid's Postulates, and that "postulate" in Aristotle has

a different meaning. He seems to base this on the alternative
description of postulate, indistinguishable from a hypothesis; but,
if we take the other description in which it is distinguished from a
hypothesis as being an assumption of something which is a proper
subject of demonstration without the assent or against the opinion of
the learner, it seems to fit Euclid's Postulates fairly well, not only the
first three (postulating three constructions), but eminently also the other
two, that all right angles are equal, and that two straight lines meeting
a third and making the internal angles on the same side of it less than
two right angles will meet on that side.   Aristotle's description also
seems to me to suit the "postulates" with which Archimedes begins
his book *On the equilibrium of planes*, namely that equal weights balance
at equal distances, and that equal weights at unequal distances do not
balance but that the weight at the longer distance will prevail.

Aristotle's distinction also between *hypothesis* and *definition*, and
between *hypothesis* and *axiom*, is clear from the following passage:
"Among immediate syllogistic principles, I call that a *thesis* which·
it is neither possible to prove nor essential for any one to hold who
is to learn anything; but that which it is necessary for any one to
hold who is to learn anything whatever is an *axiom*: for there are
some principles of this kind, and that is the most usual name by
which we speak of them.   But, of *theses*, one kind is that which
assumes one or other side of a predication, as, for instance, that
something exists or does not exist, and this is a *hypothesis*; the other,
which makes no such assumption, is a *definition*.   For a definition is
a thesis: thus the arithmetician posits ($\tau i\theta\epsilon\tau\alpha\iota$) that a unit is that
which is indivisible in respect of quantity; but this is not a hypo-
thesis, since what is meant by a unit and the fact that a unit exists
are different things[1]."

Aristotle uses as an alternative term for axioms "common (things),"
$\tau\grave{\alpha}\ \kappa o\iota\nu\acute{\alpha}$, or "common opinions" ($\kappa o\iota\nu\alpha\grave{\iota}\ \delta\acute{o}\xi\alpha\iota$), as in the following
passages.   "That, when equals are taken from equals, the remainders
are equal is (a) common (principle) in the case of all quantities, but
mathematics takes a separate department ($\grave{\alpha}\pi o\lambda\alpha\beta o\hat{v}\sigma\alpha$) and directs its
investigation to some portion of its proper subject-matter, as e.g. lines
or angles, numbers, or any of the other quantities[2]."   "The common
(principles), e.g. that one of two contradictories must be true, that
equals taken from equals etc., and the like[3]...."   "With regard to the
principles of demonstration, it is questionable whether they belong to
one science or to several.   By principles of demonstration I mean the
common opinions from which all demonstration proceeds, e.g. that one
of two contradictories must be true, and that it is impossible for the
same thing to be and not be[4]."   Similarly "every demonstrative
(science) investigates, with regard to some subject-matter, the essential
attributes, starting from the *common opinions[5]*."   We have then here,
as Heiberg says, a sufficient explanation of Euclid's term for axioms,

---

[1] *Anal. post.* I. 2, 72 a 14—24.   [2] *Metaph.* 1061 b 19—24.
[3] *Anal. post.* I. 11, 77 a 30.   [4] *Metaph.* 996 b 26—30.
[5] *Metaph.* 997 a 20—22.

viz. *common notions* (κοιναὶ ἔννοιαι), and there is no reason to suppose it to be a substitution for the original term due to the Stoics: cf. Proclus' remark that, according to Aristotle and the geometers, axiom and common notion are the same thing[1].

Aristotle discusses the *indemonstrable* character of the axioms in the *Metaphysics*. Since "all the demonstrative sciences use the axioms[2]," the question arises, to what science does their discussion belong[3]? The answer is that, like that of Being (οὐσία), it is the province of the (first) philosopher[4]. It is impossible that there should be demonstration of everything, as there would be an infinite series of demonstrations: if the axioms were the subject of a demonstrative science, there would have to be here too, as in other demonstrative sciences, a *subject-genus*, its *attributes* and corresponding *axioms*[5]; thus there would be axioms behind axioms, and so on continually. The axiom is the most firmly established of all principles[6]. It is ignorance alone that could lead any one to try to prove the axioms[7]; the supposed proof would be a *petitio principii*[8]. If it is admitted that not everything can be proved, no one can point to any principle more truly indemonstrable[9]. If any one thought he could prove them, he could at once be refuted; if he did not attempt to say anything, it would be ridiculous to argue with him: he would be no better than a vegetable[10]. The first condition of the possibility of any argument whatever is that words should signify something both to the speaker and to the hearer: without this there can be no reasoning with any one. And, if any one admits that words can mean anything to both hearer and speaker, he admits that something can be true without demonstration. And so on[11].

It was necessary to give some sketch of Aristotle's view of the first principles, if only in connexion with Proclus' account, which is as follows. As in the case of other sciences, so "the compiler of elements in geometry must give separately the principles of the science, and after that the conclusions from those principles, not giving any account of the principles but only of their consequences. No science proves its own principles, or even discourses about them: they are treated as self-evident....Thus the first essential was to distinguish the principles from their consequences. Euclid carries out this plan practically in every book and, as a preliminary to the whole enquiry, sets out the common principles of this science. Then he divides the common principles themselves into *hypotheses, postulates,* and *axioms*. For all these are different from one another: an axiom, a postulate and a hypothesis are not the same thing, as the inspired Aristotle somewhere says. But, whenever that which is assumed and ranked as a principle is both known to the learner and convincing in itself, such a thing is an *axiom*, e.g. the statement that things which are equal to the same thing are also equal to one another. When, on

---

[1] Proclus, p. 194, 8.   [2] *Metaph.* 997 a 10.
[3] *ibid.* 996 b 26.   [4] *ibid.* 1005 a 21—b 11.   [5] *ibid.* 997 a 5—8.
[6] *ibid.* 1005 b 11—17.   [7] *ibid.* 1006 a 5.   [8] *ibid.* 1006 a 17.
[9] *ibid.* 1006 a 10.   [10] *ibid.* 1006 a 11—15.   [11] *ibid.* 1006 a 18 sqq.

the other hand, the pupil has not the notion of what is told him which carries conviction in itself, but nevertheless lays it down and assents to its being assumed, such an assumption is a *hypothesis*. Thus we do not preconceive by virtue of a common notion, and without being taught, that the circle is such and such a figure, but, when we are told so, we assent without demonstration. When again what is asserted is both unknown and assumed even without the assent of the learner, then, he says, we call this a *postulate*, e.g. that all right angles are equal. This view of a postulate is clearly implied by those who have made a special and systematic attempt to show, with regard to one of the postulates, that it cannot be assented to by any one straight off. According then to the teaching of Aristotle, an axiom, a postulate and a hypothesis are thus distinguished[1]."

We observe, first, that Proclus in this passage confuses *hypotheses* and *definitions*, although Aristotle had made the distinction quite plain. The confusion may be due to his having in his mind a passage of Plato from which he evidently got the phrase about "not giving an account of" the principles. The passage is[2]: " I think you know that those who treat of geometries and calculations (arithmetic) and such things take for granted (ὑποθέμενοι) odd and even, figures, angles of three kinds, and other things akin to these in each subject, implying that they know these things, and, though using them as hypotheses, do not even condescend to give any account of them either to themselves or to others, but begin from these things and then go through everything else in order, arriving ultimately, by recognised methods, at the conclusion which they started in search of." But the hypothesis is here the assumption, e.g. ' that *there may be such a thing* as length without breadth, henceforward called a line[3],' and so on, without any attempt to show that there is such a thing; it is mentioned in connexion with the distinction between Plato's ' superior' and ' inferior' intellectual method, the former of which uses successive hypotheses as stepping-stones by which it mounts upwards to the idea of Good.

We pass now to Proclus' account of the difference between *postulates* and *axioms*. He begins with the view of Geminus, according to which "they differ from one another in the same way as theorems are also distinguished from problems. For, as in theorems we propose to see and determine what follows on the premises, while in problems we are told to find and do something, in like manner in the *axioms* such things are assumed as are manifest of themselves and easily apprehended by our untaught notions, while in the *postulates* we assume such things as are easy to find and effect (our understanding suffering no strain in their assumption), and we require no complication of machinery[4]."..." Both must have the characteristic of being simple

---

[1] Proclus, pp. 75, 10—77, 2.
[2] *Republic*, VI. 510 C.   Cf. Aristotle, *Nic. Eth.* 1151 a 17.
[3] H. Jackson, *Journal of Philology*, vol. x. p. 144.
[4] Proclus, pp. 178, 12—179, 8. In illustration Proclus contrasts the drawing of a straight line or a circle with the drawing of a " single-turn spiral " or of an equilateral triangle, the

and readily grasped, I mean both the postulate and the axiom ; but the postulate bids us contrive and find some subject-matter ($\ddot{v}\lambda\eta$) to exhibit a property simple and easily grasped, while the axiom bids us assert some essential attribute which is self-evident to the learner, just as is the fact that fire is hot, or any of the most obvious things[1]."

Again, says Proclus, " some claim that all these things are alike postulates, in the same way as some maintain that all things that are sought are problems. For Archimedes begins his first book on *In-equilibrium*[2] with the remark ' I postulate that equal weights at equal distances are in equilibrium,' though one would rather call this an axiom. Others call them all axioms in the same way as some regard as theorems everything that requires demonstration[3]."

" Others again will say that postulates are peculiar to geometrical subject-matter, while axioms are common to all investigation which is concerned with quantity and magnitude. Thus it is the geometer who knows that all right angles are equal and how to produce in a straight line any limited straight line, whereas it is a common notion that things which are equal to the same thing are also equal to one another, and it is employed by the arithmetician and any scientific person who adapts the general statement to his own subject[4]."

The third view of the distinction between a postulate and an axiom is that of Aristotle above described[5].

The difficulties in the way of reconciling Euclid's classification of postulates and axioms with any one of the three alternative views are next dwelt upon. If we accept the first view according to which an axiom has reference to something known, and a postulate to something done, then the 4th postulate (that all right angles are equal) is not a postulate ; neither is the 5th which states that, if a straight line falling on two straight lines makes the interior angles on the same side less than two right angles, the straight lines, if produced indefinitely, will meet on that side on which are the angles less than two right angles. On the second view, the assumption that two straight lines cannot enclose a space, "which even now," says Proclus, "some add as an axiom," and which is peculiar to the subject-matter of geometry, like the fact that all right angles are equal, is not an axiom. According to the third (Aristotelian) view, "everything which is confirmed ($\pi\iota\sigma\tau o\hat{v}\tau\alpha\iota$) by a sort of demonstration

spiral requiring more complex machinery and even the equilateral triangle needing a certain method. " For the geometrical intelligence will say that by conceiving a straight line fixed at one end but, as regards the other end, moving round the fixed end, and a point moving along the straight line from the fixed end, I have described the single-turn spiral ; for the end of the straight line describing a circle, and the point moving on the straight line simultaneously, when they arrive and meet at the same point, complete such a spiral. And again, if I draw equal circles, join their common point to the centres of the circles and draw a straight line from one of the centres to the other, I shall have the equilateral triangle. These things then are far from being completed by means of a single act or of a moment's thought" (p. 180, 8—21).

[1] Proclus, p. 181, 4—11.

[2] It is necessary to coin a word to render $\dot{\alpha}\nu\iota\sigma o\rho\rho o\pi\iota\hat{\omega}\nu$, which is moreover in the plural. The title of the treatise as we have it is *Equilibria of planes or centres of gravity of planes* in Book I and *Equilibria of planes* in Book II.

[3] Proclus, p. 181, 16—23.　　　[4] *ibid.* p. 182, 6—14.　　　[5] Pp. 118, 119.

will be a postulate, and what is incapable of proof will be an axiom[1]."
This last statement of Proclus is loose, as regards the axiom, because
it omits Aristotle's requirement that the axiom should be a self-
evident truth, and one that must be admitted by any one who is to
learn anything at all, and, as regards the postulate, because Aristotle
calls a postulate something assumed without proof though it is
"matter of demonstration" (ἀποδεικτὸν ὄν), but says nothing of a
quasi-demonstration of the postulates.   On the whole I think it is
from Aristotle that we get the best idea of what Euclid understood
by a postulate and an axiom or common notion.   Thus Aristotle's
account of an axiom as a principle common to all sciences, which is
self-evident, though incapable of proof, agrees sufficiently with the
contents of Euclid's *common notions* as reduced to five in the most
recent text (not omitting the fourth, that "things which coincide are
equal to one another").   As regards the *postulates*, it must be borne
in mind that Aristotle says elsewhere[2] that, "other things being equal,
that proof is the better which proceeds from the fewer postulates or
hypotheses or propositions."   If then we say that a geometer must
lay down as principles, first certain axioms or common notions, and
then an *irreducible minimum* of postulates in the Aristotelian sense
concerned only with the subject-matter of geometry, we are not far
from describing what Euclid in fact does.   As regards the postulates
we may imagine him saying : "Besides the common notions there are
a few other things which I must assume without proof, but which
differ from the common notions in that they are not self-evident.
The learner may or may not be disposed to agree to them ; but he
must accept them at the outset on the superior authority of his
teacher, and must be left to convince himself of their truth in the
course of the investigation which follows.   In the first place certain
simple constructions, the drawing and producing of a straight line,
and the drawing of a circle, must be assumed to be possible, and with
the constructions the existence of such things as straight lines and
circles ; and besides this we must lay down some postulate to form
the basis of the theory of parallels."   It is true that the admission of
the 4th postulate that all right angles are equal still presents a
difficulty to which we shall have to recur.

There is of course no foundation for the idea, which has found
its way into many text-books, that "the object of the postulates is to
declare that the only instruments the use of which is permitted in
geometry are the *rule* and *compass*[3]."

§ 4. THEOREMS AND PROBLEMS.

"Again the deductions from the first principles," says Proclus,
"are divided into **problems** and **theorems**, the former embracing the

---

[1] Proclus, pp. 182, 21—183, 13.   [2] *Anal. post.* 1. 25, 86 a 33—35.
[3] Cf. Lardner's Euclid : also Todhunter.

generation, division, subtraction or addition of figures, and generally the changes which are brought about in them, the latter exhibiting the essential attributes of each[1]."

"Now, of the ancients, some, like Speusippus and Amphinomus, thought it proper to call them all theorems, regarding the name of theorems as more appropriate than that of problems to theoretic sciences, especially as these deal with eternal objects. For there is no becoming in things eternal, so that neither could the problem have any place with them, since it promises the generation and making of what has not before existed, e.g. the construction of an equilateral triangle, or the describing of a square on a given straight line, or the placing of a straight line at a given point. Hence they say it is better to assert that all (propositions) are of the same kind, and that we regard the generation that takes place in them as referring not to actual *making* but to *knowledge*, when we treat things existing eternally as if they were subject to becoming: in other words, we may say that everything is treated by way of theorem and not by way of problem[2] (πάντα θεωρηματικῶς ἀλλ' οὐ προβληματικῶς λαμβάνεσθαι).

"Others on the contrary, like the mathematicians of the school of Menaechmus, thought it right to call them all problems, describing their purpose as twofold, namely in some cases to furnish (πορίσασθαι) the thing sought, in others to take a determinate object and see either what it is, or of what nature, or what is its property, or in what relations it stands to something else.

"In reality both assertions are correct. Speusippus is right because the problems of geometry are not like those of mechanics, the latter being matters of sense and exhibiting becoming and change of every sort. The school of Menaechmus are right also because the discoveries even of theorems do not arise without an issuing-forth into matter, by which I mean intelligible matter. Thus forms going out into matter and giving it shape may fairly be said to be like processes of becoming. For we say that the motion of our thought and the throwing-out of the forms in it is what produces the figures in the imagination and the conditions subsisting in them. It is in the imagination that constructions, divisions, placings, applications, additions and subtractions (take place), but everything in the mind is fixed and immune from becoming and from every sort of change[3]."

"Now those who distinguish the theorem from the problem say that every problem implies the possibility, not only of that which is predicated of its subject-matter, but also of its opposite, whereas every theorem implies the possibility of the thing predicated but not of its opposite as well. By the subject-matter I mean the genus which is the subject of inquiry, for example, a triangle or a square or a circle, and by the property predicated the essential attribute, as equality, section, position, and the like. When then any one

[1] Proclus, p. 77, 7—12.   [2] *ibid.* pp. 77, 15—78, 8.
[3] *ibid.* pp. 78, 8—79, 2.

enunciates thus, *To inscribe an equilateral triangle in a circle*, he states a *problem*; for it is also possible to inscribe in it a triangle which is not equilateral. Again, if we take the enunciation *On a given limited straight line to construct an equilateral triangle*, this is a *problem*; for it is possible also to construct one which is not equilateral. But, when any one enunciates that *In isosceles triangles the angles at the base are equal*, we must say that he enunciates a *theorem*; for it is not also possible that the angles at the base of isosceles triangles should be unequal. It follows that, if any one were to use the form of a problem and say *In a semicircle to describe a right angle*, he would be set down as no geometer. For every angle in a semicircle is right[1]."

"Zenodotus, who belonged to the succession of Oenopides, but was a disciple of Andron, distinguished the theorem from the problem by the fact that the theorem inquires what is the property predicated of the subject-matter in it, but the problem what is the cause of what effect (τίνος ὄντος τί ἐστιν). Hence too Posidonius defined the one (the problem) as a proposition in which it is inquired whether a thing exists or not (εἰ ἔστιν ἢ μή), the other (the theorem[2]) as a proposition in which it is inquired what (a thing) is or of what nature (τί ἐστιν ἢ ποῖόν τι); and he said that the theoretic proposition must be put in a declaratory form, e.g., *Any triangle has two sides (together) greater than the remaining side* and *In any isosceles triangle the angles at the base are equal*, but that we should state the problematic proposition as if inquiring whether it is possible to construct an equilateral triangle upon such and such a straight line. For there is a difference between inquiring absolutely and indeterminately (ἁπλῶς τε καὶ ἀορίστως) whether there exists a straight line from such and such a point at right angles to such and such a straight line and investigating which is the straight line at right angles[3]."

"That there is a certain difference between the problem and the theorem is clear from what has been said; and that the Elements of Euclid contain partly problems and partly theorems will be made manifest by the individual propositions, where Euclid himself adds at the end of what is proved in them, in some cases, 'that which it was required to do,' and in others, 'that which it was required to prove,' the latter expression being regarded as characteristic of theorems, in spite of the fact that, as we have said, demonstration is found in problems also. In problems, however, even the demonstration is for the purpose of (confirming) the construction: for we bring in the demonstration in order to show that what was enjoined has been done; whereas in theorems the demonstration is worthy of study for its own sake as being capable of putting before us the nature of the thing sought. And you will find that Euclid sometimes interweaves theorems with problems and employs them in turn, as in the first

---

[1] Proclus, pp. 79, 11—80, 5.
[2] In the text we have τὸ δὲ πρόβλημα answering to τὸ μὲν without substantive: πρόβλημα was obviously inserted in error.
[3] Proclus, pp. 80, 15—81, 4.

book, while at other times he makes one or other preponderate. For the fourth book consists wholly of problems, and the fifth of theorems[1]."

Again, in his note on Eucl. I. 4, Proclus says that Carpus, the writer on mechanics, raised the question of theorems and problems in his treatise on astronomy. Carpus, we are told, "says that the class of problems is in order prior to theorems. For the subjects, the properties of which are sought, are discovered by means of problems. Moreover in a problem the enunciation is simple and requires no skilled intelligence; it orders you plainly to do such and such a thing, *to construct an equilateral triangle*, or, *given two straight lines, to cut off from the greater* (*a straight line*) *equal to the lesser*, and what is there obscure or elaborate in these things? But the enunciation of a theorem is a matter of labour and requires much exactness and scientific judgment in order that it may not turn out to exceed or fall short of the truth; an example is found even in this proposition (I. 4), the first of the theorems. Again, in the case of problems, one general way has been discovered, that of *analysis*, by following which we can always hope to succeed; it is this method by which the more obscure problems are investigated. But, in the case of theorems, the method of setting about them is hard to get hold of since 'up to our time,' says Carpus, 'no one has been able to hand down a general method for their discovery. Hence, by reason of their easiness, the class of problems would naturally be more simple.' After these distinctions, he proceeds: 'Hence it is that in the Elements too problems precede theorems, and the Elements begin from them; the first theorem is fourth in order, not because the fifth[2] is proved from the problems, but because, even if it needs for its demonstration none of the propositions which precede it, it was necessary that they should be first because they are problems, while it is a theorem. In fact, in this theorem he uses the common notions exclusively, and in some sort takes the same triangle placed in different positions; the coincidence and the equality proved thereby depend entirely upon sensible and distinct apprehension. Nevertheless, though the demonstration of the first theorem is of this character, the problems properly preceded it, because in general problems are allotted the order of precedence[3].'"

Proclus himself explains the position of Prop. 4 after Props. 1—3 as due to the fact that a theorem about the essential properties of triangles ought not to be introduced before we know that such a thing as a triangle can be constructed, nor a theorem about the equality of sides or straight lines until we have shown, by constructing them, that there can be two straight lines which are equal to one another[4]. It is plausible enough to argue in this way that Props. 2 and 3 at all events should precede Prop. 4. And Prop. 1 is used in

---

[1] Proclus, p. 81, 5—22.

[2] τὸ πέμπτον. This should apparently be the *fourth* because in the next words it is implied that none of the first three propositions are required in proving it.

[3] Proclus, pp. 241, 19—243, 11.   [4] *ibid.* pp. 233, 21—234, 6.

Prop. 2, and must therefore precede it. But Prop. 1 showing how to construct an *equilateral* triangle on a given base is not important, in relation to Prop. 4, as dealing with the "production of triangles" in general: for it is of no use to say, as Proclus does, that the construction of the equilateral triangle is "common to the three species (of triangles)[1]," as we are not in a position to know this at such an early stage. The *existence* of triangles in general was doubtless assumed as following from the existence of straight lines and points in one plane and from the possibility of drawing a straight line from one point to another.

Proclus does not however seem to reject definitely the view of Carpus, for he goes on[2]: "And perhaps problems are in order before theorems, and especially for those who need to ascend from the arts which are concerned with things of sense to theoretical investigation. But in dignity theorems are prior to problems.... It is then foolish to blame Geminus for saying that the theorem is more perfect than the problem. For Carpus himself gave the priority to problems in respect of *order*, and Geminus to theorems in point of more perfect *dignity*," so that there was no real inconsistency between the two.

Problems were classified according to the number of their possible solutions. Amphinomus said that those which had a unique solution ($\mu o \nu a \chi \hat{\omega} \varsigma$) were called "ordered" (the word has dropped out in Proclus, but it must be $\tau \epsilon \tau a \gamma \mu \acute{\epsilon} \nu a$, in contrast to the third kind, $\check{a} \tau a \kappa \tau a$); those which had a definite number of solutions "intermediate" ($\mu \acute{\epsilon} \sigma a$); and those with an infinite variety of solutions "unordered" ($\check{a} \tau a \kappa \tau a$)[3]. Proclus gives as an example of the last the problem *To divide a given straight line into three parts in continued proportion*[4]. This is the same thing as solving the equations $x+y+z=a$, $xz=y^2$. Proclus' remarks upon the problem show that it was solved, like all quadratic equations, by the method of "application of areas." The straight line a was first divided into any two parts, $(x+z)$ and y, subject to the sole limitation that $(x+z)$ must not be less than $2y$, which limitation is the $\delta \iota o \rho \iota \sigma \mu \acute{o} \varsigma$, or condition of possibility. Then an area was applied to $(x+z)$, or $(a-y)$, "*falling short by a square figure*" ($\dot{\epsilon} \lambda \lambda \epsilon \hat{\iota} \pi o \nu$ $\epsilon \check{\iota} \delta \epsilon \iota$ $\tau \epsilon \tau \rho a \gamma \acute{\omega} \nu \psi$) and equal to the square on y. This determines x and z separately in terms of a and y. For, if z be the side of the square by which the area (i.e. rectangle) "falls short," we have $\{(a-y)-z\}z=y^2$, whence $2z=(a-y)\pm\sqrt{\{(a-y)^2-4y^2\}}$. And y may be chosen arbitrarily, provided that it is not greater than $a/3$. Hence there are an infinite number of solutions. If $y=a/3$, then, as Proclus remarks, the three parts are equal.

Other distinctions between different kinds of problems are added by Proclus. The word "problem," he says, is used in several senses. In its widest sense it may mean anything "propounded" ($\pi \rho o \tau \epsilon \iota \nu \acute{o} \mu \epsilon \nu o \nu$), whether for the purpose of instruction ($\mu a \theta \acute{\eta} \sigma \epsilon \omega \varsigma$) or construction ($\pi o \iota \acute{\eta} \sigma \epsilon \omega \varsigma$). (In this sense, therefore, it would include a theorem.)

---

[1] Proclus, p. 234, 21.
[2] *ibid.* p. 243, 12—25.
[3] *ibid.* p. 220, 7—12.
[4] *ibid.* pp. 220, 16—221, 6.

But its special sense in mathematics is that of something "propounded with a view to a theoretic construction[1]."

Again you may apply the term (in this restricted sense) even to something which is *impossible*, although it is more appropriately used of what is *possible* and neither asks too much nor contains too little in the shape of data. According as a problem has one or other of these defects respectively, it is called (1) a problem *in excess* (πλεονάζον) or (2) a *deficient* problem (ἐλλιπὲς πρόβλημα). The problem *in excess* (1) is of two kinds, (*a*) a problem in which the properties of the figure to be found are either *inconsistent* (ἀσύμβατα) or *non-existent* (ἀνύπαρκτα), in which case the problem is called impossible, or (*b*) a problem in which the enunciation is merely redundant: an example of this would be a problem requiring us to construct an equilateral triangle with its vertical angle equal to two-thirds of a right angle; such a problem is possible and is called "more than a problem" (μεῖζον ἢ πρόβλημα). The *deficient* problem (2) is similarly called "less than a problem" (ἔλασσον ἢ πρόβλημα), its characteristic being that something has to be added to the enunciation in order to convert it from indeterminateness (ἀοριστία) to order (τάξις) and scientific determinateness (ὅρος ἐπιστημονικός): such would be a problem bidding you "to construct an isosceles triangle," for the varieties of isosceles triangles are unlimited. Such "problems" are not problems in the proper sense (κυρίως λεγόμενα προβλήματα), but only equivocally[2].

§ 5. THE FORMAL DIVISIONS OF A PROPOSITION.

"Every problem," says Proclus[3], "and every theorem which is complete with all its parts perfect purports to contain in itself all of the following elements: enunciation (πρότασις), setting-out (ἔκθεσις), definition or specification (διορισμός), construction or machinery (κατασκευή), proof (ἀπόδειξις), conclusion (συμπέρασμα). Now of these the *enunciation* states what is given and what is that which is sought, the perfect *enunciation* consisting of both these parts. The *setting-out* marks off what is given, by itself, and adapts it beforehand for use in the investigation. The *definition* or *specification* states separately and makes clear what the particular thing is which is sought. The *construction* or *machinery* adds what is wanting to the datum for the purpose of finding what is sought. The *proof* draws the required inference by reasoning scientifically from acknowledged facts. The *conclusion* reverts again to the *enunciation*, confirming what has been demonstrated. These are all the parts of problems and theorems, but the most essential and those which are found in all are *enunciation, proof, conclusion*. For it is equally necessary to know beforehand what is sought, to prove this by means of the intermediate steps, and to state the proved fact as a conclusion; it is impossible to dispense with any of these three things. The remaining parts are often brought in, but are often left out as serving no purpose.

---

[1] Proclus, p. 221, 7—11.   [2] *ibid.* pp. 221, 13—222, 14.
[3] *ibid.* pp. 203, 1—204, 13; 204, 23—205, 8.

Thus there is neither *setting-out* nor *definition* in the problem of constructing an isosceles triangle having each of the angles at the base double of the remaining angle, and in most theorems there is no *construction* because the *setting-out* suffices without any addition for proving the required property from the data. When then do we say that the *setting-out* is wanting? The answer is, when there is nothing *given* in the *enunciation*; for, though the enunciation is in general divided into what is given and what is sought, this is not always the case, but sometimes it states only what is sought, i.e. what must be known or found, as in the case of the problem just mentioned. That problem does not, in fact, state beforehand with what datum we are to construct the isosceles triangle having each of the equal angles double of the remaining angle, but (simply) that we are to find such a triangle.... When, then, the enunciation contains both (what is given and what is sought), in that case we find both *definition* and *setting-out*, but, whenever the datum is wanting, they too are wanting. For not only is the *setting-out* concerned with the datum, but so is the *definition* also, as, in the absence of the datum, the *definition* will be identical with the enunciation. In fact, what could you say in defining the object of the aforesaid problem except that it is required to find an isosceles triangle of the kind referred to? But that is what the *enunciation* stated. If then the *enunciation* does not include, on the one hand, what is given and, on the other, what is sought, there is no *setting-out* in virtue of there being no datum, and the *definition* is left out in order to avoid a mere repetition of the *enunciation*."

The constituent parts of an Euclidean proposition will be readily identified by means of the above description. As regards the *definition* or *specification* (διορισμός) it is to be observed that we have here only one of its uses. Here it means a closer definition or description of the object aimed at, by means of the concrete lines or figures set out in the ἔκθεσις instead of the general terms used in the enunciation; and its purpose is to rivet the attention better, as Proclus indicates in a later passage (τρόπον τινὰ προσεχείας ἐστὶν αἴτιος ὁ διορισμός)[1].

The other technical use of the word to signify the limitations to which the possible solutions of a problem are subject is also described by Proclus, who speaks of διορισμοί determining "whether what is sought is impossible or possible, and how far it is practicable and in how many ways[2]"; and the διορισμός in this sense appears in Euclid as well as in Archimedes and Apollonius. Thus we have in Eucl. I. 22 the *enunciation* "From three straight lines which are equal to three given straight lines to construct a triangle," followed immediately by the *limiting condition* (διορισμός). "Thus two of the straight lines taken together in any manner must be greater than the remaining one." Similarly in VI. 28 the *enunciation* "To a given straight line to apply a parallelogram equal to a given rectilineal

---

[1] Proclus, p. 208, 21.   [2] *ibid.* p. 202, 3.

figure and falling short by a parallelogrammic figure similar to a given one" is at once followed by the necessary condition of possibility: "Thus the given rectilineal figure must not be greater than that described on half the line and similar to the defect."

Tannery supposed that, in giving the other description of the διορισμός as quoted above, Proclus, or rather his guide, was using the term incorrectly. The διορισμός in the better known sense of the determination of limits or conditions of possibility was, we are told, invented by Leon. Pappus uses the word in this sense only. The other use of the term might, Tannery thought, be due to a confusion occasioned by the use of the same words (δεῖ δή) in introducing the parts of a proposition corresponding to the two meanings of the word διορισμός[1]. On the other hand it is to be observed that Eutocius distinguishes clearly between the two uses and implies that the difference was well known[2]. The διορισμός in the sense of condition of possibility follows immediately on the enunciation, is even part of it; the διορισμός in the other sense of course comes immediately after the *setting-out*.

Proclus has a useful observation respecting the *conclusion* of a proposition[3]. "The conclusion they are accustomed to make double in a certain way: I mean, by proving it in the given case and then drawing a general inference, passing, that is, from the partial conclusion to the general. For, inasmuch as they do not make use of the *individuality* of the subjects taken, but only draw an angle or a straight line with a view to placing the datum before our eyes, they consider that this same fact which is established in the case of the particular figure constitutes a conclusion true of every other figure of the same kind. They pass accordingly to the general in order that we may not conceive the conclusion to be partial. And they are justified in so passing, since they use for the demonstration the particular things set out, not *quâ* particulars, but *quâ* typical of the rest. For it is not in virtue of such and such a size attaching to the angle which is set out that I effect the bisection of it, but in virtue of its being rectilineal and nothing more. Such and such size is peculiar to the angle set out, but its quality of being rectilineal is common to all rectilineal angles. Suppose, for example, that the given angle is a right angle. If then I had employed in the proof the fact of its being right, I should not have been able to pass to every species of rectilineal angle; but, if I make no use of its being right, and only consider it as rectilineal, the argument will equally apply to rectilineal angles in general."

---

[1] *La Géométrie grecque*, p. 149 note. Where δεῖ δή introduces the closer description of the problem we may translate, "it is then required" or "thus it is required" (to construct etc.): when it introduces the condition of possibility we may translate "thus it is necessary etc." Heiberg originally wrote δεῖ δὲ in the latter sense in I. 22 on the authority of Proclus and Eutocius, and against that of the MSS. Later, on the occasion of XI. 23, he observed that he should have followed the MSS. and written δεῖ δή which he found to be, after all, the right reading in Eutocius (Apollonius, ed. Heiberg, II. p. 178). δεῖ δή is also the expression used by Diophantus for introducing conditions of possibility.

[2] See the passage of Eutocius referred to in last note.   [3] Proclus, p. 207, 4—25.

§ 6. OTHER TECHNICAL TERMS.

1. Things said to be given.

Proclus attaches to his description of the formal divisions of a proposition an explanation of the different senses in which the word *given* or *datum* (δεδομένον) is used in geometry. "Everything that is given is given in one or other of the following ways, *in position, in ratio, in magnitude,* or *in species.* The point is given *in position* only, but a line and the rest may be given in all the senses[1]."

The illustrations which Proclus gives of the four senses in which a thing may be *given* are not altogether happy, and, as regards things which are given *in position, in magnitude,* and *in species,* it is best, I think, to follow the definitions given by Euclid himself in his book of *Data.* Euclid does not mention the fourth class, things given *in ratio,* nor apparently do any of the great geometers.

(1) *Given in position* really needs no definition; and, when Euclid says (*Data,* Def. 4) that "Points, lines and angles are said to be *given in position* which always occupy the same place," we are not really the wiser.

(2) *Given in magnitude* is defined thus (*Data,* Def. 1): "Areas, lines and angles are called *given in magnitude* to which we can find equals." Proclus' illustration is in this case the following: when, he says, two unequal straight lines are given from the greater of which we have to cut off a straight line equal to the lesser, the straight lines are obviously *given in magnitude,* "for greater and less, and finite and infinite are predications peculiar to magnitude." But he does not explain that part of the implication of the term is that a thing is given in magnitude *only,* and that, for example, its position is not given and is a matter of indifference

(3) *Given in species.* Euclid's definition (*Data,* Def. 3) is: "Rectilineal figures are said to be *given in species* in which the angles are severally given and the ratios of the sides to one another are given." And this is the recognised use of the term (cf. Pappus, *passim*) Proclus uses the term in a much wider sense for which I am not aware of any authority. Thus, he says, when we speak of (bisecting) a given rectilineal angle, the angle is given in species by the word *rectilineal,* which prevents our attempting, by the same method, to bisect a curvilineal angle! On Eucl. I. 9, to which he here refers, he says that an angle is given in species when e.g. we say that it is right or acute or obtuse or rectilineal or "mixed," but that the actual angle in the proposition is given in species only. As a matter of fact, we should say that the actual angle in the figure of the proposition is given *in magnitude* and not *in species,* part of the implication of *given in species* being that the actual magnitude of the thing *given in species* is indifferent; an angle cannot be *given in species* in this sense at all. The confusion in Proclus' mind is shown when, after saying that a right angle is given *in species,* he describes a third of a right angle as given *in magnitude.*

---

[1] Proclus, p. 205, 13—15.

No better example of what is meant by *given in species*, in its proper sense, as limited to rectilineal figures, can be quoted than the given parallelogram in Eucl. VI. 28, to which the required parallelogram has to be made similar; the former parallelogram is in fact *given in species*, though its actual size, or scale, is indifferent.

(4) *Given in ratio* presumably means something which is given by means of its ratio to some other given thing. This we gather from Proclus' remark (in his note on I. 9) that an angle may be given in ratio "as when we say that it is double and treble of such and such an angle or, generally, greater and less." The term, however, appears to have no authority and to serve no purpose. Proclus may have derived it from such expressions as "in a given ratio" which are common enough.

2. **Lemma.**

"The term *lemma*," says Proclus[1], "is often used of any proposition which is assumed for the construction of something else: thus it is a common remark that a proof has been made out of such and such lemmas. But the special meaning of *lemma* in geometry is a proposition requiring confirmation. For when, in either construction or demonstration, we assume anything which has not been proved but requires argument, then, because we regard what has been assumed as doubtful in itself and therefore worthy of investigation, we call it a *lemma*[2], differing as it does from the postulate and the axiom in being matter of demonstration, whereas they are immediately taken for granted, without demonstration, for the purpose of confirming other things. Now in the discovery of lemmas the best aid is a mental aptitude for it. For we may see many who are quick at solutions and yet do not work by method; thus Cratistus in our time was able to obtain the required result from first principles, and those the fewest possible, but it was his natural gift which helped him to the discovery.

---

[1] Proclus, pp. 211, 1—212, 4.

[2] It would appear, says Tannery (p. 151 n.), that Geminus understood a lemma as being simply λαμβανόμενον, something assumed (cf. the passage of Proclus, p. 73, 4, relating to Menaechmus' view of *elements*): hence we cannot consider ourselves authorised in attributing to Geminus the more technical definition of the term here given by Proclus, according to which it is only used of propositions not proved beforehand. This view of a lemma must be considered as relatively modern. It seems to have had its origin in an imperfection of method. In the course of a demonstration it was necessary to assume a proposition which required proof, but the proof of which would, if inserted in the particular place, break the thread of the demonstration: hence it was necessary either to prove it beforehand as a preliminary proposition or to postpone it to be proved afterwards (ὡς ἑξῆς δειχθήσεται). When, after the time of Geminus, the progress of original discovery in geometry was arrested, geometers occupied themselves with the study and elucidation of the works of the great mathematicians who had preceded them. This involved the investigation of propositions explicitly quoted or tacitly assumed in the great classical treatises; and naturally it was found that several such remained to be demonstrated, either because the authors had omitted them as being easy enough to be left to the reader himself to prove, or because books in which they were proved had been lost in the meantime. Hence arose a class of complementary or auxiliary propositions which were called *lemmas*. Thus Pappus gives in his Book VII a collection of lemmas in elucidation of the treatises of Euclid and Apollonius included in the so-called "Treasury of Analysis" (τόπος ἀναλυόμενος). When Proclus goes on to distinguish three methods of discovering lemmas, *analysis*, *division*, and *reductio ad absurdum*, he seems to imply that the principal business of contemporary geometers was the investigation of these auxiliary propositions.

Nevertheless certain methods have been handed down. The finest is the method which by means of *analysis* carries the thing sought up to an acknowledged principle, a method which Plato, as they say, communicated to Leodamas[1], and by which the latter, too, is said to have discovered many things in geometry. The second is the method of *division*[2], which divides into its parts the genus proposed for consideration and gives a starting-point for the demonstration by means of the elimination of the other elements in the construction of what is proposed, which method also Plato extolled as being of assistance to all sciences. The third is that by means of the *reductio ad absurdum*, which does not show what is sought directly, but refutes its opposite and discovers the truth incidentally."

3. Case.

"The *case*[3] (πτῶσις)," Proclus proceeds[4], "announces different ways of construction and alteration of positions due to the transposition of points or lines or planes or solids. And, in general, all its varieties are seen in the figure, and this is why it is called *case*, being a transposition in the construction."

4. Porism.

"The term *porism* is used also of certain problems such as the Porisms written by Euclid. But it is specially used when from what has been demonstrated some other theorem is revealed at the same time without our propounding it, which theorem has on this very account been called a *porism* (corollary) as being a sort of incidental gain arising from the scientific demonstration[5]." Cf. the note on I. 15.

---

[1] This passage and another from Diogenes Laertius (III. 24, p. 74 ed. Cobet) to the effect that "He [Plato] explained (εἰσηγήσατο) to Leodamas of Thasos the method of inquiry by analysis" have been commonly understood as ascribing to Plato the *invention* of the method of analysis; but Tannery points out forcibly (pp. 112, 113) how difficult it is to explain in what Plato's discovery could have consisted if *analysis* be taken in the sense attributed to it in Pappus, where we can see no more than a series of successive, *reductions* of a problem until it is finally reduced to a known problem. On the other hand, Proclus' words about carrying up the thing sought to "an acknowledged principle" suggest that what he had in mind was the process described at the end of Book VI of the *Republic* by which the dialectician (unlike the mathematician) uses hypotheses as stepping-stones up to a principle which is not hypothetical, and then is able to descend step by step verifying every one of the hypotheses by which he ascended. This description does not of course refer to mathematical analysis, but it may have given rise to the idea that analysis was Plato's discovery, since *analysis* and *synthesis* following each other are related in the same way as the upward and the downward progression in the dialectician's intellectual method. And it may be that Plato's achievement was to observe the importance, from the point of view of logical rigour, of the confirmatory synthesis following analysis, and to regularise in this way and elevate into a completely irrefragable method the partial and uncertain analysis upon which the works of his predecessors depended.

[2] Here again the successive bipartitions of genera into species such as we find in the *Sophist* and *Republic* have very little to say to geometry, and the very fact that they are here mentioned side by side with analysis suggests that Proclus confused the latter with the philosophical method of *Rep.* VI.

[3] Tannery rightly remarks (p. 152) that the subdivision of a theorem or problem into several cases is foreign to the really classic form; the ancients preferred, where necessary, to multiply enunciations. As, however, some omissions necessarily occurred, the writers of lemmas naturally added separate *cases*, which in some instances found their way into the text. A good example is Euclid I. 7, the second case of which, as it appears in our text-books, was interpolated. On the commentary of Proclus on this proposition Th. Taylor rightly remarks that "Euclid everywhere avoids a multitude of cases."

[4] Proclus, p. 212, 5—11.

[5] Tannery notes however that, so far from distinguishing his corollaries from the con-

5. Objection.

"The *objection* (ἔνστασις) obstructs the whole course of the argument by appearing as an obstacle (or crying 'halt,' ἀπαντῶσα) either to the construction or to the demonstration. There is this difference between the *objection* and the *case*, that, whereas he who propounds the case has to prove the proposition to be true of it, he who makes the objection does not need to prove anything: on the contrary it is necessary to destroy the objection and to show that its author is saying what is false[1]."

That is, in general the *objection* endeavours to make it appear that the demonstration is not true in every case; and it is then necessary to prove, in refutation of the objection, either that the supposed case is impossible, or that the demonstration *is* true even for that case. A good instance is afforded by Eucl. I. 7. The text-books give a second case which is not in the original text of Euclid. Proclus remarks on the proposition as given by Euclid that the objection may conceivably be raised that what Euclid declares to be impossible may after all be possible in the event of one pair of straight lines falling completely within the other pair. Proclus then refutes the objection by proving the impossibility in that case also. His proof then came to be given in the text-books as part of Euclid's proposition.

The *objection* is one of the technical terms in Aristotle's logic and its nature is explained in the *Prior Analytics*[2]. "An *objection* is a proposition contrary to a proposition.... Objections are of two sorts, general or partial.... For when it is maintained that an attribute belongs to every (member of a class), we object either that it belongs to none (of the class) or that there is some one (member of the class) to which it does not belong."

6. Reduction.

This is again an Aristotelian term, explained in the *Prior Analytics*[3]. It is well described by Proclus in the following passage:

"*Reduction* (ἀπαγωγή) is a transition from one problem or theorem to another, the solution or proof of which makes that which is propounded manifest also. For example, after the doubling of the cube had been investigated, they transformed the investigation into another upon which it follows, namely the finding of the two means; and from that time forward they inquired how between two given straight lines two mean proportionals could be discovered. And they say that the first to effect the reduction of difficult constructions was Hippocrates of Chios, who also squared a lune and discovered many other things in geometry, being second to none in ingenuity as regards constructions[4]."

clusions of his propositions, Euclid inserts them before the closing words "(being) what it was required to do" or "to prove." In fact the porism-corollary is with Euclid rather a modified form of the regular conclusion than a separate proposition.

[1] Proclus, p. 212, 18—23.
[2] *Anal. prior*. II. 26, 69 a 37.·   [3] *ibid*. II. 25, 69 a 20.
[4] Proclus, pp. 212, 24—213, 11. This passage has frequently been taken as crediting Hippocrates with the discovery of the method of geometrical reduction: cf. Taylor (Translation of Proclus, II. p. 26), Allman (p. 41 *n*., 59), Gow (pp. 169, 170). As Tannery remarks (p. 110), if the particular reduction of the duplication problem to that of the two means is

7. Reductio ad absurdum.

This is variously called by Aristotle "*reductio ad absurdum*" (ἡ εἰς τὸ ἀδύνατον ἀπαγωγή)[1], "proof *per impossibile*" (ἡ διὰ τοῦ ἀδυνάτου δεῖξις or ἀπόδειξις)[2], "proof leading to the impossible" (ἡ εἰς τὸ ἀδύνατον ἄγουσα ἀπόδειξις)[3]. It is part of "proof (starting) from a hypothesis[4]" (ἐξ ὑποθέσεως). "All (syllogisms) which reach the conclusion *per impossibile* reason out a conclusion which is false, and they prove the original contention (by the method starting) from a hypothesis, when something impossible results from assuming the contradictory of the original contention, as, for example, when it is proved that the diagonal (of a square) is incommensurable because, if it be assumed commensurable, it will follow that odd (numbers) are equal to even (numbers)[5]." Or again, "proof (leading) to the impossible differs from the direct (δεικτικῆς) in that it assumes what it desires to destroy [namely the hypothesis of the falsity of the conclusion] and then reduces it to something admittedly false, whereas the direct proof starts from premisses admittedly true[6]."

Proclus has the following description of the *reductio ad absurdum*. "Proofs by *reductio ad absurdum* in every case reach a conclusion manifestly impossible, a conclusion the contradictory of which is admitted. In some cases the conclusions are found to conflict with the common notions, or the postulates, or the hypotheses (from which we started); in others they contradict propositions previously established[7]."..."Every *reductio ad absurdum* assumes what conflicts with the desired result, then, using that as a basis, proceeds until it arrives at an admitted absurdity, and, by thus destroying the hypothesis, establishes the result originally desired. For it is necessary to understand generally that all mathematical arguments either proceed from the first principles or lead back to them, as Porphyry somewhere says. And those which proceed from the first principles are again of two kinds, for they start either from common notions and the clearness of the self-evident alone, or from results previously proved; while those which lead back to the principles are either by way of assuming the principles or by way of destroying them. Those which assume the principles are called *analyses*, and the opposite of these are *syntheses*— for it is possible to start from the said principles and to proceed in the regular order to the desired conclusion, and this process is *synthesis*—while the arguments which would destroy the principles are

the first noted in history, it is difficult to suppose that it was really the first ; for Hippocrates must have found instances of it in the Pythagorean geometry. Bretschneider, I think, comes nearer the truth when he boldly (p. 99) translates : "This reduction *of the aforesaid construction* is said to have been first given by Hippocrates." The words are πρῶτον δέ φασι τῶν ἀπορουμένων διαγραμμάτων τὴν ἀπαγωγὴν ποιήσασθαι, which must, literally, be translated as in the text above; but, when Proclus speaks vaguely of "difficult constructions," he probably means to say simply that "this first recorded instance of a reduction of a difficult construction is attributed to Hippocrates."

---

[1] Aristotle, *Anal. prior.* I. 7, 29 b 5 ; I. 44, 50 a 30.
[2] *ibid.* I. 21, 39 b 32 ; I. 29, 45 a 35.
[3] *Anal. post.* I. 24, 85 a 16 etc.   [4] *Anal. prior.* I. 23, 40 b 25.
[5] *Anal. prior.* I. 23, 41 a 24.   [6] *ibid.* II. 14, 62 b 29.
[7] Proclus, p. 254, 22—27.

called *reductiones ad absurdum*. For it is the function of this method to upset something admitted as clear[1]."

8. Analysis and Synthesis.

It will be seen from the note on Eucl. XIII. 1 that the MSS. of the *Elements* contain definitions of *Analysis* and *Synthesis* followed by alternative proofs of XIII. 1—5 after that method. The definitions and alternative proofs are interpolated, but they have great historical interest because of the possibility that they represent an ancient method of dealing with these propositions, anterior to Euclid. The propositions give properties of a line cut "in extreme and mean ratio," and they are preliminary to the construction and comparison of the five regular solids. Now Pappus, in the section of his *Collection* dealing with the latter subject[2], says that he will give the comparisons between the five figures, the pyramid, cube, octahedron, dodecahedron and icosahedron, which have equal surfaces, " not by means of the so-called *analytical* inquiry, by which some of the ancients worked out the proofs, but by the synthetical method[3]...." The conjecture of Bretschneider that the matter interpolated in Eucl. XIII. is a survival of investigations due to Eudoxus has at first sight much to commend it[4]. In the first place, we are told by Proclus that Eudoxus "greatly added to the number of the theorems which Plato originated regarding *the section*, and employed in them the method of analysis[5]." It is obvious that "*the section*" was some particular section which by the time of Plato had assumed great importance; and the one section of which this can safely be said is that which was called the "golden section," namely, the division of a straight line in extreme and mean ratio which appears in Eucl. II. 11 and is therefore most probably Pythagorean. Secondly, as Cantor points out[6], Eudoxus was the founder of the theory of proportions in the form in which we find it in Euclid V., VI., and it was no doubt through meeting, in the course of his investigations, with proportions not expressible by whole numbers that he came to realise the necessity for a new theory of proportions which should be applicable to incommensurable as well as commensurable magnitudes. The "golden section" would furnish such a case. And it is even mentioned by Proclus in this connexion. He is explaining[7] that it is only in arithmetic that all quantities bear "rational" ratios (ῥητὸς λόγος) to one another, while in geometry there are "irrational" ones (ἄρρητος) as well. "Theorems about sections like those in Euclid's second Book are common to both [arithmetic and geometry] *except that in which the straight line is cut in extreme and mean ratio*[8]."

---

[1] Proclus, p. 255, 8—26.
[2] Pappus, V. p. 410 sqq.
[3] *ibid.* pp. 410, 27—412, 2.
[4] Bretschneider, p. 168. See however Heiberg's recent suggestion (*Paralipomena zu Euklid* in *Hermes*, XXXVIII., 1903) that the author was Heron. The suggestion is based on a comparison with the remarks on analysis and synthesis quoted from Heron by an-Nairīzī (ed. Curtze, p. 89) at the beginning of his commentary on Eucl. Book II. On the whole, this suggestion commends itself to me more than that of Bretschneider.
[5] Proclus, p. 67, 6.
[6] Cantor, *Gesch. d. Math.* I₃, p. 241.
[7] Proclus, p. 60, 7—9.
[8] *ibid.* p. 60, 16—19.

The definitions of *Analysis* and *Synthesis* interpolated in Eucl. XIII. are as follows (I adopt the reading of B and V, the only intelligible one, for the second).

"**Analysis** is an assumption of that which is sought as if it were admitted < and the passage > through its consequences to something admitted (to be) true.

"**Synthesis** is an assumption of that which is admitted < and the passage > through its consequences to the finishing or attainment of what is sought."

The language is by no means clear and has, at the best, to be filled out.

Pappus has a fuller account[1]:

"The so-called ἀναλυόμενος ('Treasury of Analysis') is, to put it shortly, a special body of doctrine provided for the use of those who, after finishing the ordinary Elements, are desirous of acquiring the power of solving problems which may be set them involving (the construction of) lines, and it is useful for this alone. It is the work of three men, Euclid the author of the Elements, Apollonius of Perga, and Aristaeus the elder, and proceeds by way of analysis and synthesis.

"**Analysis** then takes that which is sought as if it were admitted and passes from it through its successive consequences to something which is admitted as the result of synthesis: for in analysis we assume that which is sought as if it were (already) done (γεγονός), and we inquire what it is from which this results, and again what is the antecedent cause of the latter, and so on, until by so retracing our steps we come upon something already known or belonging to the class of first principles, and such a method we call analysis as being solution backwards (ἀνάπαλιν λύσιν).

"But in **synthesis**, reversing the process, we take as already done that which was last arrived at in the analysis and, by arranging in their natural order as consequences what were before antecedents, and successively connecting them one with another, we arrive finally at the construction of what was sought; and this we call synthesis.

"Now analysis is of two kinds, the one directed to searching for the truth and called *theoretical*, the other directed to finding what we are told to find and called *problematical*. (1) In the *theoretical* kind we assume what is sought as if it were existent and true, after which we pass through its successive consequences, as if they too were true and established by virtue of our hypothesis, to something admitted: then (*a*), if that something admitted is true, that which is sought will also be true and the proof will correspond in the reverse order to the analysis, but (*b*), if we come upon something admittedly false, that which is sought will also be false. (2) In the *problematical* kind we assume that which is propounded as if it were known, after which we pass through its successive consequences, taking them as true, up to something admitted: if then (*a*) what is admitted is possible and obtainable, that is, what mathematicians call *given*, what was originally proposed will also be possible, and the proof will again correspond in

---

[1] Pappus, VII. pp. 634—6.

reverse order to the analysis, but if (b) we come upon something admittedly impossible, the problem will also be impossible."

The ancient Analysis has been made the subject of careful studies by several writers during the last half-century, the most complete being those of Hankel, Duhamel and Zeuthen; others by Ofterdinger and Cantor should also be mentioned[1].

The method is as follows. It is required, let us say, to prove that a certain proposition A is true. We assume as a hypothesis that A is true and, starting from this we find that, if A is true, a certain other proposition B is true ; if B is true, then C ; and so on until we arrive at a proposition K which is *admittedly* true. The object of the method is to enable us to infer, in the reverse order, that, since K is true, the proposition A originally assumed is true. Now Aristotle had already made it clear that false hypotheses might lead to a conclusion which is true. There is therefore a possibility of error unless a certain precaution is taken. While, for example, B may be a necessary consequence of A, it may happen that A is not a necessary consequence of B. Thus, in order that the reverse inference from the truth of K that A is true may be logically justified, it is necessary that each step in the chain of inferences should be unconditionally convertible. As a matter of fact, a very large number of theorems in elementary geometry are unconditionally convertible, so that in practice the difficulty in securing that the successive steps shall be convertible is not so great as might be supposed. But care is always necessary. For example, as Hankel says[2], a proposition may not be unconditionally convertible in the form in which it is generally quoted. Thus the proposition "The vertices of all triangles having a common base and constant vertical angle lie on a circle" cannot be converted into the proposition that "All triangles with common base and vertices lying on a circle have a constant vertical angle"; for this is only true if the further conditions are satisfied (1) that the circle passes through the extremities of the common base and (2) that only that part of the circle is taken as the locus of the vertices which lies on *one* side of the base. If these conditions are added, the proposition is unconditionally convertible. Or again, as Zeuthen remarks[3], K may be obtained by a series of inferences in which A or some other proposition in the series is only *apparently* used ; this would be the case e.g. when the method of modern algebra is being employed and the expressions on each side of the sign of equality have been inadvertently multiplied by some composite magnitude which is in reality equal to zero.

Although the above extract from Pappus does not make it clear that each step in the chain of argument must be convertible in the case taken, he almost implies this in the second part of the definition of Analysis where, instead of speaking of the consequences B, C...

[1] Hankel, *Zur Geschichte der Mathematik in Alterthum und Mittelalter*, 1874, pp. 137—150; Duhamel, *Des méthodes dans les sciences de raisonnement*, Part I., 3 ed., Paris, 1885, pp. 39—68; Zeuthen, *Geschichte der Mathematik im Altertum und Mittelalter*, 1896, pp. 92—104; Ofterdinger, *Beiträge zur Geschichte der griechischen Mathematik*, Ulm, 1860; Cantor, *Geschichte der Mathematik*, I_3, pp. 220—2.
[2] Hankel, p. 139.   [3] Zeuthen, p. 103.

successively following from A, he suddenly changes the expression and says that we inquire *what it is* (B) *from which* A *follows* (A being thus the consequence of B, instead of the reverse), and then what (viz. C) is the antecedent cause of B; and in practice the Greeks secured what was wanted by always insisting on the *analysis* being confirmed by subsequent *synthesis,* that is, they laboriously worked backwards the whole way from K to A, reversing the order of the analysis, which process would undoubtedly bring to light any flaw which had crept into the argument through the accidental neglect of the necessary precautions.

Reductio ad absurdum a variety of analysis.

In the process of analysis starting from the hypothesis that a proposition A is true and passing through B, C... as successive consequences we may arrive at a proposition K which, instead of being admittedly true, is either admittedly false or the contradictory of the original hypothesis A or of some one or more of the propositions B, C... intermediate between A and K. Now correct inference from a true proposition cannot lead to a false proposition; and in this case therefore we may at once conclude, without any inquiry whether the various steps in the argument are convertible or not, that the hypothesis A is false, for, if it were true, all the consequences correctly inferred from it would be true and no incompatibility could arise. This method of proving that a given hypothesis is *false* furnishes an indirect method of proving that a given hypothesis A is *true*, since we have only to take the *contradictory* of A and to prove that it is false. This is the method of *reductio ad absurdum,* which is therefore a variety of analysis. The contradictory of A, or not-A, will generally include more than one case and, in order to prove its falsity, each of the cases must be separately disposed of: e.g., if it is desired to prove that a certain part of a figure is *equal* to some other part, we take separately the hypotheses (1) that it is *greater*, (2) that it is *less*, and prove that each of these hypotheses leads to a conclusion either admittedly false or contradictory to the hypothesis itself or to some one of its consequences.

Analysis as applied to problems.

It is in relation to problems that the ancient analysis has the greatest significance, because it was the one general method which the Greeks used for solving all "the more abstruse problems" (τὰ ἀσαφέστερα τῶν προβλημάτων)[1].

We have, let us suppose, to construct a figure satisfying a certain set of conditions. If we are to proceed at all methodically and not by mere guesswork, it is first necessary to "analyse" those conditions. To enable this to be done we must get them clearly in our minds, which is only possible by assuming all the conditions to be actually fulfilled, in other words, by supposing the problem solved. Then we have to transform those conditions, by all the means which practice in such cases has taught us to employ, into other conditions which are necessarily fulfilled if the original conditions are, and to continue this

[1] Proclus, p. 242, 16, 17.

transformation until we at length arrive at conditions which we are in a position to satisfy[1]. In other words, we must arrive at some relation which enables us to *construct* a particular part of the figure which, it is true, has been hypothetically assumed and even drawn, but which nevertheless really requires to be *found* in order that the problem may be solved. From that moment the particular part of the figure becomes one of the *data*, and a fresh relation has to be found which enables a fresh part of the figure to be determined by means of the original data and the new one together. When this is done, the second new part of the figure also belongs to the data; and we proceed in this way until all the parts of the required figure are found[2]. The first part of the analysis down to the point of discovery of a relation which enables us to say that a certain new part of the figure not belonging to the original data is *given*, Hankel calls the *transformation*; the second part, in which it is proved that all the remaining parts of the figure are "given," he calls the *resolution*. Then follows the *synthesis*, which also consists of two parts, (1) the *construction*, in the order in which it has to be actually carried out, and in general following the course of the second part of the analysis, the *resolution*; (2) the *demonstration* that the figure obtained does satisfy all the given conditions, which follows the steps of the first part of the analysis, the *transformation*, but in the reverse order. The second part of the analysis, the *resolution*, would be much facilitated and shortened by the existence of a systematic collection of *Data* such as Euclid's book bearing that title, consisting of propositions proving that, if in a figure certain parts or relations are *given*, other parts or relations are also *given*. As regards the first part of the analysis, the *transformation*, the usual rule applies that every step in the chain must be unconditionally convertible; and any failure to observe this condition will be brought to light by the subsequent synthesis. The second part, the *resolution*, can be directly turned into the *construction* since that only is *given* which can be constructed by the means provided in the *Elements*.

It would be difficult to find a better illustration of the above than the example chosen by Hankel from Pappus[3].

Given a circle ABC *and two points* D, E *external to it, to draw straight lines* DB, EB *from* D, E *to a point* B *on the circle such that, if* DB, EB *produced meet the circle again in* C, A, AC *shall be parallel to* DE.

Analysis.

Suppose the problem solved and the tangent at A drawn, meeting ED produced in F.

(Part I.   *Transformation.*)

Then, since $A\overset{\frown}{C}$ is parallel to DE, the angle at C is equal to the angle CDE.

But, since FA is a tangent, the angle at C is equal to the angle FAE.

Therefore the angle FAE is equal to the angle CDE, whence A, B, D, F are concyclic.

[1] Zeuthen, p. 93.   [2] Hankel, p. 141.   [3] Pappus, VII. pp. 830—2.

Therefore the rectangle *AE, EB* is equal to the rectangle *FE, ED.*

(Part II. *Resolution.*)

But the rectangle *AE, EB* is given, because it is equal to the square on the tangent from *E*.

Therefore the rectangle *FE, ED* is given ;

and, since *ED* is given, *FE* is given (in length).   [*Data,* 57.]

But *FE* is given in position also, so that *F* is also given.   [*Data,* 27.]

Now *FA* is the tangent from a given point *F* to a circle *ABC* given in position;

therefore *FA* is given in position and magnitude.   [*Data,* 90.]

And *F* is given ; therefore *A* is given.

But *E* is also given; therefore the straight line *AE* is given in position.   [*Data,* 26.]

And the circle *ABC* is given in position ;

therefore the point *B* is also given.   [*Data,* 25.]

But the points *D, E* are also given ;

therefore the straight lines *DB, BE* are also given in position.

Synthesis.

(Part I. *Construction.*)

Suppose the circle *ABC* and the points *D, E* given.

Take a rectangle contained by *ED* and by a certain straight line *EF* equal to the square on the tangent to the circle from *E*.

From *F* draw *FA* touching the circle in *A* ; join *ABE* and then *DB*, producing *DB* to meet the circle at *C*. Join *AC*.

I say then that *AC* is parallel to *DE*.

(Part II. *Demonstration.*)

Since, by hypothesis, the rectangle *FE, ED* is equal to the square on the tangent from *E*, which again is equal to the rectangle *AE, EB*, the rectangle *AE, EB* is equal to the rectangle *FE, ED*.

Therefore *A, B, D, F* are concyclic,

whence the angle *FAE* is equal to the angle *BDE*.

But the angle *FAE* is equal to the angle *ACB* in the alternate segment ;

therefore the angle *ACB* is equal to the angle *BDE*.

Therefore *AC* is parallel to *DE*.

In cases where a διορισμός is necessary, i.e. where a solution is only possible under certain conditions, the analysis will enable those conditions to be ascertained. Sometimes the διορισμός is stated and proved at the end of the analysis, e.g. in Archimedes, *On the Sphere and Cylinder,* II. 7 ; sometimes it is stated in that place and the proof postponed till after the end of the synthesis, e.g. in the solution of the problem subsidiary to *On the Sphere and Cylinder,* II. 4, preserved in Eutocius' commentary on that proposition. The analysis should also enable us to determine the number of solutions of which the problem is susceptible.

§ 7. THE DEFINITIONS.

General. "Real" and "Nominal" Definitions.

It is necessary, says Aristotle, whenever any one treats of any whole subject, to divide the genus into its primary constituents, those which are indivisible in species respectively: e.g. number must be divided into triad and dyad; then an attempt must be made in this way to obtain definitions, e.g. of a straight line, of a circle, and of a right angle[1].

The word for definition is ὅρος. The original meaning of this word seems to have been "boundary," "landmark." Then we have it in Plato and Aristotle in the sense of standard or determining principle ("id quo alicuius rei natura constituitur vel definitur," *Index Aristotelicus*)[2]; and closely connected with this is the sense of *definition*. Aristotle uses both ὅρος and ὁρισμός for definition, the former occurring more frequently in the *Topics*, the latter in the *Metaphysics*.

Let us now first be clear as to what a definition does *not* do. There is nothing in connexion with definitions which Aristotle takes more pains to emphasise than that a definition asserts nothing as to the *existence* or *non-existence* of the thing defined. It is an answer to the question *what* a thing is (τί ἐστι), and does not say *that* it is (ὅτι ἐστι). The *existence* of the various things defined has to be *proved*, except in the case of a few primary things in each science, the existence of which is indemonstrable and must be *assumed* among the first principles of each science; e.g. points and lines in geometry must be *assumed* to exist, but the existence of everything else must be *proved*. This is stated clearly in the long passage quoted above under First Principles[3]. It is reasserted in such passages as the following. "The (answer to the question) *what is a man* and *the fact that a man exists* are different things[4]." "It is clear that, even according to the view of definitions now current, those who define things do not prove that they exist[5]." "We say that it is by *demonstration* that we must show that everything exists, except essence (εἰ μὴ οὐσία εἴη). But the *existence* of a thing is never essence; for the *existent* is not a genus. Therefore there must be demonstration that a thing exists. Thus, *what is meant by triangle* the geometer assumes, but that it exists he has to prove[6]." "Anterior knowledge of two sorts is necessary: for it is necessary to presuppose, with regard to some things, that they *exist*; in other cases it is necessary to understand *what* the thing described is, and in other cases it is necessary to do both. Thus, with the fact that one of two contradictories must be true, we must know that it exists (is true);

---

[1] *Anal. post.* II. 13, 96 b 15.

[2] Cf. *De anima*, I. 2, 404 a 9, where "breathing" is spoken of as the ὅρος of "life," and the many passages in the *Politics* where the word is used to denote that which gives its *special character* to the several forms of government (virtue being the ὅρος of aristocracy, wealth of oligarchy, liberty of democracy, 1294 a 10); Plato, *Republic*, VIII. 551 C.

[3] *Anal. post.* I. 10, 76 a 31 sqq.
[4] *ibid.* II. 7, 92 b 10.
[5] *ibid.* 92 b 19.
[6] *ibid.* 92 b 12 sqq.

of the triangle we must know that it means such and such a thing ; of the unit we must know both what it means and that it exists[1]." What is here so much insisted on is the very fact which Mill pointed out in his discussion of earlier views of Definitions, where he says that the so-called *real* definitions or definitions of *things* do not constitute a different kind of definition from *nominal* definitions, or definitions of *names* ; the former is simply the latter *plus* something else, namely a covert assertion that the thing defined exists. "This covert assertion is not a definition but a postulate. The definition is a mere identical proposition which gives information only about the use of language, and from which no conclusion affecting matters of fact can possibly be drawn. The accompanying postulate, on the other hand, affirms a fact which may lead to consequences of every degree of importance. It affirms the actual or possible existence of Things possessing the combination of attributes set forth in the definition : and this, if true, may be foundation sufficient on which to build a whole fabric of scientific truth[2]." This statement really adds nothing to Aristotle's doctrine[3] : it has even the slight disadvantage, due to the use of the word "postulate" to describe "the covert assertion" in all cases, of not definitely pointing out that there are cases where existence has to be *proved* as distinct from those where it must be *assumed*. It is true that the existence of a definiend may have to be taken for granted provisionally until the time comes for proving it ; but, so far as regards any case where existence must be proved sooner or later, the provisional assumption would be for Aristotle, not a *postulate*, but a *hypothesis*. In modern times, too, Mill's account of the true distinction between *real* and *nominal* definitions had been fully anticipated by Saccheri[4], the editor of *Euclides ab omni naevo vindicatus* (1733), famous in the history of non-Euclidean geometry). In his *Logica Demonstrativa* (to which he also refers in his Euclid) Saccheri lays down the clear distinction between what he calls *definitiones quid nominis* or *nominales*, and *definitiones quid rei* or *reales*, namely that the former are only intended to explain the meaning

---

[1] *Anal. post.* I. 1, 71 a 11 sqq.   [2] Mill's *System of Logic*, Bk. I. ch. viii.

[3] It is true that it was in opposition to "the ideas of most of the *Aristotelian logicians*" (rather than of Aristotle himself) that Mill laid such stress on his point of view. Cf. his observation : "We have already made, and shall often have to repeat, the remark, that the philosophers who overthrew Realism by no means got rid of the consequences of Realism, but retained long afterwards, in their own philosophy, numerous propositions which could only have a rational meaning as part of a Realistic system. It had been handed down from Aristotle, and probably from earlier times, as an obvious truth, that the science of geometry is deduced from definitions. This, so long as a definition was considered to be a proposition 'unfolding the nature of the thing,' did well enough. But Hobbes followed and rejected utterly the notion that a definition declares the nature of the thing, or does anything but state the meaning of a name ; yet he continued to affirm as broadly as any of his predecessors that the ἀρχαί, *principia*, or original premisses of mathematics, and even of all science, are definitions ; producing the singular paradox that systems of scientific truth, nay, all truths whatever at which we arrive by reasoning, are deduced from the arbitrary conventions of mankind concerning the signification of words." Aristotle was guilty of no such paradox ; on the contrary, he exposed it as plainly as did Mill.

[4] This has been fully brought out in two papers by G. Vailati, *La teoria Aristotelica della definizione* (*Rivista di Filosofia e scienze affini*, 1903), and *Di un' opera dimenticata del P. Gerolamo Saccheri* ("Logica Demonstrativa," 1697) (in *Rivista Filosofica*, 1903).

that is to be attached to a given term, whereas the latter, besides declaring the meaning of a word, affirm at the same time the existence of the thing defined or, in geometry, the possibility of constructing it. The *definitio quid nominis* becomes a *definitio quid rei* "by means of a *postulate*, or when we come to the question whether the thing *exists* and it is answered affirmatively[1]." *Definitiones quid nominis* are in themselves quite arbitrary, and neither require nor are capable of proof; they are merely provisional and are only intended to be turned as quickly as possible into *definitiones quid rei*, either (1) by means of a postulate in which it is asserted or conceded that what is defined exists or can be constructed, e.g. in the case of *straight lines* and *circles*, to which Euclid's first three postulates refer, or (2) by means of a demonstration reducing the construction of the figure defined to the successive carrying-out of a certain number of those elementary constructions, the possibility of which is *postulated*. Thus *definitiones quid rei* are in general obtained as the result of a series of demonstrations. Saccheri gives as an instance the construction of a square in Euclid I. 46. Suppose that it is objected that Euclid had no right to define a square, as he does at the beginning of the Book, when it was not certain that such a figure exists in nature; the objection, he says, could only have force if, before proving and making the construction, Euclid had assumed the aforesaid figure as given. That Euclid is not guilty of this error is clear from the fact that he never presupposes the existence of the square as defined until after I. 46.

Confusion between the *nominal* and the *real* definition as thus described, i.e. the use of the former in demonstration before it has been turned into the latter by the necessary proof that the thing defined exists, is according to Saccheri one of the most fruitful sources of illusory demonstration, and the danger is greater in proportion to the "complexity" of the definition, i.e. the number and variety of the attributes belonging to the thing defined. For the greater is the possibility that there may be among the attributes some that are *incompatible*, i.e. the simultaneous presence of which in a given figure can be proved, by means of *other* postulates etc. forming part of the basis of the science, to be impossible.

The same thought is expressed by Leibniz also. "If," he says, "we give any definition, and it is not clear from it that the idea, which we ascribe to the thing, is possible, we cannot rely upon the demonstrations which we have derived from that definition, because, if that idea by chance involves a contradiction, it is possible that even contradictories may be true of it at one and the same time, and thus our demonstrations will be useless. Whence it is clear that definitions are not arbitrary. And this is a secret which is hardly sufficiently known[2]." Leibniz' favourite illustration was the "regular polyhedron with ten faces," the impossibility of which is not obvious at first sight.

---

[1] "Definitio *quid nominis* nata est evadere definitio *quid rei* per *postulatum* vel dum venitur ad quaestionem *an est* et respondetur affirmative."

[2] *Opuscules et fragments inédits de Leibniz*, Paris, Alcan, 1903, p. 431. Quoted by Vailati.

It need hardly be added that, speaking generally, Euclid's definitions, and his use of them, agree with the doctrine of Aristotle that the definitions themselves say nothing as to the existence of the things defined, but that the existence of each of them must be proved or (in the case of the "principles") *assumed.* In geometry, says Aristotle, the existence of points and lines only must be assumed, the existence of the rest being proved. Accordingly Euclid's first three postulates declare the possibility of constructing straight lines and circles (the only "lines" except straight lines used in the *Elements*). Other things are defined and afterwards constructed and proved to exist: e.g. in Book I., Def. 20, it is explained what is meant by an equilateral triangle ; then (I. 1) it is proposed to construct it, and, when constructed, it is proved to agree with the definition. When a square is defined (I. Def. 22), the question whether such a thing really exists is left open until, in I. 46, it is proposed to construct it and, when constructed, it is proved to satisfy the definition[1]. Similarly with the right angle (I. Def. 10, and I. 11) and parallels (I. Def. 23, and I. 27—29). The greatest care is taken to exclude mere presumption and imagination. The transition from the subjective definition of names to the objective definition of things is made, in geometry, by means of *constructions* (the first principles of which are postulated), as in other sciences it is made by means of experience[2].

Aristotle's requirements in a definition.

We now come to the positive characteristics by which, according to Aristotle, scientific definitions must be marked.

First, the different attributes in a definition, when taken separately, cover more than the notion defined, but the combination of them does not. Aristotle illustrates this by the "triad," into which enter the several notions of number, odd and prime, and the last "in both its two senses (*a*) of not being measured by any (other) number (ὡς μὴ μετρεῖσθαι ἀριθμῷ) and (*b*) of not being obtainable by adding numbers together" (ὡς μὴ συγκεῖσθαι ἐξ ἀριθμῶν), a unit not being a number. Of these attributes some are present in all other odd numbers as well, while the last [primeness in the second sense] belongs also to the dyad, but in nothing but the triad are they *all* present[3]." The fact can be equally well illustrated from geometry. Thus, e.g. into the definition of a square (Eucl. I., Def. 22) there enter the several notions of figure, four-sided, equilateral, and right-angled, each of which covers more than the notion into which *all* enter as attributes[4].

Secondly, a definition must be expressed in terms of things which are prior to, and better known than, the things defined[5]. This is

---

[1] Trendelenburg, *Elementa Logices Aristoteleae,* § 50.

[2] Trendelenburg, *Erläuterungen zu den Elementen der aristotelischen Logik,* 3 ed. p. 107. On construction as proof of existence in ancient geometry cf. H. G. Zeuthen, *Die geometrische Construction als "Existenzbeweis" in der antiken Geometrie* (in *Mathematische Annalen,* 47. Band).

[3] *Anal. post.* II. 13, 96 a 33—b 1.

[4] Trendelenburg, *Erläuterungen,* p. 108.   [5] *Topics* VI. 4, 141 a 26 sqq.

clear, since the object of a definition is to give us knowledge of the thing defined, and it is by means of things prior and better known that we acquire fresh knowledge, as in the course of demonstrations. But the terms "prior" and "better known" are, as usual susceptible of two meanings; they may mean (1) *absolutely* or *logically* prior and better known, or (2) better known *relatively to us*. In the absolute sense, or from the standpoint of reason, a point is better known than a line, a line than a plane, and a plane than a solid, as also a unit is better known than number (for the unit is prior to, and the first principle of, any number). Similarly, in the absolute sense, a letter is prior to a syllable. But the case is sometimes different relatively to us; for example, a solid is more easily realised by the senses than a plane, a plane than a line, and a line than a point. Hence, while it is more scientific to begin with the *absolutely* prior, it may, perhaps, be permissible, in case the learner is not capable of following the scientific order, to explain things by means of what is more intelligible *to him*. "Among the definitions framed on this principle are those of the point, the line and the plane; all these explain what is prior by means of what is posterior, for the point is described as the extremity of a line, the line of a plane, the plane of a solid." But, if it is asserted that such definitions by means of things which are more intelligible relatively only to a particular individual are really definitions, it will follow that there may be many definitions of the same thing, one for each individual for whom a thing is being defined, and even different definitions for one and the same individual at different times, since at first sensible objects are more intelligible, while to a better trained mind they become less so. It follows therefore that a thing should be defined by means of the absolutely prior and not the relatively prior, in order that there may be one sole and immutable definition. This is further enforced by reference to the requirement that a good definition must state the *genus* and the *differentiae*, for these are among the things which are, in the absolute sense, better known than, and prior to, the species (τῶν ἁπλῶς γνωριμωτέρων καὶ προτέρων τοῦ εἴδους ἐστίν). For to destroy the genus and the differentia is to destroy the species, so that the former are *prior* to the species; they are also *better known*, for, when the species is known, the genus and the differentia must necessarily be known also, e.g. he who knows "man" must also know "animal" and "land-animal," but it does not follow, when the genus and differentia are known, that the species is known too, and hence the species is less known than they are[1]. It may be frankly admitted that the scientific definition will require superior mental powers for its apprehension; and the extent of its use must be a matter of discretion. So far Aristotle; and we have here the best possible explanation why Euclid supplemented his definition of a point by the statement in I. Def. 3 that *the extremities of a line are points* and his definition of a surface by I. Def. 6 to the effect that *the extremities of a surface are lines*. The supplementary expla-

---

[1] *Topics* VI. 4, 141 b 25—34.

nations do in fact enable us to arrive at a better understanding of the
formal definitions of a point and a line respectively, as is well explained by Simson in his note on Def. 1. Simson says, namely, that
we must consider a solid, that is, a magnitude which has length,
breadth and thickness, in order to understand aright the definitions of
a point, a line and a surface. Consider, for instance, the boundary
common to two solids which are contiguous or the boundary which
divides one solid into two contiguous parts; this boundary is a surface.
We can prove that it has no thickness by taking away either solid,
when it remains the boundary of the other; for, if it had thickness, the
thickness must either be a part of one solid or of the other, in which
case to take away one or other solid would take away the thickness
and therefore the boundary itself: which is impossible. Therefore
the boundary or the surface has no thickness. In exactly the same
way, regarding a line as the boundary of two contiguous surfaces, we
prove that the line has no breadth; and, lastly, regarding a point as
the common boundary or extremity of two lines, we prove that a
point has no length, breadth or thickness.

Aristotle on unscientific definitions.

Aristotle distinguishes three kinds of definition which are un-
scientific because founded on what is *not* prior ($\mu\grave{\eta}$ $\grave{\epsilon}\kappa$ $\pi\rho o\tau\acute{\epsilon}\rho\omega\nu$). The
first is a definition of a thing by means of its opposite, e.g. of "good"
by means of "bad"; this is wrong because opposites are naturally
evolved together, and the knowledge of opposites is not uncommonly
regarded as one and the same, so that one of the two opposites
cannot be better known than the other. It is true that, in some
cases of opposites, it would appear that no other sort of definition is
possible: e.g. it would seem impossible to define double apart from the
half and, generally, this would be the case with things which in their
very nature ($\kappa\alpha\theta^{\prime}$ $\alpha\grave{v}\tau\acute{a}$) are *relative* terms ($\pi\rho\acute{o}\varsigma$ $\tau\iota$ $\lambda\acute{\epsilon}\gamma\epsilon\tau\alpha\iota$), since one
cannot be known without the other, so that in the notion of either the
other must be comprised as well[1]. The *second* kind of definition
which is based on what is not prior is that in which there is a
complete circle through the unconscious use in the definition itself of
the notion to be defined though not of the name[2]. Trendelenburg
illustrates this by two current definitions, (1) that of magnitude as
that which can be increased or diminished, which is bad because the
positive and negative comparatives "more" and "less" presuppose
the notion of the positive "great," (2) the famous Euclidean definition
of a straight line as that which "lies evenly with the points on itself"
($\grave{\epsilon}\xi$ $\acute{\iota}\sigma ov$ $\tauο\hat{\iota}\varsigma$ $\grave{\epsilon}\phi^{\prime}$ $\grave{\epsilon}\alpha v\tau\hat{\eta}\varsigma$ $\sigma\eta\mu\epsilon\acute{\iota}o\iota\varsigma$ $\kappa\epsilon\hat{\iota}\tau\alpha\iota$), where "lies evenly" can only
be understood with the aid of the very notion of a straight line which is
to be defined[3]. The *third* kind of vicious definition from that which
is not prior is the definition of one of two coordinate species by means
of its coordinate ($\grave{\alpha}\nu\tau\iota\delta\iota\eta\rho\eta\mu\acute{\epsilon}\nu o\nu$), e.g. a definition of "odd" as that
which exceeds the even by a unit (the second alternative in Eucl. VII.
Def. 7); for "odd" and "even" are coordinates, being *differentiae* of

---

[1] *Topics* VI. 4, 142 a 22—31.   [2] *ibid.* 142 a 34—b 6.
[3] Trendelenburg, *Erläuterungen*, p. 115.

number[1]. This third kind is similar to the first. Thus, says Trendelenburg, it would be wrong to define a *square* as "a *rectangle*
with equal sides."

Aristotle's third requirement.

A third general observation of Aristotle which is specially relevant
to geometrical definitions is that "to know *what* a thing is (τί ἐστιν) is
the same as knowing *why* it is (διὰ τί ἐστιν)[2]." " *What* is an eclipse?
A deprivation of light from the moon through the interposition of the
earth. *Why* does an eclipse take place? Or *why* is the moon
eclipsed? Because the light fails through the earth obstructing it.
What is harmony? A ratio of numbers in high or low pitch. *Why*
does the high-pitched harmonise with the low-pitched? Because
the high and the low have a numerical ratio to one another[3]." "We
seek the *cause* (τὸ διότι) when we are already in possession of the
fact (τὸ ὅτι). Sometimes they both become evident at the same time,
but at all events the cause cannot possibly be known [as a cause]
before the fact is known[4]." " It is impossible to know *what* a thing is
if we do not know *that* it is[5]" Trendelenburg paraphrases: "The
definition of the notion does not fulfil its purpose until it is made
genetic. It is the producing cause which first reveals the essence of
the thing.... .The nominal definitions of geometry have only a
provisional significance and are superseded as soon as they are made
genetic by means of construction." E.g. the genetic definition of a
parallelogram is evolved from Eucl. I. 31 (giving the construction for
parallels) and I. 33 about the lines joining corresponding ends of two
straight lines parallel and equal in length. Where existence is proved
by *construction*, the cause and the fact appear *together*[6].

Again, "it is not enough that the defining statement should set
forth the fact, as most definitions do; it should also contain and
present the cause; whereas in practice what is stated in the definition
is usually no more than a conclusion (συμπέρασμα). For example,
what is quadrature? The construction of an equilateral right-angled
figure equal to an oblong. But such a definition expresses merely the
conclusion. Whereas, if you say that quadrature is the discovery of a
mean proportional, then you state the reason[7]." This is better understood if we compare the statement elsewhere that "the cause is the
middle term, and this is what is sought in all cases[8]," and the illustration of this by the case of the proposition that the angle in a semicircle is a right angle. Here the middle term which it is sought to
establish by means of the figure is that the angle in the semi-circle is
equal to *the half of two right angles.* We have then the syllogism:
Whatever is half of two right angles is a right angle; the angle in a
semi-circle is the half of two right angles; therefore (*conclusion*) the
angle in a semi-circle is a right angle[9]. As with the demonstration, so

[1] *Topics* VI. 4, 142 b 7—10.
[2] *Anal. post.* II. 2, 90 a 31.
[3] *Anal. post.* II. 2, 90 a 15—21.
[4] *ibid.* II. 8, 93 a 17.
[5] *ibid.* 93 a 20.
[6] Trendelenburg, *Erläuterungen*, p. 110.
[7] *De anima* II. 2, 413 a 13—20.
[8] *Anal. post.* II. 2, 90 a 6.
[9] *ibid.* II. 11, 94 a 28.

it should be with the definition. A definition which is to show the *genesis* of the thing defined should contain the middle term or cause ; otherwise it is a mere statement of a conclusion. Consider, for instance, the definition of "quadrature" as "making a square equal in area to a rectangle with unequal sides." This gives no hint as to whether a solution of the problem is possible or how it is solved : but, if you add that to find the mean proportional between two given straight lines gives another straight line such that the square on it is equal to the rectangle contained by the first two straight lines, you supply the necessary middle term or cause[1].

Technical terms not defined by Euclid.

It will be observed that what is here defined, "quadrature" or "squaring" (τετραγωνισμός), is not a geometrical figure, or an attribute of such a figure or a part of a figure, but a technical term used to describe a certain problem. Euclid does not define such things ; but the fact that Aristotle alludes to this particular definition as well as to definitions of *deflection* (κεκλάσθαι) and of *verging* (νεύειν) seems to show that earlier text-books included among definitions explanations of a number of technical terms, and that Euclid deliberately omitted these explanations from his *Elements* as surplusage. Later the tendency was again in the opposite direction, as we see from the much expanded Definitions of Heron, which, for example, actually include a definition of a *deflected line* (κεκλασμένη γραμμή)[2]. Euclid uses the passive of κλᾶν occasionally[3], but evidently considered it unnecessary to explain such terms, which had come to bear a recognised meaning.

The mention too by Aristotle of a definition of *verging* (νεύειν) suggests that the problems indicated by this term were not excluded from elementary text-books before Euclid. The type of problem (νεῦσις) was that of placing a straight line across two lines, e.g. two straight lines, or a straight line and a circle, so that it shall *verge* to a given point (i.e. pass through it if produced) and at the same time the intercept on it made by the two given lines shall be of given length.

---

[1] Other passages in Aristotle may be quoted to the like effect : e.g. *Anal. post.* I. 2, 71 b 9 "We consider that we know a particular thing in the absolute sense, as distinct from the sophistical and incidental sense, when we consider that we know the cause on account of which the thing is, in the sense of knowing that it is the cause of that thing and that it cannot be otherwise," *ibid.* I. 13, 79 a 2 "For here to know the *fact* is the function of those who are concerned with sensible things, to know the *cause* is the function of the mathematician ; it is he who possesses the proofs of the causes, and often he does not know the fact." In view of such passages it is difficult to see how Proclus came to write (p. 202, 11) that Aristotle was the originator (Ἀριστοτέλους κατάρξαντος) of the idea of Amphinomus and others that geometry does not investigate the cause and the *why* (τὸ διὰ τί). To this Geminus replied that the investigation of the cause does, on the contrary, appear in geometry. "For how can it be maintained that it is not the business of the geometer to inquire for what reason, on the one hand, an infinite number of equilateral polygons are inscribed in a circle, but, on the other hand, it is not possible to inscribe in a sphere an infinite number of polyhedral figures, equilateral, equiangular, and made up of similar plane figures ? Whose business is it to ask this question and find the answer to it if it is not that of the geometer? Now when geometers reason *per impossibile* they are content to discover the property, but when they argue by direct proof, if such proof be only partial (ἐπὶ μέρους), this does not suffice for showing the cause ; if however it is general and applies to all like cases, the why (τὸ διὰ τί) is at once and concurrently made evident."

[2] Heron, Def. 12 (vol. IV. Heib. pp. 22–24).   [3] e.g. in III. 20 and in *Data* 89.

In general, the use of conics is required for the theoretical solution of these problems, or a mechanical contrivance for their practical solution[1]. Zeuthen, following Oppermann, gives reasons for supposing, not only that mechanical constructions were *practically* used by the older Greek geometers for solving these problems, but that they were *theoretically* recognised as a permissible means of solution when the solution could not be effected by means of the straight line and circle, and that it was only in later times that it was considered necessary to use conics in *every* case where that was possible[2]. Heiberg[3] suggests that the allusion of Aristotle to νεύσεις perhaps confirms this supposition, as Aristotle nowhere shows the slightest acquaintance with conics. I doubt whether this is a safe inference, since the problems of this type included in the elementary text-books might easily have been limited to those which could be solved by "plane" methods (i.e. by means of the straight line and circle). We know, e.g., from Pappus that Apollonius wrote two Books on *plane* νεύσεις[4]. But one thing is certain, namely that Euclid deliberately excluded this class of problem, doubtless as not being essential in a book of Elements.

Definitions not afterwards used.

Lastly, Euclid has definitions of some terms which he never afterwards uses, e.g. oblong (ἐτερόμηκες), rhombus, rhomboid. The "oblong" occurs in Aristotle; and it is certain that all these definitions are survivals from earlier books of Elements.

---

[1] Cf. the chapter on νεύσεις in *The Works of Archimedes*, pp. c—cxxii.
[2] Zeuthen, *Die Lehre von den Kegelschnitten im Altertum*, ch. 12, p. 262.
[3] Heiberg, *Mathematisches zu Aristoteles*, p. 16.
[4] Pappus VII. pp. 670—2.

BOOK I.

DEFINITIONS.

1. A **point** is that which has no part.
2. A **line** is breadthless length.
3. The extremities of a line are points.
4. A **straight line** is a line which lies evenly with the points on itself.
5. A **surface** is that which has length and breadth only.
6. The extremities of a surface are lines.
7. A **plane surface** is a surface which lies evenly with the straight lines on itself.
8. A **plane angle** is the inclination to one another of two lines in a plane which meet one another and do not lie in a straight line.
9. And when the lines containing the angle are straight, the angle is called **rectilineal.**
10. When a straight line set up on a straight line makes the adjacent angles equal to one another, each of the equal angles is **right,** and the straight line standing on the other is called a **perpendicular** to that on which it stands.
11. An **obtuse angle** is an angle greater than a right angle.
12. An **acute angle** is an angle less than a right angle.
13. A **boundary** is that which is an extremity of anything.
14. A **figure** is that which is contained by any boundary or boundaries.
15. A **circle** is a plane figure contained by one line such that all the straight lines falling upon it from one point among those lying within the figure are equal to one another ;

16.   And the point is called the **centre** of the circle.

17.   A **diameter** of the circle is any straight line drawn through the centre and terminated in both directions by the circumference of the circle, and such a straight line also bisects the circle.

18.   A **semicircle** is the figure contained by the diameter and the circumference cut off by it.   And the centre of the semicircle is the same as that of the circle.

19.   **Rectilineal figures** are those which are contained by straight lines, **trilateral** figures being those contained by three, **quadrilateral** those contained by four, and **multi-lateral** those contained by more than four straight lines.

20.   Of trilateral figures, an **equilateral triangle** is that which has its three sides equal, an **isosceles triangle** that which has two of its sides alone equal, and a **scalene triangle** that which has its three sides unequal.

21.   Further, of trilateral figures, a **right-angled tri-angle** is that which has a right angle, an **obtuse-angled triangle** that which has an obtuse angle, and an **acute-angled triangle** that which has its three angles acute.

22.   Of quadrilateral figures, a **square** is that which is both equilateral and right-angled ; an **oblong** that which is right-angled but not equilateral ; a **rhombus** that which is equilateral but not right-angled ; and a **rhomboid** that which has its opposite sides and angles equal to one another but is neither equilateral nor right-angled.   And let quadrilaterals other than these be called **trapezia**.

23.   **Parallel** straight lines are straight lines which, being in the same plane and being produced indefinitely in both directions, do not meet one another in either direction.

POSTULATES.

Let the following be postulated :

1.   To draw a straight line from any point to any point.

2.   To produce a finite straight line continuously in a straight line.

3.   To describe a circle with any centre and distance.

4.   That all right angles are equal to one another.

5. That, if a straight line falling on two straight lines make the interior angles on the same side less than two right angles, the two straight lines, if produced indefinitely, meet on that side on which are the angles less than the two right angles.

COMMON NOTIONS.

1. Things which are equal to the same thing are also equal to one another.

2. If equals be added to equals, the wholes are equal.

3. If equals be subtracted from equals, the remainders are equal.

[7] 4. Things which coincide with one another are equal to one another.

[8] 5. The whole is greater than the part.

DEFINITION 1.

Σημεῖόν ἐστιν, οὗ μέρος οὐθέν.
A point *is that which has no part.*

An exactly parallel use of μέρος (ἐστί) in the singular is found in Aristotle, *Metaph.* 1035 b 32 μέρος μὲν οὖν ἐστι καὶ τοῦ εἴδους, literally "There is a *part* even of the form"; Bonitz translates as if the plural were used, "Theile giebt es," and the meaning is simply "even the form is *divisible* (into parts)." Accordingly it would be quite justifiable to translate in this case "A point is that which is *indivisible into parts*."

Martianus Capella (5th C. A.D.) alone or almost alone translated differently, "Punctum est cuius pars *nihil* est," "a point is that a part of which is *nothing.*" Notwithstanding that Max Simon (*Euclid und die sechs planimetrischen Bücher,* 1901) has adopted this translation (on grounds which I shall presently mention), I cannot think that it gives any sense. If a part of a point is *nothing,* Euclid might as well have said that a point is *itself* "nothing," which of course he does not do.

Pre-Euclidean definitions.

It would appear that this was not the definition given in earlier text-books; for Aristotle (*Topics* VI. 4, 141 b 20), in speaking of "*the* definitions" of point, line, and surface, says that they *all* define the prior by means of the posterior, a point as an extremity of a line, a line of a surface, and a surface of a solid.

The first definition of a point of which we hear is that given by the Pythagoreans (cf. Proclus, p. 95, 21), who defined it as a "monad having position" or "with position added" (μονὰς προσλαβοῦσα θέσιν). It is frequently used by Aristotle, either in this exact form (cf. *De anima* I. 4, 409 a 6) or its equivalent: e.g. in *Metaph.* 1016 b 24 he says that that which is indivisible every way in respect of magnitude and *quâ* magnitude but has not position is a *monad,* while that which is similarly indivisible and has position is a *point.*

Plato appears to have objected to this definition. Aristotle says (*Metaph.*

992 a 20) that he objected "to this genus [that of points] as being a geometrical fiction (γεωμετρικὸν δόγμα), and called a point the beginning of a line (ἀρχὴ γραμμῆς), while again he frequently spoke of 'indivisible lines.'" To which Aristotle replies that even "indivisible lines" must have extremities, so that the same argument which proves the existence of *lines* can be used to prove that *points* exist. It would appear therefore that, when Aristotle objects to the definition of a point as the extremity of a line (πέρας γραμμῆς) as unscientific (*Topics* VI. 4, 141 b 21), he is aiming at Plato. Heiberg conjectures (*Mathematisches zu Aristoteles*, p. 8) that it was due to Plato's influence that the word for "point" generally used by Aristotle (στιγμή) was replaced by σημεῖον (the regular term used by Euclid, Archimedes and later writers), the latter term (= *nota*, a conventional mark) probably being considered more suitable than στιγμή (a *puncture*) which might appear to claim greater *reality* for a point.

Aristotle's conception of a point as that which is indivisible and has position is further illustrated by such observations as that a point is not a *body* (*De caelo* II. 13, 296 a 17) and has no *weight* (*ibid.* III. 1, 299 a 30); again, we can make no distinction between a point and the *place* (τόπος) where it is (*Physics* IV. 1, 209 a 11). He finds the usual difficulty in accounting for the transition from the indivisible, or infinitely small, to the finite or divisible magnitude. A point being *indivisible*, no accumulation of points, however far it may be carried, can give us anything divisible, whereas of course a line is a divisible magnitude. Hence he holds that points cannot make up anything continuous like a line, point cannot be continuous with point (οὐ γάρ ἐστιν ἐχόμενον σημεῖον σημείου ἢ στιγμὴ στιγμῆς, *De gen. et corr.* I. 2, 317 a 10), and a line is not *made up* of points (οὐ σύγκειται ἐκ στιγμῶν, *Physics* IV. 8, 215 b 19). A point, he says, is like the *now* in time: *now* is indivisible and is not a *part* of time, it is only the beginning or end, or a division, of time, and similarly a point may be an extremity, beginning or division of a line, but is not *part* of it or of magnitude (cf. *De caelo* III. 1, 300 a 14, *Physics* IV. 11, 220 a 1—21, VI. 1, 231 b 6 sqq.). It is only by *motion* that a point can generate a line (*De anima* I. 4, 409 a 4) and thus be the origin of magnitude.

Other ancient definitions.

According to an-Nairīzī (ed. Curtze, p. 3) one "Herundes" (not so far identified) defined a point as "the indivisible beginning of all magnitudes," and Posidonius as "an extremity which has no dimension, or an extremity of a line."

Criticisms by commentators.

Euclid's definition itself is of course practically the same as that which Aristotle's frequent allusions show to have been then current, except that it omits to say that the point must have position. Is it then sufficient, seeing that there are other things which are without parts or indivisible, e.g. the *now* in time, and the *unit* in number? Proclus answers (p. 93, 18) that the point is the only thing *in the subject-matter of geometry* that is indivisible. Relatively therefore to the particular science the definition is sufficient. Secondly, the definition has been over and over again criticised because it is purely negative. Proclus' answer to this is (p. 94, 10) that negative descriptions are appropriate to first principles, and he quotes Parmenides as having described his first and last cause by means of negations merely. Aristotle too admits that it may sometimes be necessary for one framing a definition to use negations, e.g. in defining privative terms such as "blind"; and he seems to accept as proper the negative element in the definition of a point, since he says (*De anima* III. 6, 430 b 20) that "the point and every division [e.g. in a length or in a period of time], and that which is indivisible in this sense, is exhibited as privation (δηλοῦται ὡς στέρησις)."

Simplicius (quoted by an-Nairīzī) says that "a point is the beginning of magnitudes and that from which they grow; it is also the only thing which, having position, is not divisible." He, like Aristotle, adds that it is by its *motion* that a point can generate a magnitude: the particular magnitude can only be "of one dimension," viz. a line, since the point does not "spread itself" (dimittat). Simplicius further observes that Euclid defined a point negatively because it was arrived at by detaching surface from body, line from surface, and finally point from line. "Since then body has three dimensions it follows that a point [arrived at after successively eliminating all three dimensions] has *none of the dimensions*, and has no part." This of course reappears in modern treatises (cf. Rausenberger, *Elementar-geometrie des Punktes, der Geraden und der Ebene*, 1887, p. 7).

An-Nairīzī adds an interesting observation. "If any one seeks to know the essence of a point, a thing more simple than a line, let him, in the sensible world, think of the centre of the universe and the *poles*." But there is nothing new under the sun: the same idea is mentioned, in an Aristotelian treatise, in controverting those who imagine that the poles have some influence in the motion of the sphere, "when the poles have no magnitude but are extremities and points" (*De motu animalium* 3, 699 a 21).

Modern views.

In the new geometry represented by the excellent treatises which start from new systems of postulates or axioms, the result of the profound study of the fundamental principles of geometry during recent years (I need only mention the names of Pasch, Veronese, Enriques and Hilbert), points come before lines, but the vain effort to define them *a priori* is not made; instead of this, the nearest material things in nature are mentioned as illustrations, with the remark that it is from them that we can get the abstract idea. Cf. the full statement as regards the notion of a point in Weber and Wellstein, *Encyclopädie der elementaren Mathematik*, II., 1905, p. 9. "This notion is evolved from the notion of the real or supposed *material* point by the process of limits, i.e. by an act of the mind which sets a term to a series of presentations in itself unlimited. Suppose a grain of sand or a mote in a sunbeam, which continually becomes smaller and smaller. In this way vanishes more and more the possibility of determining still smaller atoms in the grain of sand, and there is evolved, so we say, with growing certainty, the presentation of the point as a definite position in space which is one and is incapable of further division. But this view is untenable; we have, it is true, some idea how the grain of sand gets smaller and smaller, but only so long as it remains just visible; after that we are completely in the dark, and we cannot see or imagine the further diminution. That this procedure comes to an end is unthinkable; that nevertheless there exists a term beyond which it cannot go, we must believe or postulate without ever reaching it. . . . It is a pure act of *will*, not of the understanding." Max Simon observes similarly (*Euclid*, p. 25) "The notion 'point' belongs to the limit-notions (Grenzbegriffe), the necessary conclusions of continued, and in themselves unlimited, series of presentations." He adds, "The point is the limit of localisation; if this is more and more energetically continued, it leads to the limit-notion 'point,' better 'position,' which at the same time involves a change of notion. Content of *space* vanishes, relative *position* remains. 'Point' then, according to our interpretation of Euclid, is the extremest limit of that which we can still think of (not observe) as a *spatial* presentation, and if we go further than that, not only does extension cease but even relative *place*, and in this sense the 'part' is *nothing*." I confess I think that even the meaning which Simon intends to convey is better expressed by "it has *no* part" than by "the part is nothing," since to take a "part" of a thing in Euclid's sense of the result of a simple division, corresponding to an arithmetical fraction, would not be to change the *notion* from that of the thing divided to an entirely different one.

DEFINITION 2.

Γραμμὴ δὲ μῆκος ἀπλατές.

A line is breadthless length.

This definition may safely be attributed to the Platonic School, if not to Plato himself. Aristotle (*Topics* VI. 6, 143 b 11) speaks of it as open to objection because it " divides the genus by negation," length being necessarily either breadthless or possessed of breadth ; it would seem however that the objection was only taken in order to score a point against the Platonists, since he says (*ibid.* 143 b 29) that the argument is " of service *only* against those who assert that the genus [sc. length] is one numerically, that is, those who assume *ideas*," e.g. the idea of length (αὐτὸ μῆκος) which they regard as a genus : for if the genus, being one and self-existent, could be divided into two species, one of which asserts what the other denies, it would be self-contradictory (Waitz).

Proclus (pp. 96, 21—97, 3) observes that, whereas the definition of a point is merely negative, the line introduces the first "dimension," and so its definition is to this extent positive, while it has also a negative element which denies to it the other "dimensions" (διαστάσεις). The negation of both breadth and depth is involved in the single expression "breadthless" (ἀπλατές), since everything that is without breadth is also destitute of depth, though the converse is of course not true.

Alternative definitions.

The alternative definition alluded to by Proclus, μέγεθος ἐφ' ἓν διαστατόν " magnitude in one dimension " or, better perhaps, " magnitude extended one way " (since διάστασις as used with reference to line, surface and solid scarcely corresponds to our use of " dimension " when we speak of " one," " two," or " three dimensions "), is attributed by an-Nairīzī to " Heromides," who must presumably be the same as " Herundes," to whom he attributes a certain definition of a point. It appears however in substance in Aristotle, though Aristotle does not use the adjective διαστατόν, nor does he apparently use διάστασις except of *body* as having *three* " dimensions " or "having dimension (or extension) *all* ways (πάντῃ)," the "dimensions" being in his view (1) up and down, (2) before and behind, and (3) right and left, and "up" being the principle or beginning of *length*, "right" of *breadth*, and "before" of *depth* (*De caelo* II. 2, 284 b 24). A line is, according to Aristotle, a magnitude " *divisible* in one way only " (μοναχῇ διαιρετόν), in contrast to a magnitude divisible in *two* ways (διχῇ διαιρετόν), or a surface, and a magnitude divisible "in all or in three ways " (πάντῃ καὶ τριχῇ διαιρετόν), or a body (*Metaph.* 1016 b 25—27); or it is a magnitude "*continuous* one way (or in one direction)," as compared with magnitudes continuous *two* ways or *three* ways, which curiously enough he describes as "breadth" and "depth" respectively (μέγεθος δὲ τὸ μὲν ἐφ᾽ ἓν συνεχὲς μῆκος, τὸ δ᾽ ἐπὶ δύο πλάτος, τὸ δ᾽ ἐπὶ τρία βάθος, *Metaph.* 1020 a 11), though he immediately adds that "length" means a line, "breadth" a surface, and "depth" a body.

Proclus gives another alternative definition as *"flux of a point"* (ῥύσις σημείου), i.e. the path of a point when moved. This idea is also alluded to in Aristotle (*De anima* I. 4, 409 a 4 above quoted): "they say that a line by its motion produces a surface, and a point by its motion a line." "This definition," says Proclus (p. 97, 8—13), "is a perfect one as showing the essence of the line: he who called it the flux of a point seems to define it from its genetic cause, and it is not every line that he sets before us, but only the immaterial line; for it is this that is produced by the point, which, though itself indivisible, is the cause of the existence of things divisible."

Proclus (p. 100, 5—19) adds the useful remark, which, he says, was current in the school of Apollonius, that we have the notion of a line when we ask for the length of a road or a wall measured merely as length; for in that case we mean something irrespective of breadth, viz. distance in one "dimension." Further we can obtain sensible perception of a line if we look at the division between the light and the dark when a shadow is thrown on the earth or the moon; for clearly the division is without breadth, but has length.

Species of "lines."

After defining the "line" Euclid only mentions *one* species of line, the straight line, although of course another classification appears in the definition of a circle later. He doubtless omitted all *classification* of lines as unnecessary for his purpose, whereas, for example, Heron follows up his definition of a line by a division of lines into (1) those which are "straight" and (2) those which are not, and a further division of the latter into (*a*) "circular circumferences," (*b*) "spiral-shaped" (ἑλικοειδεῖς) lines and (*c*) "curved" (καμπύλαι) lines generally, and then explains the four terms. Aristotle tells us (*Metaph.* 986 a 25) that the Pythagoreans distinguished straight (εὐθύ) and curved (καμπύλον), and this distinction appears in Plato (cf. *Republic* x. 602 c) and in Aristotle (cf. "to a line belong the attributes straight or curved," *Anal. post.* I. 4, 73 b 19; "as in mathematics it is useful to know what is meant by the terms straight and curved," *De anima* I. 1, 402 b 19). But from the class of "curved" lines Plato and Aristotle separate off the περιφερής or "circular" as a distinct species often similarly contrasted with straight. Aristotle seems to recognise broken lines forming an angle as one line: thus "a line, if it be bent (κεκαμμένη), but yet continuous, is called one" (*Metaph.* 1016 a 2); "the straight line is more one than the bent line" (*ibid.* 1016 a 12). Cf. Heron, Def. 12, "A broken line (κεκλασμένη γραμμή) so-called is a line which, when produced, does not meet *itself*."

When Proclus says that both Plato and Aristotle divided lines into those which are "straight," "circular" (περιφερής) or "a mixture of the two," adding, as regards Plato, that he included in the last of these classes "those which are called helicoidal among plane (curves) and (curves) formed about solids, and such species of curved lines as arise from sections of solids" (p. 104, 1—5), he appears to be not quite exact. The reference as regards Plato seems to be to *Parmenides* 145 B: "At that rate it would seem that the one must have shape, either straight or round (στρογγύλον) or some combination of the two"; but this scarcely amounts to a formal classification of lines. As regards

Aristotle, Proclus seems to have in mind the passage (*De caelo* I. 2, 268 b 17) where it is stated that "all *motion* in space, which we call translation (φορά), is (in) a straight line, a circle, or a combination of the two; for the first two are the only simple (*motions*)."

For completeness it is desirable to add the substance of Proclus' account of the classification of lines, for which he quotes Geminus as his authority.

Geminus' first classification of lines.

This begins (p. 111, 1—9) with a division of lines into *composite* (σύνθετος) and *incomposite* (ἀσύνθετος). The only illustration given of the *composite* class is the "broken line which forms an angle" (ἡ κεκλασμένη καὶ γωνίαν ποιοῦσα); the subdivision of the *incomposite* class then follows (in the text as it stands the word "composite" is clearly an error for "incomposite"). The subdivisions of the incomposite class are repeated in a later passage (pp. 176, 27—177, 23) with some additional details. The following diagram reproduces the effect of both versions as far as possible (all the illustrations mentioned by Proclus being shown in brackets).

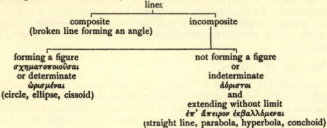

The additional details in the second version, which cannot easily be shown in the diagram, are as follows:

(1)   Of the lines which extend without limit, some do not *form a figure* at all (viz. the straight line, the parabola and the hyperbola); but some first "come together and form a figure" (i.e. have a loop), "and, for the rest, extend without limit" (p. 177, 8).

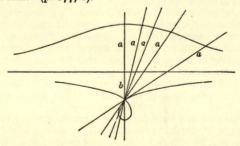

As the only other curve, besides the parabola and the hyperbola, which has been mentioned as proceeding to infinity is the *conchoid* (of Nicomedes), we can hardly avoid the conclusion of Tannery[1] that the curve which has a loop and then proceeds to infinity is a variety of the *conchoid* itself. As is

---

[1] *Notes pour l'histoire des lignes et surfaces courbes dans l'antiquité* in *Bulletin des sciences mathém. et astronom.* 2 sér. VIII. (1884), pp. 108—9 (*Mémoires scientifiques*, II. p. 23).

well known, the ordinary conchoid (which was used both for doubling the cube and for trisecting the angle) is obtained in this way. Suppose any number of rays passing through a fixed point (the *pole*) and intersecting a fixed straight line; and suppose that points are taken on the rays, beyond the fixed straight line, such that the portions of the rays intercepted between the fixed straight line and the point are equal to a constant *distance* (διάστημα), the locus of the points is a conchoid which has the fixed straight line for asymptote. If the "distance" *a* is measured from the intersection of the ray with the given straight line, not in the direction away from the pole, but towards the pole, we obtain three other curves according as *a* is less than, equal to, or greater than *b*, the distance of the pole from the fixed straight line, which is an asymptote in each case. The case in which *a* > *b* gives a curve which forms a loop and then proceeds to infinity in the way Proclus describes. Now we know both from Eutocius (*Comm. on Archimedes*, ed. Heiberg, III. p. 98) and Proclus (p. 272, 3—7) that Nicomedes wrote on conchoids (in the plural), and Pappus (IV. p. 244, 18) says that besides the "first" (used as above stated) there were "the second, the third and the fourth which are useful for other theorems."

(2) Proclus next observes (p. 177, 9) that, of the lines which extend without limit, some are "*asymptotic*" (ἀσύμπτωτοι), namely "those which never meet, however they are produced," and some are "*symptotic*," namely "those which will meet sometime"; and, of the "asymptotic" class, some are in one plane, and others not. Lastly, of the "asymptotic" lines in one plane, some preserve always the same distance from one another, while others continually "lessen the distance, like the hyperbola with reference to the straight line, and the conchoid with reference to the straight line."

Geminus' second classification.

This (from Proclus, pp. 111, 9—20 and 112, 16—18) can be shown in a diagram thus:

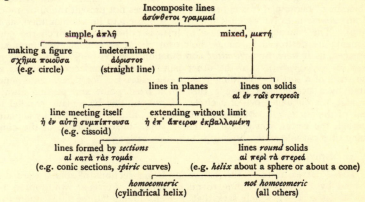

Notes on classes of "lines" and on particular curves.

We will now add the most interesting notes found in Proclus with reference to the above classifications or the particular curves mentioned.

1. Homoeomeric lines.

By this term (ὁμοιομερεῖς) are meant lines which are alike in all parts, so that in any one such curve any part can be made to coincide with any other part. Proclus observes that these lines are only three in number, two being "simple" and in a plane (the straight line and the circle), and the third "mixed," (subsisting) "about a solid," namely the cylindrical helix. The latter curve was also called the *cochlias* or *cochlion*, and its *homoeomeric* property was proved by Apollonius in his work περὶ τοῦ κοχλίου (Proclus, p. 105, 5). The fact that there are only three *homoeomeric* lines was proved by Geminus, "who proved, as a preliminary proposition, that, if from a point (ἀπό του σημείου, but on p. 251, 4 ἀφ' ἑνὸς σημείου) two straight lines be drawn to a homoeomeric line making equal angles with it, the straight lines are equal" (pp. 112, 1—113, 3, cf. p. 251, 2—19).

2. Mixed lines.

It might be supposed, says Proclus (p. 105, 11), that the cylindrical helix, being *homoeomeric*, like the straight line and the circle, must like them be *simple*. He replies that it is not simple, but *mixed*, because it is generated by *two unlike* motions. Two *like* motions, said Geminus, e.g. two motions at the same speed in the directions of two adjoining sides of a square, produce a *simple* line, namely a straight line (the diagonal); and again, if a straight line moves with its extremities upon the two sides of a right angle respectively, this same motion gives a *simple* curve (a circle) for the locus of the middle point of the straight line, and a *mixed* curve (an ellipse) for the locus of any other point on it (p. 106, 3—15).

Geminus also explained that the term "mixed," as applied to curves, and as applied to surfaces, respectively, is used in different senses. As applied to curves, "mixing" neither means simple "putting together" (σύνθεσις) nor "blending" (κρᾶσις). Thus the helix (or spiral) is a "mixed" line, but (1) it is not "mixed" in the sense of "putting together," as it would be if, say, part of it were straight and part circular, and (2) it is not mixed in the sense of "blending," because, if it is cut in any way, it does not present the appearance of any *simple* lines (of which it might be supposed to be compounded, as it were). The "mixing" in the case of lines is rather that in which the constituents are destroyed so far as their own character is concerned, and are replaced, as it were, by a *chemical* combination (ἔστιν ἐν αὐτῇ συνεφθαρμένα τὰ ἄκρα καὶ συγκεχυμένα). On the other hand "mixed" surfaces are mixed in the sense of a sort of "blending" (κατά τινα κρᾶσιν). For take a cone generated by a straight line passing through a fixed point and passing always through the circumference of a circle: if you cut this by a plane parallel to that of the circle, you obtain a circular section, and if you cut it by a plane through the vertex, you obtain a triangle, the "mixed" surface of the cone being thus cut into *simple* lines (pp. 117, 22—118, 23).

3. Spiric curves.

These curves, classed with conics as being sections of solids, were discovered by Perseus, according to an epigram of Perseus' own quoted by Proclus (p. 112, 1), which says that Perseus found "three lines upon (or, perhaps, in addition to) five sections" (τρεῖς γραμμὰς ἐπὶ πέντε τομαῖς). Proclus throws some light on these in the following passages:

"Of the spiric sections, one is interlaced, resembling the horse-fetter (ἵππου πέδη); another is widened out in the middle and contracts on each side (of the middle), a third is elongated and is narrower in the middle, broadening out on each side of it" (p. 112, 4—8).

"This is the case with the *spiric surface*; for it is conceived as generated by the revolution of a circle remaining at right angles [to a plane] and turning about a point which is not its centre [in other words, generated by the revolution of a circle about a straight line in its plane not passing through the centre]. Hence the *spire* takes three forms, for the centre [of rotation] is either on the circumference, or within it, or without it. And if the centre of rotation is on the circumference, we have the *continuous* spire (συνεχής), if within, the *interlaced* (ἐμπεπλεγμένη), and if without, the *open* (διεχής). And the spiric sections are three according to these three differences" (p. 119, 8—17).

"When the *hippopede*, which is one of the spiric curves, forms an angle with itself, this angle also is contained by mixed lines" (p. 127, 1—3).

"Perseus showed for spirics what was their property (σύμπτωμα)" (p. 356, 12).

Thus the spiric surface was what we call a *tore*, or (when open) an *anchor-ring*. Heron (Def. 97) says it was called alternatively *spire* (σπεῖρα) or *ring* (κρίκος); he calls the variety in which "the circle cuts itself," not "interlaced," but " crossing-itself " (ἐπαλλάττουσα).

Tannery[1] has discussed these passages, as also did Schiaparelli[2]. It is clear that Proclus' remark that the difference in the three curves which he mentions corresponds to the difference between the three surfaces is a slip, due perhaps to too hurried transcribing from Geminus : all three arise from plane sections of the *open* anchor-ring. If r is the radius of the revolving circle, a the distance of its centre from the axis of rotation, d the distance of the plane section (supposed to be parallel to the axis) from the axis, the three curves described in the first extract correspond to the following cases :

(1) $d = a - r$. In this case the curve is the *hippopede*, of which the lemniscate of Bernoulli is a particular case, namely that in which $a = 2r$.

The name *hippopede* was doubtless adopted for this one of Perseus' curves on the ground of its resemblance to the *hippopede* of Eudoxus, which seems to have been the curve of intersection of a sphere with a cylinder touching it internally.

(2) $a + r > d > a$. Here the curve is an oval.

(3) $a > d > a - r$. The curve is now narrowest in the middle.

Tannery explains the "three lines upon (in addition to) five sections" thus. He points out that with the *open tore* there are two other sections corresponding to

(4) $d = a$: transition from (2) to (3).

(5) $a - r > d > 0$, in which case the section consists of two symmetrical ovals.

He then shows that the sections of the *closed* or *continuous tore*, corresponding to $a = r$, give curves corresponding to (2), (3) and (4) only. Instead of (1) and (5) we have only a section consisting of two equal circles touching one another.

On the other hand, the *third* spire (the *interlaced* variety) gives three *new* forms, which make a group of three in addition to the first group of *five* sections.

[1] *Pour l'histoire des lignes et surfaces courbes dans l'antiquité* in *Bulletin des sciences mathém. et astronom.* VIII. (1884), pp. 25—27 (*Mémoires scientifiques*, II. pp. 24—28).

[2] *Die homocentrischen Sphären des Eudoxus, des Kallippus und des Aristoteles* (*Abhandlungen zur Gesch. der Math.* I. Heft, 1877, pp. 149—152).

The difficulty which I see in this interpretation is the fact that, just after "three lines on five sections" are mentioned, Proclus describes three curves which were evidently the most important; but these three belong to three of the five sections of the open tore, and are not separate from them.

4. The cissoid.

This curve is assumed to be the same as that by means of which, according to Eutocius (*Comm. on Archimedes*, III. p. 66 sqq.), Diocles in his book περὶ πυρίων (*On burning-glasses*) solved the problem of doubling the cube. It is the locus of points which he found by the following construction. Let *AC*, *BD* be diameters at right angles in a circle with centre *O*.

Let *E*, *F* be points on the quadrants *BC*. *BA* respectively such that the arcs *BE*, *BF* are equal.

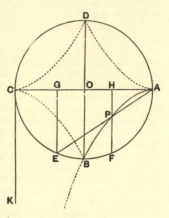

Draw *EG*, *FH* perpendicular to *CA*. Join *AE*, and let *P* be its intersection with *FH*.

The cissoid is the locus of all the points *P* corresponding to different positions of *E* on the quadrant *BC* and of *F* at an equal distance from *B* along the arc *BA*.

A is the point on the curve corresponding to the position *C* for the point *E*, and *B* the point on the curve corresponding to the position of *E* in which it coincides with *B*.

It is easy to see that the curve extends in the direction *AB* beyond *B*, and that *CK* drawn perpendicular to *CA* is an asymptote. It may be regarded also as having a branch *AD* symmetrical with *AB*, and, beyond *D*, approaching *KC* produced as asymptote.

If *OA*, *OD* are coordinate axes, the equation of the curve is obviously

$$y^2(a+x)=(a-x)^3,$$

where *a* is the radius of the circle.

There is a cusp at *A*, and it agrees with this that Proclus should say (p. 126, 24) that "cissoidal lines converging to one point like the leaves of ivy—for this is the origin of their name—form an angle." He makes the slight correction (p. 128, 5) that it is not two *parts* of a curve, but *one* curve, which in this case makes an angle.

But what is surprising is that Proclus seems to have no idea of the curve passing outside the circle and having an asymptote, for he several times speaks of it as a *closed* curve (forming a figure and including an area): cf. p. 152, 7, "the plane (area) cut off by the cissoidal line has one bounding (line), but it has not in it a centre such that all (straight lines drawn to the curve) from it are equal." It would appear as if Proclus regarded the cissoid as formed by the *four* symmetrical cissoidal arcs shown in the figure.

Even more peculiar is Proclus' view of the

5. "Single-turn Spiral."

This is really the spiral of Archimedes traced by a point starting from the fixed extremity of a straight line and moving uniformly along it, while simultaneously the straight line itself moves uniformly in a plane about its fixed extremity.   In Archimedes the spiral has of course any number of turns, the straight line making the same number of complete revolutions.   Yet Proclus, while giving the same account of the generation of the spiral (p. 180, 8—12), regards the *single-turn spiral* as actually *stopping short* at the point reached after one complete revolution of the straight line: "it is necessary to know that extending without limit is not a property of all lines; for it, neither belongs to the circle nor to the cissoid, nor in general to lines which form figures; nor even to those which do not form figures.   For even the single-turn spiral does not extend without limit—*for it is constructed between two points*—nor does any of the other lines so generated do so" (p. 187, 19—25). It is curious that Pappus (VIII. p. 1110 sqq.) uses the same term μονόστροφος ἕλιξ to denote one turn, not of the spiral, but of the *cylindrical helix*.

DEFINITION 3.

Γραμμῆς δὲ πέρατα σημεῖα.
The extremities of a line are points.

It being unscientific, as Aristotle said, to define a point as the "extremity of a line" (πέρας γραμμῆς), thereby explaining the prior by the posterior, Euclid defined a point differently; then, as it was necessary to connect a point with a line, he introduced this explanation after the definitions of both had been given.   This compromise is no doubt his own idea; the same thing occurs with reference to a surface and a line as its extremity in Def. 6, and with reference to a solid and a surface as its extremity in XI. Def. 2.

We miss a statement of the facts, equally requiring to be known, that a "division" (διαίρεσις) of a line, no less than its "beginning" or "end," is a point (this is brought out by Aristotle: cf. *Metaph.* 1060 b 15), and that the *intersection* of two lines is also a point.   If these additional explanations had been given, Proclus would have been saved the difficulty which he finds in the fact that some of the lines used in Euclid (namely infinite straight lines on the one hand, and circles on the other) have no "extremities."   So also the ellipse, which Proclus calls by the old name θυρεός ("shield").   In the case of the circle and ellipse we can, he observes (p. 103, 7), take a portion bounded by points, and the definition applies to that portion.   His rather far-fetched distinction between two aspects of a circle or ellipse as a *line* and as a *closed figure* (thus, while you are *describing* a circle, you have two extremities at any moment, but they disappear when it is finished) is an unnecessarily elaborate attempt to establish the literal universality of the "definition," which is really no more than an explanation that, if a line *has* extremities, those extremities are points.

DEFINITION 4.

Εὐθεῖα γραμμή ἐστιν, ἥτις ἐξ ἴσου τοῖς ἐφ' ἑαυτῆς σημείοις κεῖται.
A straight line is a line which lies evenly with the points on itself.

The only definition of a straight line authenticated as pre-Euclidean is that of Plato, who defined it as "*that of which the middle covers the ends*" (relatively, that is, to an eye placed at either end and looking along the straight line).   It appears in the *Parmenides* 137 E: "straight is whatever has its middle in front of (i.e. so placed as to obstruct the view of) both its ends" (εὐθύ γε οὗ ἂν τὸ μέσον ἀμφοῖν τοῖν ἐσχάτοιν ἐπίπροσθεν ᾖ). Aristotle quotes it in equivalent terms (*Topics* VI. 11, 148 b 27), οὗ τὸ μέσον ἐπιπροσθεῖ τοῖς πέρασιν; and, as he does not mention the name of its author, but states it in combination with the definition of a line as the extremity of a surface, we may assume that he used it as being well known. Proclus also quotes the definition as Plato's in almost identical terms, ἧς τὰ μέσα τοῖς ἄκροις ἐπιπροσθεῖ (p. 109, 21). This definition is ingenious, but implicitly appeals to the sense of sight and involves the postulate that the line of sight is straight. (Cf. the Aristotelian *Problems* 31, 20, 959 a 39, where the question is why we can better observe straightness in a row, say, of letters with one eye than with two.) As regards the straightness of "visual rays," ὄψεις, cf. Euclid's own *Optics*, Deff. 1, 2, assumed as *hypotheses*, in which he first speaks of the "straight lines" drawn from the eye, avoiding the word ὄψεις, and then says that the figure contained by the *visual rays* (ὄψεις) is a cone with its vertex in the eye.

As Aristotle mentions no definition of a straight line resembling Euclid's, but gives only Plato's definition and the other explaining it as the "extremity of a surface," the latter being evidently the current definition in contemporary textbooks, we may safely infer that Euclid's definition was a new departure of his own.

Proclus on Euclid's definition.

Coming now to the interpretation of Euclid's definition, εὐθεῖα γραμμή ἐστιν, ἥτις ἐξ ἴσου τοῖς ἐφ᾿ ἑαυτῆς σημείοις κεῖται, we find any number of slightly different versions, but none that can be described as quite satisfactory; some authorities, e.g. Savile, have confessed that they could make nothing of it. It is natural to appeal to Proclus first; and we find that he does in fact give an interpretation which at first sight seems plausible. He says (p. 109, 8 sq.) that Euclid "shows by means of this that the straight line alone [of all lines] occupies a distance (κατέχειν διάστημα) equal to that between the points on it. For, as far as one of the points is distant from another, so great is the length (μέγεθος) of the straight line of which they are the extremities; and this is the meaning of lying ἐξ ἴσου to (or with) the points on it" [ἐξ ἴσου being thus, apparently, interpreted as "at" (or "over") "an equal distance"]. "But if you take two points on the circumference (of a circle) or any other line, the distance cut off between them along the line is greater than the interval separating them. And this is the case with every line except the straight line. Hence the ordinary remark, based on a common notion, that those who journey in a straight line only travel the necessary distance, while those who do not go straight travel more than the necessary distance." (Cf. Aristotle, *De caelo* I. 4, 271 a 13, "we always call the distance of anything the straight line" drawn to it.) Thus Proclus would interpret somewhat in this way: "a straight line is that which represents extension equal with (the distances separating) the points on it." This explanation seems to be an attempt to graft on to Euclid's definition the *assumption* (it is a λαμβανόμενον, not a definition) of Archimedes (*On the sphere and cylinder* I. *ad init.*) that "of all the lines which have the same extremities the straight line is least." For this purpose ἐξ ἴσου has apparently to be taken as meaning "at an equal distance," and again "lying at an equal distance" as equivalent to "extending over (or representing) an equal distance." This is difficult enough in itself, but is seen to be an impossible interpretation when applied to the similar definition of a plane by Euclid (Def. 7) as a surface "which lies evenly with the straight lines on itself." In that connexion Proclus tries to make the same words ἐξ ἴσου κεῖται mean "extends over an equal *area* with." He says namely (p. 117, 2) that, "if two straight lines are set out" on the plane, the plane surface "occupies a space equal to that between the straight lines." But two straight lines do not determine by themselves any space at all; it would be necessary to have a *closed* figure with its boundaries in the plane before we could arrive at the equivalent of the other assumption of Archimedes that "of surfaces which have the same extremities, if those extremities are in a plane, the plane is the least [in area]." This seems to be an impossible sense for ἐξ ἴσου even on the assumption that it means "at an equal distance" in the present definition. The necessity therefore of interpreting ἐξ ἴσου similarly in both definitions makes it impossible to regard it as referring to *distance* or *length* at all. It should be added that Simplicius gave the same explanations as Proclus (an-Nairīzī, p. 5).

The language and construction of the definition.

Let us now consider the actual wording and grammar of the phrase ἥτις ἐξ ἴσου τοῖς ἐφ᾽ ἑαυτῆς σημείοις κεῖται. As regards the expression ἐξ ἴσου we note that Plato and Aristotle (whose use of it seems typical) commonly have it in the sense of "on a footing of equality": cf. οἱ ἐξ ἴσου in Plato's *Laws* 777 D, 919 D; Aristotle, *Politics* 1259 b 5 ἐξ ἴσου εἶναι βούλεται τὴν φύσιν, "tend to be on an equality in nature," *Eth. Nic.* VIII. 12, 1161 a 8 ἐνταῦθα πάντες ἐξ ἴσου, "there all are on a footing of equality." Slightly different are the uses in Aristotle, *Eth. Nic.* X. 8, 1178 a 25 τῶν μὲν γὰρ ἀναγκαίων χρεία καὶ ἐξ ἴσου ἔστω, "both need the necessaries of life *to the same extent*, let us say"; *Topics* IX. 15, 174 a 32 ἐξ ἴσου ποιοῦντα τὴν ἐρώτησιν, "asking the question indifferently" (i.e. without showing any expectation of one answer being given rather than another). The natural meaning would therefore appear to be "evenly placed" (or balanced), "in equal measure," "indifferently" or "without bias" one way or the other. Next, is the dative τοῖς ἐφ᾽ ἑαυτῆς σημείοις constructed with ἐξ ἴσου or with κεῖται? In the first case the phrase must mean "that which lies *evenly with* (or in respect to) the points on it," in the second apparently "that which, in (or by) the points on it, lies (or is placed) evenly (or uniformly)." Max Simon takes the first construction to give the sense "die Gerade liegt in gleicher Weise wie ihre Punkte." If the last words mean "in the same way as (or in like manner as) its points," I cannot see that they tell us anything, although Simon attaches to the words the notion of *distance* (Abstand) like Proclus. The second construction he takes as giving "die Gerade liegt für (durch) ihre Punkte gleichmässig," "the straight line lies symmetrically for (or through) its points"; or, if κεῖται is taken as the passive of τίθημι, "die Gerade ist durch ihre Punkte gleichmässig gegeben worden," "the straight line is symmetrically determined by its points." He adds that the idea is here *direction*, and that both *direction* and *distance* (as between two different given points simply) would be to Euclid, as later to Bolzano (*Betrachtungen über einige Gegenstände der Elementargeometrie*, 1804, quoted by Schotten, *Inhalt und Methode des planimetrischen Unterrichts*, II. p. 16), primary irreducible notions.

While the language is thus seen to be hopelessly obscure, we can safely say that the sort of idea which Euclid wished to express was that of a line which presents the same shape at and relatively to all points on it, without any irregular or unsymmetrical feature distinguishing one part or side of it from another. Any such irregularity could, as Saccheri points out (Engel and Stäckel, *Die Theorie der Parallellinien von Euklid bis Gauss*, 1895, p. 109), be at once made perceptible by keeping the ends fixed and turning the line about them right round; if any two positions were distinguishable, e.g. one being to the left or right relatively to another, "it would not lie in a uniform manner between its points."

A conjecture as to its origin and meaning.

The question arises, what· was the origin of Euclid's definition, or, how was it suggested to him? It seems to me that the basis of it was really Plato's definition of a straight line as "that line the middle of which covers the ends." Euclid was a Platonist, and what more natural than that he should have adopted Plato's definition in substance, while regarding it as essential to change the form of words in order to make it independent of any implied appeal to vision, which, as a physical fact, could not properly find a place in a purely geometrical definition? I believe therefore that Euclid's definition is simply an attempt (albeit unsuccessful, from the nature of the case) to express, in terms to which a geometer could not object as not being part of geometrical subject-matter, the same thing as the Platonic definition.

The truth is that Euclid was attempting the impossible. As Pfleiderer says (Scholia to Euclid), "It seems as though the notion of a *straight line*, owing to its simplicity, cannot be explained by any regular definition which does not introduce words already containing in themselves, by implication, the notion to be defined (such e.g. are direction, equality, uniformity or evenness of position, unswerving course), and as though it were impossible, if a person does not already know what the term *straight* here means, to teach it to him unless by putting before him in some way a picture or a drawing of it." This is accordingly done in such books as Veronese's *Elementi di geometria* (Part I., 1904, p. 10): "A stretched string, e.g. a plummet, a ray of light entering by a small hole into a dark room, are *rectilineal* objects. The image of them gives us the abstract idea of the limited line which is called a *rectilineal segment.*"

Other definitions.

We will conclude this note with some other famous definitions of a straight line. The following are given by Proclus (p. 110, 18—23).

1. *A line stretched to the utmost,* ἐπ' ἄκρον τεταμένη γραμμή. This appears in Heron also, with the words "towards the ends" (ἐπὶ τὰ πέρατα) added. (Heron, Def. 4).

2. *Part of it cannot be in the assumed plane while part is in one higher up* (ἐν μετεωροτέρῳ). This is a *proposition* in Euclid (XI. 1).

3. *All its parts fit on all (other parts) alike,* πάντα αὐτῆς τὰ μέρη πᾶσιν ὁμοίως ἐφαρμόζει. Heron has this too (Def. 4), but instead of "alike" he says παντοίως, "in all ways," which is better as indicating that the applied part may be applied one way or the *reverse* way, with the same result.

4. *That line which, when its ends remain fixed, itself remains fixed,* ἡ τῶν περάτων μενόντων καὶ αὐτὴ μένουσα. Heron's addition to this, "*when it is, as it were, turned round in the same plane*" (οἷον ἐν τῷ αὐτῷ ἐπιπέδῳ στρεφομένη), and his next variation, "and about the same ends having always the same position," show that the definition of a straight line as "that which does not change its position when it is turned about its extremities (or any two points in it) as poles" was no original discovery of Leibniz, or Saccheri, or Krafft, or Gauss, but goes back at least to the beginning of the Christian era. Gauss' form of this definition was: "The line in which lie all points that, during the revolution of a body (a part of space) about two fixed points, maintain their position unchanged is called a straight line." Schotten

(I. p. 315) maintains that the notion of a straight line and its property of being determined by two points are unconsciously assumed in this definition, which is therefore a logical "circle."

5. *That line which with one other of the same species cannot complete a figure*, ἡ μετὰ τῆς ὁμοειδοῦς μιᾶς σχῆμα μὴ ἀποτελοῦσα. This is an obvious ὕστερον-πρότερον, since it assumes the notion of a *figure*.

Lastly Leibniz' definition should be mentioned: *A straight line is one which divides a plane into two halves identical in all but position.* Apart from the fact that this definition introduces the plane, it does not seem to have any advantages over the definition last but one referred to.

Legendre uses the Archimedean property of a straight line as *the shortest distance between two points.* Van Swinden observes (*Elemente der Geometrie*, 1834, p. 4), that to take this as the definition involves *assuming* the proposition that any two sides of a triangle are greater than the third and *proving* that straight lines which have two points in common coincide throughout their length (cf. Legendre, *Éléments de Géométrie* I. 3, 8).

The above definitions all illustrate the observation of Unger (*Die Geometrie des Euklid*, 1833): "*Straight* is a simple notion, and hence all definitions of it must fail.... But if the proper idea of a straight line has once been grasped, it will be recognised in all the various definitions usually given of it; all the definitions must therefore be regarded as *explanations*, and among them that one is the best from which further inferences can immediately be drawn as to the essence of the straight line."

DEFINITION 5.

Ἐπιφάνεια δέ ἐστιν, ὃ μῆκος καὶ πλάτος μόνον ἔχει.

A surface is that which has length and breadth only.

The word ἐπιφάνεια was used by Euclid and later writers to denote *surface* in general, while they appropriated the word ἐπίπεδον for *plane surface*, thus making ἐπίπεδον a *species* of the *genus* ἐπιφάνεια. A solitary use of ἐπιφάνεια by Euclid when a plane is meant (XI. Def. 11) is probably due to the fact that the particular definition came from an earlier textbook. Proclus (p. 116, 17) remarks that the older philosophers, including Plato and Aristotle, used the words ἐπιφάνεια and ἐπίπεδον indifferently for any kind of surface. Aristotle does indeed use both words for a surface, with perhaps a tendency to use ἐπιφάνεια more than ἐπίπεδον for a surface not plane. Cf. *Categories* 6, 5 a 1 sq., where both words are used in one sentence: "You can find a common boundary at which the parts fit together, a point in the case of a line, and a line in the case of a surface (ἐπιφάνεια); for the parts of the surface (ἐπιπέδου) do fit together at some common boundary. Similarly also in the case of a body you can find a common boundary, a line or a surface (ἐπιφάνεια), at which the parts of the body fit together." Plato however does not use ἐπιφάνεια at all in the sense of surface, but only ἐπίπεδον for both *surface* and *plane surface*. There is reason therefore for doubting the correctness of the notice in Diogenes Laertius, III. 24, that Plato "was the first philosopher to name, among extremities, the *plane* surface" (ἐπίπεδος ἐπιφάνεια).

ἐπιφάνεια of course means literally the feature of a body which is *apparent* to the eye (ἐπιφανής), namely the surface.

Aristotle tells us (*De sensu* 3, 439 a 31) that the Pythagoreans called a surface χροιά, which seems to have meant *skin* as well as *colour*. Aristotle explains the term with reference to colour (χρῶμα) as a thing inseparable from the extremity (πέρας) of a body.

Alternative definitions.

The definitions of a surface correspond to those of a line. As in Aristotle a line is a magnitude "(extended) one way, or in one 'dimension'" (ἐφ' ἕν), "continuous one way" (ἐφ' ἓν συνεχές), or "divisible in one way" (μοναχῇ διαιρετόν), so a surface is a magnitude extended or continuous *two ways* (ἐπὶ δύο), or divisible *in two ways* (διχῇ). As in Euclid a surface has "length and breadth" only, so in Aristotle "breadth" is characteristic of the surface and is once used as synonymous with it (*Metaph.* 1020 a 12), and again "lengths are made up of long and short, *surfaces of broad and narrow*, and solids (ὄγκοι) of deep and shallow" (*Metaph.* 1085 a 10).

Aristotle mentions the common remark that *a line by its motion produces a surface* (*De anima* I. 4, 409 a 4). He also gives the *a posteriori* description of a surface as the "extremity of a solid" (*Topics* VI. 4, 141 b 22), and as "the section (τομή) or division (διαίρεσις) of a body" (*Metaph.* 1060 b 14).

Proclus remarks (p. 114, 20) that we get a notion of a surface when we measure areas and mark their boundaries in the sense of length and breadth; and we further get a sort of perception of it by looking at shadows, since these have no depth (for they do not penetrate the earth) but only have length and breadth.

Classification of surfaces.

Heron gives (Def. 74, p. 50, ed. Heiberg) two alternative divisions of surfaces into two classes, corresponding to Geminus' alternative divisions of lines, viz. into (1) *incomposite* and *composite* and (2) *simple* and *mixed*.

(1) *Incomposite* surfaces are "those which, when produced, fall into (or coalesce with) themselves" (ὅσαι ἐκβαλλόμεναι αὐταὶ καθ' ἑαυτῶν πίπτουσιν), i.e. are of continuous curvature, e.g. the sphere.

Composite surfaces are "those which, when produced, cut one another." Of composite surfaces, again, some are (*a*) made up of non-homogeneous (elements) (ἐξ ἀνομοιογενῶν) such as cones, cylinders and hemispheres, others (*b*) made up of homogeneous (elements), namely the rectilineal (or polyhedral) surfaces.

(2) Under the alternative division, *simple* surfaces are the plane and the spherical surfaces, but no others; the *mixed* class includes all other surfaces whatever and is therefore infinite in variety.

Heron specially mentions as belonging to the mixed class (*a*) the surface of cones, cylinders and the like, which are a mixture of plane and circular (μικταὶ ἐξ ἐπιπέδου καὶ περιφερείας) and (*b*) *spiric* surfaces, which are "a mixture of two circumferences" (by which he must mean a mixture of two circular elements, namely the generating circle and its circular motion about an axis in the same plane).

Proclus adds the remark that, curiously enough, *mixed* surfaces may arise by the revolution either of *simple* curves, e.g. in the case of the *spire*, or of *mixed* curves, e.g. the "right-angled conoid" from a parabola, "another conoid" from the hyperbola, the "oblong" (ἐπίμηκες, in Archimedes παρα-μᾶκες) and "flat" (ἐπιπλατύ) spheroids from an ellipse according as it revolves about the major or minor axis respectively (pp. 119, 6—120, 2). The *homoeomeric* surfaces, namely those any part of which will coincide with any other part, are *two* only (the plane and the spherical surface), not three as in the case of lines (p. 120, 7).

DEFINITION 6.

Ἐπιφανείας δὲ πέρατα γραμμαί.

The extremities of a surface are lines.

It being unscientific, as Aristotle says, to define a line as the extremity of a surface, Euclid avoids the error of defining the prior by means of the posterior in this way, and gives a different definition not open to this objection. Then, by way of compromise, and in order to show the connexion between a line and a surface, he adds the equivalent of the definition of a line previously current as an explanation.

As in the corresponding Def. 3 above, he omits to add what is made clear by Aristotle (*Metaph.* 1060 b 15) that a "division" (διαίρεσις) or "section" (τομή) of a solid or body is also a surface, or that the common boundary at which two parts of a solid fit together (*Categories* 6, 5 a 2) may be a surface.

Proclus discusses how the fact stated in Def. 6 can be said to be true of surfaces like that of the sphere "which is bounded (πεπέρασται), it is true, but not by lines." His explanation (p. 116, 8—14) is that, "if we take the surface (of a sphere), so far as it is extended two ways (διχῇ διαστατή), we shall find that it is bounded by lines as to length and breadth; and if we consider the spherical surface as possessing a form of its own and invested with a fresh quality, we must regard it as having fitted end on to beginning and made the two ends (or extremities) one, being thus one potentially only, and not in actuality."

DEFINITION 7.

Ἐπίπεδος ἐπιφάνειά ἐστιν, ἥτις ἐξ ἴσου ταῖς ἐφ' ἑαυτῆς εὐθείαις κεῖται.

A plane surface *is a surface which lies evenly with the straight lines on itself.*

The Greek follows exactly the definition of a straight line *mutatis mutandis,* i.e. with ταῖς...εὐθείαις for τοῖς...σημείοις. Proclus remarks that, in general, all the definitions of a straight line can be adapted to the plane surface by merely changing the *genus.* Thus, for instance, a plane surface is "a *surface* the middle of which covers the ends" (this being the adaptation of Plato's definition of a straight line). Whether Plato actually gave this as the definition of a plane surface or not, I believe that Euclid's definition of a plane surface as *lying evenly with the straight lines on itself* was intended simply to express the same idea without any implied appeal to vision (just as in the corresponding case of the definition of a straight line).

As already noted under Def. 4, Proclus tries to read into Euclid's definition the Archimedean assumption that "of surfaces which have the same extremities, if those extremities are in a plane, the plane is the least." But, as I have stated, his interpretation of the words seems impossible, although it is adopted by Simplicius also (see an-Nairīzī).

Ancient alternatives.

The other ancient definitions recorded are as follows.

1. *The surface which is stretched to the utmost* (ἐπ' ἄκρον τεταμένη): a definition which Proclus describes as equivalent to Euclid's definition (on Proclus' own view of that definition). Cf. Heron, Def. 9, "(a surface) which is right (and) stretched out" (ὀρθὴ οὖσα ἀποτεταμένη), words which he adds to Euclid's definition.

2. *The least surface among all those which have the same extremities.*
Proclus is here (p. 117, 9) obviously quoting the Archimedean *assumption.*

3. *A surface all the parts of which have the property of fitting on (each other)* (Heron, Def. 9).

4. *A surface such that a straight line fits on all parts of it* (Proclus, p. 117, 8), or *such that the straight line fits on it all ways,* i.e. however placed (Proclus, p. 117, 20).

With this should be compared :

5. "*(A plane surface is) such that, if a straight line pass through two points on it, the line coincides wholly with it at every spot, all ways,*" i.e. however placed (one way or the reverse, no matter how), ἧς ἐπειδὰν δύο σημείων ἄψηται εὐθεῖα, καὶ ὅλη αὐτῇ κατὰ πάντα τόπον παντοίως ἐφαρμόζεται (Heron, Def. 9). This appears, with the words κατὰ πάντα τόπον παντοίως omitted, in Theon of Smyrna (p. 112, 5, ed. Hiller), so that it goes back at least as far as the 1st C. A.D. It is of course the same as the definition commonly attributed to Robert Simson, and very widely adopted as a substitute for Euclid's.

This same definition appears also in an-Nairīzī (ed. Curtze, p. 10) who, after quoting Simplicius' explanation (on the same lines as Proclus') of the meaning of Euclid's definition, goes on to say that "others defined the plane surface as that in which it is possible to draw a straight line from any point to any other."

Difficulties in ordinary definitions.

Gauss observed in a letter to Bessel that the definition of a plane surface as *a surface such that, if any two points in it be taken, the straight line joining them lies wholly in the surface* (which, for short, we will call "Simson's" definition) contains more than is necessary, in that a plane can be obtained by simply projecting a straight line lying in it from a point outside the line but also lying on the plane; in fact the definition includes a theorem, or postulate, as well. The same is true of Euclid's definition of a plane as the surface which "lies evenly with (*all*) the straight lines on itself," because it is sufficient for a definition of a plane if the surface "lies evenly" with those lines only which pass through a fixed point on it and each of the several points of a straight line also lying in it but not passing through the point. But from Euclid's point of view it is immaterial whether a definition contains more than the necessary minimum *provided* that the *existence* of a thing possessing all the attributes contained in the definition is afterwards proved. This however is not done in regard to the plane. No proposition about the nature of a plane as such appears before Book XI., although its existence is presupposed in all the geometrical Books I.—IV. and VI.; nor in Book XI. is there any attempt to prove, e.g. by construction, the existence of a surface conforming to the definition. The explanation may be that the existence of the plane as defined was deliberately assumed from the beginning like that of points and lines, the existence of which, according to Aristotle, must be assumed as principles unproved, while the existence of everything else must be proved; and it may well be that Aristotle would have included plane surfaces with points and lines in this statement had it not been that he generally took his illustrations from *plane* geometry (excluding solid).

But, whatever definition of a plane is taken, the evolution of its essential properties is extraordinarily difficult. Crelle, who wrote an elaborate article *Zur Theorie der Ebene* (read in the Academie der Wissenschaften in 1834) of which account must be taken in any full history of the subject, observes that, since the plane is the field, as it were, of almost all the rest of geometry, while a proper conception of it is necessary to enable Eucl. I. 1 to be understood, it might have been expected that the theory of the plane would have been the subject of at least the same amount of attention as, say, that of parallels. This however was far from being the case, perhaps because the subject of parallels (which, for the rest, presuppose the notion of a plane) is *much easier* than that of the plane. The nature of the difficulties as regards the plane have also been pointed out recently by Mr Frankland (*The First Book of Euclid's Elements*, Cambridge, 1905): it would appear that, whatever definition is taken, whether the simplest (as containing the minimum necessary to determine a plane) or the more complex, e.g. Simson's, some postulate has to be assumed in addition before the fundamental properties, or the truth of the other definitions, can be established. Crelle notes the same thing as regards Simson's definition, containing *more* than is necessary. Suppose a plane in which lies the triangle ABC. Let AD join the vertex A to any point D on BC, and BE the vertex B to any point E on CA. Then, according to the definition, AD lies wholly in the plane of the triangle; so does BE. But, if both AD and BE are to lie wholly in the one plane, AD, BE must intersect, say at F: if they did not, there would be two planes in question, not one. But the fact that the lines intersect and that, say, AD does not pass above or below BE, is by no means self-evident.

Mr Frankland points out the similar difficulty as regards the simpler definition of a plane as the surface generated by a straight line passing always through a fixed point and always intersecting a fixed straight line. Let OPP', OQQ' drawn from O intersect the straight line X at P, Q respectively. Let R be any third point on X: then it needs to be proved that OR intersects $P'Q'$ in some point, say R'. Without some postulate, however, it is not easy to see how to prove this, or even to prove that $P'Q'$ intersects X.

Crelle's essay. Definitions by Fourier, Deahna, Becker.

Crelle takes as the standard of a good definition that it shall be, not only as simple as possible, but also the best adapted for deducing, with the aid of the simplest possible principles, further properties belonging to the thing defined. He was much attracted by a very lucid definition, due, he says, to Fourier, according to which *a plane is formed by the aggregate of all the straight lines which, passing through one point on a straight line in space, are perpendicular to that straight line.* (This is really no more than an adaptation from Euclid's proposition XI. 5, to the effect that, if one of four concurrent straight lines is at right angles to each of the other three, those three are in one plane, which proposition is also used in Aristotle, *Meteorologica* III. 3, 373 a 13.) But Crelle confesses that he had not been able to deduce the necessary properties from this and had had to substitute the definition, already mentioned, of a plane as *the surface containing, throughout their whole length, all the straight lines passing through a fixed point and also intersecting a straight line in space*; and he only claims to have proved, after a long series of propositions, that the "Fourier"- or "perpendicular"-surface and the *plane* of the other definition just given are identical, after which the properties of the "Fourier"-surface can be used along with those of the plane. The advantage of the Fourier definition is that it leads easily, by means of the two propositions that triangles are equal in all respects (1) when two sides and the included angle are respectively equal and (2) when all three sides are respectively equal, to the property expressed in Simson's definition. But Crelle uses to establish these two congruence-theorems a number of propositions about *equal angles, supplementary* angles, *right* angles, *greater* and *less* angles; and it is difficult to question the soundness of Schotten's criticism that these notions in themselves really presuppose that of a plane. The difficulty due to Fourier's use of the word "perpendicular," if that were all, could no doubt be got over. Thus Deahna in a dissertation (Marburg, 1837) constructed a plane as follows. Presupposing the notions of a straight line and a sphere, he observes that, if a sphere revolve about a diameter, all the points of its surface which move describe closed curves (circles). Each of these circles, during the revolution, moves along itself, and one of them divides the surface of the sphere into two congruent parts. The aggregate then of the lines joining the centre to the points of this circle forms the *plane*. Again, J. K. Becker (*Die Elemente der Geometrie*, 1877) pointed out that the revolution of a right angle about one side of it produces a conical surface which differs from all other conical surfaces generated by the revolution of other angles in the fact that *the particular cone coincides with the cone vertically opposite to it*: this characteristic might therefore be taken in order to get rid of the use of the *right angle*.

W. Bolyai and Lobachewsky.

Very similar to Deahna's equivalent for Fourier's definition is the device of W. Bolyai and Lobachewsky (described by Frischauf, *Elemente der absoluten Geometrie*, 1876). They worked upon a fundamental idea first suggested, apparently, by Leibniz. Briefly stated, their way of evolving a *plane* and a *straight line* was as follows. Conceive an infinite number of pairs of concentric spheres described about two fixed points in space, O, O', as centres, and with equal radii, gradually increasing: these pairs of equal spherical surfaces intersect respectively in homogeneous curves (circles), and the "Inbegriff" or aggregate of these curves of intersection forms a *plane*. If A be a point on one of these circles (k say), suppose points M, M' to start simultaneously from A and to move in opposite directions *at the same speed* till they meet at B, say: B then is "opposite" to A, and A, B divide the circumference into two equal halves. If the points A, B be held fast and the whole system be turned about them until O takes the place of O', and O' of O, the circle k will occupy the same position as before (though turned a different way). Two opposite points, P, Q say, of each of the other circles will remain stationary during the motion as well as A, B: the "Inbegriff" or aggregate of all such points which remain stationary forms a *straight line*. It is next observed that the *plane* as defined can be generated by the revolution of the straight line about OO', and this suggests the following construction for a plane. Let a circle as one of the curves of intersection of the pairs of spherical surfaces be divided as before into two equal halves at A, B. Let the arc ADB be similarly bisected at D, and let C be the middle point of AB. This determines a straight line CD which is then *defined* as "perpendicular" to AB. The revolution of CD about AB generates a *plane*. The property stated in Simson's definition is then proved by means of the congruence-theorems proved in Eucl. I. 8 and I. 4. The first is taken as proved, practically by considerations of symmetry and homogeneity. If two spherical surfaces, not necessarily equal, with centres O, O' intersect, A and its "opposite" point B are taken as

before on the curve of intersection (a circle) and, relatively to OO', the point
A is taken to be convertible with B or any other point on the homogeneous
curve. The second (that of Eucl. I. 4) is established by simple application.
Rausenberger objects to these proofs on the grounds that the first *assumes*
that the two spherical surfaces intersect in one single curve, not in several,
and that the second compares *angles* : a comparison which, he says, is possible
only in a *plane*, so that a plane is really presupposed. Perhaps as regards
the particular comparison of angles Rausenberger is hypercritical; but it is
difficult to regard the supposed proof of the theorem of Eucl. I. 8 as sufficiently
rigorous (quite apart from the use of the uniform *motion* of points for the
purpose of bisecting lines).

Simson's property is proved from the two congruence-theorems thus.
Suppose that AB is "perpendicular" (as defined by Bolyai) to two generators
CM, CN of a plane, or suppose CM, CN respectively to make with AB two
angles congruent with one another. It is enough to prove that, if P be any
point on the straight line MN, then CP, just as
much as CM, CN respectively, makes with AB two
angles congruent with one another and is therefore
a generator. We prove successively the congruence
of the following pairs of triangles :

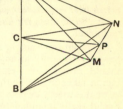

$$ACM, \quad BCM$$
$$ACN, \quad BCN$$
$$AMN, \quad BMN$$
$$AMP, \quad BMP$$
$$ACP, \quad BCP,$$

whence the angles ACP, BCP are congruent.

Other views.

Enriques and Amaldi (*Elementi di geometria*, Bologna, 1905), Veronese
(in his *Elementi*) and Hilbert all assume as a *postulate* the property stated in
Simson's definition. But G. Ingrami (*Elementi di geometria*, Bologna, 1904)
proves it in the course of a remarkable series of closely argued proposition
based upon a much less comprehensive postulate. He evolves the theory of
the plane from that of a triangle, beginning with a triangle as a mere *three-side*
(trilatero), i.e. a *frame*, as it were. His postulate relates to the *three-side* and
is to the effect that each "(rectilineal) segment" joining a vertex to a point of
the opposite side meets every segment similarly joining each of the other two
vertices to the points of the sides opposite to them respectively, and, conversely, if a point be taken on a segment joining a vertex to a point of the
opposite side, and if a straight line be drawn from another vertex to the point
on the segment so taken, it will if produced meet the opposite side. A
triangle is then defined as the figure formed by the aggregate of all the
segments joining the respective vertices of a *three-side* to points on the
opposite sides. After a series of propositions, Ingrami evolves a plane as *the
figure formed by the "half straight-lines" which project from an internal point
of the triangle the points of the perimeter*, and then, after two more theorems,
proves that a plane is determined by any three of its points which are not in
a straight line, and that *a straight line which has two points in a plane has all
its points in it*.

The argument by which Bolyai and Lobachewsky evolved the plane is
of course equivalent to the definition of a plane as *the locus of all points
equidistant from two fixed points in space*.

Leibniz in a letter to Giordano defined a plane as *that surface which divides space into two congruent parts*. Adverting to Giordano's criticism that you could conceive of surfaces and lines which divided space or a plane into two congruent parts without being *plane* or *straight* respectively, Beez (*Über Euklidische und Nicht-Euklidische Geometrie*, 1888) pointed out that what was wanted to complete the definition was the further condition that the two congruent spaces could be *slid along each other* without the surfaces ceasing to coincide, and claimed priority for his completion of the definition in this way. But the idea of *all the parts* of a plane fitting exactly on *all other parts* is ancient, appearing, as we have seen, in Heron, Def. 9.

DEFINITIONS 8, 9.

8. Ἐπίπεδος δὲ γωνία ἐστὶν ἡ ἐν ἐπιπέδῳ δύο γραμμῶν ἀπτομένων ἀλλήλων καὶ μὴ ἐπ' εὐθείας κειμένων πρὸς ἀλλήλας τῶν γραμμῶν κλίσις.

9. Ὅταν δὲ αἱ περιέχουσαι τὴν γωνίαν γραμμαὶ εὐθεῖαι ὦσιν, εὐθύγραμμος καλεῖται ἡ γωνία.

8. *A plane angle is the inclination to one another of two lines in a plane which meet one another and do not lie in a straight line.*

9. *And when the lines containing the angle are straight, the angle is called* rectilineal.

The phrase "not in a *straight line*" is strange, seeing that the definition purports to apply to angles formed by *curves* as well as straight lines. We should rather have expected *continuous* (συνεχής) with one another; and Heron takes this to be the meaning, since he at once adds an explanation as to what is meant by lines not being *continuous* (οὐ συνεχεῖς). It looks as though Euclid really intended to define a *rectilineal* angle, but on second thoughts, as a concession to the then common recognition of curvilineal angles, altered "straight lines" into "lines" and separated the definition into two.

I think all our evidence suggests that Euclid's definition of an angle as *inclination* (κλίσις) was a new departure. The word does not occur in Aristotle; and we should gather from him that the idea generally associated with an angle in his time was rather *deflection* or *breaking* of lines (κλάσις): cf. his common use of κεκλάσθαι and other parts of the verb κλᾶν, and also his reference to *one bent line* forming an angle (τὴν κεκαμμένην καὶ ἔχουσαν γωνίαν, *Metaph.* 1016 a 13)

Proclus has a long and elaborate note on this definition, much of which (pp. 121, 12—126, 6) is apparently taken direct from a work by his master Syrianus (ὁ ἡμέτερος καθηγεμών). Two criticisms contained in the note need occasion no difficulty. One of these asks how, if an angle be an inclination, one inclination can produce two angles. The other (p. 128, 2) is to the effect that the definition seems to exclude an angle formed by one and the same curve with itself, e.g. the complete *cissoid* [at what we call the "cusp"] or the curve known as the *hippopede* (horse-fetter) [shaped like a lemniscate]. But such an "angle" as this belongs to higher geometry, which Euclid may well be excused for leaving out of account in any case.

Other ancient definitions: Apollonius, Plutarch, Carpus.

Proclus' note records other definitions of great interest. Apollonius defined an angle as *a contracting of a surface or a solid at one point under a broken line or surface* (συναγωγὴ ἐπιφανείας ἢ στερεοῦ πρὸς ἑνὶ σημείῳ ὑπὸ κεκλασμένῃ γραμμῇ ἢ ἐπιφανείᾳ), where again an angle is supposed to be formed by *one* broken line or surface. Still more interesting, perhaps, is the definition by "those who say that *the first distance under the point* (τὸ πρῶτον διάστημα ὑπὸ τὸ σημεῖον) *is the angle.* Among these is Plutarch, who insists that Apollonius meant the same thing; for, he says, there must be *some* first distance under the breaking (or deflection) of the including lines or surfaces, though, the distance under the point being continuous, it is impossible to obtain the actual *first,* since every distance is divisible without limit" (ἐπ' ἄπειρον). There is some vagueness in the use of the word "distance" (διάστημα); thus it was objected that "if we anyhow separate off the *first*" (*distance* being apparently the word understood) "and draw a straight line *through it,* we get a *triangle* and not one angle." In spite of the objection, I cannot but see in the idea of Plutarch and the others the germ of a valuable conception in infinitesimals, an attempt (though partial and imperfect) to get at the *rate of divergence* between the lines at their point of meeting as a measure of the angle between them.

A third view of an angle was that of Carpus of Antioch, who said "that the angle was a *quantity* (ποσόν), namely a *distance* (διάστημα) between the lines or surfaces containing it. This means that it would be a distance (or divergence) in one *sense* (ἐφ' ἓν διεστώς), although the angle is not on that account a straight line. For it is not everything *extended in one sense* (τὸ ἐφ' ἓν διαστατόν) that is a line." This very phrase "extended one way" being held to define a *line,* it is natural that Carpus' idea should have been described as the greatest possible paradox (πάντων παραδοξότατον). The difficulty seems to have been caused by the want of a different technical term to express a new idea; for Carpus seems undoubtedly to have been anticipating the more modern idea of an angle as representing *divergence* rather than distance, and to have meant by ἐφ' ἓν *in one sense* (*rotationally*) as distinct from *one way* or *in one dimension* (linearly).

To what category does an angle belong?

There was much debate among philosophers as to the particular category (according to the Aristotelian scheme) in which an angle should be placed; is it, namely, a *quantum* (ποσόν), *quale* (ποιόν) or *relation* (πρός τι)?

1. Those who put it in the category of *quantity* argued from the fact that a plane angle is divided by a line and a solid angle by a surface. Since, then, it is a surface which is divided by a line, and a solid which is divided by a surface, they felt obliged to conclude that an angle *is* a surface or a solid, and therefore a magnitude. But homogeneous finite magnitudes, e.g. plane angles, must bear a ratio to one another, or one must be capable of being multiplied until it exceeds the other. This is, however, not the case with a rectilineal angle and the *horn-like* angle (κερατοειδής), by which latter is meant the "angle" between a circle and a tangent to it, since (Eucl. III. 16) the latter "angle" is less than any rectilineal angle whatever. The objection, it will be observed, assumes that the two sorts of angles *are* homogeneous. Plutarch and Carpus are classed among those who, in one way or other, placed an angle among *magnitudes*; and, as above noted, Plutarch claimed Apollonius as a supporter of his view, although the word *contraction* (of a surface or solid) used by the latter does not in itself suggest magnitude much more than Euclid's *inclination.* It was this last consideration which doubtless led "Aganis," the "friend" (socius) apparently of Simplicius, to substitute for Apollonius' wording "*a quantity which has dimensions and the extremities of which arrive at one point*" (an-Nairīzī, p. 13).

2. Eudemus the Peripatetic, who wrote a whole work on the angle, maintained that it belonged to the category of *quality.* Aristotle had given as his fourth variety of *quality* "figure and the shape subsisting in each thing, and,

besides these, straightness, curvature, and the like " (*Categories* 8, 10 a 11).
He says that each individual thing is spoken of as *quale* in respect of its form,
and he instances a triangle and a square, using them again later on (*ibid.* 11 a 5)
to show that it is not all qualities which are susceptible of *more* and *less*; again,
in *Physics* I. 5, 188 a 25 *angle, straight, circular* are called kinds of *figure*.
Aristotle would no doubt have regarded *deflection* (κεκλάσθαι) as belonging to
the same category with straightness and curvature (καμπυλότης). At all events,
Eudemus took up an angle as having its origin in the *breaking* or *deflection*
(κλάσις) of lines: deflection, he argued, was quality if straightness was, and that
which has its origin in quality is itself quality. Objectors to this view argued
thus. If an angle be a quality (ποιότης) like heat or cold, how can it be bisected,
say? It can in fact be divided; and, if things of which divisibility is an
essential attribute are varieties of *quantum* and not qualities, an angle cannot
be a quality. Further, the *more* and the *less* are the appropriate attributes of
quality, not the equal and the unequal; if therefore an angle were a quality,
we should have to say of angles, not that one is greater and another smaller,
but that one is more an angle and another less an angle, and that two angles
are not unequal but *dissimilar* (ἀνόμοιοι). As a matter of fact, we are told by
Simplicius, 538, 21, on Arist. *De caelo* that those who brought the angle under
the category of *quale* did call equal angles *similar* angles; and Aristotle
himself speaks of *similar* angles in this sense in *De caelo* 296 b 20, 311 b 34.

3. Euclid and all who called an angle an inclination are held by Syrianus
to have classed it as a *relation* (πρός τι). Yet Euclid certainly regarded angles
as magnitudes; this is clear both from the earliest propositions dealing
specifically with angles, e.g. I. 9, 13, and also (though in another way) from
his describing an angle in the very next definition and always as *contained*
(περιεχομένη) by the two lines forming it (Simon, *Euclid*, p. 28).

Proclus (i.e. in this case Syrianus) adds that the truth lies between these
three views. The angle partakes in fact of all those categories: it needs the
quantity involved in magnitude, thereby becoming susceptible of equality,
inequality and the like; it needs the *quality* given it by its *form*, and lastly
the *relation* subsisting between the lines or planes bounding it.

Ancient classification of " angles."

An elaborate classification of angles given by Proclus (pp. 126, 7—127, 16)
may safely be attributed to Geminus. In order to show it by a diagram it

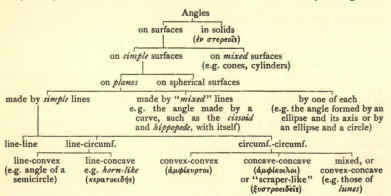

will be necessary to make a convention about terms. Angles are to be understood under each class, "line-circumference" means an angle contained by a straight line and an arc of a circle, "line-convex" an angle contained by a straight line and a circular arc with convexity *outwards*, and so on in every case.

Definitions of angle classified.

As for the point, straight line, and plane, so for the *angle*, Schotten gives a valuable summary, classification and criticism of the different modern views up to date (*Inhalt und Methode des planimetrischen Unterrichts*, II., 1893, pp. 94—183); and for later developments represented by Veronese reference may be made to the third article (by Amaldi) in *Questioni riguardanti le matematiche elementari*, I. (Bologna, 1912).

With one or two exceptions, says Schotten, the definitions of an angle may be classed in three groups representing generally the following views:

1.   *The angle is the difference of direction between two straight lines.* (With this group may be compared Euclid's definition of an angle as an inclination.)

2.   *The angle is the quantity or amount (or the measure) of the rotation necessary to bring one of its sides from its own position to that of the other side without its moving out of the plane containing both.*

3.   *The angle is the portion of a plane included between two straight lines in the plane which meet in a point (or two rays issuing from the point).*

It is remarkable however that nearly all of the text-books which give definitions different from those in group 2 add to them something pointing to a connexion between an angle and rotation: a striking indication that the essential nature of an angle is closely connected with rotation, and that a good definition must take account of that connexion.

The definitions in the first group must be admitted to be tautologous, or *circular*, inasmuch as they really presuppose some conception of an angle. *Direction (as between two given points)* may no doubt be regarded as a primary notion; and it may be defined as "the immediate relation of two points which the *ray* enables us to realise" (Schotten). But "a *direction* is no intensive magnitude, and therefore two directions cannot have any quantitative difference" (Bürklen). Nor is direction susceptible of differences such as those between qualities, e.g. colours. Direction is a *singular* entity: there cannot be different *sorts* or *degrees* of direction. If we speak of "a *different* direction," we use the word equivocally; what we mean is simply "another" direction. The fact is that these definitions of an angle as a difference of direction unconsciously appeal to something outside the notion of direction altogether, to some conception equivalent to that of the angle itself.

Recent Italian views.

The second group of definitions are (says Amaldi) based on the idea of the rotation of a straight line or ray in a plane about a point: an idea which, logically formulated, may lead to a convenient method of introducing the angle. But it must be made independent of *metric* conceptions, or of the conception of *congruence*, so as to bring out *first* the notion of an angle, and *afterwards* the notion of *equal* angles.

The third group of definitions satisfy the condition of not including metric conceptions; but they do not entirely correspond to our intuitive conception of an angle, to which we attribute the character of an entity in *one* dimension (as Veronese says) with respect to the *ray* as element, or an entity in *two* dimensions with reference to *points* as elements, which may be called an *angular sector*. The defect is however easily remedied by considering the angle as "the aggregate of the rays issuing from the vertex and comprised in the angular sector."

Proceeding to consider the principal methods of arriving at the logical formulation of the first superficial properties of the *plane* from which a definition of the angle may emerge, Amaldi distinguishes two points of view (1) the *genetic*, (2) the *actual*.

(1) From the first point of view we consider the *cluster of straight lines* or *rays* (the aggregate of all the straight lines in a plane passing through a point, or of all the rays with their extremities in that point) as generated by the movement of a straight line or ray in the plane, about a point. This leads to the *postulation* of a *closed order*, or *circular disposition*, of the straight lines or rays in a cluster. Next comes the connexion subsisting between the disposition of any two clusters whatever in one plane, and so on.

(2) Starting from the point of view of the *actual*, we lay the foundation of the definition of an angle in *the division of the plane into two parts* (half-planes) *by the straight line*. Next, two straight lines (*a*, *b*) in the plane, intersecting at a point *O*, divide the plane into four *regions* which are called *angular sectors* (convex) ; and finally the *angle* (*ab*) or (*ba*) may be defined as *the aggregate of the rays issuing from* O *and belonging to the angular sector which has* a *and* b *for sides*.

Veronese's procedure (in his *Elementi*) is as follows. He begins with the first properties of the plane introduced by the following definition.

The figure given by all the straight lines joining the points of a straight line *r* to a point *P* outside it and by the parallel to *r* through *P* is called a *cluster of straight lines*, a *cluster of rays*, or a *plane*, according as we consider the *element* of the figure itself to be the *straight line*, the *ray* terminated at *P*, or a *point*.

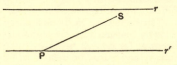

[It will be observed that this method of producing a plane involves using the *parallel* to *r*. This presents no difficulty to Veronese because he has previously defined parallels, without reference to the plane, by means of *reflex* or *opposite* figures, with respect to a point *O*: "two straight lines are called *parallel*, if one of them contains two points opposite to (or the reflex of) two points of the other with respect to the middle point of a common transversal (of the two lines)." He proves by means of a postulate that the parallel *r'* does belong to the plane *Pr*. Ingrami avoids the use of the parallel by defining a *plane* as "the figure formed by the half straight lines which project from an internal point of a triangle (i.e. a point on a line joining any vertex of a *three-side* to a point of the opposite side) the points of its perimeter," and then defining a *cluster* of rays as "the aggregate of the half straight lines in a plane starting from a given point of the plane and passing through the points of the perimeter of a triangle containing the point."]

Veronese goes on to the definition of an angle. "*We call an* angle *a part of a cluster of rays, bounded by two rays* (*as the segment is a part of a straight line bounded by two points*).

"*An angle of the cluster, the bounding rays of which are opposite, is called a* flat angle."

Then, after a postulate corresponding to postulates which he lays down for a *rectilineal segment* and for a *straight line*, Veronese proves that *all flat angles are equal to one another.*

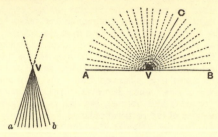

Hence he concludes that "the cluster of rays is a homogeneous linear system in which the element is the *ray* instead of the *point.* The cluster being a homogeneous linear system, all the propositions deduced from [Veronese's] Post. 1 for the straight line apply to it, e.g. that relative to the sum and difference of the segments : it is only necessary to substitute the ray for the point, and the angle for the segment."

DEFINITIONS 10, 11, 12.

10. Ὅταν δὲ εὐθεῖα ἐπ᾽ εὐθεῖαν σταθεῖσα τὰς ἐφεξῆς γωνίας ἴσας ἀλλήλαις ποιῇ, ὀρθὴ ἑκατέρα τῶν ἴσων γωνιῶν ἐστί, καὶ ἡ ἐφεστηκυῖα εὐθεῖα κάθετος καλεῖται, ἐφ᾽ ἣν ἐφέστηκεν.

11. Ἀμβλεῖα γωνία ἐστὶν ἡ μείζων ὀρθῆς.

12. Ὀξεῖα δὲ ἡ ἐλάσσων ὀρθῆς.

10. *When a straight line set up on a straight line makes the adjacent angles equal to one another, each of the equal angles is* right, *and the straight line standing on the other is called a* perpendicular *to that on which it stands.*

11. *An obtuse angle is an angle greater than a right angle.*

12. *An acute angle is an angle less than a right angle.*

ἐφεξῆς is the regular term for *adjacent* angles, meaning literally "(next) in order." I do not find the term used in Aristotle of *angles,* but he explains its meaning in such passages as *Physics* VI. 1, 231 b 8 : "those things are (next) in order which have nothing of the same kind (συγγενές) between them."

κάθετος, *perpendicular,* means literally *let fall* : the full expression is *perpendicular straight line,* as we see from the enunciation of Eucl. I. 11, and the notion is that of a straight line let fall *upon the surface of the earth,* a *plumb-line.* Proclus (p. 283, 9) tells us that in ancient times the perpendicular was called *gnomon-wise* (κατὰ γνώμονα), because the gnomon (an upright stick) was set up at right angles to the horizon.

The three kinds of angles are among the things which according to the Platonic Socrates (*Republic* VI. 510 c) the geometer assumes and argues from, declining to give any account of them because they are obvious. Aristotle discusses the *priority* of the right angle in comparison with the acute (*Metaph.* 1084 b 7): in one way the right angle is prior, i.e. *in being defined* (ὅτι ὥρισται) and by its *notion* (τῷ λόγῳ), in another way the acute is prior, i.e. as being a *part,* and because the right angle is divided into acute angles ; the acute angle is prior as *matter,* the right angle in respect of *form*; cf. also *Metaph.* 1035 b 6, "the notion of the right angle is not divided into that of an acute angle, but the reverse; for, when defining an acute angle, you make use of the right angle." Proclus (p. 133, 15) observes that it is by the *perpendicular* that we measure the heights of figures, and that it is by reference to the right angle that we distinguish the other rectilineal angles, which are otherwise undistinguished the one from the other.

The Aristotelian *Problems* (16, 4, 913 b 36) contain an expression perhaps worth quoting. The question discussed is why things which fall on the ground and rebound make "similar" angles with the surface on both sides of the point of impact; and it is observed that "the right angle is the *limit* (ὅρος) of the opposite angles," where however "opposite" seems to mean, not "supplementary" (or acute and obtuse), but the equal angles made with the surface on opposite sides of the perpendicular.

Proclus, after his manner, remarks that the statement that an angle less than a right angle is acute is not true without qualification, for (1) the *horn-like* angle (between the circumference of a circle and a tangent) is less than a right angle, since it is less than an *acute* angle, but is not an acute angle, while (2) the "angle of a semicircle" (between the arc and a diameter) is also less than a right angle, but is not an acute angle.

The *existence* of the right angle is of course proved in I. 11.

DEFINITION 13.

Ὅρος ἐστίν, ὅ τινός ἐστι πέρας.

A boundary is that which is an extremity of anything.

Aristotle also uses the words ὅρος and πέρας as synonymous. Cf. *De gen. animal.* II. 6, 745 a 6, 9, where in the expression "limit of magnitude" first one and then the other word is used.

Proclus (p. 136, 8) remarks that the word boundary is appropriate to the origin of geometry, which began from the measurement of areas of ground and involved the marking of boundaries.

DEFINITION 14.

Σχῆμά ἐστι τὸ ὑπό τινος ἤ τινων ὅρων περιεχόμενον.

A figure is that which is contained by any boundary or boundaries.

Plato in the *Meno* observes that *roundness* (στρογγυλότης) or the *round* is a "figure," and that *the straight* and many other things are so too; he then inquires what there is common to all of them, in virtue of which we apply the term "figure" to them. His answer is (76 A): "with reference to every figure I say that *that in which the solid terminates* (τοῦτο, εἰς ὅ τὸ στερεὸν περαίνει) *is a figure*, or, to put it briefly, *a figure is an extremity of a solid.*" The first observation is similar to Aristotle's in the *Physics* I. 5, 188 a 25, where *angle*, *straight*, and *circular* are mentioned as genera of figure. In the *Categories* 8, 10 a 11, "figure" is placed with straightness and curvedness in the category of quality. Here however "figure" appears to mean *shape* (μορφή) rather than "figure" in our sense. Coming nearer to "figure" in our sense, Aristotle admits that figure is "*a sort of* magnitude" (*De anima* III. 1, 425 a 18), and he distinguishes *plane figures* of two kinds, in language not unlike Euclid's, as *contained* by straight and circular lines respectively: "every plane figure is either rectilineal or formed by circular lines (περιφερόγραμμον), and the rectilineal figure is contained by several lines, the circular by one line" (*De caelo* II. 4, 286 b 13). He is careful to explain that a plane is not a figure, nor a figure a plane, but that a plane figure constitutes one notion and is a *species* of the *genus* figure (*Anal. post.* II. 3, 90 b 37). Aristotle does not attempt to define figure in general, in fact he says it would be useless : " From this it is clear that there is one definition of soul in the same way as there is one definition of *figure*; for in the one case there is no figure except the triangle, quadrilateral, and so on, nor is there any soul other than those above mentioned. A definition might be constructed which should apply to all figures but not specially to any particular figure, and similarly with the species of soul referred to. [But such a general definition would serve no purpose.] Hence it is absurd here as elsewhere to seek a general definition which will not be properly a definition of anything in existence and will not be applicable to the particular irreducible species before us, to the neglect of the definition which is so applicable " (*De anima* II. 3, 414 b 20—28).

Comparing Euclid's definition with the above, we observe that by introducing *boundary* (ὅρος) he at once excludes the *straight* which Aristotle classed as figure; he doubtless excluded *angle* also, as we may judge by (1) Heron's statement that "neither one nor two straight lines can complete a figure," (2) the alternative definition of a straight line as "that which cannot with another line of the same species form a figure," (3) Geminus' distinction between the line which *forms a figure* (σχηματοποιοῦσα) and the line which *extends indefinitely* (ἐπ᾽ ἄπειρον ἐκβαλλομένη), which latter term includes a hyperbola and a parabola. Instead of calling figure an *extremity* as Plato did in the expression "extremity (or limit) of a solid," Euclid describes a figure as *that which has* a boundary or boundaries. And lastly, in spite of Aristotle's objection, he does attempt a general definition to cover all kinds of figure, solid and plane. It appears certain therefore that Euclid's definition is entirely his own.

Another view of a figure, recalling that of Plato in *Meno* 76 A, is attributed by Proclus (p. 143, 8) to Posidonius. The latter regarded the *figure* as the *confining extremity* or *limit* (πέρας συγκλεῖον), " separating the notion of figure from *quantity* (or magnitude) and making it the cause of *definition, limitation,* and *inclusion* (τοῦ ὡρίσθαι καὶ πεπεράσθαι καὶ τῆς περιοχῆς)...Posidonius thus seems to have in view only the boundary placed round from outside, Euclid the whole content, so that Euclid will speak of the circle as a figure in respect of its whole plane (surface) and of its inclusion (from) without, whereas Posidonius (makes it a figure) in respect of its circumference...Posidonius wished to explain the notion of figure as itself *limiting* and *confining* magnitude."

Proclus observes that a logical and refining critic might object to Euclid's definition as defining the genus from the species, since that which is enclosed by one boundary and that which is enclosed by several are both species of figure. The best answer to this seems to be supplied by the passage of Aristotle's *De anima* quoted above.

DEFINITIONS 15, 16.

15. Κύκλος ἐστὶ σχῆμα ἐπίπεδον ὑπὸ μιᾶς γραμμῆς περιεχόμενον [ἣ καλεῖται περιφέρεια], πρὸς ἣν ἀφ᾽ ἑνὸς σημείου τῶν ἐντὸς τοῦ σχήματος κειμένων πᾶσαι αἱ προσπίπτουσαι εὐθεῖαι [πρὸς τὴν τοῦ κύκλου περιφέρειαν] ἴσαι ἀλλήλαις εἰσίν.

16. Κέντρον δὲ τοῦ κύκλου τὸ σημεῖον καλεῖται.

15. *A circle is a plane figure contained by one line such that all the straight lines falling upon it from one point among those lying within the figure are equal to one another;*

16. *And the point is called the* centre *of the circle.*

The words ἡ καλεῖται περιφέρεια, "which is called the circumference," and πρὸς τὴν τοῦ κύκλου περιφέρειαν, "to the circumference of the circle," are bracketed by Heiberg because, although the MSS. have them, they are omitted in other ancient sources, viz. Proclus, Taurus, Sextus Empiricus and Boethius, and Heron also omits the second gloss. The recently discovered papyrus Herculanensis No. 1061 also quotes the definition without the words in question, confirming Heiberg's rejection of them (see Heiberg in *Hermes* XXXVIII., 1903, p. 47). The words were doubtless added in view of the occurrence of the word "circumference" in Deff. 17, 18 immediately following, without any explanation. But no explanation was needed. Though the word περιφέρεια does not occur in Plato, Aristotle uses it several times (1) in the general sense of *contour* without any special mathematical signification, (2) mathematically, with reference to the rainbow and the circumference, as well as an arc, of a circle. Hence Euclid was perfectly justified in employing the word in Deff. 17, 18 and elsewhere, but leaving it undefined as being a word universally understood and not involving in itself any mathematical conception. It may be added that an-Nairīzī had not the bracketed words in his text; for he comments on and tries to explain Euclid's omission to define the circumference.

The definition itself contained nothing new in substance. Plato (*Parmenides* 137 E) says: "*Round* is, I take it, that the extremes of which are every way equally distant from the middle" (στρογγύλον γέ πού ἐστι τοῦτο, οὗ ἂν τὰ ἔσχατα πανταχῇ ἀπὸ τοῦ μέσου ἴσον ἀπέχῃ). In Aristotle we find the following expressions: "the circular (περιφερόγραμμον) plane figure bounded by one line" (*De caelo* II. 4, 286 b 13—16); "the plane equal (i.e. extending equally all ways) from the middle" (ἐπίπεδον τὸ ἐκ τοῦ μέσου ἴσον), meaning a circle (*Rhetoric* III. 6, 1407 b 27); he also contrasts with the circle "any other figure which has not the lines from the middle equal, as for example an egg-shaped figure" (*De caelo* II. 4, 287 a 19). The word "centre" (κέντρον) was also regularly used: cf. Proclus' quotation from the "oracles" (λόγια), "the centre from which all (lines extending) as far as the rim are equal."

The definition as it stands has no *genetic* character. It says nothing as to the existence or non-existence of the thing defined or as to the method of constructing it. It simply explains what is meant by the word "circle," and is a provisional definition which cannot be used until the existence of circles is proved or assumed. Generally, in such a case, existence is proved by actual construction; but here the possibility of constructing the circle as defined, and consequently its existence, are *postulated* (Postulate 3). A *genetic* definition might state that a circle is the figure described when a straight line, always remaining in one plane, moves about one extremity as a fixed point until it returns to its first position (so Heron, Def. 27).

Simplicius indeed, who points out that the distance between the feet of a pair of compasses is a straight line from the centre to the circumference, will have it that Euclid intended by this definition to show how to construct a circle by the revolution of a straight line about one end as centre; and an-Nairīzī points to this as the explanation (1) of Euclid's definition of a circle as a *plane figure*, meaning the whole surface bounded by the circumference, and not the circumference itself, and (2) of his omission to mention the "circumference," since with this construction the circumference is not drawn separately as a *line*. But it is not necessary to suppose that Euclid himself did more than follow the traditional view; for the same conception of the circle as a *plane figure* appears, as we have seen, in Aristotle. While, however,

Euclid is generally careful to say the "*circumference* of a circle" when he means the circumference, or an arc, only, there are cases where "circle" means "circumference of a circle," e.g. in III. 10: "A circle does not cut a circle in more points than two."

Heron, Proclus and Simplicius are all careful to point out that the centre is not the only point which is equidistant from all points of the circumference. The centre is the only point *in the plane of the circle* ("lying within the figure," as Euclid says) of which this is true; any point not in the same plane which is equidistant from all points of the circumference is a *pole.* If you set up a "gnomon" (an upright stick) at the centre of a circle (i.e. a line through the centre perpendicular to the plane of the circle), its upper extremity is a pole (Proclus, p. 153, 3); the perpendicular is the locus of all such poles.

DEFINITION 17.

Διάμετρος δὲ τοῦ κύκλου ἐστὶν εὐθεῖά τις διὰ τοῦ κέντρου ἠγμένη καὶ περατου-
μένη ἐφ' ἑκάτερα τὰ μέρη ὑπὸ τῆς τοῦ κύκλου περιφερείας, ἥτις καὶ δίχα τέμνει τὸν
κύκλον.

A diameter *of the circle is any straight line drawn through the centre and terminated in both directions by the circumference of the circle, and such a straight line also bisects the circle.*

The last words, literally "which (straight line) also bisects the circle," are omitted by Simson and the editors who followed him. But they are necessary even though they do not "belong to the definition" but only express a property of the diameter as defined. For, without this explanation, Euclid would not have been justified in describing as a *semi*-circle a portion of a circle bounded by a diameter and the circumference cut off by it.

Simplicius observes that the *diameter* is so called because it passes *through* the whole surface of a circle as if *measuring* it, and also because it divides the circle into two equal parts. He might however have added that, in general, it is a line passing through a figure where it is *widest*, as well as dividing it equally: thus in Aristotle τὰ κατὰ διάμετρον κείμενα, "things diametrically situated" in space, are at their maximum distance apart. *Diameter* was the regular word in Euclid and elsewhere for the diameter of a *square*, and also of a parallelogram; *diagonal* (διαγώνιος) was a later term, defined by Heron (Def. 67) as the straight line drawn from an angle to an angle.

Proclus (p. 157, 10) says that Thales was the first to prove that a circle is bisected by its diameter; but we are not told how he proved it. Proclus gives as the *reason* of the property "the undeviating course of the straight line through the centre" (a simple appeal to symmetry), but adds that, if it is desired to prove it mathematically, it is only necessary to imagine the diameter drawn and one part of the circle applied to the other; it is then clear that they must coincide, for, if they did not, and one fell inside or outside the other, the straight lines from the centre to the circumference would not all be equal: which is absurd.

Saccheri's proof is worth quoting. It depends on three "Lemmas" immediately preceding, (1) that two straight lines cannot enclose a space, (2) that two straight lines cannot have one and the same segment common, (3) that, if two straight lines meet at a point, they do not touch, but cut one another, at it.

"Let *MDHNKM* be a circle, *A* its centre, *MN* a diameter. Suppose the portion *MNKM* of the circle turned about the fixed points *M*, *N*, so that it ultimately comes near to or coincides with the remaining portion *MNHDM*.

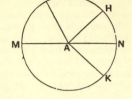

"Then (i) the whole diameter *MAN*, with all its points, clearly remains in the same position, since otherwise two straight lines would enclose a space (contrary to the first Lemma).

"(ii) Clearly no point *K* of the circumference *NKM* falls within or outside the surface enclosed by the diameter *MAN* and the other part, *NHDM*, of the circumference, since otherwise, contrary to the nature of the circle, a radius as *AK* would be less or greater than another radius as *AH*.

"(iii) Any radius *MA* can clearly be rectilineally produced only along a single other radius *AN*, since otherwise (contrary to the second Lemma) two lines assumed straight, e.g. *MAN*, *MAH*, would have one and the same common segment.

"(iv) All diameters of the circle obviously cut one another in the centre (Lemma 3 preceding), and they bisect one another there, by the general properties of the circle.

"From all this it is manifest that the diameter *MAN* divides its circle and the circumference of it just exactly into two equal parts, and the same may be generally asserted for every diameter whatsoever of the same circle; which was to be proved."

Simson observes that the property is easily deduced from III. 31 and 24; for it follows from III. 31 that the two parts of the circle are "similar segments" of a circle (segments containing equal angles, III. Def. 11), and from III. 24 that they are equal to one another.

· DEFINITION 18.

Ἡμικύκλιον δέ ἐστι τὸ περιεχόμενον σχῆμα ὑπό τε τῆς διαμέτρου καὶ τῆς ἀπολαμβανομένης ὑπ᾿ αὐτῆς περιφερείας. κέντρον δὲ τοῦ ἡμικυκλίου τὸ αὐτό, ὃ καὶ τοῦ κύκλου ἐστίν.

A semicircle is the figure contained by the diameter and the circumference cut off by it. And the centre of the semicircle is the same as that of the circle.

The last words, "And the centre of the semicircle is the same as that of the circle," are added from Proclus to the definition as it appears in the MSS. Scarburgh remarks that a semicircle has no centre, properly speaking, and thinks that the words are not Euclid's, but only a note by Proclus. I am however inclined to think that they are genuine, if only because of the very futility of an observation added by Proclus. He explains, namely, that the semicircle is the only plane figure that has its centre on its perimeter (!), "so that you may conclude that the centre has three positions, since it may be within the figure, as in the case of a circle, or on the perimeter, as with the semicircle, or outside, as with some conic lines (the single-branch hyperbola presumably)"!

Proclus and Simplicius point out that, in the order adopted by Euclid for these definitions of figures, the first figure taken is that bounded by *one* line (the circle), then follows that bounded by *two* lines (the semicircle), then the triangle, bounded by *three* lines, and so on. Proclus, as usual, distinguishes different kinds of figures bounded by two lines (pp. 159, 14—160, 9). Thus they may be formed

(1) by circumference and circumference, e.g. (*a*) those forming angles, as a *lune* (τὸ μηνοειδές) and the figure included by two arcs with convexities outward, and (*b*) the *angle-less* (ἀγώνιον), as the figure included between two concentric circles (the *coronal*) ;

(2) by circumference and straight line, e.g. the semicircle or segments of circles (ἁψῖδες is a name given to those less than a semicircle);

(3) by "mixed" line and "mixed" line, e.g. two ellipses cutting one another ;

(4) by "mixed" line and circumference, e.g. intersecting ellipse and circle ;

(5) by "mixed" line and straight line, e.g. half an ellipse.

Following Def. 18 in the MSS. is a definition of a *segment of a circle* which was obviously interpolated from III. Def. 6. Proclus, Martianus Capella and Boethius do not give it in this place, and it is therefore properly omitted.

DEFINITIONS 19, 20, 21.

19. Σχήματα εὐθύγραμμά ἐστι τὰ ὑπὸ εὐθειῶν περιεχόμενα, τρίπλευρα μὲν τὰ ὑπὸ τριῶν, τετράπλευρα δὲ τὰ ὑπὸ τεσσάρων, πολύπλευρα δὲ τὰ ὑπὸ πλειόνων ἢ τεσσάρων εὐθειῶν περιεχόμενα.

20. Τῶν δὲ τριπλεύρων σχημάτων ἰσόπλευρον μὲν τρίγωνόν ἐστι τὸ τὰς τρεῖς ἴσας ἔχον πλευράς, ἰσοσκελὲς δὲ τὸ τὰς δύο μόνας ἴσας ἔχον πλευράς, σκαληνὸν δὲ τὸ τὰς τρεῖς ἀνίσους ἔχον πλευράς.

21. Ἔτι δὲ τῶν τριπλεύρων σχημάτων ὀρθογώνιον μὲν τρίγωνόν ἐστι τὸ ἔχον ὀρθὴν γωνίαν, ἀμβλυγώνιον δὲ τὸ ἔχον ἀμβλεῖαν γωνίαν, ὀξυγώνιον δὲ τὸ τὰς τρεῖς ὀξείας ἔχον γωνίας.

19. Rectilineal figures *are those which are contained by straight lines*, trilateral *figures being those contained by three*, quadrilateral *those contained by four, and* multilateral *those contained by more than four straight lines.*

20. *Of trilateral figures, an* equilateral triangle *is that which has its three sides equal, an* isosceles triangle *that which has two of its sides alone equal, and a* scalene triangle *that which has its three sides unequal.*

21. *Further, of trilateral figures, a* right-angled triangle *is that which has a right angle, an* obtuse-angled triangle *that which has an obtuse angle, and an* acute-angled triangle *that which has its three angles acute.*

19.

The latter part of this definition, distinguishing *three-sided, four-sided* and *many-sided* figures, is probably due to Euclid himself, since the words τρίπλευρον, τετράπλευρον and πολύπλευρον do not appear in Plato or Aristotle (only in one passage of the *Mechanics* and of the *Problems* respectively does even τετράπλευρον, *quadrilateral,* occur). By his use of τετράπλευρον, quadrilateral, Euclid seems practically to have put an end to any ambiguity in the use by mathematicians of the word τετράγωνον, literally "four-angled (figure)," and to have got it restricted to the *square.* Cf. note on Def. 22.

20.

Isosceles (ἰσοσκελής, with equal legs) is used by Plato as well as Aristotle. *Scalene* (σκαληνός, with the variant σκαληνής) is used by Aristotle of a triangle with no two sides equal : cf. also Tim. Locr. 98 B. Plato, *Euthyphro* 12 D, applies the term "scalene" to an *odd* number in contrast to "isosceles" used of an even number. Proclus (p. 168, 24) seems to connect it with σκάζω, to *limp*; others make it akin to σκολιός, *crooked, aslant.* Apollonius uses the same word "scalene" of an *oblique* circular cone.

Triangles are classified, first with reference to their sides, and then with reference to their angles. Proclus points out that seven distinct species of triangles emerge: (1) the *equilateral* triangle, (2) three species of *isosceles* triangles, the right-angled, the obtuse-angled and the acute-angled, (3) the same three varieties of *scalene* triangles.

Proclus gives an odd reason for the dual classification according to sides and angles, namely that Euclid was mindful of the fact that it is not every *triangle* that is *trilateral* also. He explains this statement by reference (p. 165, 22) to a figure which some called *barb-like* (ἀκιδοειδής) while Zenodorus called it *hollow-angled* (κοιλογώνιος). Proclus mentions it again in his note on I. 22 (p. 328, 21 sqq.) as one of the paradoxes of geometry, observing that it is seen in the figure of that proposition. This "triangle" is merely a *quadrilateral* with a re-entrant angle; and the idea that it has only three angles is due to the non-recognition of the fourth angle (which is greater than two right angles) as being an angle at all. Since Proclus speaks of the *four-sided triangle* as "one of the paradoxes in geometry," it is perhaps not safe to assume that the misconception underlying the expression existed in the mind of Proclus alone; but there does not seem to be any evidence that Zenodorus called the figure in question a triangle (cf. Pappus, ed. Hultsch, pp. 1154, 1206).

DEFINITION 22.

Τῶν δὲ τετραπλεύρων σχημάτων τετράγωνον μέν ἐστιν, ὃ ἰσόπλευρόν τέ ἐστι καὶ ὀρθογώνιον, ἑτερόμηκες δέ, ὃ ὀρθογώνιον μέν, οὐκ ἰσόπλευρον δέ, ῥόμβος δέ, ὃ ἰσόπλευρον μέν, οὐκ ὀρθογώνιον δέ, ῥομβοειδὲς δὲ τὸ τὰς ἀπεναντίον πλευράς τε καὶ γωνίας ἴσας ἀλλήλαις ἔχον, ὃ οὔτε ἰσόπλευρόν ἐστιν οὔτε ὀρθογώνιον· τὰ δὲ παρὰ ταῦτα τετράπλευρα τραπέζια καλείσθω.

Of quadrilateral figures, a square *is that which is both equilateral and right-angled; an* oblong *that which is right-angled but not equilateral; a* rhombus *that which is equilateral but not right-angled; and a* rhomboid *that which has its opposite sides and angles equal to one another but is neither equilateral nor right-angled. And let quadrilaterals other than these be called* trapezia.

τετράγωνον was already a *square* with the Pythagoreans (cf. Aristotle, *Metaph.* 986 a 26), and it is so most commonly in Aristotle; but in *De anima* II. 3, 414 b 31 it seems to be a quadrilateral, and in *Metaph.* 1054 b 2, "equal and equiangular τετράγωνα," it cannot be anything else but quadrilateral if "equiangular" is to have any sense. Though, by introducing τετράπλευρον for any quadrilateral, Euclid enabled ambiguity to be avoided, there seem to be traces of the older vague use of τετράγωνον in much later writers. Thus Heron (Def. 100) speaks of a cube as "contained by six equilateral and *equiangular* τετράγωνα" and Proclus (p. 166, 10) adds to his remark about the "four-sided triangle" that "you might have τετράγωνα with more than the four sides," where τετράγωνα can hardly mean squares.

ἑτερόμηκες, oblong (with sides of *different length*), is also a Pythagorean term.

The word *right-angled* (ὀρθογώνιον) as here applied to quadrilaterals must mean *rectangular* (i.e., practically, having all its angles right angles); for, although it is tempting to take the word in the same sense for a square as for a triangle (i.e. "having *one* right angle"), this will not do in the case of the oblong, which, unless it were stated that *three* of its angles are right angles, would not be sufficiently defined.

If it be objected, as it was by Todhunter for example, that the definition of a square assumes more than is necessary, since it is sufficient that, being equilateral, it should have one right angle, the answer is that, as in other cases, the superfluity does not matter from Euclid's point of view; on the contrary, the more of the essential attributes of a thing that could be included in its definition the better, provided that the existence of the thing defined and its possession of all those attributes is proved before the definition is, actually used; and Euclid does this in the case of the square by construction in I. 46, making no use of the definition before that proposition.

The word *rhombus* (ῥόμβος) is apparently derived from ῥέμβω, to *turn round and round*, and meant among other things a *spinning-top*. Archimedes uses the term *solid rhombus* to denote a solid figure made up of two right cones with a common circular base and vertices turned in opposite directions. We can of course easily imagine this solid generated by *spinning*; and, if the cones were equal, the section through the common axis would be a *plane* rhombus, which would also be the *apparent* form of the spinning solid to the eye. The difficulty in the way of supposing the plane figure to have been named after the solid figure is that in Archimedes the cones forming the solid are not necessarily equal. It is however possible that the solid to which the name was originally given was made up of two equal cones, that the plane rhombus then received its name from that solid, and that Archimedes, in taking up the old name again, extended its signification (cf. J. H. T. Müller, *Beiträge zur Terminologie der griechischen Mathematiker*, 1860, p. 20). Proclus, while he speaks of a rhombus as being like a shaken, i.e. deformed, square, and of a rhomboid as an oblong that has been moved, tries to explain the rhombus by reference to the appearance of a *spinning* square (τετράγωνον ῥομβούμενον).

It is true that the definition of a rhomboid says more than is necessary in describing it as having its opposite sides *and angles* equal to one another. The answer to the objection is the same as the answer to the similar objection to the definition of a square.

Euclid makes no use in the *Elements* of the *oblong*, the *rhombus* and the *rhomboid*. The explanation of his inclusion of definitions of these figures is no doubt that they were taken from earlier text-books. From the words "*let* quadrilaterals other than these *be called* trapezia" we may perhaps infer that *trapezium* was a new name or a new application of an old name.

As Euclid has not yet defined parallel lines and does not anywhere define a *parallelogram*, he is not in a position to make the more elaborate classification of quadrilaterals attributed by Proclus to Posidonius and appearing also in Heron's Definitions. It may be shown by the following diagram, distinguishing seven species of quadrilaterals.

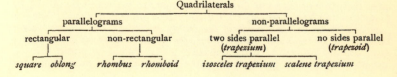

It will be observed that, while Euclid in the above definition classes as *trapezia* all quadrilaterals other than squares, oblongs, rhombi, and rhomboids, the word is in this classification restricted to quadrilaterals having two sides (only) parallel, and *trapezoid* is used to denote the rest. Euclid appears to have used *trapezium* in the restricted sense of a quadrilateral with two sides parallel in his book περὶ διαιρέσεων (on divisions of figures). Archimedes uses it in the same sense, but in one place describes it more precisely as a trapezium with its two sides parallel.

DEFINITION 23.

Παράλληλοί εἰσιν εὐθεῖαι, αἵτινες ἐν τῷ αὐτῷ ἐπιπέδῳ οὖσαι καὶ ἐκβαλλόμεναι εἰς ἄπειρον ἐφ' ἑκάτερα τὰ μέρη ἐπὶ μηδέτερα συμπίπτουσιν ἀλλήλαις.

Parallel straight lines are straight lines which, being in the same plane and being produced indefinitely in both directions, do not meet one another in either direction.

Παράλληλος (alongside one another) written in one word does not appear in Plato; but with Aristotle it was already a familiar term.

εἰς ἄπειρον cannot be translated "to infinity" because these words might seem to suggest a *region* or *place* infinitely distant, whereas εἰς ἄπειρον, which seems to be used indifferently with ἐπ' ἄπειρον, is adverbial, meaning "without limit," i.e. "indefinitely." Thus the expression is used of a magnitude being "infinitely divisible," or of a series of terms extending without limit.

In both directions, ἐφ' ἑκάτερα τὰ μέρη, literally "towards both the parts" where "parts" must be used in the sense of "regions" (cf Thuc. II. 96).

It is clear that with Aristotle the general notion of parallels was that of straight lines *which do not meet*, as in Euclid: thus Aristotle discusses the question whether to think that parallels do meet should be called a geometrical or an ungeometrical error (*Anal. post.* I. 12, 77 b 22), and (more interesting still in relation to Euclid) he observes that there is nothing surprising in different hypotheses leading to the same error, as one might conclude that parallels meet by starting from the assumption, either (*a*) that the interior (angle) is greater than the exterior, or (*b*) that the angles of a triangle make up more than two right angles (*Anal. prior.* II. 17, 66 a 11).

Another definition is attributed by Proclus to Posidonius, who said that "*parallel lines are those which, (being) in one plane, neither converge nor diverge, but have all the perpendiculars equal which are drawn from the points of one line to the other*," while such (straight lines) as make the perpendiculars less and less continually do converge to one another; for the perpendicular is enough to define (ὁρίζειν δύναται) the heights of areas and the distances between lines. For this reason, when the perpendiculars are equal, the distances between the straight lines are equal, but when they become greater and less, the interval is lessened, and the straight lines converge to one another in the direction in which the less perpendiculars are" (Proclus, p. 176, 6—17).

Posidonius' definition, with the explanation as to distances between straight lines, their convergence and divergence, amounts to the definition quoted by Simplicius (an-Nairīzī, p. 25, ed. Curtze) which described straight lines as parallel *if, when they are produced indefinitely both ways, the distance between them, or the perpendicular drawn from either of them to the other, is always equal and not different*. To the objection that it should be *proved* that the distance between two parallel lines is the perpendicular to them Simplicius replies that the definition will do equally well if all mention of the *perpendicular* be omitted and it be merely stated that the *distance* remains equal, although "for *proving* the matter in question it is necessary to say that one straight line is perpendicular to both" (an-Nairīzī, ed. Besthorn-Heiberg, p. 9). He then quotes the definition of "the philosopher Aganis": "*Parallel straight lines are straight lines, situated in the same plane, the distance between which, if they are produced indefinitely in both directions at the same time, is everywhere the same.*" (This definition forms the basis of the attempt of "Aganis" to prove the Postulate of Parallels.) On the definition Simplicius remarks that the words "situated in the same plane" are perhaps unnecessary, since, if the distance between the lines is everywhere the same, and one does not incline at all towards the other, they must for that reason be in the same plane. He adds that the "distance" referred to in the definition is the shortest line which joins things disjoined. Thus, between point and point, the distance is the straight line joining them; between a point and a straight line or between a point and a plane it is the perpendicular drawn from the point to the line or plane; "as regards the distance between two lines, that distance is, if the lines are parallel, one and the same, equal to itself at all places on the lines, it is the *shortest* distance and, at all places on the lines, perpendicular to both" (*ibid.* p. 10).

The same idea occurs in a quotation by Proclus (p. 177, 11) from Geminus. As part of a classification of lines which do not meet he observes: "Of lines which do not meet, some are in one plane with one another, others not. Of those which meet and are in one plane, *some are always the same distance from one another*, others lessen the distance continually, as the hyperbola (approaches) the straight line, and the conchoid the straight line (i.e. the asymptote in each case). For these, while the distance is being continually lessened, are continually (in the position of) not meeting, though they converge to one another; they never converge entirely, and this is the most paradoxical theorem in geometry, since it shows that the convergence of some lines is non-convergent. But of lines which are always an equal distance apart, those which are straight and never make the (distance) between them smaller, and which are in one plane, are parallel."

Thus the *equidistance*-theory of parallels (to which we shall return) is very fully represented in antiquity. I seem also to see traces in Greek writers of a conception equivalent to the vicious *direction*-theory which has been adopted in so many modern text-books. Aristotle has an interesting, though obscure, allusion in *Anal. prior.* II. 16, 65 a 4 to a *petitio principii* committed by "those who think that they draw parallels" (or "establish the theory of parallels," which is a possible translation of τὰς παραλλήλους γράφειν): ".for they unconsciously assume such things as it is not possible to demonstrate if parallels do not exist." It is clear from this that there was a vicious circle in the then current theory of parallels; something which depended for its truth on the properties of parallels was assumed in the actual proof of those properties, e.g. that the three angles of a triangle make up two right angles. This is not the case in Euclid, and the passage makes it clear that it was Euclid himself who got rid of the *petitio principii* in earlier text-books by formulating and premising before I. 29 the famous Postulate 5, which must ever be regarded as among the most epoch-making achievements in the domain of geometry. But one of the commentators on Aristotle, Philoponus, has a note on the above passage purporting to give the specific character of the *petitio principii* alluded to; and it is here that a *direction*-theory of parallels may be hinted at, whether Philoponus is or is not right in supposing that this was what Aristotle had in mind. Philoponus says: "The same thing is done by those who draw parallels, namely begging the original question; for they will have it that it is possible to draw parallel straight lines from the meridian circle, and they assume a point, so to say, falling on the plane of that circle and thus they draw the straight lines. And what was sought is thereby assumed; for he who does not admit the genesis of the parallels will not admit the point referred to either." What is meant is, I think, somewhat as follows. Given a straight line and a point through which a parallel to it is to be drawn, we are to suppose the given straight line placed in the plane of the meridian. Then we are told to draw through the given point another straight line in the plane of the meridian (strictly speaking it should be drawn in a plane parallel to the plane of the meridian, but the idea is that, compared with the size of the meridian circle, the distance between the point and the straight line is negligible); and this, as I read Philoponus, is supposed to be equivalent to assuming a very distant point in the meridian plane and joining the given point to it. But obviously no ruler would stretch to such a point, and the objector would say that we cannot really direct a straight line to the assumed distant point except by drawing it, without more ado, *parallel* to the given straight line. And herein is the *petitio principii*. I am confirmed in seeing in Philoponus an allusion to a *direction*-theory by a remark of Schotten on a similar reference to the meridian plane supposed to be used by advocates of that theory. Schotten is arguing that direction is not in itself a conception such that you can predicate *one* direction of *two* different lines. "If any one should reply that nevertheless many lines can be conceived *which all have the direction from north to south,*" he replies that this represents only a nominal, not a real, identity of direction.

Coming now to modern times, we may classify under three groups practically all the different definitions that have been given of parallels (Schotten, *op. cit.* II. p. 188 sqq.).

(1) *Parallel straight lines have no point common*, under which general conception the following varieties of statement may be included:

(a)   *they do not cut one another,*

(b)   *they meet at infinity,* or

(c)   *they have a common point at infinity.*

(2) *Parallel straight lines have the same, or like, direction or directions,* under which class of definitions must be included all those which introduce transversals and say that the parallels *make equal angles with a transversal.*

(3) *Parallel straight lines have the distance between them constant*; with which group we may connect the attempt to explain a parallel as *the geometrical locus of all points which are equidistant from a straight line.*

But the three points of view have a good deal in common; some of them lead easily to the others. Thus the idea of the lines having no point common led to the notion of their having a common point at infinity, through the influence of modern geometry seeking to embrace different cases under one conception; and then again the idea of the lines having a common point at infinity might suggest their having the same direction. The "non-secant" idea would also naturally lead to that of equidistance (3), since our observation shows that it is things which come nearer to one another that tend to meet, and hence, if lines are not to meet, the obvious thing is to see that they shall not come nearer, i.e. shall remain the same distance apart.

We will now take the three groups in order.

(1) The first observation of Schotten is that the varieties of this group which regard parallels as (a) meeting at infinity or (b) having a common point at infinity (first mentioned apparently by Kepler, 1604, as a "façon de parler" and then used by Desargues, 1639) are at least unsuitable definitions for elementary text-books. How do we know that the lines cut or meet at infinity? We are not entitled to assume either that they do or that they do not, because "infinity" is outside our field of observation and we cannot verify either. As Gauss says (letter to Schumacher), "Finite man cannot claim to be able to regard the infinite as something to be grasped by means of ordinary methods of observation." Steiner, in speaking of the rays passing through a point and successive points of a straight line, observes that as the point of intersection gets further away the ray moves continually in one and the same direction ("nach einer und derselben Richtung hin"); only in one position, that in which it is parallel to the straight line, "there is *no real cutting*" between the ray and the straight line; what we have to say is that the ray is "*directed towards the infinitely distant point on the straight line.*" It is true that higher geometry has to assume that the lines do meet at infinity: whether such lines exist in nature or not does not matter (just as we deal with "straight lines" although there is no such thing as a straight line). But if two lines do not cut at any finite distance, may not the same thing be true at infinity also? Are lines conceivable which would not cut even at infinity but always remain at the same distance from one another even there? Take the case of a line of railway. Must the two rails meet at infinity so that a train could not stand on them there (whether we could *see* it or not makes no difference)? It seems best therefore to leave to higher geometry the conception of infinitely distant points on a line and of two straight lines meeting at infinity, like *imaginary* points of intersection, and, for the purposes of elementary geometry, to rely on the plain distinction between "parallel" and "cutting" which average human intelligence can readily grasp. This is the method adopted by Euclid in his definition, which of course belongs to the group (1) of definitions regarding parallels as non-secant.

It is significant, I think, that such authorities as Ingrami (*Elementi di geometria*, 1904) and Enriques and Amaldi (*Elementi di geometria*, 1905), after all the discussion of principles that has taken place of late years, give definitions of parallels equivalent to Euclid's: "those straight lines in a plane which have not any point in common are called parallels." Hilbert adopts the same point of view. Veronese, it is true, takes a different line. In his great work *Fondamenti di geometria*, 1891, he had taken a ray to be parallel to another when a point at infinity on the second is situated on the first; but he appears to have come to the conclusion that this definition was unsuitable for his *Elementi*. He avoids however giving the Euclidean definition of parallels as "straight lines in a plane which, though produced indefinitely, never meet," because "no one has ever seen two straight lines of this sort," and because the postulate generally used in connexion with this definition is not evident in the way that, in the field of our experience, it is evident that only one straight line can pass through two points. Hence he gives a different definition, for which he claims the advantage that it is independent of the plane. It is based on a definition of figures "opposite to one another with respect to a point" (or *reflex* figures). "Two figures are opposite to one another with respect to a point *O*, e.g. the figures *ABC* ... and *A'B'C'* ..., if to every point of the one there corresponds one sole point of the other, and if the segments

OA, *OB*, *OC*, ... joining the points of one figure to *O* are respectively equal and opposite to the segments *OA'*, *OB'*, *OC'*, ... joining to *O* the corresponding points of the second": then, a *transversal* of two straight lines being any segment having as its extremities one point of one line and one point of the other, "*two straight lines are called parallel if one of them contains two points opposite to two points of the other with respect to the middle point of a common transversal.*" It is true, as Veronese says, that the parallels so defined and the parallels of Euclid are in substance the same; but it can hardly be said that the definition gives as good an idea of the essential nature of parallels as does Euclid's. Veronese has to *prove*, of course, that his parallels have no point in common, and his "Postulate of Parallels" can hardly be called more evident than Euclid's: "If two straight lines are parallel, they are figures opposite to one another with respect to the middle points of all their transversal segments."

(2) The *direction*-theory.

The fallacy of this theory has nowhere been more completely exposed than by C. L. Dodgson (*Euclid and his modern Rivals*, 1879). According to Killing (*Einführung in die Grundlagen der Geometrie*, I. p. 5) it would appear to have originated with no less a person than Leibniz. In the text-books which employ this method the notion of *direction* appears to be regarded as a primary, not a derivative notion, since no definition is given. But we ought at least to know how the same direction or like directions can be recognised when two different straight lines are in question. But no answer to this question is forthcoming. The fact is that the whole idea as applied to non-coincident straight lines is derived from knowledge of the properties of *parallels*; it is a case of explaining a thing by itself. The idea of parallels being in the *same* direction perhaps arose from the conception of an angle as a *difference* of direction (the hollowness of which has already been exposed); sameness of direction for parallels follows from the same "difference of direction" which both exhibit relatively to a third line. But this is not enough. As Gauss said (*Werke*, IV. p. 365), "If it [identity of direction] is recognised by the equality of the angles formed with *one* third straight line, we do not yet know without an antecedent proof whether this same equality will also be found in the angles formed with a *fourth* straight line" (and any number of other transversals); and in order to make this theory of parallels valid, so far from getting rid of axioms such as Euclid's, you would have to assume as an axiom what is much less axiomatic, namely that "straight lines which make equal corresponding angles with a certain transversal do so with *any* transversal" (Dodgson, p. 101).

(3) In modern times the conception of parallels as *equidistant* straight lines was practically adopted by Clavius (the editor of Euclid, born at Bamberg, 1537) and (according to Saccheri) by Borelli (*Euclides restitutus*, 1658) although they do not seem to have *defined* parallels in this way. Saccheri points out that, before such a definition can be used, it has to be *proved* that "the geometrical locus of points equidistant from a straight line is a straight line." To do him justice, Clavius saw this and tried to prove it: he makes out that the locus is a straight line according to the definition of Euclid, because "it lies evenly with respect to all the points on it"; but there is a confusion here, because such "evenness" as the locus has is with respect to the straight line from which its points are equidistant, and there is nothing to show that it possesses this property with respect to itself. In fact the theorem cannot be proved without a postulate.

POSTULATE I.

'Ηιτήσθω ἀπὸ παντὸς σημείου ἐπὶ πᾶν σημεῖον εὐθεῖαν γραμμὴν ἀγαγεῖν.

Let the following be postulated: to draw a straight line from any point to any point.

From any point to any point. In general statements of this kind the Greeks did not say, as we do, "*any* point," "*any* triangle" etc., but "*every* point," "*every* triangle" and the like. Thus the words are here literally "from every point to every point." Similarly the first words of Postulate 3 are "with *every* centre and distance," and the enunciation, e.g., of I. 18 is "In *every* triangle the greater side subtends the greater angle."

It will be remembered that, according to Aristotle, the geometer must in general assume *what* a thing is, or its definition, but must prove *that* it is, i.e. the *existence* of the thing corresponding to the definition : only in the case of the two most primary things, points and lines, does he assume, without proof, both the definition and the existence of the thing defined. Euclid has indeed no separate assumption affirming the existence of *points* such as we find nowadays in text-books like those of Veronese, Ingrami, Enriques, "there exist distinct points" or "there exist an infinite number of points." But, as regards the only lines dealt with in the *Elements,* straight lines and circles, existence is asserted in Postulates 1 and 3 respectively. Postulate 1 however does much more than (1) postulate the existence of straight lines. It is (2) an answer to a possible objector who should say that you cannot, with the imperfect instruments at your disposal, draw a mathematical straight line at all, and consequently (in the words of Aristotle, *Anal. post.* I. 10, 76 b 41) that the geometer uses false hypotheses, since he calls a line a foot long when it is not or straight when it is not straight. It would seem (if Gherard's translation is right) that an-Nairīzī saw that one purpose of the Postulate was to refute this criticism : "the utility of the first three postulates is (to ensure) that the weakness of our equipment shall not prevent (scientific) demonstration" (ed. Curtze, p. 30). The fact is, as Aristotle says, that the geometer's demonstration is not concerned with the particular imperfect straight line which he has drawn, but with the ideal straight line of which it is the imperfect representation. Simplicius too indicates that the object of the Postulate is rather to enable the drawing of a mathematical straight line to be *imagined* than to assert that it can actually be realised in practice : "he would be a rash person who, taking things as they actually are, should postulate the drawing of a straight line from Aries to Libra."

There is still something more that must be inferred from the Postulate combined with the definition of a straight line, namely (3) that the straight line joining two points is *unique* : in other words that, *if two straight lines* ("rectilineal segments," as Veronese would call them) *have the same extremities, they must coincide throughout their length.* The omission of Euclid to state this in so many words, though he assumes it in I. 4, is no doubt answerable for the interpolation in the text of the equivalent assumption that *two straight lines cannot enclose a space,* which has constantly appeared in MSS. and editions of Euclid, either among Axioms or Postulates. That Postulate 1 included it, by conscious implication, is even clear from Proclus' words in his note on I. 4 (p. 239, 16): "therefore two straight lines do not enclose a space, and it was with knowledge of this fact that the writer of the Elements said in the first of his Postulates, *to draw a straight line from any point to any point,* implying that it is *one* straight line which would always join the two points, not *two.*"

Proclus attempts in the same note (p. 239) to *prove* that two straight lines cannot enclose a space, using as his basis the definition of the diameter of a circle and the theorem, stated in it, that any diameter divides the circle into two equal parts.

Suppose, he says, *ACB*, *ADB* to be two straight lines enclosing a space. Produce them (beyond *B*) indefinitely. With centre *B* and distance *AB* describe a circle, cutting the lines so produced in *F*, *E* respectively.

Then, since *ACBF*, *ADBE* are both diameters cutting off semi-circles, the arcs *AE*, *AEF* are equal: which is impossible. Therefore etc.

It will be observed, however, that the straight lines produced are assumed to meet the circle given in two *different* points *E*, *F*, whereas, for anything we know, *E*, *F* might coincide and the straight lines have *three* common points. The proof is therefore delusive.

Saccheri gives a different proof. From Euclid's definition of a straight line as that which lies evenly with its points he infers that, when such a line is turned about its two extremities, which remain fixed, all the points on it must remain throughout in the same position, and cannot take up different positions as the revolution proceeds. "In this view of the straight line the truth of the assertion that two straight lines do not enclose a space is obviously involved. In fact, if two lines are given which enclose a space, and of which the two points *A* and *X* are the common extremities, it is easily shown that neither, or else only one, of the two lines is straight."

It is however better to assume as a *postulate* the fact, inseparably connected with the idea of a straight line, that *there exists only one straight line containing two given points*, or, *if two straight lines have two points in common, they coincide throughout.*

POSTULATE 2.

Καὶ πεπερασμένην εὐθεῖαν κατὰ τὸ συνεχὲς ἐπ' εὐθείας ἐκβαλεῖν.
To produce a finite straight line continuously in a straight line.

I translate πεπερασμένην by *finite*, because that is the received equivalent, and because any alternative word such as *limited, terminated*, if applied to a straight line, would equally fail to express what modern Italian geometers aptly call a *rectilineal segment*, that is, a straight line having *two* extremities.

Just as Post. 1 asserting the possibility of drawing a straight line from any one point to another must be held to declare at the same time that the straight line so drawn is unique, so Post. 2 maintaining the possibility of producing a finite straight line (a "rectilineal segment") continuously in a straight line must also be held to assert that the straight line can only be produced *in one way* at either end, or that the produced part in either direction is *unique*; in other words, that *two straight lines cannot have a common segment*. This latter assumption is not expressly appealed to by Euclid until XI. 1. But it is needed at the very beginning of Book I. Proclus (p. 214, 18) says that Zeno of Sidon, an Epicurean, maintained that the very first proposition I. 1 requires it to be admitted that "two straight lines cannot have the same segments"; otherwise *AC*, *BC* might meet before they arrive at *C* and have the rest of their length common, in which case the actual triangle formed by them and *AB* would not be equilateral. The assumption that two straight lines cannot have a common segment is certainly necessary in I. 4, where one side of one triangle is placed on that side of the other triangle which is equal to it, and it is inferred that the two coincide throughout their length : this would by no means follow if two straight lines could have a common segment. Proclus (p. 215, 24), while observing that Post. 2 clearly indicates that the produced portion must be *one*, attempts to prove it, but unsuccessfully. Both he and Simplicius practically use the same argument. Suppose, says Proclus, that the straight lines *AC, AD* have *AB* as a common segment. With centre *B* and radius *BA* describe a circle (Post. 3) meeting *AC, AD* in *C, D.* Then, since *ABC* is a straight line through the centre, *AEC* is a semi-circle. Similarly, *ABD* being a straight line through the centre, *AED* is a semi-circle. Therefore *AEC* is equal to *AED*: which is impossible.

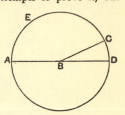

Proclus observes that Zeno would object to this proof as really depending on the assumption that "two circumferences (of circles) cannot have one portion common"; for this, he would say, is assumed in the common proof by superposition of the fact that a circle is bisected by a diameter, since that proof takes it for granted that, if one part of the circumference cut off by the diameter, when applied to the other, does not coincide with it, it must necessarily fall either *entirely* outside or *entirely* inside it, whereas there is nothing to prevent their coinciding, not altogether, but in part only ; and, until you really prove the bisection of a circle by its diameter, the above proof is not valid. Posidonius is represented as having derided Zeno for not seeing that the proof of the bisection of a circle by its diameter goes on just as well if the circumferences fail to coincide *in part* only. But the true objection to the proof above given is that the proof of the bisection of a circle by any diameter *itself* assumes that two straight lines cannot have a common segment; for, if we wish to draw the diameter of a circle which has its extremity at a given point of the circumference we have to join the latter point to the centre (Post. 1) and then to *produce* the straight line so drawn till it meets the circle again (Post. 2), and it is necessary for the proof that the produced part shall be *unique*.

Saccheri adopted the proper order when he gave, first the proposition that two straight lines cannot have a common segment, and after that the proposition that any diameter of a circle bisects the circle and its circumference.

Saccheri's proof of the former is very interesting as showing the thoroughness of his method, if not at the end entirely convincing. It is in five stages which I shall indicate shortly, giving the full argument of the first only.

Suppose, if possible, that *AX* is a common segment of both the straight lines *AXB, AXC*, in one plane, produced beyond *X*. Then describe about *X* as centre, with radius *XB* or *XC*, the arc *BMC*, and draw through *X* to any point on it the straight line *XM*.

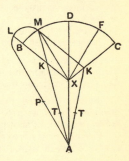

(i) I maintain that, with the assumption made, *the line* AXM *is also a straight line which is drawn from the point* A *to the point* X *and produced beyond* X.

For, if this line were not straight, we could draw another straight line *AM* which for its part would be straight. This straight line will either (*a*) cut one of the two straight lines *XB, XC* in a certain point *K* or (*b*) enclose one of them, for instance *XB*, in the area bounded by *AX, XM* and *APLM*.

But the first alternative (*a*) obviously contradicts the foregoing lemma [that two straight lines cannot enclose a space], since in that case the two lines *AXK*, *ATK*, which by hypothesis are straight, would enclose a space.

The second possibility (*b*) is at once seen to involve a similar absurdity. For the straight line *XB* must, when produced beyond *B*, ultimately meet *APLM* in a point *L*. Consequently the two lines *AXBL*, *APL*, which by hypothesis are straight, would again enclose a space. If however we were to assume that the straight line *XB* produced beyond *B* will ultimately meet either the straight line *XM* or the straight line *XA* in another point, we should in the same way arrive at a contradiction.

From this it obviously follows that, on the assumption made, the line *AXM* is itself the straight line which was drawn from the point *A* to the point *M*; and that is what was maintained.

The remaining stages are in substance these.

(ii)   *If the straight line* AXB, *regarded as rigid, revolves about* AX *as axis, it cannot assume two more positions in the same plane, so that, for example, in one position* XB *should coincide with* XC, *and in the other with* XM.

[This is proved by considerations of symmetry. *AXB* cannot be altogether "similar or equal to" *AXC*, if viewed from the same side (left or right) of both : otherwise they would coincide, which by hypothesis they do not. But there is nothing to prevent *AXB* viewed from one side (say the left) being "similar or equal to" *AXC* viewed from the other side (i.e. the right), so that *AXB can*, without any change, be brought into the position *AXC*.

AXB cannot however take the position of the other straight line *AXM* as well. If they were like on one side, they would coincide; if they were like on opposite sides, *AXM*, *AXC* would be like on the same side and therefore coincide.]

(iii)   The other positions of *AXB* during the revolution must be above or below the original plane.

(iv)   It is next maintained that *there* is *a point* D *on the arc* BC *such that, if* XD *is drawn*, AXD *is not only a straight line but is such that viewed from the left side it is exactly "similar or equal" to what it is when viewed from the right side.*

[*First*, it is proved that points *M*, *F* can be found on the arc, corresponding in the same way as *B*, *C* do, but nearer together, and of course *AXM*, *AXF* are both straight lines.

Secondly, similar corresponding points can be found still nearer together, and so on continually, until either (*a*) we come to *one* point *D* such that *AXD is exactly like itself when the right and left sides are compared*, or (*b*) there are *two* ultimate points of this sort *M*, *F*, so that *both AXM, AXF have this property.*

Thirdly, (*b*) is ruled out by reference to the definition of a straight line.

Hence (*a*) only is true, and there is only *one* point *D* such as described.]

(v)   Lastly, Saccheri concludes that the straight line *AXD* so determined "is *alone* a straight line, and the *immediate* prolongation from *A* beyond *X* to *D*," relying again on the definition of a straight line as "lying evenly."

Simson deduced the proposition that *two straight lines cannot have a common segment* as a corollary from I. 11; but his argument is a complete *petitio principii*, as shown by Todhunter in his note on that proposition.

Proclus (p. 217, 10) records an ancient proof also based on the proposition I. 11. Zeno, he says, propounded this proof and then criticised it.

Suppose that two straight lines AC, AD have a common segment AB, and let BE be drawn at right angles to AC.

Then the angle EBC is right.

If then the angle EBD is also right, the two angles will be equal : which is impossible.

If the angle EBD is not right, draw BF at right angles to AD; therefore the angle FBA is right.

But the angle EBA is right.

Therefore the angles EBA, FBA are equal : which is impossible.

Zeno objected to this, says Proclus, because it assumed the later proposition I. 11 for its proof. Posidonius said that there was no trace of such a proof to be found in the text-books of Elements, and that it was only invented by Zeno for the purpose of slandering contemporary geometers. Posidonius maintains further that even this proof has something to be said for it. There must be some straight line at right angles to each of the two straight lines AC, AD (the very definition of right angles assumes this): "*suppose then it happens to be the straight line we have set up.*" Here then we have an ancient instance of a defence of *hypothetical construction*, but in such apologetic terms ("it is possible to say *something* even for this proof") that we may conclude that in general it would not have been accepted by geometers of that time as a legitimate means of proving a proposition.

Todhunter proposed to deduce that *two straight lines cannot have a common segment* from I. 13. But this will not serve either, since, as before mentioned, the assumption is really required for I. 4.

It is best to make it a postulate.

POSTULATE 3.

Καὶ παντὶ κέντρῳ καὶ διαστήματι κύκλον γράφεσθαι.

To describe a circle with any centre and distance.

In this case Euclid's text has the passive of the verb: "a circle can be drawn"; Proclus however has the active (γράψαι) as Euclid has in the first two Postulates.

Distance, διαστήματι. This word, meaning "distance" quite generally (cf. Arist. *Metaph.* 1055 a 9 "it is between extremities that distance is greatest," *ibid.* 1056 a 36 "things which have something between them, that is, a certain distance"), and also "distance" in the sense of "dimension" (as in "space has three dimensions, length, breadth and depth," Arist. *Physics* IV. 1, 209 a 4), was the regular word used for describing a circle with a certain *radius*, the idea being that each point of the circumference was at that *distance* from the centre (cf. Arist *Meteorologica* III. 5, 376 b 8 : "if a circle be drawn...with distance MII"). The Greeks had no word corresponding to *radius*: if they had to express it, they said "(straight lines) drawn from the centre" (αἱ ἐκ τοῦ κέντρου, Eucl. III. Def. 1 and Prop. 26; *Meteorologica* II. 5, 362 b 1 has the full phrase αἱ ἐκ τοῦ κέντρου ἀγόμεναι γραμμαί).

Mr Frankland observes that it would be remarkable if, unlike Postulates 1 and 2, this Postulate implied *merely* what it says, that a circle can be drawn with any centre and distance. We may regard it, if we please, as helping to the complete delineation of the Space which Euclid's geometry is to investigate formally. The Postulate has the effect of removing any restriction upon the size of the circle. It may (1) be indefinitely small, and this implies that space is *continuous*, not discrete, with an irreducible minimum distance between contiguous points in it. (2) The circle may be indefinitely large, which implies the fundamental hypothesis of *infinitude* of space. This last assumed characteristic of space is essential to the proof of I. 16, a theorem not universally valid in a space which is unbounded in extent but finite in size. It would however be unsafe to suppose that Euclid foresaw the use to which his Postulate might thus be put, or formulated it with such an intention.

POSTULATE 4.

Καὶ πάσας τὰς ὀρθὰς γωνίας ἴσας ἀλλήλαις εἶναι.
That all right angles are equal to one another.

While this Postulate asserts the essential truth that a right angle is a *determinate magnitude* so that it really serves as an invariable standard by which other (acute and obtuse) angles may be measured, much more than this is implied, as will easily be seen from the following consideration. If the statement is to be *proved*, it can only be proved by the method of applying one pair of right angles to another and so arguing their equality. But this method would not be valid unless on the assumption of the *invariability of figures*, which would therefore have to be asserted as an antecedent postulate. Euclid preferred to assert as a postulate, directly, the fact that all right angles are equal; and hence his postulate must be taken as equivalent to the principle of *invariability of figures* or its equivalent, the *homogeneity of space*.

According to Proclus, Geminus held that this Postulate should not be classed as a postulate but as an axiom, since it does not, like the first three Postulates, assert the possibility of some *construction* but expresses an essential property of right angles. Proclus further observes (p. 188, 8) that it is not a postulate in Aristotle's sense either. (In this I think he is wrong, as explained above.) Proclus himself, while regarding the assumption as axiomatic ("the equality of right angles suggests itself even by virtue of our common notions"), is prepared with a proof, if such is asked for.

Let *ABC*, *DEF* be two right angles.

If they are not equal, one of them must be the greater, say *ABC*.

Then, if we apply *DE* to *AB*, *EF* will fall within *ABC*, as *BG*.

Produce *CB* to *H*. Then, since *ABC* is a right angle, so is *ABH*, and the two angles are equal (a right angle being by definition equal to its adjacent angle).

Therefore the angle *ABH* is *greater* than the angle *ABG*.

Producing *GB* to *K*, we have similarly the two angles *ABK*, *ABG* both right and equal to one another; whence the angle *ABH* is *less* than the angle *ABG*.

But it is also greater: which is impossible.

Therefore etc.

A defect in this proof is the assumption that *CB*, *GB* can each be produced only in one way, and that *BK* falls outside the angle *ABH*.

Saccheri's proof is more careful in that he premises a third lemma in addition to those asserting (1) that two straight lines cannot enclose a space and (2) that two straight lines cannot have a common segment. The third lemma is: *If two straight lines* AB, CXD *meet one another at an intermediate point* X, *they do not touch at that point, but* cut one another.

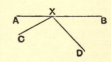

Suppose now that *DA* standing on *BAC* makes the two angles *DAB*, *DAC* equal, so that each is a right angle by the definition.

Similarly, let *LH* form with the straight line *FHM* the right angles *LHF*, *LHM*.

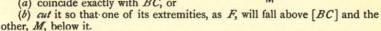

Let *DA*, *HL* be equal; and suppose the whole of the second figure so laid upon the first that the point *H* falls on *A*, and *L* on *D*.

Then the straight line *FHM* will (by the third lemma) not *touch* the straight line *BC* at *A*; it will either

(*a*) coincide exactly with *BC*, or

(*b*) *cut* it so that one of its extremities, as *F*, will fall above [*BC*] and the other, *M*, below it.

If the alternative (*a*) is true, we have already proved the exact equality of all rectilineal right angles.

Under alternative (*b*) we prove that the angle *LHF*, being equal to the angle *DAF*, is less than the angle *DAB* or *DAC*, and *a fortiori* less than the angle *DAM* or *LHM*: which is contrary to the hypothesis.

[Hence (*a*) is the only possible alternative, so that all right angles are equal.]

Saccheri adds that it makes no difference if the angle *DAF* diverges *infinitely little* from the angle *DAB*. This would equally lead to a conclusion contradicting the hypothesis.

It will be observed that Saccheri speaks of "the exact equality of all *rectilineal* right angles." He may have had in mind the remark of Pappus, quoted by Proclus (p. 189, 11), that the converse of this postulate, namely that an angle which is equal to a right angle is also right, is not necessarily true, unless the former angle is *rectilineal*. Suppose two equal straight lines *BA*, *BC* at right angles to one another, and semi-circles described on *BA*, *BC* respectively as *AEB*, *BDC* in the figure. Then, since the semi-circles are equal, they coincide if applied to one another. Hence the "angles" *EBA*, *DBC* are equal. Add to each the "angle"

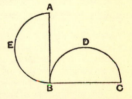

ABD; and it follows that the *lunular* angle *EBD* is equal to the right angle *ABC*. (Similarly, if *BA*, *BC* be inclined at an acute or obtuse angle, instead of at a right angle, we find a *lunular* angle equal to an acute or obtuse angle.) This is one of the curiosities which Greek commentators delighted in.

Veronese, Ingrami, and Enriques and Amaldi deduce the fact that *all right angles are equal* from the equivalent fact that *all flat angles are equal*, which is either itself assumed as a postulate or immediately deduced from some other postulate.

Hilbert takes quite a different line. He considers that Euclid did wrong in placing Post. 4 among "axioms." He himself, after his Group III. of Axioms containing six relating to congruence, proves several theorems about the congruence of triangles and angles, and then deduces our Postulate.

As to the *raison d'être* and the place of Post. 4 one thing is quite certain. It was essential from Euclid's point of view that it should come before Post. 5, since the condition in the latter that a certain pair of angles are together less than two right angles would be useless unless it were first made clear that right angles are angles of determinate and invariable magnitude.

POSTULATE 5.

Καὶ ἐὰν εἰς δύο εὐθείας εὐθεία ἐμπίπτουσα τὰς ἐντὸς καὶ ἐπὶ τὰ αὐτὰ μέρη γωνίας δύο ὀρθῶν ἐλάσσονας ποιῇ, ἐκβαλλομένας τὰς δύο εὐθείας ἐπ' ἄπειρον συμπίπτειν, ἐφ' ἃ μέρη εἰσὶν αἱ τῶν δύο ὀρθῶν ἐλάσσονες.

That, if a straight line falling on two straight lines make the interior angles on the same side less than two right angles, the two straight lines, if produced indefinitely, meet on that side on which are the angles less than the two right angles.

Although Aristotle gives a clear idea of what he understood by a *postulate*, he does not give any instances from geometry; still less has he any allusion recalling the particular postulates found in Euclid. We naturally infer that the formulation of these postulates was Euclid's own work. There is a more positive indication of the originality of Postulate 5, since in the passage (*Anal. prior.* II. 16, 65 a 4) quoted above in the note on the definition of parallels he alludes to some *petitio principii* involved in the theory of parallels current in his time. This reproach was removed by Euclid when he laid down this epoch-making Postulate. When we consider the countless successive attempts made through more than twenty centuries to prove the Postulate, many of them by geometers of ability, we cannot but admire the genius of the man who concluded that such a hypothesis, which he found necessary to the validity of his whole system of geometry, was really indemonstrable.

From the very beginning, as we know from Proclus, the Postulate was attacked as such, and attempts were made to prove it as a theorem or to get rid of it by adopting some other definition of parallels; while in modern times the literature of the subject is enormous. Riccardi (*Saggio di una bibliografia Euclidea*, Part IV., Bologna, 1890) has twenty quarto pages of titles of monographs relating to Post. 5 between the dates 1607 and 1887. Max Simon (*Ueber die Entwicklung der Elementar-geometrie im XIX. Jahrhundert*, 1906) notes that he has seen three new attempts, as late as 1891 (a century after Gauss laid the foundation of non-Euclidean geometry), to prove the theory of parallels independently of the Postulate. Max Simon himself (pp. 53—61) gives a large number of references to books or articles on the subject and refers to the copious information, as to contents as well as names, contained in Schotten's *Inhalt und Methode des planimetrischen Unterrichts*, II. pp. 183—332.

This note will include some account of or allusion to a few of the most noteworthy attempts to prove the Postulate. Only those of ancient times, as being less generally accessible, will be described at any length; shorter references must suffice in the case of the modern geometers who have made the most important contributions to the discussion of the Postulate and have thereby, in particular, contributed most towards the foundation of the non-Euclidean geometries, and here I shall make use principally of the valuable Article 8, *Sulla teoria delle parallele e sulle geometrie non-euclidee* (by Roberto Bonola), in *Questioni riguardanti le matematiche elementari*, I. pp. 247—363.

Proclus (p. 191, 21 sqq.) states very clearly the nature of the first objections taken to the Postulate.

"This ought even to be struck out of the Postulates altogether; for it is a theorem involving many difficulties, which Ptolemy, in a certain book, set himself to solve, and it requires for the demonstration of it a number of definitions as well as theorems. And the converse of it is actually proved by Euclid himself as a theorem. It may be that some would be deceived and would think it proper to place even the assumption in question among the postulates as affording, in the lessening of the two right angles, ground for an instantaneous belief that the straight lines converge and meet. To such as these Geminus correctly replied that we have learned from the very pioneers of this science not to have any regard to mere plausible imaginings when it is a question of the reasonings to be included in our geometrical doctrine. For Aristotle says that it is as justifiable to ask scientific proofs of a rhetorician as to accept mere plausibilities from a geometer; and Simmias is made by Plato to say that he recognises as quacks those who fashion for themselves proofs from probabilities. So in this case the fact that, when the right angles are lessened, the straight lines converge is true and necessary; but the statement that, since they converge more and more as they are produced, they will sometime meet is plausible but not necessary, in the absence of some argument showing that this is true in the case of straight lines. For the fact that some lines exist which approach indefinitely, but yet remain non-secant (ἀσύμπτωτοι), although it seems improbable and paradoxical, is nevertheless true and fully ascertained with regard to other species of lines. May not then the same thing be possible in the case of straight lines which happens in the case of the lines referred to? Indeed, until the statement in the Postulate is clinched by proof, the facts shown in the case of other lines may direct our imagination the opposite way. And, though the controversial arguments against the meeting of the straight lines should contain much that is surprising, is there not all the more reason why we should expel from our body of doctrine this merely plausible and unreasoned (hypothesis)?

"It is then clear from this that we must seek a proof of the present theorem, and that it is alien to the special character of postulates. But how it should be proved, and by what sort of arguments the objections taken to it should be removed, we must explain at the point where the writer of the Elements is actually about to recall it and use it as obvious. It will be necessary at that stage to show that its obvious character does not appear independently of proof, but is turned by proof into matter of knowledge."

Before passing to the attempts of Ptolemy and Proclus to prove the Postulate, I should note here that Simplicius says (in an-Nairīzī, ed. Besthorn-Heiberg, p. 119, ed. Curtze, p. 65) that this Postulate is by no means manifest, but requires proof, and accordingly "Abthiniathus" and Diodorus had already proved it by means of many different propositions, while Ptolemy also had explained and proved it, using for the purpose Eucl. I. 13, 15 and 16 (or 18). The Diodorus here mentioned may be the author of the *Analemma* on which Pappus wrote a commentary. It is difficult even to frame a conjecture as to who "Abthiniathus" is. In one place in the Arabic text the name appears to be written "Anthisathus" (H. Suter in *Zeitschrift für Math. und Physik*, XXXVIII., hist. litt. Abth. p. 194). It has occurred to me whether he might be Peithon, a friend of Serenus of Antinoeia (Antinoupolis) who was long known as Serenus of *Antissa*. Serenus says (*De sectione cylindri*, ed. Heiberg, p. 96): "Peithon the geometer, explaining parallels in a work of his, was not satisfied with what Euclid said, but showed their nature more cleverly by an example; for he says that parallel straight lines are such a thing as we see on walls or on the ground in the shadows of pillars which are made when either a torch or a lamp is burning behind them. And, although this has only been matter of merriment to every one, I at least must not deride it, for the respect I have for the author, who is my friend." If Peithon was known as "of Antinoeia" or "of Antissa," the two forms of the mysterious name might perhaps be an attempt at an equivalent; but this is no more than a guess.

Simplicius adds in full and word for word the attempt of his "friend" or his "master Aganis" to prove the Postulate.

Proclus returns to the subject (p. 365, 5) in his note on Eucl. I. 29. He says that before his time a certain number of geometers had classed as a theorem this Euclidean postulate and thought it matter for proof, and he then proceeds to give an account of Ptolemy's argument.

Noteworthy attempts to prove the Postulate.

Ptolemy.

We learn from Proclus (p. 365, 7—11) that Ptolemy wrote a book on the proposition that "straight lines drawn from angles less than two right angles meet if produced," and that he used in his "proof" many of the theorems in Euclid preceding I. 29. Proclus excuses himself from reproducing the early part of Ptolemy's argument, only mentioning as one of the propositions proved in it the theorem of Eucl. I. 28 that, if two straight lines meeting a transversal make the two interior angles on the same side equal to two right angles, the straight lines do not meet, however far produced.

I. From Proclus' note on I. 28 (p. 362, 14 sq.) we know that Ptolemy proved this somewhat as follows.

Suppose that there are two straight lines AB, CD, and that $EFGH$, meeting them, makes the angles BFG, FGD equal to two right angles. I say that AB, CD are parallel, that is, they are non-secant.

For, if possible, let FB, GD meet at K.

Now, since the angles BFG, FGD are equal to two right angles, while the four angles AFG, BFG, FGD, FGC are together equal to four right angles, the angles AFG, FGC are equal to two right angles.

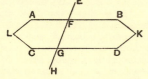

"*If therefore* FB, GD, *when the interior angles are equal to two right angles, meet at* K, *the straight lines* FA, GC *will also meet if produced;* for the angles AFG, CGF are also equal to two right angles.

"Therefore the straight lines will either meet in both directions or in neither direction, if the two pairs of interior angles are both equal to two right angles.

"Let, then, FA, GC meet at L.

"Therefore the straight lines $LABK$, $LCDK$ enclose a space: which is impossible.

"Therefore it is not possible for two straight lines to meet when the interior angles are equal to two right angles. Therefore they are parallel."

[The argument in the words italicised would be clearer if it had been shown that the two interior angles on one side of EH are *severally* equal to the two interior angles on the other, namely BFG to CGF and FGD to AFG; whence, assuming FB, GD to meet in K, we can take the triangle KFG and place it (e.g. by rotating it in the plane about O the middle point of FG) so that FG falls where GF is in the figure and GD falls on FA, in which case FB must also fall on GC; hence, since FB, GD meet at K, GC and FA must meet at a corresponding point L. Or, as Mr Frankland does, we may substitute for FG a straight line MN through O the middle point of FG drawn perpendicular to one of the parallels, say AB. Then, since the two triangles OMF, ONG have two angles equal respectively, namely FOM to

GON (I. 15) and *OFM* to *OGN*, and one side *OF* equal to one side *OG*, the triangles are congruent, the angle *ONG* is a right angle, and *MN* is perpendicular to both *AB* and *CD*. Then, by the same method of application, *MA*, *NC* are shown to form with *MN* a triangle *MALCN* congruent with the triangle *NDKBM*, and *MA*, *NC* meet at a point *L* corresponding to *K*. Thus the two straight lines would meet at the *two* points *K*, *L*. This is what happens under the Riemann hypothesis, where the axiom that two straight lines cannot enclose a space does not hold, but all straight lines meeting in one point have another point common also, and e.g. in the particular figure just used *K*, *L* are points common to all perpendiculars to *MN*. If we suppose that *K*, *L* are not distinct points, but *one* point, the axiom that two straight lines cannot enclose a space is *not* contradicted.]

II. Ptolemy now tries to prove I. 29 without using our Postulate, and then deduces the Postulate from it (Proclus, pp. 365, 14—367, 27).

The argument to prove I. 29 is as follows.

The straight line which cuts the parallels must make the sum of the interior angles on the same side equal to, greater than, or less than, two right angles.

"Let *AB*, *CD* be parallel, and let *FG* meet them. I say (1) that *FG* does not make the interior angles on the same side greater than two right angles.

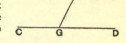

"For, if the angles *AFG*, *CGF* are greater than two right angles, the remaining angles *BFG*, *DGF* are less than two right angles.

"But the same two angles are also greater than two right angles; *for* AF, CG *are no more parallel than* FB, GD, *so that, if the straight line falling on* AF, CG *makes the interior angles greater than two right angles, the straight line falling on* FB, GD *will also make the interior angles greater than two right angles.*

"But the same angles are also less than two right angles; for the four angles *AFG*, *CGF*, *BFG*, *DGF* are equal to four right angles: which is impossible.

"Similarly (2) we can show that the straight line falling on the parallels does not make the interior angles on the same side less than two right angles.

"But (3), if it makes them neither greater nor less than two right angles, it can only make the interior angles on the same side *equal* to two right angles."

III. Ptolemy deduces Post. 5 thus:

Suppose that the straight lines making angles with a transversal less than two right angles do not meet on the side on which those angles are.

Then, *a fortiori*, they will not meet on the other side on which are the angles *greater* than two right angles.

Hence the straight lines will not meet in either direction; they are therefore *parallel*.

But, if so, the angles made by them with the transversal are equal to two right angles, by the preceding proposition (= I. 29).

Therefore the same angles will be both equal to and less than two right angles: which is impossible.

Hence the straight lines will meet.

IV.   Ptolemy lastly enforces his conclusion that the straight lines will meet *on the side on which are the angles less than two right angles* by recurring to the *a fortiori* step in the foregoing proof.

Let the angles *AFG*, *CGF* in the accompanying figure be together less than two right angles.

Therefore the angles *BFG*, *DGF* are greater than two right angles.

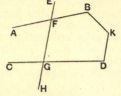

We have proved that the straight lines are not non-secant.

If they meet, they must meet either towards *A*, *C*, or towards *B*, *D*.

(1)   Suppose they meet towards *B*, *D*, at *K*.

Then, since the angles *AFG*, *CGF* are less than two right angles, and the angles *AFG*, *GFB* are equal to two right angles, take away the common angle *AFG*, and

the angle *CGF* is less than the angle *BFG*;

that is, the exterior angle of the triangle *KFG* is less than the interior and opposite angle *BFG*:
which is impossible.

Therefore *AB*, *CD* do not meet towards *B*, *D*.

(2)   But they do meet, and therefore they must meet in one direction or the other:

therefore they meet towards *A*, *B*, that is, on the side where are the angles less than two right angles.

The flaw in Ptolemy's argument is of course in the part of his proof of I. 29 which I have italicised.   As Proclus says, he is not entitled to assume that, if *AB*, *CD* are parallel, whatever is true of the interior angles on one side of *FG* (i.e. that they are together equal to, greater than, or less than, two right angles) is necessarily true at the same time of the interior angles on the other side.   Ptolemy justifies this by saying that *FA*, *GC* are no more parallel in one direction than *FB*, *GD* are in the other: which is equivalent to the assumption that *through any point only one parallel can be drawn to a given straight line.*   That is, he assumes an equivalent of the very Postulate he is endeavouring to prove.

Proclus.

Before passing to his own attempt at a proof, Proclus (p. 368, 26 sqq.) examines an ingenious argument (recalling somewhat the famous one about Achilles and the tortoise) which appeared to show that it was *impossible* for the lines described in the Postulate to meet.

Let *AB*, *CD* make with *AC* the angles *BAC*, *ACD* together less than two right angles.

Bisect *AC* at *E* and along *AB*, *CD* respectively measure *AF*, *CG* so that each is equal to *AE*.

Bisect *FG* at *H* and mark off *FK*, *GL* each equal to *FH*; and so on.

Then *AF*, *CG* will not meet at any point on *FG*; for, if that were the case, two sides of a triangle would be together equal to the third: which is impossible.

Similarly, *AB, CD* will not meet at any point on *KL*; and "proceeding like this indefinitely, joining the non-coincident points, bisecting the lines so drawn, and cutting off from the straight lines portions equal to the half of these, they say they thereby prove that the straight lines *AB, CD* will not meet anywhere."

It is not surprising that Proclus does not succeed in exposing the fallacy here (the fact being that the process will indeed be endless, and yet the straight lines will intersect within a finite distance). But Proclus' criticism contains nevertheless something of value. He says that the argument will prove too much, since we have only to join *AG* in order to see that straight lines making *some* angles which are together less than two right angles do in fact meet, namely *AG, CG*. "Therefore it is not possible to assert, without some definite limitation, that the straight lines produced from angles less than two right angles do not meet. On the contrary, it is manifest that *some* straight lines, when produced from angles less than two right angles, do meet, although the argument seems to require it to be proved that this property belongs to *all* such straight lines. For one might say that, the lessening of the two right angles being subject to no limitation, *with such and such an amount of lessening the straight lines remain non-secant, but with an amount of lessening in excess of this they meet* (p. 371, 2—10)."

[Here then we have the germ of such an idea as that worked out by Lobachewsky, namely that the straight lines issuing from a point in a plane can be divided with reference to a straight line lying in that plane into two classes, "secant" and "non-secant," and that we may define as *parallel* the two straight lines which divide the secant from the non-secant class.]

Proclus goes on (p. 371, 10) to base his own argument upon "an axiom such as Aristotle too used in arguing that the universe is finite. For, *if from one point two straight lines forming an angle be produced indefinitely, the distance* (διάστασις, Arist. διάστημα) *between the said straight lines produced indefinitely will exceed any finite magnitude.* Aristotle at all events showed that, if the straight lines drawn from the centre to the circumference are infinite, the interval between them is infinite. For, if it is finite, it is impossible to increase the distance, so that the straight lines (the radii) are not infinite. Hence the straight lines, when produced indefinitely, will be at a distance from one another greater than any assumed finite magnitude."

This is a fair representation of Aristotle's argument in *De caelo* I. 5, 271 b 28, although of course it is not a proof of what Proclus assumes as an axiom.

This being premised, Proclus proceeds (p. 371, 24):

I. "I say that, *if any straight line cuts one of two parallels, it will cut the other also.*

"For let *AB, CD* be parallel, and let *EFG* cut *AB*; I say that it will cut *CD* also.

"For, since *BF, FG* are two straight lines from one point *F,* they have, when produced indefinitely, a distance greater than any magnitude, so that it will also be greater than the interval between the parallels. Whenever therefore they are at a distance from one another greater than the distance between the parallels, *FG* will cut *CD*.

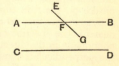

"Therefore etc."

II. "Having proved this, we shall prove, as a deduction from it, the theorem in question.

"For let AB, CD be two straight lines, and let EF falling on them make the angles BEF, DFE less than two right angles.

"I say that the straight lines will meet on that side on which are the angles less than two right angles.

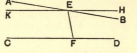

"For, since the angles BEF, DFE are less than two right angles, let the angle HEB be equal to the excess of two right angles (over them), and let HE be produced to K.

"Since then EF falls on KH, CD and makes the two interior angles HEF, DFE equal to two right angles,

the straight lines HK, CD are parallel.

"And AB cuts KH; therefore it will also cut CD, by what was before shown.

"Therefore AB, CD will meet on that side on which are the angles less than two right angles.

"Hence the theorem is proved."

Clavius criticised this proof on the ground that the axiom from which it starts, taken from Aristotle, itself requires proof. He points out that, just as you cannot assume that two lines which continually approach one another will meet (witness the hyperbola and its asymptote), so you cannot assume that two lines which continually diverge will ultimately be so far apart that a perpendicular from a point on one let fall on the other will be greater than any assigned distance; and he refers to the *conchoid* of Nicomedes, which continually approaches its asymptote, and therefore continually gets farther away from the tangent at the vertex; yet the perpendicular from any point on the curve to that tangent will always be less than the distance between the tangent and the asymptote.   Saccheri supports the objection.

Proclus' first proposition is open to the objection that it assumes that two "parallels" (in the Euclidean sense) or, as we may say, two straight lines which have a common perpendicular, are (not necessarily equidistant, but) so related that, when they are produced indefinitely, the perpendicular from a point of one upon the other remains finite.

This last assumption is incorrect on the hyperbolic hypothesis; the "axiom" taken from Aristotle does not hold on the elliptic hypothesis.

Naṣīraddīn aṭ-Ṭūsī.

The Persian-born editor of Euclid, whose date is 1201—1274, has three lemmas leading up to the final proposition.   Their content is substantially as follows, the first lemma being apparently assumed as evident.

I. (*a*) If AB, CD be two straight lines such that successive perpendiculars, as EF, GH, KL, from points on AB to CD always make with AB unequal angles, which are always acute on the side towards B and always obtuse on the side towards A, then the lines AB, CD, so long as they do not cut, approach continually nearer in the direction of the acute angles and diverge continually in the direction of the obtuse angles, and the perpendiculars diminish towards B, D, and increase towards A, C.

(*b*) Conversely, if the perpendiculars so drawn continually become shorter in the direction of B, D, and longer in the direction of *A*, *C*, the straight lines *AB*, *CD* approach continually nearer in the direction of *B*, *D* and diverge continually in the other direction; also each perpendicular will make with *AB* two angles one of which is acute and the other is obtuse, and all the acute angles will lie in the direction towards *B*, *D*, and the obtuse angles in the opposite direction.

[Saccheri points out that even the first part (*a*) requires proof.   As regards the converse (*b*) he asks, why should not the successive acute angles made by the perpendiculars with *AB*, while remaining acute, become greater and greater as the perpendiculars become smaller until we arrive at last at a perpendicular which is a common perpendicular to *both* lines? If that happens, all the author's efforts are in vain.   And, if you are to assume the truth of the statement in the lemma without proof, would it not, as Wallis said, be as easy to assume as axiomatic the statement in Post. 5 without more ado?]

II.   *If* AC, BD *be drawn from the extremities of* AB *at right angles to it and on the same side, and if* AC, BD *be made equal to one another and* CD *be joined, each of the angles* ACD, BDC *will be right, and* CD *will be equal to* AB.

The first part of this lemma is proved by *reductio ad absurdum* from the preceding lemma. If, e.g., the angle *ACD* is not right, it must either be acute or obtuse.

Suppose it is acute; then, by lemma 1, *AC* is greater than *BD*, which is contrary to the hypothesis.   And so on.

The angles *ACD*, *BDC* being proved to be right angles, it is easy to prove that *AB*, *CD* are equal.

[It is of course assumed in this "proof" that, if the angle *ACD* is acute, the angle *BDC* is obtuse, and *vice versa*.]

III.   *In any triangle the three angles are together equal to two right angles.*

This is proved for a *right-angled* triangle by means of the foregoing lemma, the four angles of the quadrilateral *ABCD* of that lemma being all right angles. The proposition is then true for *any* triangle, since any triangle can be divided into two right-angled triangles.

IV.   Here we have the final "proof" of Post. 5.   Three cases are distinguished, but it is enough to show the case where one of the interior angles is right and the other acute.

Suppose *AB*, *CD* to be two straight lines met by *FCE* making the angle *ECD* a right angle and the angle *CEB* an acute angle.

Take any point *G* on *EB*, and draw *GH* perpendicular to *EC*.

Since the angle *CEG* is acute, the perpendicular *GH* will fall on the side of *E* towards *D*, and will either coincide with *CD* or not coincide with it. In the former case the proposition is proved.

If *GH* does not coincide with *CD* but falls on the side of it towards *F*, *CD*, being within the triangle formed by the perpendicular and by *CE*, *EG*, must cut *EG*. [An axiom is here used, namely that, if *CD* be produced far enough, it must pass *outside* the triangle and therefore cut *some* side, which must be *EB*, since it cannot be the perpendicular (I. 27), or *CE*.]

Lastly, let *GH* fall on the side of *CD* towards *E*.

Along *HC* set off *HK*, *KL* etc., each equal to *EH*, until we get the first point of division, as *M*, beyond *C*.

Along *GB* set off *GN*, *NO* etc., each equal to *EG*, until *EP* is the same multiple of *EG* that *EM* is of *EH*.

Then we can prove that the perpendiculars from *N*, *O*, *P* on *EC* fall on the points *K*, *L*, *M* respectively.

For take the first perpendicular, that from *N*, and call it *NS*.

Draw *EQ* at right angles to *EH* and equal to *GH*, and set off *SR* along *SN* also equal to *GH*. Join *QG*, *GR*.

Then (second lemma) the angles *EQG*, *QGH* are right, and *QG* = *EH*. Similarly the angles *SRG*, *RGH* are right, and *RG* = *SH*.

Thus *RGQ* is one straight line, and the vertically opposite angles *NGR*, *EGQ* are equal. The angles *NRG*, *EQG* are both right, and *NG* = *GE*, by construction.

Therefore (I. 26) *RG* = *GQ*;

whence *SH* = *HE* = *KH*, and *S* coincides with *K*.

We may proceed similarly with the other perpendiculars.

Thus *PM* is perpendicular to *FE*. Hence *CD*, being parallel to *MP* and within the triangle *PME*, must cut *EP*, if produced far enough.

John Wallis.

As is well known, the argument of Wallis (1616—1703) assumed as a postulate that, *given a figure, another figure is possible which is similar to the given one and of any size whatever*. In fact Wallis assumed this for *triangles* only. He first proved (1) that, if a finite straight line is placed on an infinite straight line, and is then moved in its own direction as far as we please, it will always lie on the same infinite straight line, (2) that, if an angle be moved so that one leg always slides along an infinite straight line, the angle will remain the same, or equal, (3) that, if two straight lines, cut by a third, make the interior angles on the same side less than two right angles, each of the exterior angles is greater than the opposite interior angle (proved by means of I. 13).

(4) If *AB*, *CD* make, with *AC*, the interior angles less than two right angles, suppose *AC* (with *AB* rigidly attached to it) to move along *AF* to the position *αγ*, such that *α* coincides with *C*. If *AB* then takes the position *αβ*, *αβ* *lies entirely outside* CD (proved by means of (3) above).

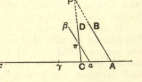

(5) With the same hypotheses, *the straight line αβ, or* AB, *during its motion, and before α reaches* C, *must* cut *the straight line* CD.

(6) Here is enunciated the postulate stated above.

(7) Postulate 5 is now proved thus.

Let *AB*, *CD* be the straight lines which make, with the infinite straight line *ACF* meeting them, the interior angles *BAC*, *DCA* together less than two right angles.

Suppose *AC* (with *AB* rigidly attached to it) to move along *ACF* until *AB* takes the position of *αβ* cutting *CD* in *π*.

Then, *αCπ* being a triangle, we can, by the above postulate, suppose a triangle drawn on the base *CA* similar to the triangle *αCπ*.

Let it be *ACP*.

[Wallis here interposes a defence of the hypothetical construction.]

Thus CP and AP meet at P; and, as by the definition of similar figures the angles of the triangles PCA, πCa are respectively equal, the angle PCA being equal to the angle πCa and the angle PAC to the angle πaC or BAC, it follows that CP, AP lie on CD, AB produced respectively.

Hence AB, CD meet on the side on which are the angles less than two right angles.

[The whole gist of this proof lies in the assumed postulate as to the existence of similar figures; and, as Saccheri points out, this is equivalent to unconditionally assuming the "hypothesis of the right angle," and consequently Euclid's Postulate 5.]

Gerolamo Saccheri.

The book *Euclides ab omni naevo vindicatus* (1733) by Gerolamo Saccheri (1667—1733), a Jesuit, and professor at the University of Pavia, is now accessible (1) edited in German by Engel and Stäckel, *Die Theorie der Parallellinien von Euklid bis auf Gauss*, 1895, pp. 41—136, and (2) in an Italian version, abridged but annotated, *L'Euclide emendato del P. Gerolamo Saccheri*, by G. Boccardini (Hoepli, Milan, 1904). It is of much greater importance than all the earlier attempts to prove Post. 5 because Saccheri was the first to contemplate the possibility of hypotheses other than that of Euclid, and to work out a number of consequences of those hypotheses. He was therefore a true precursor of Legendre and of Lobachewsky, as Beltrami called him (1889), and, it might be added, of Riemann also. For, as Veronese observes (*Fondamenti di geometria*, p. 570), Saccheri obtained a glimpse of the theory of parallels in all its generality, while Legendre, Lobachewsky and G. Bolyai excluded *a priori*, without knowing it, the "hypothesis of the obtuse angle," or the Riemann hypothesis. Saccheri, however, was the victim of the preconceived notion of his time that the sole possible geometry was the Euclidean, and he presents the curious spectacle of a man laboriously erecting a structure upon new foundations for the very purpose of demolishing it afterwards; he sought for contradictions in the heart of the systems which he constructed, in order to prove thereby the falsity of his hypotheses.

For the purpose of formulating his hypotheses he takes a plane quadrilateral $ABDC$, two opposite sides of which, AC, BD, are equal and perpendicular to a third AB. Then the angles at C and D are easily proved to be equal. On the Euclidean hypothesis they are both right angles; but apart from this hypothesis they might be both obtuse or both acute. To the three possibilities, which Saccheri distinguishes by the names (1) *the hypothesis of the right angle*, (2) *the hypothesis of the obtuse angle* and

(3) *the hypothesis of the acute angle* respectively, there corresponds a certain group of theorems; and Saccheri's point of view is that the Postulate will be completely proved if the consequences which follow from the last two hypotheses comprise results inconsistent with one another.

Among the most important of his propositions are the following:

(1) *If the hypothesis of the right angle, or of the obtuse angle, or of the acute angle is proved true in a single case, it is true in every other case.* (Props. v., vi., vii.)

(2) *According as the hypothesis of the right angle, the obtuse angle, or the acute angle is true, the sum of the three angles of a triangle is equal to, greater than, or less than two right angles.* (Prop. ix.)

(3) *From the existence of a single triangle in which the sum of the angles is equal to, greater than, or less than two right angles the truth of the hypothesis of the right angle, obtuse angle, or acute angle respectively follows.* (Prop. XV.)

These propositions involve the following : *If in a single triangle the sum of the angles is equal to, greater than, or less than two right angles, then* any triangle has the sum of its angles equal to, greater than, or less than two right angles respectively, which was proved about a century later by Legendre for the two cases only where the sum is *equal to* or *less than* two right angles.

The proofs are not free from imperfections, as when, in the proofs of Prop. XII. and the part of Prop. XIII. relating to the hypothesis of the *obtuse angle*, Saccheri uses Eucl. I. 18 depending on I. 16, a proposition which is only valid on the assumption that *straight lines are infinite in length* ; for this assumption itself does not hold under the hypothesis of the obtuse angle (the Riemann hypothesis).

The hypothesis of the acute angle takes Saccheri much longer to dispose of, and this part of the book is less satisfactory ; but it contains the following propositions afterwards established anew by Lobachewsky and Bolyai, viz.:

(4) *Two straight lines in a plane* (even on the hypothesis of the acute angle) *either have a common perpendicular, or must, if produced in one and the same direction, either intersect once at a finite distance or at least continually approach one another.* (Prop. XXIII.)

(5) *In a cluster of rays issuing from a point there exist always* (on the hypothesis of the acute angle) *two determinate straight lines which separate the straight lines which intersect a fixed straight line from those which do not intersect it, ending with and including the straight line which has a common perpendicular with the fixed straight line.* (Props. XXX., XXXI., XXXII.)

Lambert.

A dissertation by G. S. Klügel, *Conatuum praecipuorum theoriam parallelarum demonstrandi recensio* (1763), contained an examination of some thirty "demonstrations" of Post. 5 and is remarkable for its conclusion expressing, apparently for the first time, *doubt as to its demonstrability* and observing that the certainty which we have in us of the truth of the Euclidean hypothesis is not the result of a series of rigorous deductions but rather of experimental observations. It also had the greater merit that it called the attention of Johann Heinrich Lambert (1728—1777) to the theory of parallels. His *Theory of Parallels* was written in 1766 and published after his death by G. Bernoulli and C. F. Hindenburg; it is reproduced by Engel and Stäckel (*op. cit.* pp. 152—208).

The third part of Lambert's tract is devoted to the discussion of the same three hypotheses as Saccheri's, the hypothesis of the *right angle* being for Lambert the *first*, that of the *obtuse angle* the *second*, and that of the *acute angle* the *third*, hypothesis; and, with reference to a quadrilateral with *three right angles* from which Lambert starts (that is, one of the halves into which the median divides Saccheri's quadrilateral), the three hypotheses are the assumptions that the fourth angle is a right angle, an obtuse angle, or an acute angle respectively.

Lambert goes much further than Saccheri in the deduction of new propositions from the *second* and *third* hypotheses. The most remarkable is the following.

The area of a plane triangle, under the second *and* third *hypotheses, is proportional to the difference between the sum of the three angles and two right angles.*

Thus the numerical expression for the area of a triangle is, under the *third* hypothesis

$$\Delta = k\,(\pi - A - B - C) \dots\dots\dots\dots\dots(1),$$

and under the *second* hypothesis

$$\Delta = k\,(A + B + C - \pi) \dots\dots\dots\dots\dots(2),$$

where k is a positive constant.

A remarkable observation is appended (§ 82): "In connexion with this it seems to be remarkable that the *second* hypothesis holds if *spherical* instead of plane triangles are taken, because in the former also the sum of the angles is greater than two right angles, and the excess is proportional to the area of the triangle.

"It appears still more remarkable that what I here assert of spherical triangles can be proved independently of the difficulty of parallels."

This discovery that the *second* hypothesis is realised on the surface of a sphere is important in view of the development, later, of the Riemann hypothesis (1854).

Still more remarkable is the following prophetic sentence: "*I am almost inclined to draw the conclusion that the* third *hypothesis arises with an imaginary spherical surface*" (cf. Lobachewsky's *Géométrie imaginaire*, 1837).

No doubt Lambert was confirmed in this by the fact that, in the formula (2) above, which, for $k = r^2$, represents the area of a spherical triangle, if $r \sqrt{-1}$ is substituted for r, and $r^2 = k$, we obtain the formula (1).

Legendre.

No account of our present subject would be complete without a full reference to what is of permanent value in the investigations of Adrien Marie Legendre (1752—1833) relating to the theory of parallels, which extended over the space of a generation. His different attempts to prove the Euclidean hypothesis appeared in the successive editions of his *Éléments de Géométrie* from the first (1794) to the twelfth (1823), which last may be said to contain his last word on the subject. Later, in 1833, he published, in the *Mémoires de l'Académie Royale des Sciences*, XII. p. 367 sqq., a collection of his different proofs under the title *Réflexions sur différentes manières de démontrer la théorie des parallèles*. His exposition brought out clearly, as Saccheri had done, and kept steadily in view, the essential connexion between the theory of parallels and the sum of the angles of a triangle. In the first edition of the *Éléments* the proposition that *the sum of the angles of a triangle is equal to two right angles* was proved analytically on the basis of the assumption that the choice of a *unit of length* does not affect the correctness of the proposition to be proved, which is of course equivalent to Wallis' assumption of *the existence of similar figures*. A similar analytical proof is given in the notes to the twelfth edition. In his second edition Legendre proved Postulate 5 by means of the assumption that, *given three points not in a straight line, there exists a circle passing through all three.* In the third edition (1800) he gave the proposition that *the sum of the angles of a triangle is not greater than two right angles*; this proof, which was geometrical, was replaced later by another, the best known, depending on a construction like that of Euclid I. 16, the continued application of which enables any number of successive triangles to be evolved in which, while the sum of the angles in each remains always equal to the sum of the angles of the original triangle, one of the angles increases and the sum of the other two diminishes continually. But Legendre found the proof of the equally necessary proposition that the sum of the angles of a triangle is *not less* than two right angles to present great difficulties. He first observed that, as in the case of spherical triangles (in which the sum of the angles is greater than two right angles) the excess of the sum of the angles over two right angles is proportional to the area of the triangle, so in the case of rectilineal triangles, if the sum of the angles is less than two right angles by a certain *deficit*, the *deficit* will be proportional to the area of the triangle. Hence if, starting from a given triangle, we could construct another triangle in which the original triangle is contained at least *m* times, the *deficit* of this new triangle will be equal to at least *m* times that of the original triangle, so that the sum of the angles of the greater triangle will diminish progressively as *m* increases, until it becomes zero or negative: which is absurd. The whole difficulty was thus reduced to that of the construction of a triangle containing the given triangle at least twice; but the solution of even this simple problem requires it to be assumed (or proved) that *through a given point within a given angle less than two-thirds of a right angle we can always draw a straight line which shall meet both sides of the angle.* This is however really equivalent to Euclid's Postulate. The proof in the course of which the necessity for the assumption appeared is as follows.

It is required to prove that the sum of the angles of a triangle cannot be *less* than two right angles.

Suppose A is the least of the three angles of a triangle ABC. Apply to the opposite side BC a triangle DBC, equal to the triangle ACB, and such that the angle DBC is equal to the angle ACB, and the angle DCB to the angle ABC; and *draw any straight line through D cutting AB, AC produced in E, F.*

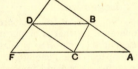

If now the sum of the angles of the triangle ABC is less than two right angles, being equal to $2R - \delta$ say, the sum of the angles of the triangle DBC, equal to the triangle ABC, is also $2R - \delta$.

Since the sum of the three angles of the remaining triangles DEB, FDC respectively cannot at all events be *greater* than two right angles [for Legendre's proofs of this see below], the sum of the twelve angles of the four triangles in the figure *cannot be greater* than

$$4R + (2R - \delta) + (2R - \delta), \text{ i.e. } 8R - 2\delta.$$

Now the sum of the three angles at each of the points B, C, D is $2R$.

Subtracting these nine angles, we have the result that the three angles of the triangle AEF cannot *be greater* than $2R - 2\delta$.

Hence, if the sum of the angles of the triangle ABC is less than two right angles by δ, the sum of the angles of the larger triangle AEF is less than two right angles by *at least* 2δ.

We can continue the construction, making a still larger triangle from AEF, and so on.

But, however small δ is, we can arrive at a multiple $2^n\delta$ which shall exceed any given angle and therefore $2R$ itself; so that the sum of the three angles of a triangle sufficiently large would be zero or even less than zero: which is absurd.

Therefore etc.

The difficulty caused by the necessity of making the above-mentioned assumption made Legendre abandon, in his ninth edition, the method of the editions from the third to the eighth and return to Euclid's method pure and simple.

But again, in the twelfth, he returned to the plan of constructing any number of successive triangles such that the sum of the three angles in all of them remains equal to the sum of the three angles of the original triangle, but two of the angles of the new triangles become smaller and smaller, while the third becomes larger and larger; and this time he claims to prove in one proposition that the sum of the three angles of the original triangle is *equal* to two right angles by continuing the construction of new triangles *indefinitely* and compressing the two smaller angles of the ultimate triangle into nothing, while the third angle is made to become a *flat* angle at the same time. The construction and attempted proof are as follows.

Let *ABC* be the given triangle; let *AB* be the greatest side and *BC* the least; therefore *C* is the greatest angle and *A* the least.

From *A* draw *AD* to the middle point of *BC*, and produce *AD* to *C′*, making *AC′* equal to *AB*.

Produce *AB* to *B′*, making *AB′* equal to twice *AD*.

The triangle *AB′C′* is then such that the sum of its three angles is equal to the sum of the three angles of the triangle *ABC*.

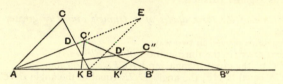

For take *AK* along *AB* equal to *AD*, and join *C′K*.

Then the triangles *ABD*, *AC′K* have two sides and the included angles respectively equal, and are therefore equal in all respects; and *C′K* is equal to *BD* or *DC*.

Next, in the triangles *B′C′K*, *ACD*, the angles *B′KC′*, *ADC* are equal, being respectively supplementary to the equal angles *AKC′*, *ADB*; and the two sides about the equal angles are respectively equal;

therefore the triangles *B′C′K*, *ACD* are equal in all respects.

Thus the angle *AC′B′* is the sum of two angles respectively equal to the angles *B*, *C* of the original triangle; and the angle *A* in the original triangle is the sum of two angles respectively equal to the angles at *A* and *B′* in the triangle *AB′C′*.

It follows that the sum of the three angles of the new triangle *AB′C′* is equal to the sum of the angles of the triangle *ABC*.

Moreover, the side *AC′*, being equal to *AB*, and therefore greater than *AC*, is greater than *B′C′* which is equal to *AC*.

Hence the angle *C′AB′* is less than the angle *AB′C′*; so that the angle *C′AB′* is less than $\frac{1}{2}A$, where *A* denotes the angle *CAB* of the original triangle.

[It will be observed that the triangle *AB′C′* is really the same triangle as the triangle *AEB* obtained by the construction of Eucl. I. 16, but differently placed so that the longest side lies along *AB*.]

By taking the middle point *D′* of the side *B′C′* and repeating the same construction, we obtain a triangle *AB″C″* such that (1) the sum of its three angles is equal to the sum of the three angles of *ABC*, (2) the sum of the two angles $C''AB''$, $AB''C''$ is equal to the angle $C'AB'$ in the preceding triangle, and is therefore less than $\frac{1}{2}A$, and (3) the angle $C''AB''$ is less than half the angle $C'AB'$, and therefore less than $\frac{1}{4}A$.

Continuing in this way, we shall obtain a triangle Abc such that the sum of two angles, those at A and b, is less than $\frac{1}{2^n}A$, and the angle at c is greater than the corresponding angle in the preceding triangle.

If, Legendre argues, the construction be continued indefinitely so that $\frac{1}{2^n}A$ becomes smaller than any assigned angle, the point c ultimately lies on Ab, and the sum of the three angles of the triangle (which is equal to the sum of the three angles of the original triangle) becomes identical with the angle at c, which is then a *flat* angle, and therefore equal to two right angles.

This proof was however shown to be unsound (in respect of the final inference) by J. P. W. Stein in Gergonne's *Annales de Mathématiques* XV., 1824, pp. 77—79.

We will now reproduce shortly the substance of the theorems of Legendre which are of the most permanent value as not depending on a particular hypothesis as regards parallels.

I. *The sum of the three angles of a triangle cannot be* greater *than two right angles.*

This Legendre proved in two ways.

(1) *First proof* (in the third edition of the *Éléments*).

Let ABC be the given triangle, and ACJ a straight line.

Make CE equal to AC, the angle DCE equal to the angle BAC, and DC equal to AB. Join DE.

Then the triangle DCE is equal to the triangle BAC in all respects.

If then the sum of the three angles of the triangle ABC is greater than

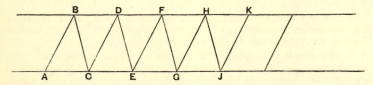

$2R$, the said sum must be greater than the sum of the angles BCA, BCD, DCE, which sum is *equal* to $2R$.

Subtracting the equal angles on both sides, we have the result that

the angle ABC is *greater* than the angle BCD.

But the two sides AB, BC of the triangle ABC are respectively equal to the two sides DC, CB of the triangle BCD.

Therefore the base AC is *greater* than the base BD (Eucl. I. 24).

Next, make the triangle FEG (by the same construction) equal in all respects to the triangle BAC or DCE; and we prove in the same way that CE (or AC) is *greater* than DF.

And, at the same time, BD is equal to DF, because the angles BCD, DEF are equal.

Continuing the construction of further triangles, however small the difference between AC and BD is, we shall ultimately reach some multiple of this difference, represented in the figure by (say) the difference between the straight line AJ and the composite line $BDFHK$, which will be greater than any assigned length, and greater therefore than the sum of AB and JK.

Hence, on the assumption that the sum of the angles of the triangle ABC is greater than $2R$, the broken line $ABDFHKJ$ may be less than the straight line AJ: which is impossible.

Therefore etc.

(2) *Proof substituted later.*

If possible, let $2R + a$ be the sum of the three angles of the triangle ABC, of which A is not greater than either of the others.

Bisect BC at H, and produce AH to D, making HD equal to AH; join BD.

Then the triangles AHC, DHB are equal in all respects (I. 4); and the angles CAH, ACH are respectively equal to the angles BDH, DBH.

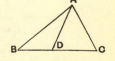

It follows that the sum of the angles of the triangle ABD is equal to the sum of the angles of the original triangle, i.e. to $2R + a$.

And one of the angles DAB, ADB is either equal to or less than half the angle CAB.

Continuing the same construction with the triangle ADB, we find a third triangle in which the sum of the angles is still $2R + a$, while one of them is equal to or less than $(\angle CAB)/4$.

Proceeding in this way, we arrive at a triangle in which the sum of the angles is $2R + a$, and one of them is not greater than $(\angle CAB)/2^n$.

And, if n is sufficiently large, this will be less than a; in which case we should have a triangle in which two angles are together greater than two right angles: which is absurd.

Therefore a is equal to or less than zero.

(It will be noted that in both these proofs, as in Eucl. I. 16, it is taken for granted that *a straight line is infinite in length* and does not return into itself, which is not true under the Riemann hypothesis.)

II. On the assumption that the sum of the angles of a triangle is *less* than two right angles, *if a triangle is made up of two others, the "deficit" of the former is equal to the sum of the "deficits" of the others.*

In fact, if the sums of the angles of the component triangles are $2R - a$, $2R - \beta$ respectively, the sum of the angles of the whole triangle is

$$(2R - a) + (2R - \beta) - 2R = 2R - (a + \beta).$$

III. *If the sum of the three angles of a triangle is* equal *to two right angles, the same is true of all triangles obtained by subdividing it by straight lines drawn from a vertex to meet the opposite side.*

Since the sum of the angles of the triangle ABC is equal to $2R$, if the sum of the angles of the triangle ABD were $2R - a$, it would follow that the sum of the angles of the triangle ADC must be $2R + a$, which is absurd (by I. above).

IV. *If in a triangle the sum of the three angles is* equal *to two right angles, a quadrilateral can always be constructed with four right angles and four equal sides exceeding in length any assigned rectilineal segment.*

Let ABC be a triangle in which the sum of the angles is equal to two right angles. We can assume ABC to be an *isosceles right-angled* triangle because we can reduce the case to this by making subdivisions of ABC by straight lines through vertices (as in Prop. III. above).

Taking two equal triangles of this kind and placing their hypotenuses together, we obtain a quadrilateral with four right angles and four equal sides.

Putting four of these quadrilaterals together, we obtain a new quadrilateral of the same kind but with its sides double of those of the first quadrilateral.

After n such operations we have a quadrilateral with four right angles and four equal sides, each being equal to 2^n times the side AB.

The diagonal of this quadrilateral divides it into two equal isosceles right-angled triangles in each of which the sum of the angles is equal to two right angles.

Consequently, from the existence of *one* triangle in which the sum of the three angles is equal to two right angles it follows that there exists an isosceles right-angled triangle with sides greater than any assigned rectilineal segment and such that the sum of its three angles is also equal to two right angles.

V. *If the sum of the three angles of* one *triangle is equal to two right angles, the sum of the three angles of* any other *triangle is also equal to two right angles.*

It is enough to prove this for a *right-angled* triangle, since any triangle can be divided into two right-angled triangles.

Let ABC be any right-angled triangle.

If then the sum of the angles of any one triangle is equal to two right angles, we can construct (by the preceding Prop.) an isosceles right-angled triangle with the same property and with its perpendicular sides greater than those of ABC.

Let $A'B'C'$ be such a triangle, and let it be applied to ABC, as in the figure.

Applying then Prop. III. above, we deduce first that the sum of the three angles of the triangle $AB'C'$ is equal to two right angles, and next, for the same reason, that the sum of the three angles of the original triangle ABC is equal to two right angles.

VI. *If in any one triangle the sum of the three angles is less than two right angles, the sum of the three angles of any other triangle is also less than two right angles.*

This follows from the preceding theorem.

(It will be observed that the last two theorems are included among those of Saccheri, which contain however in addition the corresponding theorem touching the case where the sum of the angles is *greater* than two right angles.)

We come now to the bearing of these propositions upon Euclid's Postulate 5 ; and the next theorem is

VII. *If the sum of the three angles of a triangle is equal to two right angles, through any point in a plane there can only be drawn one parallel to a given straight line.*

For the proof of this we require the following

LEMMA. *It is always possible, through a point* P, *to draw a straight line which shall make, with a given straight line* (r), *an angle less than any assigned angle.*

Let Q be the foot of the perpendicular from P upon r.

Let a segment QR be taken on r, on either side of Q, such that QR is equal to PQ.

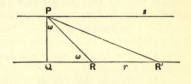

Join PR, and mark off the segment RR' equal to PR; join PR'.

If ω represents the angle QPR or the angle QRP, each of the equal angles RPR', $RR'P$ is not greater than ω/2.

Continuing the construction, we obtain, after the requisite number of operations, a triangle $PR_{n-1} R_n$ in which each of the equal angles is equal to or less than $\omega/2^n$.

Hence we shall arrive at a straight line PR_n which, starting from P and meeting r, makes with r an angle as small as we please.

To return now to the Proposition. Draw from P the straight line s perpendicular to PQ.

Then any straight line drawn from P which meets r in R will form equal angles with r and s, since, by hypothesis, the sum of the angles of the triangle PQR is equal to two right angles.

And since, by the Lemma, it is always possible to draw through P straight lines which form with r angles as small as we please, it follows that all the straight lines through P, except s, will meet r. Hence s is the only parallel to r that can be drawn through P.

The history of the attempts to prove Postulate 5 or something equivalent has now been brought down to the parting of the ways. The further developments on lines independent of the Postulate, beginning with Schweikart (1780—1857), Taurinus (1794—1874), Gauss (1777—1855), Lobachewsky (1793—1856), J. Bolyai (1802—1860), Riemann (1826—1866), belong to the history of non-Euclidean geometry, which is outside the scope of this work. I may refer the reader to the full article *Sulla teoria delle parallele e sulle geometrie non-euclidee* by R. Bonola in *Questioni riguardanti le matematiche elementari*, I., of which I have made considerable use in the above, to the same author's *La geometria non-euclidea*, Bologna, 1906, to the first volume of Killing's *Einführung in die Grundlagen der Geometrie*, Paderborn, 1893, to P. Mansion's *Premiers principes de métagéométrie*, and P. Barbarin's *La géométrie non-Euclidienne*, Paris, 1902, to the historical summary in Veronese's *Fondamenti di geometria*, 1891, p. 565 sqq., and (for original sources) to Engel and Stäckel's *Die Theorie der Parallellinien von Euklid bis auf Gauss*, 1895, and *Urkunden zur Geschichte der nicht-Euklidischen Geometrie*, I. (Lobachewsky), 1899, and II. (Wolfgang und Johann Bolyai). I will only add that it was Gauss who first expressed a conviction that the Postulate could never be proved; he indicated this in reviews in the *Göttingische gelehrte Anzeigen*, 20 Apr. 1816 and 28 Oct. 1822, and affirmed it in a letter to Bessel of 27 January, 1829. The actual indemonstrability of the Postulate was proved by Beltrami (1868) and by Hoüel (*Note sur l'impossibilité de démontrer par une construction plane le principe de la théorie des parallèles dit Postulatum d'Euclide* in Battaglini's *Giornale di matematiche*, VIII., 1870, pp. 84—89).

Alternatives for Postulate 5.

It may be convenient to collect here a few of the more noteworthy substitutes which have from time to time been formally suggested or tacitly assumed.

(1) *Through a given point only one parallel can be drawn to a given straight line* or, *Two straight lines which intersect one another cannot both be parallel to one and the same straight line.*

This is commonly known as "Playfair's Axiom," but it was of course not a new discovery.   It is distinctly stated in Proclus' note to Eucl. I. 31.

(1 *a*) *If a straight line intersect one of two parallels, it will intersect the other also* (Proclus).

(1 *b*) *Straight lines parallel to the same straight line are parallel to one another.*

The forms (1 *a*) and (1 *b*) are exactly equivalent to (1).

(2) *There exist straight lines everywhere equidistant from one another* (Posidonius and Geminus); with which may be compared Proclus' tacit assumption that *Parallels remain, throughout their length, at a finite distance from one another.*

(3) *There exists a triangle in which the sum of the three angles is equal to two right angles* (Legendre).

(4) *Given any figure, there exists a figure similar to it of any size we please* (Wallis, Carnot, Laplace).

Saccheri points out that it is not necessary to assume so much, and that it is enough to postulate that *there exist two unequal triangles with equal angles.*

(5) *Through any point within an angle less than two-thirds of a right angle a straight line can always be drawn which meets both sides of the angle* (Legendre).

With this may be compared the similar axiom of Lorenz (*Grundriss der reinen und angewandten Mathematik*, 1791): *Every straight line through a point within an angle must meet one of the sides of the angle.*

(6) *Given any three points not in a straight line, there exists a circle passing through them* (Legendre, W. Bolyai).

(7) "*If I could prove that a rectilineal triangle is possible the content of which is greater than any given area, I am in a position to prove perfectly rigorously the whole of geometry*" (Gauss, in a letter to W. Bolyai, 1799).

Cf. the proposition of Legendre numbered IV. above, and the axiom of Worpitzky: *There exists no triangle in which every angle is as small as we please.*

(8) *If in a quadrilateral three angles are right angles, the fourth angle is a right angle also* (Clairaut, 1741).

(9) *If two straight lines are parallel, they are figures opposite to (or the reflex of) one another with respect to the middle points of all their transversal segments* (Veronese, *Elementi*, 1904).

Or, *Two parallel straight lines intercept, on every transversal which passes through the middle point of a segment included between them, another segment the middle point of which is the middle point of the first* (Ingrami, *Elementi*, 1904).

Veronese and Ingrami deduce immediately Playfair's Axiom.

AXIOMS OR *COMMON NOTIONS.*

In a paper *Sur l'authenticité des axiomes d'Euclide* in the *Bulletin des sciences math. et astron.* 1884, p. 162 sq. (*Mémoires scientifiques*, II., pp. 48—63), Paul Tannery maintained that the *Common Notions* (including the first three) were not in Euclid's work but were interpolated later.   The following are his main arguments.   (1) If Euclid had set about distinguishing between indemonstrable principles (*a*) common to all demonstrative sciences and (*b*) peculiar to geometry, he would, says Tannery, certainly not have placed the common principles second and the special principles (the Postulates) first.   (2) If the *Common Notions* are Euclid's, this designation of them must be his too ; for he must have used *some* name to distinguish them from the Postulates and, if he had used another name, such as *Axioms*, it is impossible to imagine why that name was changed afterwards for a less suitable one.   The word ἔννοια (*notion*), says Tannery, never signified a notion in the sense of a *proposition*, but a notion of some *object*; nor is it found in any technical sense in Plato and Aristotle.   (3) Tannery's own view was that the formulation of the *Common Notions* dates from the time of Apollonius, and that it was inspired by his work relating to the Elements (we know from Proclus that Apollonius tried to prove the *Common Notions*).   This idea, Tannery thought, was confirmed by a "fortunate coincidence" furnished by the occurrence of the word ἔννοια (*notion*) in a quotation by Proclus (p. 100, 6): "we shall agree with Apollonius when he says that we have a *notion* (ἔννοιαν) of a line when we order the lengths, only, of roads or walls to be measured."

In reply to argument (1) that it is an unnatural order to place the purely geometrical Postulates first, and the *Common Notions*, which are not peculiar to geometry, last, it may be pointed out that it would surely have been a still more awkward arrangement to give the Definitions first and then to separate from them, by the interposition of the *Common Notions*, the Postulates, which are so closely connected with the Definitions in that they proceed to postulate the *existence* of certain of the things defined, namely straight lines and circles.

(2) Though it is true that ἔννοια in Plato and Aristotle is generally a notion of an *object*, not of a *fact* or proposition, there are instances in Aristotle where it does mean a notion of a fact : thus in the *Eth. Nic.* IX. 11, 1171 a 32 he speaks of "the notion (or consciousness) *that friends sympathise*" (ἡ ἔννοια τοῦ συναλγεῖν τοὺς φίλους) and again, b 14, of "the *notion* (or consciousness) *that they are pleased* at his good fortune."   It is true that Plato and Aristotle do not use the word in a technical sense ; but neither was there apparently in Aristotle's time any fixed technical term for what we call "axioms," since he speaks of them variously as "the so-called axioms in mathematics," "the so-called common axioms," "the common (things)" (τὰ κοινά), and even "the common *opinions*" (κοιναὶ δόξαι). I see therefore no reason why Euclid should not himself have given a technical sense to "Common Notions," which is at least a distinct improvement upon "common opinions."

(3) The use of ἔννοια in Proclus' quotation from Apollonius seems to me to be an unfortunate, rather than a fortunate, coincidence from Tannery's point of view, for it is there used precisely in the old sense of the notion of an *object* (in that case a line).

No doubt it is difficult to feel certain that Euclid did himself use the term *Common Notions*, seeing that Proclus' commentary generally speaks of *Axioms*. But even Proclus (p. 194, 8), after explaining the meaning of the word "axiom," first as used by the Stoics, and secondly as used by "Aristotle and the geometers," goes on to say : " For in their view (that of Aristotle and the geometers) *axiom* and *common notion* are the same thing." This, as it seems to me, may be a sort of apology for using the word "axiom" exclusively in what has gone before, as if Proclus had suddenly bethought himself that he had described both Aristotle and the geometers as using the one term "axiom," whereas he should have said that Aristotle spoke of "axioms," while "the geometers" (in fact Euclid), though meaning the same thing, called them *Common Notions*. It may be for a like reason that in another passage (p. 76, 16), after quoting Aristotle's view of an "axiom," as distinct from a postulate and a hypothesis, he proceeds : "For it is not by virtue of a *common notion* that, without being taught, we preconceive the circle to be such and such a figure." If this view of the two passages just quoted is correct, it would strengthen rather than weaken the case for the genuineness of *Common Notions* as the Euclidean term.

Again, it is clear from Aristotle's allusions to the "common axioms in mathematics" that more than one axiom of this kind had a place in the text-books of his day ; and as he constantly quotes the particular axiom that, *if equals be taken from equals, the remainders are equal,* which is Euclid's *Common Notion* 3, it would seem that at least the first three *Common Notions* were adopted by Euclid from earlier text-books. It is, besides, scarcely credible that, if the *Common Notions* which Apollonius tried to prove had not been introduced earlier (e.g. by Euclid), they would then have been interpolated as axioms and not as propositions to be proved. The line taken by Apollonius is much better explained on the assumption that he was directly attacking axioms which he found already admitted into the *Elements*.

Proclus, who recognised the five *Common Notions* given in the text, warns us, not only against the error of unnecessarily multiplying the axioms, but against the contrary error of reducing their number unduly (p. 196, 15), "as Heron does in enunciating three only; for it is also an axiom that *the whole is greater than the part*, and indeed the geometer employs this in many places for his demonstrations, and again that *things which coincide are equal*."

Thus Heron recognised the first three of the *Common Notions* ; and this fact, together with Aristotle's allusions to "common axioms" (in the plural), and in particular to our *Common Notion* 3, may satisfy us that at least the first three *Common Notions* were contained in the *Elements* as they left Euclid's hands.

COMMON NOTION 1.

Τὰ τῷ αὐτῷ ἴσα καὶ ἀλλήλοις ἐστὶν ἴσα.
Things which are equal to the same thing are also equal to one another.

Aristotle throughout emphasises the fact that axioms are self-evident truths, which it is impossible to demonstrate. If, he says, any one should attempt to prove them, it could only be through ignorance. Aristotle therefore would undoubtedly have agreed in Proclus' strictures on Apollonius for attempting to prove the axioms. Proclus gives (p. 194, 25), as a specimen of these attempted proofs by Apollonius, that of the first of the *Common Notions*. "Let *A* be equal to *B*, and the latter to *C* ; I say that *A* is also equal to *C*. For, since *A* is equal to *B*, it occupies the same space with it ; and since *B* is equal to *C*, it occupies the same space with it.

Therefore *A* also occupies the same space with *C*."

Proclus rightly remarks (p. 194, 22) that "the middle term is no more intelligible (better known, γνωριμώτερον) than the conclusion, if it is not actually more disputable." Again (p. 195, 6), the proof assumes two things, (1) that things which "occupy the same space" (τόπος) are equal to one another, and (2) that things which occupy the same space with one and the same thing occupy the same space with one another; which is to explain the obvious by something much more obscure, for space is an entity more unknown to us than the things which exist in space.

Aristotle would also have objected to the proof that it is partial and not general (καθόλου), since it refers only to things which can be supposed to occupy a space (or take up room), whereas the axiom is, as Proclus says (p. 196, 1), true of numbers, speeds, and periods of time as well, though of course each science uses axioms in relation to its own subject-matter only.

COMMON NOTIONS 2, 3.

2. Καὶ ἐὰν ἴσοις ἴσα προστεθῇ, τὰ ὅλα ἐστὶν ἴσα.
3. Καὶ ἐὰν ἀπὸ ἴσων ἴσα ἀφαιρεθῇ, τὰ καταλειπόμενά ἐστιν ἴσα.
2. *If equals be added to equals, the wholes are equal.*
3. *If equals be subtracted from equals, the remainders are equal.*

These two Common Notions are recognised by Heron and Proclus as genuine. The latter is the axiom which is so favourite an illustration. with Aristotle.

Following them in the MSS. and editions there came four others of the same type as 1—3. Three of these are given by Heiberg in brackets; the fourth he omits altogether.

The three are :

(*a*) *If equals be added to unequals, the wholes are unequal.*

(*b*) *Things which are double of the same thing are equal to one another.*

(*c*) *Things which are halves of the same thing are equal to one another.*

The fourth, which was placed between (*a*) and (*b*), was :

(*d*) *If equals be subtracted from unequals, the remainders are unequal.*

Proclus, in observing that axioms ought not to be multiplied, indicates that all should be rejected which follow from the five admitted by him and appearing in the text above (p. 155). He mentions the second of those just quoted (*b*) as one of those to be excluded, since it follows from *Common Notion* 1. Proclus does not mention (*a*), (*c*) or (*d*); an-Nairīzī gives (*a*), (*d*), (*b*) and (*c*), in that order, as Euclid's, adding a note of Simplicius that "three axioms (sententiae acceptae) only are extant in the ancient manuscripts, but the number was increased in the more recent."

(*a*) stands self-condemned because "unequal" tells us nothing. It is easy to see what is wanted if we refer to I. 17, where the same angle is added to a *greater* and a *less*, and it is inferred that the first sum is greater than the second. So far however as the wording of (*a*) is concerned, the addition of equal to *greater* and *less* might be supposed to produce *less* and *greater* respectively. If therefore such an axiom were given at all, it should be divided into two. Heiberg conjectures that this axiom may have been taken from the commentary of Pappus, who had the axiom about equals added to unequals quoted below (*e*); if so, it can only be an unskilful adaptation of some remark of Pappus, for his axiom (*e*) has some point, whereas (*a*) is useless.

As regards (*b*), I agree with Tannery in seeing no sufficient reason why, if we reject it (as we certainly must), the words in I. 47 "But things which are double of equals are equal to one another" should be condemned as an interpolation. If they were interpolated, we should have expected to find the same interpolation in I. 42, where the axiom is *tacitly* assumed. I think it quite possible that Euclid may have inserted such words in one case and left them out in another, without necessarily implying either that he was quoting a formal *Common Notion* of his own or that he had *not* included among his Common Notions the particular fact stated as obvious.

The corresponding axiom (*c*) about the *halves* of equals can hardly be genuine if (*b*) is not, and Proclus does not mention it. Tannery acutely observes however that, when Heiberg, in I. 37, 38, brackets words stating that "the halves of equal things are equal to one another" on the ground that axiom (*c*) was interpolated (although before Theon's time), and explains that Euclid used *Common Notion* 3 in making his inference, he is clearly mistaken. For, while axiom (*b*) is an obvious inference from *Common Notion* 2, axiom (*c*) is not an inference from *Common Notion* 3. Tannery says, in a note, that (*c*) would have to be established by *reductio ad absurdum* with the help of axiom (*b*), that is to say, of *Common Notion* 2. But, as the hypothesis in the *reductio ad absurdum* would be that one of the halves is *greater* than the other, and it would therefore be necessary to prove that the one whole is *greater* than the other, while axiom (*b*) or *Common Notion* 2 only refers to *equals*, a little argument would be necessary in addition to the reference to *Common Notion* 2. I think Euclid would not have gone through this process in order to prove (*c*), but would have assumed it as equally obvious with (*b*).

Proclus (pp. 197, 6—198, 5) definitely rejects two other axioms of the above kind given by Pappus, observing that, as they follow from the genuine axioms, they are rightly omitted in most copies, although Pappus said that they were "on record" with the others (συναναγράφεσθαι):

(*e*) *If unequals be added to equals, the difference between the wholes is equal to the difference between the added parts* ; and

(*f*) *If equals be added to unequals, the difference between the wholes is equal to the difference between the original unequals.*

Proclus and Simplicius (in an-Nairīzī) give proofs of both. The proof of the former, as given by Simplicius, is as follows :

Let *AB*, *CD* be equal magnitudes ; and let *EB*, *FD* be added to them respectively, *EB* being greater than *FD*.

I say that *AE* exceeds *CF* by the same difference as that by which *BE* exceeds *DF*.

Cut off from *BE* the magnitude *BG* equal to *DF*.

Then, since *AE* exceeds *AG* by *GE*, and *AG* is equal to *CF* and *BG* to *DF*,

AE exceeds *CF* by the same difference as that by which *BE* exceeds *DF*.

COMMON NOTION 4.

Καὶ τὰ ἐφαρμόζοντα ἐπ᾽ ἄλληλα ἴσα ἀλλήλοις ἐστίν.

Things which coincide with one another are equal to one another.

The word ἐφαρμόζειν, as a geometrical term, has a different meaning according as it is used in the active or in the passive. In the passive, ἐφαρμόζεσθαι, it means "to be *applied* to" without any implication that the applied figure will exactly fit, or coincide with, the figure to which it is applied; on the other hand the active ἐφαρμόζειν is used intransitively and means "to fit exactly," "to coincide with." In Euclid and Archimedes ἐφαρμόζειν is constructed with ἐπί and the accusative, in Pappus with the dative.

On *Common Notion* 4 Tannery observes that it is incontestably geometrical in character, and should therefore have been excluded from the *Common Notions*; again, it is difficult to see why it is not accompanied by its converse, at all events for straight lines (and, it might be added, angles also), which Euclid makes use of in I. 4. As it is, says Tannery, we have here a definition of geometrical equality more or less sufficient, but not a real axiom.

It is true that Proclus seems to recognise this *Common Notion* and the next as proper axioms in the passage (p. 196, 15—21) where he says that we should not cut down the axioms to the minimum, as Heron does in giving only three axioms; but the statement seems to rest, not upon authority, but upon an assumption that Euclid would state explicitly at the beginning all axioms subsequently used and not reducible to others unquestionably included. Now in I. 4 this *Common Notion* is not quoted; it is simply inferred that "the base *BC* will coincide with *EF*, *and will be equal* to it." The position is therefore the same as it is in regard to the statement in the same proposition that, "if... the base *BC* does not coincide with *EF*, *two straight lines will enclose a space*: which is impossible"; and, if we do not admit that Euclid had the axiom that "two straight lines cannot enclose a space," neither need we infer that he had *Common Notion* 4. I am therefore inclined to think that the latter is more likely than not to be an interpolation.

It seems clear that the Common Notion, as here formulated, is intended to assert that superposition is a legitimate way of proving the equality of two figures which have the necessary parts respectively equal, or, in other words, to serve as an *axiom of congruence*.

The phraseology of the propositions, e.g. I. 4 and I. 8, in which Euclid employs the method indicated, leaves no room for doubt that he regarded one figure as actually *moved* and *placed upon* the other. Thus in I. 4 he says, "The triangle *ABC* being applied (ἐφαρμοζομένου) to the triangle *DEF*, and the point *A* being *placed* (τιθεμένου) upon the point *D*, and the straight line *AB* on *DE*, the point *B* will also coincide with *E* because *AB* is equal to *DE*"; and in I. 8, "If the sides *BA*, *AC* do not coincide with *ED*, *DF*, but *fall beside them* (take a different position, παραλλάξουσιν), then" etc. At the same time, it is clear that Euclid disliked the method and avoided it wherever he could, e.g. in I. 26, where he proves the equality of two triangles which have two angles respectively equal to two angles and one side of the one equal to the corresponding side of the other. It looks as though he found the method handed down by tradition (we can hardly suppose that, if Thales proved that the diameter of a circle divides it into two equal parts, he would do so by any other method than that of superposition), and followed it, in ·the few cases where he does so, only because he had not been able to see his way to a satisfactory substitute. But seeing how much of the *Elements* depends on I. 4, directly or indirectly, the method can hardly be regarded as being, in Euclid, of only subordinate importance; on the contrary, it is fundamental. Nor, as a matter of fact, do we find in the ancient geometers any expression of doubt as to the legitimacy of the method. Archimedes uses it to prove that any spheroidal figure cut by a plane through the centre is divided into two equal parts in respect of both its surface and its volume; he also postulates in *Equilibrium of Planes* I. that "when equal and similar plane figures coincide if applied to one another, their centres of gravity coincide also."

Killing (*Einführung in die Grundlagen der Geometrie*, II. pp. 4, 5) contrasts the attitude of the Greek geometers with that of the philosophers, who, he says, appear to have agreed in banishing motion from geometry altogether. In support of this he refers to the view frequently expressed by Aristotle that mathematics has to do with *immovable* objects (ἀκίνητα), and that only where astronomy is admitted as part of mathematical science is motion mentioned as a subject for mathematics. Cf. *Metaph.* 989 b 32 "For mathematical objects are among things which exist apart from motion, except such as relate to astronomy"; *Metaph.* 1064 a 30 "Physics deals with things which have in themselves the principle of motion; mathematics is a theoretical science and one concerned with things which are *stationary* (μένοντα) but not separable" (sc. from matter); in *Physics* II. 2, 193 b 34 he speaks of the subjects of mathematics as "in thought separable from motion."

But I doubt whether in Aristotle's use of the words "immovable," "without motion" etc. as applied to the subjects of mathematics there is any implication such as Killing supposes. We arrive at mathematical concepts by abstraction from material objects; and just as we, in thought, eliminate the matter, so according to Aristotle we eliminate the attributes of matter as such, e.g. qualitative change and *motion*. It does not appear to me that the use of "immovable" in the passages referred to means more than this. I do not think that Aristotle would have regarded it as illegitimate to *move* a geometrical figure from one position to another; and I infer this from a passage in *De caelo* III. 1 where he is criticising "those who make up every body that has an origin by putting together *planes*, and resolve it again into *planes*." The reference must be to the *Timaeus* (54 B sqq.) where Plato evolves the four elements in this way. He begins with a right-angled triangle in which the hypotenuse is double of the smaller side; six of these put together in the proper way produce one equilateral triangle. Making solid angles with (*a*) three, (*b*) four, and (*c*) five of these equilateral triangles respectively, and taking the requisite number of these solid angles, namely four of (*a*), six of (*b*) and twelve of (*c*) respectively, and putting them together so as to form regular solids, he obtains (α) a tetrahedron, (β) an octahedron, (γ) an icosahedron respectively. For the fourth element (earth), four isosceles right-angled triangles are first put together so as to form a square, and then six of these squares are put together to form a cube. Now, says Aristotle (299 b 23), "it is absurd that planes should only admit of being put together so as to touch in a *line*; for just as a line and a line are put together in both ways, lengthwise and breadthwise, so must a plane and a plane. A line can be combined with a line in the sense of being a line *superposed*, and not *added*"; the inference being that a *plane* can be superposed on a *plane*. Now this is precisely the sort of motion in question here; and Aristotle, so far from denying its permissibility, seems to blame Plato for not using it. Cf. also *Physics* V. 4, 228 b 25, where Aristotle speaks of "the spiral or other magnitude in which any part will not coincide with any other part," an where superposition is obviously contemplated.

Motion without deformation.

It is well known that Helmholtz maintained that geometry requires us to assume the actual existence of rigid bodies and their free mobility in space, whence he inferred that geometry is dependent on mechanics.

Veronese exposed the fallacy in this (*Fondamenti di geometria*, pp. xxxv—xxxvi, 239—240 note, 615—7), his argument being as follows. Since geometry is concerned with empty space, which is immovable, it would be at least strange if it was necessary to have recourse to the real motion of bodies for a definition, and for the proof of the properties, of immovable space. We must distinguish the intuitive principle of motion in itself from that of motion *without deformation*. Every point of a figure which moves is transferred to another point in space. "Without deformation" means that the mutual relations between the points of the figure do not change, but the relations between them and other figures do change (for if they did not, the figure could not move). Now consider what we mean by saying that, when the figure A has moved from the position A_1 to the position A_2, the relations between the points of A in the position A_2 are unaltered from what they were in the position A_1, are the same in fact as if A had not moved but remained at A_1. We can only say that, judging of the figure (or the body with its physical qualities eliminated) by the impressions it produces in us during its movement, the impressions produced in us in the two different positions (which are in time distinct) *are equal*. In fact, we are making use of the notion of *equality* between two distinct figures. Thus, if we say that two bodies are equal when they can be superposed by means of *movement without deformation*, we are committing a *petitio principii*. The notion of the equality of spaces is really prior to that of rigid bodies or of motion without deformation. Helmholtz supported his view by reference to the process of measurement in which the measure must be, at least approximately, a rigid body, but the existence of a rigid body as a standard to measure by, and the question how we discover two equal spaces to be equal, are matters of no concern to the geometer. The method of superposition, depending on motion without deformation, is only of use as a *practical* test; it has nothing to do with the *theory* of geometry.

Compare an acute observation of Schopenhauer (*Die Welt als Wille*, 2 ed. 1844, II. p. 130) which was a criticism in advance of Helmholtz' theory: "I am surprised that, instead of the eleventh axiom [the Parallel-Postulate], the eighth is not rather attacked: 'Figures which coincide (sich decken) are equal to one another.' For *coincidence* (das Sichdecken) is either mere tautology, or something entirely empirical, which belongs, not to pure intuition (Anschauung), but to external sensuous experience. It presupposes in fact the mobility of figures; but that which is movable in space is matter and nothing else. Thus this appeal to coincidence means leaving pure space, the sole element of geometry, in order to pass over to the material and empirical."

Mr Bertrand Russell observes (*Encyclopaedia Britannica*, Suppl. Vol. 4, 1902, Art. "Geometry, non-Euclidean") that the apparent use of motion here is deceptive; what in geometry is called a motion is merely the transference of our attention from one figure to another. Actual superposition, which is nominally employed by Euclid, is not required; all that is required is the transference of our attention from the original figure to a new one defined by the position of some of its elements and by certain properties which it shares with the original figure.

If the method of superposition is given up as a means of defining theoretically the equality of two figures, some other definition of equality is necessary. But such a definition can be evolved out of *empirical* or *practical* observation of the result of superposing two material representations of figures. This is done by Veronese (*Elementi di geometria*, 1904) and Ingrami (*Elementi di geometria*, 1904). Ingrami says, namely (p. 66):

"If a sheet of paper be folded double, and a triangle be drawn upon it and then cut out, we obtain two triangles *superposed* which we in practice call *equal*. If points A, B, C, D ... be marked on one of the triangles, then, when we place this triangle upon the other (so as to coincide with it), we see that *each* of the particular points taken on the first is superposed on one particular point of the second in such a way that the segments *AB*, *AC*, *AD*, *BC*, *BD*, *CD*, ... are respectively superposed on as many segments in the second triangle and are therefore equal to them respectively. In this way we justify the following

"Definition of equality.

"Any two figures whatever will be called *equal* when to the points of one the points of the other can be made to correspond *univocally* [i.e. every *one* point in one to *one distinct* point in the other and *vice versa*] in such a way that the segments which join the points, two and two, in one figure are respectively equal to the segments which join, two and two, the corresponding points in the other."

Ingrami has of course previously postulated as known the signification of the phrase *equal (rectilineal) segments*, of which we get a *practical* notion when we can place one upon the other or can place a third movable segment successively on both.

New systems of Congruence-Postulates.

In the fourth Article of *Questioni riguardanti le matematiche elementari*, I., pp. 93—122, a review is given of three different systems: (1) that of Pasch in *Vorlesungen über neuere Geometrie*, 1882, p. 101 sqq., (2) that of Veronese according to the *Fondamenti di geometria*, 1891, and the *Elementi* taken together, (3) that of Hilbert (see *Grundlagen der Geometrie*, 1903, pp. 7—15).

These systems differ in the particular conceptions taken by the three authors as primary. (1) Pasch considers as primary the notion of *congruence* or *equality* between *any figures which are made up of a finite number of points only*. The definitions of congruent *segments* and of congruent *angles* have to be *deduced* in the way shown on pp. 102—103 of the Article referred to, after which Eucl. I. 4 follows immediately, and Eucl. I. 26 (1) and I. 8 by a method recalling that in Eucl. I. 7, 8.

(2) Veronese takes as primary the conception of congruence between *segments* (rectilineal). The transition to congruent *angles*, and thence to *triangles* is made by means of the following postulate:

"Let *AB*, *AC* and *A'B'*, *A'C'* be two pairs of straight lines intersecting at *A*, *A'*, and let there be determined upon them the congruent segments *AB*, *A'B'* and the congruent segments *AC*, *A'C'*;

then, if *BC*, *B'C'* are congruent, the two *pairs of straight lines* are congruent."

(3) Hilbert takes as primary the notions of congruence between *both segments and angles*.

It is observed in the Article referred to that, from the theoretical standpoint, Veronese's system is an advance upon that of Pasch, since the idea of congruence between *segments* is more simple than that of congruence between *any figures*; but, didactically, the development of the theory is more complicated when we start from Veronese's system than when we start from that of Pasch.

The system of Hilbert offers advantages over both the others from the point of view of the teaching of geometry, and I shall therefore give a short account of his system only, following the Article above quoted.

Hilbert's system.

The following are substantially the Postulates laid down.

(1) *If one segment is congruent with another, the second is also congruent with the first.*

(2) *If an angle is congruent with another angle, the second angle is also congruent with the first.*

(3) *Two segments congruent with a third are congruent with one another.*

(4) *Two angles congruent with a third are congruent with one another.*

(5) *Any segment* AB *is congruent with itself, independently of its sense.* This we may express symbolically thus:

$$AB \equiv AB \equiv BA.$$

(6) *Any angle* (ab) *is congruent with itself, independently of its sense.* This we may express symbolically thus:

$$(ab) \equiv (ab) \equiv (ba).$$

(7) *On any straight line* r′, *starting from any one of its points* A′, *and on each side of it respectively, there exists one and only one segment congruent with a segment* AB *belonging to the straight line* r.

(8) *Given a ray* a, *issuing from a point* O, *in any plane which contains it and on each of the two sides of it, there exists one and only one ray* b *issuing from* O *such that the angle* (ab) *is congruent with a given angle* (a′b′).

(9) *If* AB, BC *are two consecutive segments of the same straight line* r (*segments, that is, having an extremity and no other point common*), *and* A′B′, B′C′ *two consecutive segments on another straight line* r′, *and if* AB ≡ A′B′, BC ≡ B′C′, *then*

$$AC \equiv A'C'.$$

(10) *If* (ab), (bc) *are two consecutive angles in the same plane* π (*angles, that is, having the vertex and one side common*), *and* (a′b′), (b′c′) *two consecutive angles in another plane* π′, *and if* (ab) ≡ (a′b′), (bc) = (b′c′), *then*

$$(ac) = (a'c').$$

(11) *If two triangles have two sides and the included angles respectively congruent, they have also their third sides congruent as well as the angles opposite to the congruent sides respectively.*

As a matter of fact, Hilbert's postulate corresponding to (11) does not assert the equality of the third sides in each, but only the equality of the two remaining angles in one triangle to the two remaining angles in the other respectively.   He proves the equality of the third sides (thereby completing the theorem of Eucl. I. 4) by *reductio ad absurdum* thus. Let *ABC*, *A′B′C′* be the two triangles which have the sides *AB*, *AC* respectively congruent with the sides *A′B′*, *A′C′* and the included angle at *A* congruent with the included angle at *A′*.

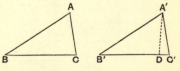

Then, by Hilbert's own postulate, the angles *ABC*, *A′B′C′* are congruent, as also the angles *ACB*, *A′C′B′*.

If *BC* is not congruent with *B′C′*, let *D* be taken on *B′C′* such that *BC*, *B′D* are congruent. and join *A′D*.

Then the two triangles ABC, $A'B'D$ have two sides and the included angles congruent respectively; therefore, by the same postulate, the angles BAC, $B'A'D$ are congruent.

But the angles BAC, $B'A'C'$ are congruent; therefore, by (4) above, the angles $B'A'C'$, $B'A'D$ are congruent: which is impossible, since it contradicts (8) above.

Hence BC, $B'C'$ cannot but be congruent.

Eucl. I. 4 is thus proved; but it seems to be as well to include all of that theorem in the postulate, as is done in (11) above, since the two parts of it are equally suggested by empirical observation of the result of one superposition.

A proof similar to that just given immediately establishes Eucl. I. 26 (1), and Hilbert next proves that

If two angles ABC, A'B'C' *are congruent with one another, their supplementary angles* CBD, C'B'D' *are also congruent with one another.*

We choose A, D on one of the straight lines forming the first angle, and A', D' on one of those forming the second angle, and again C, C' on the other

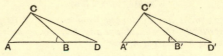

straight lines forming the angles, so that $A'B'$ is congruent with AB, $C'B'$ with CB, and $D'B'$ with DB.

The triangles ABC, $A'B'C'$ are congruent, by (11) above; and AC is congruent with $A'C'$, and the angle CAB with the angle $C'A'B'$.

Thus, AD, $A'D'$ being congruent, by (9), the triangles CAD, $C'A'D'$ are also congruent, by (11);

whence CD is congruent with $C'D'$, and the angle ADC with the angle $A'D'C'$.

Lastly, by (11), the triangles CDB, $C'D'B'$ are congruent, and the angles CBD, $C'B'D'$ are thus congruent.

Hilbert's next proposition is that

Given that the angle (h, k) *in the plane* α *is congruent with the angle* (h', k') *in the plane* α', *and that* l *is a half-ray in the plane* α *starting from the vertex of the angle* (h, k) *and lying within that angle, there always exists a half-ray* l' *in the second plane* α', *starting from the vertex of the angle* (h', k') *and lying within that angle, such that*

$$(h, l) \equiv (h', l'), \quad and \quad (k, l) \equiv (k', l').$$

If O, O' are the vertices, we choose points A, B on h, k, and points A', B' on h', k' respectively, such that OA, $O'A'$ are congruent and also OB, $O'B'$.

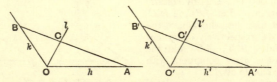

The triangles OAB, $O'A'B'$ are then congruent; and, if l meets AB in C, we can determine C' on $A'B'$ such that $A'C'$ is congruent with AC.

Then l' drawn from O' through C' is the half-ray required.

The congruence of the angles (h, l), (h', l') follows from (11) directly, and that of (k, l) and (k', l') follows in the same way after we have inferred by means of (9) that, AB, AC being respectively congruent with $A'B'$, $A'C'$, the difference BC is congruent with the difference $B'C'$.

It is by means of the two propositions just given that Hilbert proves that

All right angles are congruent with one another.

Let the angle BAD be congruent with its adjacent angle CAD, and likewise the angle $B'A'D'$ congruent with its adjacent angle $C'A'D'$ All four angles are then right angles.

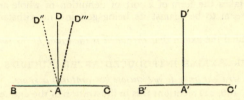

If the angle $B'A'D'$ is not congruent with the angle BAD, let the angle with AB for one side and congruent with the angle $B'A'D'$ be the angle BAD'', so that AD'' falls either within the angle BAD or within the angle DAC. Suppose the former.

By the last proposition but one (about adjacent angles), the angles $B'A'D'$, BAD'' being congruent, the angles $C'A'D'$, CAD'' are congruent.

Hence, by the hypothesis and postulate (4) above, the angles BAD'', CAD'' are also congruent.

And, since the angles BAD, CAD are congruent, we can find within the angle CAD a half-ray CAD''' such that the angles BAD'', CAD''' are congruent, and likewise the angles DAD'', DAD''' (by the last proposition).

But the angles BAD'', CAD'' were congruent (see above); and it follows, by (4), that the angles CAD'', CAD''' are congruent: which is impossible, since it contradicts postulate (8).

Therefore etc.

Euclid I. 5 follows directly by applying the postulate (11) above to ABC, ACB as distinct triangles.

Postulates (9), (10) above give in substance the proposition that "the sums or differences of segments, or of angles, respectively equal, are equal."

Lastly, Hilbert proves Eucl. I. 8 by means of the theorem of Eucl. I. 5 and the proposition just stated as applied to angles.

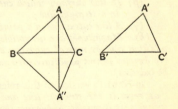

ABC, $A'B'C$ being the given triangles with three sides respectively congruent, we suppose an angle CBA'' to be determined, on the side of BC opposite to A, congruent with the angle $A'B'C'$, and we make BA'' equal to $A'B'$.

The proof is obvious, being equivalent to the alternative proof often given in our text-books for Eucl. I. 8.

COMMON NOTION 5.

καὶ τὸ ὅλον τοῦ μέρους μεῖζόν [ἐστιν].

The whole is greater than the part.

Proclus includes this "axiom" on the same ground as the preceding one. I think however there is force in the objection which Tannery takes to it, namely that it replaces a *different* expression in Eucl. I. 6, where it is stated that "the triangle *DBC* will be equal to the triangle *ACB, the less to the greater: which is absurd.*" The axiom appears to be an abstraction or generalisation substituted for an immediate inference from a geometrical figure, but it takes the form of a sort of definition of whole and part. The probabilities seem to be against its being genuine, notwithstanding Proclus' approval of it.

Clavius added the axiom that *the whole is the equal to the sum of its parts.*

OTHER AXIOMS INTRODUCED AFTER EUCLID'S TIME.

[9]   *Two straight lines do not enclose* (or *contain*) *a space.*

Proclus (p. 196, 21) mentions this in illustration of the undue multiplication of axioms, and he points out, as an objection to it, that it belongs to the subject matter of geometry, whereas axioms are of a general character, and not peculiar to any one science. The real objection to the axiom is that it is unnecessary, since the fact which it states is included in the meaning of Postulate 1. It was no doubt taken from the passage in I. 4, "if...the base *BC* does not coincide with the base *EF, two straight lines will enclose a space*: *which is impossible*"; and we must certainly regard it as an interpolation, notwithstanding that two of the best MSS. have it after Postulate 5, and one gives it as *Common Notion* 9.

Pappus added some others which Proclus objects to (p. 198, 5) because they are either anticipated in the definitions or follow from them.

(*g*)   *All the parts of a plane, or of a straight line, coincide with one another.*

(*h*)   *A point divides a line, a line a surface, and a surface a solid*; on which Proclus remarks that everything is *divided* by the same things as those by which it is *bounded*.

An-Nairīzī (ed. Besthorn-Heiberg, p. 31, ed. Curtze, p. 38) in his version of this axiom, which he also attributes to Pappus, omits the reference to solids, but mentions planes as a particular case of surfaces.

"(α)   *A surface cuts a surface in a line;*

(β)   *If two surfaces which cut one another are plane, they cut one another in a straight line;*

(γ)   *A line cuts a line in a point* (this last we need in the first proposition)."

(*k*)   *Magnitudes are susceptible of the infinite* (or *unlimited*) *both by way of addition and by way of successive diminution, but in both cases potentially only* (τὸ ἄπειρον ἐν τοῖς μεγέθεσίν ἐστιν καὶ τῇ προσθέσει καὶ τῇ ἐπικαθαιρέσει, δυνάμει δὲ ἑκάτερον).

An-Nairīzī's version of this refers to straight lines and plane surfaces only: "*as regards the straight line and the plane surface, in consequence of their evenness, it is possible to produce them indefinitely.*"

This "axiom" of Pappus, as quoted by Proclus, seems to be taken directly from the discussion of τὸ ἄπειρον in Aristotle, *Physics* III. 5—8, even to the wording, for, while Aristotle uses the term *division* (διαίρεσις) most frequently as the antithesis of *addition* (σύνθεσις), he occasionally speaks of *subtraction* (ἀφαίρεσις) and *diminution* (καθαίρεσις). Hankel (*Zur Geschichte der Mathematik im Alterthum und Mittelalter*, 1874, pp. 119—120) gave an admirable summary of Aristotle's views on this subject; and they are stated in greater detail in Görland, *Aristoteles und die Mathematik*, Marburg, 1899, pp. 157—183. The infinite or unlimited (ἄπειρον) only exists potentially (δυνάμει), not in actuality (ἐνεργείᾳ). The infinite is so in virtue of its endlessly changing into something else, like day or the Olympic Games (*Phys.* III. 6, 206 a 15—25). The infinite is manifested in different forms in time, in Man, and in the division of magnitudes. For, in general, the infinite consists in something new being continually taken, that something being itself always finite but always different. Therefore the infinite must not be regarded as a particular thing (τόδε τι), as man, house, but as being always in course of becoming or decay, and, though finite at any moment, always different from moment to moment. But there is the distinction between the forms above referred to that, whereas in the case of magnitudes what is once taken remains, in the case of time and Man it passes or is destroyed but the succession is unbroken. The case of addition is in a sense the same as that of division; in the finite magnitude the former takes place in the converse way to the latter; for, as we see the finite magnitude divided *ad infinitum*, so we shall find that addition gives a sum tending to a definite limit. I mean that, in the case of a finite magnitude, you may take a definite fraction of it and add to it (continually) in the same ratio; if now the successive added terms do not include one and the same magnitude whatever it is [i.e. if the successive terms diminish in geometrical progression], you will not come to the end of the finite magnitude, but, if the ratio is increased so that each term does include one and the same magnitude whatever it is, you will come to the end of the finite magnitude, for every finite magnitude is exhausted by continually taking from it any definite fraction whatever. Thus in no other sense does the infinite exist, but only in the sense just mentioned, that is, potentially and by way of diminution (206 a 25—b 13). And in this sense you may have potentially infinite addition, the process being, as we say, in a manner, the same as with division *ad infinitum*: for in the case of addition you will always be able to find something outside the total for the time being, but the total will never exceed every definite (or assigned) magnitude in the way that, in the direction of division, the result will pass every definite magnitude, that is, by becoming smaller than it. The infinite therefore cannot exist even potentially in the sense of exceeding every finite magnitude as the result of successive addition (206 b 16—22). It follows that the correct view of the infinite is the opposite of that commonly held: it is not that which has nothing outside it, but that which always has something outside it (206 b 33—207 a 1).

Contrasting the case of number and magnitude, Aristotle points out that (1) in number there is a limit in the direction of smallness, namely unity, but none in the other direction: a number may exceed any assigned number however great; but (2) with magnitude the contrary is the case: you can find a magnitude smaller than any assigned magnitude, but in the other direction there is no such thing as an infinite magnitude (207 b 1—5). The latter assertion he justified by the following argument. However large a thing can be potentially, it can be as large actually. But there is no magnitude perceptible to sense that is infinite. Therefore excess over every assigned magnitude is an impossibility; otherwise there would be something larger than the universe (οὐρανός) (207 b 17—21).

Aristotle is aware that it is essentially of physical magnitudes that he is speaking. He had observed in an earlier passage (*Phys.* III. 5, 204 a 34) that it is perhaps a more general inquiry that would be necessary to determine whether the infinite is possible in mathematics, and in the domain of thought and of things which have no magnitude; but he excuses himself from entering upon this inquiry on the ground that his subject is physics and sensible objects. He returns however to the bearing of his conclusions on mathematics in III. 7, 207 b 27 : "my argument does not even rob mathematicians of their study, although it denies the existence of the infinite in the sense of actual existence as something increased to such an extent that it cannot be gone through (ἀδιεξίτητον); for, as it is, they do not even need the infinite or use it, but only require that the finite (straight line) shall be as long *as they please*; and another magnitude of any size whatever can be cut in the same ratio as the greatest magnitude. Hence it will make no difference to them for the purpose of demonstration."

Lastly, if it should be urged that the infinite exists in *thought*, Aristotle replies that this does not involve its existence in *fact*. A thing is not greater than a certain size because it is conceived to be so, but because it *is*; and magnitude is not infinite in virtue of increase in thought (208 a 16—22).

Hankel and Görland do not quote the passage about an infinite series of magnitudes (206 b 3—13) included in the above paraphrase; but I have thought that mathematicians would be interested in the distinct expression of Aristotle's view that the existence of an infinite series the terms of which are *magnitudes* is impossible unless it is convergent, and (with reference to Riemann's developments) in the statement that it does not matter to geometry if the straight line is not infinite in length, provided that it is as long as we please.

Aristotle's denial of even the potential existence of a sum of magnitudes which shall exceed every definite magnitude was, as he himself implies, in conflict with the lemma or assumption used by Eudoxus (as we infer from Archimedes) to prove the theorem about the volume of a pyramid. The lemma is thus stated by Archimedes (*Quadrature of a parabola*, preface): "The excess by which the greater of two unequal areas exceeds the less can, if it be continually added to itself, be made to exceed any assigned finite area." We can therefore well understand why, a century later, Archimedes felt it necessary to justify his own use of the lemma as he does in the same preface : "The earlier geometers too have used this lemma : for it is by its help that they have proved that circles have to one another the duplicate ratio of their diameters, that spheres have to one another the triplicate ratio of their diameters, and so on. And, in the result, each of the said theorems has been accepted no less than those proved without the aid of this lemma."

Principle of continuity.

The use of actual construction as a method of proving the existence of figures having certain properties is one of the characteristics of the *Elements*. Now constructions are effected by means of straight lines and circles drawn in accordance with Postulates 1—3; the essence of them is that such straight lines and circles determine by their intersections other points in addition to those given, and these points again are used to determine new lines, and so on. This being so, the *existence* of such points of intersection must be postulated or proved in the same way as that of the lines which determine them. Yet there is no postulate of this character expressed in Euclid except Post. 5. This postulate asserts that two straight lines meet if they satisfy a certain condition. The condition is of the nature of a διορισμός (*discrimination*, or condition of possibility) in a problem ; and, if the existence of the point of intersection were not granted, the solutions of problems in which the points of intersection of straight lines are used would not in general furnish the required proofs of the existence of the figures to be constructed.

But, equally with the intersections of straight lines, the intersections of circle with straight line, and of circle with circle, are used in constructions. Hence, in addition to Postulate 5, we require postulates asserting the actual existence of points of intersection of circle with straight line and of circle with circle. In the very first proposition the vertex of the required equilateral triangle is determined as one of the intersections of two circles, and we need therefore to be assured that the circles will intersect. Euclid seems to assume it as obvious, although it is not so; and he makes a similar assumption in I. 22. It is true that in the latter case Euclid adds to the enunciation that two of the given straight lines must be together greater than the third; but there is nothing to show that, if this condition is satisfied, the construction is always possible. In I. 12, in order to be sure that the circle with a given centre will intersect a given straight line, Euclid makes the circle pass through a point on the side of the line opposite to that where the centre is. It appears therefore as if, in this case, he based his inference in some way upon the definition of a circle combined with the fact that the point within it called the centre is on one side of the straight line and one point of the circumference on the other, and, in the case of two intersecting circles, upon similar considerations. But not even in Book III., where there are several propositions about the relative positions of two circles, do we find any discussion of the conditions under which two circles have two, one, or no point common.

The deficiency can only be made good by the *Principle of Continuity.*

Killing (*Einführung in die Grundlagen der Geometrie*, II. p. 43) gives the following forms as sufficient for most purposes.

(*a*)   Suppose a line belongs entirely to a figure which is divided into two parts; then, if the line has at least one point common with each part, it must also meet the boundary between the parts; or

(*b*)   If a point moves in a figure which is divided into two parts, and if it belongs at the beginning of the motion to one part and at the end of the motion to the other part, it must during the motion arrive at the boundary between the two parts.

In the *Questioni riguardanti le matematiche elementari*, I., Art. 5, pp. 123—143, the principle of continuity is discussed with special reference to the Postulate of Dedekind, and it is shown, first, how the Postulate may be led up to and, secondly, how it may be applied for the purposes of elementary geometry.

Suppose that in a segment AB of a straight line a point C determines two segments AC, CB. If we consider the point C as belonging to only one of the two segments AC, CB, we have a division of the segment AB into two parts with the following properties.

1.   Every point of the segment AB belongs to *one* of the two parts.

2.   The point A belongs to one of the two parts (which we will call the *first*) and the point B to the other; the point C may belong indifferently to one or the other of the two parts according as we choose to premise.

3.   Every point of the first part precedes every point of the second in the order AB of the segment.

(For generality we may also suppose the case in which the point C falls at A or at B. Considering C, in these cases respectively, as belonging to the first or second part, we still have a division into parts which have the properties above enunciated, one part being then a single point A or B.)

Now, considering carefully the inverse of the above proposition, we see that it agrees with the idea which we have of the continuity of the straight line. Consequently we are induced to admit as a *postulate* the following.

If a segment of a straight line AB *is divided into two parts so that*

(1) *every point of the segment* AB *belongs to one of the parts,*

(2) *the extremity* A *belongs to the first part and* B *to the second, and*

(3) *any point whatever of the first part precedes any point whatever of the second part, in the order* AB *of the segment,*

there exists a point C *of the segment* AB (*which may belong either to one part or to the other*) *such that every point of* AB *that precedes* C *belongs to the first part, and every point of* AB *that follows* C *belongs to the second part in the division originally assumed.*

(If one of the two parts consists of the single point A or B, the point C is the said extremity A or B of the segment.)

This is the Postulate of Dedekind, which was enunciated by Dedekind himself in the following slightly different form (*Stetigkeit und irrationale Zahlen*, 1872, new edition 1905, p. 11).

" *If all points of a straight line fall into two classes such that every point of the first class lies to the left of every point of the second class, there exists one and only one point which produces this division of all the points into two classes, this division of the straight line into two parts.*"

The above enunciation may be said to correspond to the intuitive notion which we have that, if in a segment of a straight line two points start from the ends and describe the segment in opposite senses, they meet in a point. The point of meeting might be regarded as belonging to both parts, but for the present purpose we must regard it as belonging to one only and subtracted from the other part.

Application of Dedekind's postulate to angles.

If we consider an angle less than two right angles bounded by two rays a, b, and draw the straight line connecting A, a point on a, with B, a point on b, we see that all points on the finite segment AB correspond univocally to all the rays of the angle, the point corresponding to any ray being the point in which the ray cuts the segment AB; and if a ray be supposed to move about the vertex of the angle from the position a to the position b, the corresponding points of the segment AB are seen to follow in the same order as the corresponding rays of the angle (ab).

Consequently, if the angle (ab) is divided into two parts so that

(1) each ray of the angle (ab) belongs to one of the two parts,

(2) the outside ray a belongs to the first part and the ray b to the second,

(3) any ray whatever of the first part precedes any ray whatever of the second part,

the corresponding points of the segment AB determine two parts of the segments such that

(1) every point of the segment AB belongs to one of the two parts,

(2) the extremity A belongs to the first part and B to the second,

(3) any point whatever of the first part precedes any point whatever of the second.

But in that case there exists a point C of AB (which may belong to one or the other of the two parts) such that every point of AB that precedes C belongs to the first part and every point of AB that follows C belongs to the second part.

Thus exactly the same thing holds of c, the ray corresponding to C, with reference to the division of the angle (ab) into two parts.

It is not difficult to extend this to an angle (ab) which is either *flat* or greater than two right angles ; this is done (Vitali, *op. cit.* pp. 126—127) by supposing the angle to be divided into two, (ad), (db), each less than two right angles, and considering the three cases in which

(1) the ray d is such that all the rays that precede it belong to the first part and those which follow it to the second part,
(2) the ray d is followed by some rays of the first part,
(3) the ray d is preceded by some rays of the second part.

Application to circular arcs.

If we consider an arc AB of a circle with centre O, the points of the arc correspond univocally, and in the same order, to the rays from the point O passing through those points respectively, and the same argument by which we passed from a segment of a straight line to an angle can be used to make the transition from an angle to an arc.

Intersections of a straight line with a circle.

It is possible to use the Postulate of Dedekind to prove that

If a straight line has one point inside and one point outside a circle, it has two points common with the circle.

For this purpose it is necessary to assume (1) the proposition with reference to the perpendicular and obliques drawn from a given point to a given straight line, namely that of all straight lines drawn from a given point to a given straight line the perpendicular is the shortest, and of the rest (the obliques) that is the longer which has the longer projection upon the straight line, while those are equal the projections of which are equal, so that for any given length of projection there are two equal obliques and two only, one on each side of the perpendicular, and (2) the proposition that any side of a triangle is less than the sum of the other two.

Consider the circle (C) with centre O, and a straight line (r) with one point A inside and one point B outside the circle.

By the definition of the circle, if R is the radius,

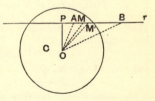

$$OA < R, \quad OB > R.$$

Draw OP perpendicular to the straight line r.

Then $OP < OA$, so that OP is always less than R, and P is therefore within the circle C.

Now let us fix our attention on the finite segment AB of the straight line r. It can be divided into two parts, (1) that containing all the points H for which $OH < R$ (i.e. points inside C), and (2) that containing all the points K for which $OK \geqq R$ (points outside C or on the circumference of C).

Thus, remembering that, of two obliques from a given point to a given straight line, that is greater the projection of which is greater, we can assert that all the points of the segment PB which *precede* a point inside C are inside C, and those which *follow* a point on the circumference of C or outside C are outside C.

Hence, by the Postulate of Dedekind, there exists on the segment PB a point M such that all the points which precede it belong to the first part and those which follow it to the second part.

I say that M is common to the straight line r and the circle C, or

$$OM = R.$$

For suppose, e.g., that $OM < R$.

There will then exist a segment (or length) σ less than the difference between R and OM.

Consider the point M', one of those which *follow* M, such that MM' is equal to σ.

Then, because any side of a triangle is less than the sum of the other two,

$$OM' < OM + MM'.$$

But $OM + MM' = OM + \sigma < R,$

whence $OM' < R,$

which is absurd.

A similar absurdity would follow if we suppose that $OM > R$.

Therefore OM must be equal to R.

It is immediately obvious that, corresponding to the point M on the segment PB which is common to r and C, there is another point on r which has the same property, namely that which is symmetrical to M with respect to P.

And the proposition is proved.

Intersections of two circles.

We can likewise use the Postulate of Dedekind to prove that

If in a given plane a circle C has one point X inside and one point Y outside another circle C', the two circles intersect in two points.

We must first prove the following

Lemma.

If O, O' are the centres of two circles C, C', and R, R' their radii respectively, the straight line OO' meets the circle C in two points A, B, one of which is inside C' and the other outside it.

Now one of these points must fall (1) on the prolongation of $O'O$ beyond O or (2) on OO' itself or (3) on the prolongation of OO' beyond O'.

(1)   First, suppose A to lie on $O'O$ produced.

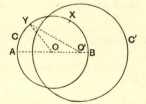

Then $AO' = AO + OO' = R + OO'$ ......(a).

But, in the triangle $OO'Y$,

$$O'Y < OY + OO',$$

and, since $O'Y > R'$, $OY = R$,

$$R' < R + OO'.$$

It follows from (a) that $AO' > R'$; and A therefore lies *outside* C'.

(2)   Secondly, suppose A to lie on OO'.

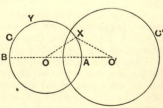

Then $OO' = OA + AO' = R + AO'$ ...(β).

From the triangle $OO'X$ we have

$$OO' < OX + O'X,$$

and, since $OX = R$, $O'X < R'$, it follows that

$$OO' < R + R',$$

whence, by (β), $AO' < R'$, so that A lies *inside* C'.

(3) Thirdly, suppose A to lie on OO' produced.

Then $\qquad R = OA = OO' + O'A \ldots\ldots\ldots(\gamma)$.

And, in the triangle $OO'X$,

$$OX < OO' + O'X,$$

that is $\qquad R < OO' + O'X,$

whence, by (γ),

$$OO' + O'A < OO' + O'X,$$

or $\qquad O'A < O'X,$

and A lies *inside* C'.

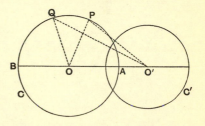

It is to be observed that one of the two points A, B is in the position of case (1) and the other in the position of either case (2) or case (3) : whence we must conclude that one of the two points A, B is *inside* and the other *outside* the circle C'.

Proof of theorem.

The circle C is divided by the points A, B into two semicircles. Consider one of them, and suppose it to be described by a point moving from A to B.

Take two separate points P, Q on it and, to fix our ideas, suppose that P precedes Q.

Comparing the triangles $OO'P$, $OO'Q$, we observe that one side OO' is common, OP is equal to OQ, and the angle POO' is less than the angle QOO'.

Therefore $O'P < O'Q$.

Now, considering the semicircle $APQB$ as divided into two parts, so that the points of the first part are inside the circle C', and those of the second part on the circumference of C' or outside it, we have the conditions necessary for the applicability of the Postulate of Dedekind (which is true for arcs of circles as for straight lines) ; whence *there exists a point* M *separating the two parts*.

I say that $O'M = R'$.

For, if not, suppose $O'M < R'$.

If then σ signifies the difference between R' and $O'M$, suppose a point M', which *follows* M, taken on the semicircle such that the chord MM' is not greater than σ (for a way of doing this see below).

Then, in the triangle $O'MM'$,

$$O'M' < O'M + MM' < O'M + \sigma,$$

and therefore $\qquad O'M' < R'$.

It follows that M', a point on the arc MB, is inside the circle C' : which is absurd.

Similarly it may be proved that $O'M$ is not greater than R.

Hence $O'M = R$.

[To find a point M' such that the chord MM' is not greater than σ, we may proceed thus.

Draw from M a straight line MP distinct from OM, and cut off MP on it equal to $\sigma/2$.

Join *OP*, and draw another radius *OQ* such that the angle *POQ* is equal
to the angle *MOP*.

The intersection, *M'*, of *OQ* with the
circle satisfies the required condition.

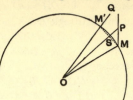

For *MM'* meets *OP* at right angles
in *S*.

Therefore, in the right-angled triangle
MSP, *MS* is not greater than *MP* (it is
less, unless *MP* coincides with *MS*, when
it is equal).

Therefore *MS* is not greater than σ/2, so that *MM'* is not greater than σ.]

PROPOSITION I.

On a given finite straight line to construct an equilateral triangle.

Let AB be the given finite straight line.

Thus it is required to con-
5 struct an equilateral triangle on
the straight line AB.

With centre A and distance
AB let the circle BCD be
described ;   [Post. 3]
10 again, with centre B and distance BA let the circle ACE
be described ;   [Post. 3]

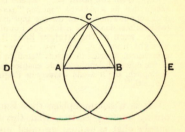

and from the point C, in which the circles cut one another, to
the points A, B let the straight lines CA, CB be joined.

[Post. 1]

15   Now, since the point A is the centre of the circle CDB,
AC is equal to AB.   [Def. 15]

Again, since the point B is the centre of the circle CAE,
BC is equal to BA.   [Def. 15]

But CA was also proved equal to AB ;
20 therefore each of the straight lines CA, CB is equal to AB.

And things which are equal to the same thing are also
equal to one another ;   [C. N. 1]

therefore CA is also equal to CB.

Therefore the three straight lines CA, AB, BC are
25 equal to one another.

Therefore the triangle *ABC* is equilateral; and it has been constructed on the given finite straight line *AB*.

(Being) what it was required to do.

1. **On a given finite straight line.** The Greek usage differs from ours in that the definite article is employed in such a phrase as this where we have the indefinite. ἐπὶ τῆς δοθείσης εὐθείας πεπερασμένης, "on *the* given finite straight line," i.e. the finite straight line which we choose to take.

·3. **Let AB be the given finite straight line.** To be strictly literal we should have to translate in the reverse order "let the given finite straight line be the (straight line) *AB*"; but this order is inconvenient in other cases where there is more than one datum, e.g. in the *setting-out* of I. 2, "let the given point be *A*, and the given straight line *BC*," the awkwardness arising from the omission of the verb in the second clause. Hence I have, for clearness' sake, adopted the other order throughout the book.

8. **let the circle BCD be described.** Two things are here to be noted, (1) the elegant and practically universal use of the perfect passive imperative in constructions, γεγράφθω meaning of course "let it *have been* described " or "suppose it described," (2) the impossibility of expressing shortly in a translation the force of the words in their original order. κύκλος γεγράφθω ὁ ΒΓΔ means literally "let a circle have been described, the (circle, namely, which I denote by) *BCD*." Similarly we have lower down "let straight lines, (namely) the (straight lines) *CA, CB*, be joined," ἐπεζεύχθωσαν εὐθεῖαι αἱ ΓΑ, ΓΒ. There seems to be no practicable alternative, in English, but to translate as I have done in the text.

13. **from the point C**.... Euclid is careful to adhere to the phraseology of Postulate 1 except that he speaks of "joining" (ἐπεζεύχθωσαν) instead of "drawing" (γράφειν). He does not allow himself to use the shortened expression "let the straight line *FC* be joined" (without mention of the points *F, C*) until I. 5.

20. **each of the straight lines CA, CB,** ἑκατέρα τῶν ΓΑ, ΓΒ and 24. **the three straight lines CA, AB, BC,** αἱ τρεῖς αἱ ΓΑ, ΑΒ, ΒΓ. I have, here and in all similar expressions, inserted the words "straight lines" which are not in the Greek. The possession of the inflected definite article enables the Greek to omit the words, but this is not possible in English, and it would scarcely be English to write "each of *CA, CB*" or "the three *CA, AB, BC*."

It is a commonplace that Euclid has no right to assume, without premising some postulate, that the two circles *will* meet in a point *C*. To supply what is wanted we must invoke the Principle of Continuity (see note thereon above, p. 235). It is sufficient for the purpose of this proposition and of I. 22, where there is a similar tacit assumption, to use the form of postulate suggested by Killing. "*If a line* [in this case e.g. the circumference *ACE*] *belongs entirely to a figure* [in this case a plane] *which is divided into two parts* [namely the part enclosed within the circumference of the circle *BCD* and the part outside that circle], *and if the line has at least one point common with each part, it must also meet the boundary between the parts* [i.e. the circumference *ACE* must meet the circumference *BCD*]."

Zeno's remark that the problem is not solved unless it is taken for granted that two straight lines cannot have a common segment has already been mentioned (note on Post. 2, p. 196). Thus, if *AC, BC* meet at *F* before reaching *C*, and have the part *FC* common, the triangle obtained, namely *FAB*, will not be equilateral, but *FA, FB* will each be less than *AB*. But Post. 2 has already laid it down that two straight lines cannot have a common segment.

Proclus devotes considerable space to this part of Zeno's criticism, but satisfies himself with the bare mention of the other part, to the effect that it is also necessary to assume that two *circumferences* (with different centres) cannot have a common part. That is, for anything we know, there may be any number of points *C* common to the two circumferences *ACE, BCD*. It is not until III. 10 that it is proved that two circles cannot intersect in more points than two, so that we are not entitled to assume it here. The most we can say is that it is enough for the purpose of this proposition if *one* equilateral triangle can be found with the given base; that the construction only gives *two* such triangles has to be left over to be proved subsequently. And indeed we have not long to wait; for I. 7 clearly shows that on either side of the base *AB* only *one* equilateral triangle can be described. Thus I. 7 gives us the *number of solutions* of which the present problem is susceptible, and it supplies the same want in I. 22 where a triangle has to be described with three sides of given length; that is, I. 7 furnishes us, in both cases, with one of the essential parts of a complete διορισμός, which includes not only the determination of the conditions of possibility but also the number of solutions (ποσαχῶς ἐγχωρεῖ, Proclus, p. 202, 5). This view of I. 7 as supplying an equivalent for III. 10 absolutely needed in I. 1 and I. 22 should serve to correct the idea so common among writers of text-books that I. 7 is merely of use as a lemma to Euclid's proof of I. 8, and therefore may be left out if an alternative proof of that proposition is adopted.

Agreeably to his notion that it is from I. 1 that we must satisfy ourselves that isosceles and scalene triangles actually exist, as well as equilateral triangles, Proclus shows how to draw, first a particular isosceles triangle, and then a scalene triangle, by means of the figure of the proposition. To make an isosceles triangle he produces *AB* in both directions to meet the respective circles in *D, E,* and then describes circles with *A, B* as centres and *AE, BD* as radii respectively. The result is an isosceles triangle with each of two sides double of the third side. To make an isosceles triangle in which the equal sides are not so related to the third side but have any given length would require the use of I. 3; and there is no object in treating the question at all in advance of I. 22. An easier way of satisfying ourselves of the existence of some isosceles triangles would surely be to conceive any two radii of a circle drawn and their extremities joined.

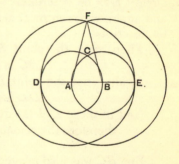

There is more point in Proclus' construction of a *scalene* triangle. Suppose *AC* to be a radius of one of the two circles, and *D* a point on *AC* lying in that portion of the circle with centre *A* which is outside the circle with centre *B*. Then, joining *BD*, as in the figure, we have a triangle which obviously has all its sides unequal, that is, a *scalene* triangle.

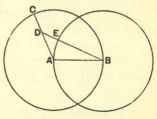

The above two constructions appear in an-Nairīzī's commentary under the name of Heron; Proclus does not mention his source.

In addition to the above construction for a scalene triangle (producing a triangle in which the "given" side is greater than one and less than the other of the two remaining sides), Heron has two others showing the other two possible cases, in which the "given" side is (1) less than, (2) greater than, either of the other two sides.

PROPOSITION 2.

To place at a given point (as an extremity) a straight line equal to a given straight line.

Let A be the given point, and BC the given straight line.

Thus it is required to place at the point A (as an extremity)
5 a straight line equal to the given straight line BC.

From the point A to the point B let the straight line AB be joined;
                  [Post. 1]

and on it let the equilateral triangle
10 DAB be constructed.   [I. 1]

Let the straight lines AE, BF be produced in a straight line with DA, DB;   [Post. 2]

with centre B and distance BC let the
15 circle CGH be described;   [Post. 3]

and again, with centre D and distance DG let the circle GKL be described.   [Post. 3]

Then, since the point B is the centre of the circle CGH, BC is equal to BG.

20   Again, since the point D is the centre of the circle GKL, DL is equal to DG.

And in these DA is equal to DB;
therefore the remainder AL is equal to the remainder BG.   [C.N. 3]

25   But BC was also proved equal to BG;
therefore each of the straight lines AL, BC is equal to BG.

And things which are equal to the same thing are also equal to one another;   [C.N. 1]
30   therefore AL is also equal to BC.

Therefore at the given point A the straight line AL is placed equal to the given straight line BC.

(Being) what it was required to do.

---

1. **(as an extremity).** I have inserted these words because " to place a straight line *at* a given point " ($\pi\rho\grave{o}\varsigma \ \tau\hat{\omega} \ \delta o\theta\acute{\epsilon}\nu\tau\iota \ \sigma\eta\mu\epsilon\acute{\iota}\omega$) is not quite clear enough, at least in English.

11. **Let the straight lines AE, BF be produced....** It will be observed that in this first application of Postulate 2, and again in I. 5, Euclid speaks of the *continuation* of the straight line as that which is produced in such cases, $\dot{\epsilon}\kappa\beta\epsilon\beta\lambda\acute{\eta}\sigma\theta\omega\sigma\alpha\nu$ and $\pi\rho o\sigma\epsilon\kappa\beta\epsilon\beta\lambda\acute{\eta}\sigma\theta\omega\sigma\alpha\nu$ meaning little more than *drawing* straight lines " in a straight line with " the given straight lines. The first place in which Euclid uses phraseology exactly corresponding to ours when

speaking of a straight line being produced is in I. 16: "let one side of it, *BC*, be produced to *D* " (προσεκβεβλήσθω αὐτοῦ μία πλευρά ἡ ΒΓ ἐπὶ τὸ Δ).

23. **the remainder AL...the remainder BG.** The Greek expressions are λοιπὴ ἡ ΑΛ and λοιπῇ τῇ ΒΗ, and the literal translation would be "*AL* (or *BG*) *remaining*," but the shade of meaning conveyed by the position of the definite article can hardly be expressed in English.

This proposition gives Proclus an opportunity, such as the Greek commentators revelled in, of distinguishing a multitude of *cases*. After explaining that those theorems and problems are said to have *cases* which have the same force, though admitting of a number of different figures, and preserve the same method of demonstration while admitting variations of position, and that cases reveal themselves in the *construction*, he proceeds to distinguish the cases in this problem arising from the different positions which the given point may occupy relatively to the given straight line. It may be (he says) either (1) outside the line or (2) on the line, and, if (1), it may be either (*a*) on the line produced or (*b*) situated obliquely with regard to it; if (2), it may be either (*a*) one of the extremities of the line or (*b*) an intermediate point on it. It will be seen that Proclus' anxiety to subdivide leads him to give a "case," (2) (*a*), which is useless, since in that "case" we are given what we are required to find, and there is really no problem to solve. As Savile says, " qui quaerit ad *β* punctum ponere rectam aequalem *τῇ* *βγ* rectae, quaerit quod datum est, quod nemo faceret nisi forte insaniat."

Proclus gives the construction for (2) (*b*) following Euclid's way of taking *G* as the point in which the circle with centre *B* intersects *DB produced*, and then proceeds to "cases," of which there are still more, which result from the different ways of drawing the equilateral triangle and of producing its sides.

This last class of "cases" he subdivides into three according as *AB* is (1) equal to, (2) greater than or (3) less than *BC*. Here again "case" (1) serves no purpose, since, if *AB* is equal to *BC*, the problem is already solved. But Proclus' figures for the other two cases are worth giving, because in one of them the point *G* is on *BD* produced beyond *D*, and in the other it lies on *BD* itself and there is no need to produce any side of the equilateral triangle.

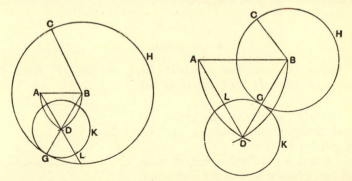

A glance at these figures will show that, if they were used in the proposition, each of them would require a slight modification in the wording (1) of the construction, since *BD* is in one case produced beyond *D* instead of *B* and in the other case not produced at all, (2) of the proof, since *BG*, instead of being the difference between *DG* and *DB*, is in one case the sum of *DG* and *DB* and in the other the difference between *DB* and *DG*.

Modern editors generally seem to classify the cases according to the possible variations in the construction rather than according to differences in the data. Thus Lardner, Potts, and Todhunter distinguish eight cases due to the three possible alternatives, (1) that the given point may be joined to either end of the given straight line, (2) that the equilateral triangle may then be described on either side of the joining line, and (3) that the side of the equilateral triangle which is produced may be produced in either direction. (But it should have been observed that, where AB is greater than BC, the third alternative is between producing DB and not producing it at all.) Potts adds that, when the given point lies either on the line or on the line produced, the distinction which arises from joining the two ends of the line with the given point no longer exists, and there are only four cases of the problem (I think he should rather have said *solutions*).

To distinguish a number of cases in this way was foreign to the really classical manner. Thus, as we shall see, Euclid's method is to give one case only, for choice the most difficult, leaving the reader to supply the rest for himself. Where there was a real distinction between cases, sufficient to necessitate a substantial difference in the proof, the practice was to give separate *enunciations* and proofs altogether, as we may see, e.g., from the *Conics* and the *De sectione rationis* of Apollonius.

Proclus alludes, in conclusion, to the error of those who proposed to solve I. 2 by describing a circle with the given point as centre and with a distance *equal* to BC, which, as he says, is a *petitio principii*. De Morgan puts the matter very clearly (*Supplementary Remarks on the first six Books of Euclid's Elements* in the *Companion to the Almanac*, 1849, p. 6). We should "insist," he says, "here upon the restrictions imposed by the first three postulates, which do not allow a circle to be drawn with a compass-carried distance; suppose the compasses to close of themselves the moment they cease to touch the paper. These two propositions [I. 2, 3] extend the power of construction to what it would have been if *all* the usual power of the compasses had been assumed; they are mysterious to all who do not see that postulate iii does not ask for *every use of the compasses*."

PROPOSITION 3.

Given two unequal straight lines, to cut off from the greater a straight line equal to the less.

Let AB, C be the two given un- equal straight lines, and let AB be the greater of them.

Thus it is required to cut off from AB the greater a straight line equal to C the less.

At the point A let AD be placed equal to the straight line C;   [I. 2]
and with centre A and distance AD let the circle DEF be described.

[Post. 3]

Now, since the point A is the centre of the circle DEF,
AE is equal to AD.   [Def. 15]

But C is also equal to AD.

Therefore each of the straight lines AE, C is equal to AD ; so that AE is also equal to C.   [C.N. 1]

Therefore, given the two straight lines AB, C, from AB the greater AE has been cut off equal to C the less.

(Being) what it was required to do.

Proclus contrives to make a number of "cases" out of this proposition also, and gives as many as eight figures. But he only produces this variety by practically incorporating the construction of the preceding proposition, instead of assuming it as we are entitled to do. If Prop. 2 is assumed, there is really only one "case" of the present proposition, for Potts' distinction between two cases according to the particular extremity of the straight line from which the given length has to be cut off scarcely seems to be worth making.

PROPOSITION 4.

If two triangles have the two sides equal to two sides respectively, and have the angles contained by the equal straight lines equal, they will also have the base equal to the base, the triangle will be equal to the triangle, and the remaining angles
5 *will be equal to the remaining angles respectively, namely those which the equal sides subtend.*

Let ABC, DEF be two triangles having the two sides AB, AC equal to the two sides DE, DF respectively, namely AB to DE and AC to DF, and the angle BAC equal to the
10 angle EDF.

I say that the base BC is also equal to the base EF, the triangle ABC will be equal to the triangle DEF, and the remaining angles will be equal to the remaining angles respectively, namely those which the equal sides subtend, that
15 is, the angle ABC to the angle DEF, and the angle ACB to the angle DFE.

For, if the triangle ABC be applied to the triangle DEF, and if the point A be placed
20 on the point D and the straight line AB on DE,

then the point B will also coincide with E, because AB is equal to DE.

25   Again, *AB* coinciding with *DE*,
the straight line *AC* will also coincide with *DF*, because the
angle *BAC* is equal to the angle *EDF*;
        hence the point *C* will also coincide with the point *F*,
because *AC* is again equal to *DF*.

30   But *B* also coincided with *E*;
hence the base *BC* will coincide with the base *EF*.
        [For if, when *B* coincides with *E* and *C* with *F*, the base
BC does not coincide with the base *EF*, two straight lines
will enclose a space: which is impossible.

35   Therefore the base *BC* will coincide with

                EF] and will be equal to it.   [*C.N.* 4]

        Thus the whole triangle *ABC* will coincide with the
whole triangle *DEF*,

                and will be equal to it.

40   And the remaining angles will also coincide with the
remaining angles and will be equal to them,

the angle *ABC* to the angle *DEF*,

and the angle *ACB* to the angle *DFE*.

        Therefore etc.

45   (Being) what it was required to prove.

1—3. It is a fact that Euclid's enunciations not infrequently leave something to be
desired in point of clearness and precision. Here he speaks of the triangles having "the
angle equal to the angle, namely the angle contained by the equal straight lines" (τὴν γωνίαν
τῇ γωνίᾳ ἴσην ἔχῃ τὴν ὑπὸ τῶν ἴσων εὐθειῶν περιεχομένην), only one of the two angles being
described in the latter expression (in the accusative), and a similar expression in the dative
being left to be understood of the other angle. It is curious too that, after mentioning two
"*sides*," he speaks of the angles contained by the equal "*straight lines*," not "*sides*." It
may be that he wished to adhere scrupulously, at the outset, to the phraseology of the
definitions, where the angle is the inclination to one another of two *lines* or *straight lines*.
Similarly in the enunciation of I. 5 he speaks of producing the equal "*straight lines*" as if to
keep strictly to the wording of Postulate 2.

2. **respectively.** I agree with Mr H. M. Taylor (*Euclid*, p. ix) that it is best to
abandon the traditional translation of "each to each," which would naturally seem to imply
that all the four magnitudes are equal rather than (as the Greek ἑκατέρα ἑκατέρᾳ does) that
one is equal to one and the other to the other.

3. **the base.** Here we have the word *base* used for the first time in the *Elements*.
Proclus explains it (p. 236, 12—15) as meaning (1), when no side of a triangle has been
mentioned before, the side "which is on a level with the sight" (τὴν πρὸς τῇ ὄψει κειμένην),
and (2), when two sides have already been mentioned, the third side. Proclus thus avoids
the mistake made by some modern editors who explain the term exclusively with reference to
the case where two sides have been mentioned before. That this is an error is proved (1) by
the occurrence of .the term in the enunciations of I. 37 etc. about triangles on the same base
and equal bases, (2) by the application of the same term to the bases of parallelograms in
I. 35 etc. The truth is that the use of the term must have been suggested by the practice of
drawing the particular side horizontally, as it were, and the rest of the figure above it. The
base of a figure was therefore spoken of, primarily, in the same sense as the base of anything

else, e.g. of a pedestal or column ; but when, as in I. 5, two triangles were compared occupying other than the normal positions which gave rise to the name, and when two sides had been previously mentioned, the base was, as Proclus says, necessarily the third side.

6. **subtend.** ὑποτείνειν ὑπό, "to stretch under," with accusative.

9. **the angle BAC.** The full Greek expression would be ἡ ὑπὸ τῶν ΒΑ, ΑΓ περιεχομένη γωνία, "the angle contained by the (straight lines) *BA, AC*." But it was a common practice of Greek geometers, e.g. of Archimedes and Apollonius (and Euclid too in Books x.—xiii.), to use the abbreviation αἱ ΒΑΓ for αἱ ΒΑ, ΑΓ, "the (straight lines) *BA, AC*." Thus, on περιεχομένη being dropped, the expression would become first ἡ ὑπὸ τῶν ΒΑΓ γωνία, then ἡ ὑπὸ ΒΑΓ γωνία, and finally ἡ ὑπὸ ΒΑΓ, without γωνία, as we regularly find it in Euclid.

17. **if the triangle be applied to..., 23. coincide.** The difference between the technical use of the passive ἐφαρμόζεσθαι "to be *applied* (to)," and of the active ἐφαρμόζειν "to *coincide* (with)" has been noticed above (note on *Common Notion* 4, pp. 224—5).

32. [For if, when B coincides...36. coincide with EF]. Heiberg (*Paralipomena zu Euklid* in *Hermes*, XXXVIII., 1903, p. 56) has pointed out, as a conclusive reason for regarding these words as an early interpolation, that the text of an-Nairīzī (*Codex Leidensis* 399, 1, ed. Besthorn-Heiberg, p. 55) does not give the words in this place but after the conclusion Q.E.D., which shows that they constitute a *scholium* only. They were doubtless added by some commentator who thought it necessary to explain the immediate inference that, since *B* coincides with *E* and *C* with *F*, the straight line *BC* coincides with the straight line *EF*, an inference which really follows from the definition of a straight line and Post. 1; and no doubt the Postulate that "Two straight lines cannot enclose a space" (afterwards placed among the *Common Notions*) was interpolated at the same time.

44. **Therefore etc.** Where (as here) Euclid's *conclusion* merely repeats the enunciation word for word, I shall avoid the repetition and write "Therefore etc." simply.

In the note on *Common Notion* 4 I have already mentioned that Euclid obviously used the method of superposition with reluctance, and I have given, after Veronese for the most part, the reason for holding that that method is not admissible as a *theoretical* means of proving equality, although it may be of use as a *practical* test, and may thus furnish an empirical basis on which to found a postulate.   Mr Bertrand Russell observes (*Principles of Mathematics* I. p. 405) that Euclid would have done better to assume I. 4 as an axiom, as is practically done by Hilbert (*Grundlagen der Geometrie*, p. 9). It may be that Euclid himself was as well aware of the objections to the method as are his modern critics ; but at all events those objections were stated, with almost equal clearness, as early as the middle of the 16th century.   Peletarius (Jacques Peletier) has a long note on this proposition (*In Euclidis Elementa geometrica demonstrationum libri sex*, 1557), in which he observes that, if superposition of lines and figures could be assumed as a method of proof, the whole of geometry would be full of such proofs, that it could equally well have been used in I. 2, 3 (thus in I. 2 we could simply have supposed the line taken up and *placed* at the point), and that in short it is obvious how far removed the method is from the dignity of geometry.   The theorem, he adds, is obvious in itself and does not require proof ; although it is introduced as a theorem, it would seem that Euclid intended it rather as a *definition* than a theorem, "for I cannot think that two angles are equal unless I have a conception of what equality of angles is."   Why then did Euclid include the proposition among theorems, instead of placing it among the axioms ?   Peletarius makes the best excuse he can, but concludes thus : " Huius itaque propositionis veritatem non aliunde quam a communi iudicio petemus ; cogitabimusque figuras figuris superponere, Mechanicum quippiam esse : intelligere verò, id demum esse Mathematicum."

Expressed in terms of the modern systems of Congruence-Axioms referred to in the note on *Common Notion* 4, what Euclid really assumes amounts to the following :

(1)   On the line *DE*, there is a point *E*, on either side of *D*, such that *AB* is equal to *DE*.

(2) On either side of the ray *DE* there is a ray *DF* such that the angle *EDF* is equal to the angle *BAC*.

It now follows that on *DF* there is a point *F* such that *DF* is equal to *AC*.

And lastly (3), we require an axiom from which to infer that the two remaining angles of the triangles are respectively equal and that the bases are equal.

I have shown above (pp. 229—230) that Hilbert has an axiom stating the equality of the remaining angles simply, but proves the equality of the bases.

Another alternative is that of Pasch (*Vorlesungen über neuere Geometrie*, p. 109) who has the following "Grundsatz":

If two figures *AB* and *FGH* are given (*FGH* not being contained in a straight length), and *AB*, *FG* are congruent, and if a plane surface be laid through *A* and *B*, we can specify in this plane surface, produced if necessary, two points *C*, *D*, neither more nor less, such that the figures *ABC* and *ABD* are congruent with the figure *FGH*, and the straight line *AB* or with *AB* produced one point common.

I pass to two points of detail in Euclid's proof:

(1) The inference that, since *B* coincides with *E*, and *C* with *F*, the bases of the triangles are wholly coincident rests, as expressly stated, on the impossibility of two straight lines enclosing a space, and therefore presents no difficulty.

But (2) most editors seem to have failed to observe that at the very beginning of the proof a much more serious assumption is made without any explanation whatever, namely that, if *A* be placed on *D*, and *AB* on *DE*, the point *B* will coincide with *E*, because *AB* is equal to *DE*. That is, the *converse* of *Common Notion* 4 is assumed for straight lines. Proclus merely observes, with regard to the converse of this Common Notion, that it is only true in the case of things "of the same form" (ὁμοειδῆ), which he explains as meaning straight lines, arcs of one and the same circle, and angles "contained by lines similar and similarly situated" (p. 241, 3—8).

Savile however saw the difficulty and grappled with it in his note on the Common Notion. After stating that all straight lines with two points common are congruent between them (for otherwise two straight lines would enclose a space), he argues thus. Let there be two straight lines *AB*, *DE*, and let *A* be placed on *D*, and *AB* on *DE*. Then *B* will coincide with *E*. For, if not, let *B* fall somewhere short of *E* or beyond *E*; and in either case it will follow that the less is equal to the greater, which is impossible.

Savile seems to assume (and so apparently does Lardner who gives the same proof) that, if the straight lines be "applied," *B* will fall somewhere on *DE* or *DE* produced. But the ground for this assumption should surely be stated; and it seems to me that it is necessary to use, not Postulate 1 alone, nor Postulate 2 alone, but both, for this purpose (in other words to assume, not only that *two straight lines cannot enclose a space*, but also that *two straight lines cannot have a common segment*). For the only safe course is to place *A* upon *D* and then turn *AB* about *D* until *some* point on *AB* intermediate between *A* and *B* coincides with *some* point on *DE*. In this position *AB* and *DE* have two points common. Then Postulate 1 enables us to infer that the straight lines coincide *between* the two common points, and Postulate 2 that they coincide beyond the second common point towards *B* and *E*. Thus the straight lines coincide throughout so far as *both* extend; and Savile's argument then proves that *B* coincides with *E*.

PROPOSITION 5.

In isosceles triangles the angles at the base are equal to one another, and, if the equal straight lines be produced further, the angles under the base will be equal to one another.

Let *ABC* be an isosceles triangle having the side *AB*
5 equal to the side *AC*;

and let the straight lines *BD*, *CE* be produced further in a straight line with *AB*, *AC*.   [Post. 2]

I say that the angle *ABC* is equal to the angle *ACB*, and the angle *CBD* to the angle *BCE*.

10   Let a point *F* be taken at random
on *BD*;

from *AE* the greater let *AG* be cut off
equal to *AF* the less;   [I. 3]

and let the straight lines *FC*, *GB* be joined.
                                  [Post. 1]

15   Then, since *AF* is equal to *AG* and
AB to *AC*,

the two sides *FA*, *AC* are equal to the
two sides *GA*, *AB*, respectively;

and they contain a common angle, the angle *FAG*.

20   Therefore the base *FC* is equal to the base *GB*,

and the triangle *AFC* is equal to the triangle *AGB*,

and the remaining angles will be equal to the remaining angles
respectively, namely those which the equal sides subtend,

that is, the angle *ACF* to the angle *ABG*,

25   and the angle *AFC* to the angle *AGB*.   [I. 4]

And, since the whole *AF* is equal to the whole *AG*,

and in these *AB* is equal to *AC*,

the remainder *BF* is equal to the remainder *CG*.

But *FC* was also proved equal to *GB*;

30 therefore the two sides *BF*, *FC* are equal to the two sides
CG, *GB* respectively;

and the angle *BFC* is equal to the angle *CGB*,

while the base *BC* is common to them;

therefore the triangle *BFC* is also equal to the triangle *CGB*,
35 and the remaining angles will be equal to the remaining

angles respectively, namely those which the equal sides subtend;

   therefore the angle *FBC* is equal to the angle *GCB*,

      and the angle *BCF* to the angle *CBG*.

40   Accordingly, since the whole angle *ABG* was proved equal to the angle *ACF*,

   and in these the angle *CBG* is equal to the angle *BCF*,

the remaining angle *ABC* is equal to the remaining angle *ACB*;

45   and they are at the base of the triangle *ABC*.

But the angle *FBC* was also proved equal to the angle *GCB*;

      and they are under the base.

Therefore etc.   Q. E. D.

2. **the equal straight lines** (meaning the equal *sides*).   Cf. note on the similar expression in Prop. 4, lines 2, 3.

10. **Let a point F be taken at random on BD**, εἰλήφθω ἐπὶ τῆς ΒΔ τυχὸν σημεῖον τὸ Z, where τυχὸν σημεῖον means "a chance point."

17. **the two sides FA, AC are equal to the two sides GA, AB respectively**, δύο αἱ ZA, ΑΓ δυσὶ ταῖς ΗΑ, ΑΒ ἴσαι εἰσὶν ἑκατέρα ἑκατέρᾳ.   Here, and in numberless later passages, I have inserted the word "sides" for the reason given in the note on I. 1, line 20. It would have been permissible to supply either "straight lines" or "sides"; but on the whole "sides" seems to be more in accordance with the phraseology of I. 4.

33. **the base BC is common to them**, i.e., apparently, common to the *angles*, as the αὐτῶν in βάσις αὐτῶν κοινὴ can only refer to γωνίᾳ and γωνίᾳ preceding.   Simson wrote "and the base *BC* is common to the two triangles *BFC*, *CGB*"; Todhunter left out these words as being of no use and tending to perplex a beginner.   But Euclid evidently chose to quote the conclusion of I. 4 exactly; the first phrase of that conclusion is that the bases (of the two triangles) are equal, and, as the equal bases are here the *same* base, Euclid naturally substitutes the word "common" for "equal."

48. As "(Being) what it was required to prove" (or "do") is somewhat long, I shall henceforth write the time-honoured "Q. E. D." and "Q. E. F." for ὅπερ ἔδει δεῖξαι and ὅπερ ἔδει ποιῆσαι.

According to Proclus (p. 250, 20) the discoverer of the fact that in any isosceles triangle the angles at the base are equal was Thales, who however is said to have spoken of the angles as being *similar*, and not as being *equal*. (Cf. Arist. *De caelo* IV. 4, 311 b 34 πρὸς ὁμοίας γωνίας φαίνεται φερόμενον where *equal* angles are meant.)

A pre-Euclidean proof of I. 5.

One of the most interesting of the passages in Aristotle indicating differences between Euclid's proofs and those with which Aristotle was familiar, in other words, those of the text-books immediately preceding Euclid's, has reference to the theorem of I. 5.   The passage (*Anal. Prior.* I. 24, 41 b 13—22) is so important that I must quote it in full.   Aristotle is illustrating the fact that in any syllogism one of the propositions must be affirmative and *universal* (καθόλου).   "This," he says, "is better shown in the case of geometrical propositions" (ἐν τοῖς διαγράμμασιν), e.g. the proposition that *the angles at the base of an isosceles triangle are equal*.

"For let *A, B* be drawn [i.e. joined] to the centre.

"If, then, we assumed (1) that the angle AC [i.e. $A + C$] is equal to the angle BD [i.e. $B + D$] without asserting generally that *the angles of semicircles are equal*, and again (2) that the angle C is equal to the angle D without making the further assumption that *the two angles of all segments are equal*, and if we then inferred, lastly, that, since the whole angles are equal, and equal angles are subtracted from them, the angles which remain, namely E, F, are equal. we should commit a *petitio principii*, unless we assumed [generally] that, *when equals are subtracted from equals, the remainders are equal*."

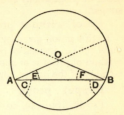

The language is noteworthy in some respects.

(1) A, B are said to be *drawn* ($\mathring{\eta}\gamma\mu\acute{\epsilon}\nu\alpha\iota$) to the centre (of the circle of which the two equal sides are radii) as if A, B were not the angular points but the sides or the radii themselves. (There is a parallel for this in Eucl. IV. 4.)

(2)   "The angle AC" is the angle which is the sum of A and C, and A means here the angle at A of the *isosceles triangle* shown in the figure, and afterwards spoken of by Aristotle as E, while C is the "mixed" angle between AB and the circumference of the smaller segment cut off by it.

(3)   The "angle *of* a semicircle" (i.e. the "angle" between the diameter and the circumference, at the extremity of the diameter) and the "angle *of* a segment" appear in Euclid III. 16 and III. Def. 7 respectively, obviously as survivals from earlier text-books.

But the most significant facts to be gathered from the extract are that in the text-books which preceded Euclid's "mixed" angles played a much more important part than they do with Euclid, and, in particular, that at least two propositions concerning such angles appeared quite at the beginning, namely the propositions that *the (mixed) angles of semicircles are equal* and that *the two (mixed) angles of any segment of a circle are equal*. The wording of the first of the two propositions is vague, but it does not necessarily mean more than that the two (mixed) angles in *one* semicircle are equal, and I know of no evidence going to show that it asserts that the angle of any one semicircle is equal to the angle of any other semicircle (of different size). It is quoted in the same form, "because the angles of semicircles are equal," in the Latin translation from the Arabic of Heron's *Catoptrica*, Prop. 9 (Heron, Vol. II., Teubner, p. 334), but it is only inferred that the different radii of *one* circle make equal "angles" with the circumference ; and in the similar proposition of the Pseudo-Euclidean *Catoptrica* (Euclid, Vol. VII., p. 294) angles of the same sort in *one* circle are said to be equal "because they are (angles) of a semicircle." Therefore the first of the two propositions may be only a particular case of the second.

But it is remarkable enough that the second proposition (that *the two "angles of" any segment of a circle are equal*) should, in earlier text-books, have been placed before the theorem of Eucl. I. 5. We can hardly suppose it to have been proved otherwise than by the superposition of the semicircles into which the circle is divided by the diameter which bisects at right angles the base of the segment; and no doubt the proof would be closely connected with that of Thales' other proposition that any diameter of a circle bisects it, which must also (as Proclus indicates) have been proved by superposing one of the two parts upon the other.

It is a natural inference from the passage of Aristotle that Euclid's proof of

I. 5 was his own, and it would thus appear that his innovations as regards order of propositions and methods of proof began at the very threshold of the subject.

Proof without producing the sides.

In this proof, given by Proclus (pp. 248, 22—249, 19), *D* and *E* are taken on *AB*, *AC*, instead of on *AB*, *AC produced*, so that *AD*, *AE* are equal. The method of proof is of course exactly like Euclid's, but it does not establish the equality of the angles beyond the base as well.

Pappus' proof.

Proclus (pp. 249, 20—250, 12) says that Pappus proved the theorem in a still shorter manner without the help of any construction whatever.

This very interesting proof is given as follows :

" Let *ABC* be an isosceles triangle, and *AB* equal to *AC*.

Let us conceive this one triangle as two triangles, and let us argue in this way.

Since *AB* is equal to *AC*, and *AC* to *AB*,

the two sides *AB*, *AC* are equal to the two sides *AC*, *AB*.

And the angle *BAC* is equal to the angle *CAB*, for it is the same.

Therefore all the corresponding parts (in the triangles) are equal, namely

BC to BC,

the triangle *ABC* to the triangle *ABC* (i.e. *ACB*),

the angle *ABC* to the angle *ACB*,

and the angle *ACB* to the angle *ABC*,

(for these are the angles subtended by the equal sides *AB*, *AC*.

Therefore in isosceles triangles the angles at the base are equal."

This will no doubt be recognised as the foundation of the alternative proof frequently given by modern editors, though they do not refer to Pappus. But they state the proof in a different form, the common method being to suppose the triangle to be taken up, turned over, and placed again upon *itself*, after which the same considerations of congruence as those used by Euclid in I. 4 are used over again. There is the obvious difficulty that it supposes the triangle to be taken up and at the same time to remain where it is. (Cf. Dodgson's humorous remark upon this, *Euclid and his modern Rivals*, p. 47.) Whatever we may say in justification of the proceeding (e.g. that the triangle may be supposed to leave a *trace*), it is really equivalent to assuming the construction (hypothetical, if you will) of another triangle equal in all respects to the given triangle ; and such an assumption is not in accordance with Euclid's principles and practice.

It seems to me that the form given to the proof by Pappus himself is by far the best, for the reasons (1) that it assumes no construction of a second triangle, real or hypothetical, (2) that it avoids the distinct awkwardness involved by a proof which, instead of merely quoting and applying the *result* of a previous proposition, repeats, with reference to a new set of data, the *process* by which that result was established. If it is asked how we are to realise Pappus' idea of *two* triangles, surely we may answer that we keep to one triangle and merely view it in two aspects. If it were a question of helping a beginner to understand this, we might say that one triangle is the triangle looked at in front and that the other triangle is the *same* triangle looked at from *behind*; but even this is not really necessary.

Pappus' proof, of course, does not include the proof of the second part of the proposition about the angles under the base, and we should still have to establish this much in the same way as Euclid does.

Purpose of the second part of the theorem.

An interesting question arises as to the reason for Euclid's insertion of the second part, to which, it will be observed, the converse proposition I. 6 has nothing corresponding. As a matter of fact, it is not necessary for any subsequent demonstration that is to be found in the original text of Euclid, but only for the interpolated second case of I. 7; and it was perhaps not unnatural that the undoubted genuineness of the second part of I. 5 convinced many editors that the second case of I. 7 must necessarily be Euclid's also. Proclus' explanation, which must apparently be the right one, is that the second part of I. 5 was inserted for the purpose of fore-arming the learner against a possible *objection* (ἔνστασις), as it was technically called, which might be raised to I. 7 as given in the text, with one case only. The *objection* would, as we have seen, take the specific ground that, as demonstrated, the theorem was not conclusive, since it did not cover all possible cases. From this point of view, the second part of I. 5 is useful not only for I. 7 but, according to Proclus, for I. 9 also. Simson does not seem to have grasped Proclus' meaning, for he says: "And Proclus acknowledges, that the second part of Prop. 5 was added upon account of Prop. 7 but gives a ridiculous reason for it, 'that it might afford an answer to objections made against the 7th,' as if the case of the 7th which is left out were, as he expressly makes it, an objection against the proposition itself."

PROPOSITION 6.

If in a triangle two angles be equal to one another, the sides which subtend the equal angles will also be equal to one another.

Let *ABC* be a triangle having the angle *ABC* equal to the angle *ACB*;

I say that the side *AB* is also equal to the side *AC*.

For, if *AB* is unequal to *AC*, one of them is greater.

Let *AB* be greater; and from *AB* the greater let *DB* be cut off equal to *AC* the less;

let *DC* be joined.

Then, since *DB* is equal to *AC*, and *BC* is common,

the two sides *DB*, *BC* are equal to the two sides *AC*, *CB* respectively;

and the angle *DBC* is equal to the angle *ACB* ;
therefore the base *DC* is equal to the base *AB*,
and the triangle *DBC* will be equal to the triangle *ACB*,
the less to the greater :
which is absurd.
Therefore *AB* is not unequal to *AC* ;
it is therefore equal to it.
Therefore etc.

Q. E. D.

Euclid assumes that, because *D* is between *A* and *B*, the triangle *DBC* is less than the triangle *ABC*. Some postulate is necessary to justify this tacit assumption; considering an angle less than two right angles, say the angle *ACB* in the figure of the proposition, as a cluster of rays issuing from *C* and bounded by the rays *CA*, *CB*, and joining *AB* (where *A*, *B* are any two points on *CA*, *CB* respectively), we see that to each successive ray taken in the direction from *CA* to *CB* there corresponds one point on *AB* in which the said ray intersects *AB*, and that all the points on *AB* taken in order from *A* to *B* correspond univocally to all the rays taken in order from *CA* to *CB*, each point namely to the ray intersecting *AB* in the point.

We have here used, for the first time in the *Elements*, the method of *reductio ad absurdum*, as to which I would refer to the section above (pp. 136, 140) dealing with this among other technical terms.

This proposition also, being the *converse* of the preceding proposition, brings us to the subject of

Geometrical Conversion.

This must of course be distinguished from the *logical* conversion of a proposition. Thus, from the proposition that all isosceles triangles have the angles opposite to the equal sides equal, *logical* conversion would only enable us to conclude that *some* triangles with two angles equal are isosceles. Thus I. 6 is the geometrical, but not the logical, converse of I. 5. On the other hand, as De Morgan points out (*Companion to the Almanac*, 1849, p. 7), I. 6 is a purely *logical* deduction from I. 5 and I. 18 taken together, as is I. 19 also. For the general argument see the note on I. 19. For the present proposition it is enough to state the matter thus. Let *X* denote the class of triangles which have the two sides other than the base equal, *Y* the class of triangles which have the base angles equal ; then we may call non-*X* the class of triangles having the sides other than the base unequal, non-*Y* the class of triangles having the base angles unequal.

Thus we have
All *X* is *Y*,   [I. 5]
All non-*X* is non-*Y*;   [I. 18]
and it is a purely logical deduction that
All *Y* is *X*.   [I. 6]

According to Proclus (p. 252, 5 sqq.) two forms of *geometrical conversion* were distinguished.

(1)   The leading form (προηγουμένη). the conversion *par excellence* (ἡ κυρίως

ἀντιστροφή), is the complete or simple conversion in which the hypothesis and the conclusion of a theorem change places exactly, the conclusion of the theorem being the hypothesis of the converse theorem, which again establishes, as its conclusion, the hypothesis of the original theorem. The relation between the first part of I. 5 and I. 6 is of this character. In the former the hypothesis is that two sides of a triangle are equal and the conclusion is that the angles at the base are equal, while the converse (I. 6) starts from the hypothesis that two angles are equal and proves that the sides subtending them are equal.

(2) The other form of conversion, which we may call *partial*, is seen in cases where a theorem starts from two or more hypotheses combined into one enunciation and leads to a certain conclusion, after which the converse theorem takes this conclusion in substitution for one of the hypotheses of the original theorem and from the said conclusion along with the rest of the original hypotheses obtains, as its conclusion, the omitted hypothesis of the original theorem. I. 8 is in this sense a converse proposition to I. 4; for I. 4 takes as hypotheses (1) that two sides in two triangles are respectively equal, (2) that the included angles are equal, and proves (3) that the bases are equal, while I. 8 takes (1) and (3) as hypotheses and proves (2) as its conclusion. It is clear that a conversion of the *leading* type must be unique, while there may be many *partial* conversions of a theorem according to the number of hypotheses from which it starts.

Further, of convertible theorems, those which took as their hypothesis the *genus* and proved a *property* were distinguished as the *leading* theorems (προηγούμενα), while those which started from the property as hypothesis and described, as the conclusion, the genus possessing that property were the *converse* theorems. I. 5 is thus the leading theorem and I. 6 its converse, since the genus is in this case taken to be the isosceles triangle.

Converse of second part of I. 5.

Why, asks Proclus, did not Euclid convert the *second* part of I. 5 as well? He suggests, properly enough, two reasons: (1) that the second part of I. 5 itself is not wanted for any proof occurring in the original text, but is only put in to enable *objections* to the existing form of later propositions to be met, whereas the converse is not even wanted for this purpose; (2) that the converse could be deduced from I. 6, if wanted, at any time after we have passed I. 13, which can be used to prove that, if the angles formed by producing two sides of a triangle beyond the base are equal, the base angles themselves are equal.

Proclus adds a proof of the converse of the second part of I. 5, i.e. of the proposition that, if the angles formed by producing two sides of a triangle beyond the base are equal, the triangle is isosceles; but it runs to some length and then only effects a reduction to the theorem of I. 6 as we have it. As the result of this should hardly be assumed, a better proof would be an independent one *adapting* Euclid's own method in I. 6. Thus, with the construction of I. 5, we first prove by means of I. 4 that the triangles *BFC*, *CGB* are equal in all respects, and therefore that *FC* is equal to *GB*, and the angle *BFC* equal to the angle *CGB*. Then we have to prove that *AF*, *AG* are equal. If they are not, let *AF* be the greater, and from *FA* cut off *FH* equal to *GA*. Join *CH*.

Then we have, in the two triangles *HFC*, *AGB*,

two sides *HF*, *FC* equal to two sides *AG*, *GB*

and the angle *HFC* equal to the angle *AGB*.

Therefore (I. 4) the triangles *HFC*, *AGB* are equal. But the triangles *BFC*, *CGB* are also equal.

Therefore (if we take away these equals respectively) the triangles *HBC*, *ACB* are equal: which is impossible.

Therefore *AF*, *AG* are not unequal.

Hence *AF* is equal to *AG* and, if we subtract the equals *BF*, *CG* respectively, *AB* is equal to *AC*.

This proof is found in the commentary of an-Nairīzī (ed. Besthorn-Heiberg, p. 61 ; ed. Curtze, p. 50).

Alternative proofs of I. 6.

Todhunter points out that I. 6, not being wanted till II. 4, could be postponed till later and proved by means of I. 26. Bisect the angle *BAC* by a straight line meeting the base at *D*. Then the triangles *ABD*, *ACD* are equal in all respects.

Another method depending on I. 26 is given by an-Nairīzī after that proposition.

Measure equal lengths *BD*, *CE* along the sides *BA*, *CA*. Join *BE*, *CD*.

Then [I. 4] the triangles *DBC*, *ECB* are equal in all respects ;

therefore *EB*, *DC* are equal, and the angles *BEC*, *CDB* are equal.

The supplements of the latter angles are equal [I. 13], and hence the triangles *ABE*, *ACD* have two angles equal respectively and the side *BE* equal to the side *CD*.

Therefore [I. 26] *AB* is equal to *AC*.

Proposition 7.

Given two straight lines constructed on a straight line (from its extremities) and meeting in a point, there cannot be constructed on the same straight line (from its extremities),
5 *and on the same side of it, two other straight lines meeting in another point and equal to the former two respectively, namely each to that which has the same extremity with it.*

For, if possible, given two straight lines *AC*, *CB* constructed on the straight line *AB* and meeting at the point *C*, let two other straight lines
10 *AD*, *DB* be constructed on the same straight line *AB*, on the same side of it, meeting in another point *D* and equal to the former two respectively, namely each to that which has the same extremity with it, so that *CA* is equal to *DA* which has the same extremity
15 equal to *DA* which has the same extremity *A* with it, and

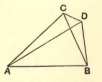

CB to *DB* which has the same extremity *B* with it; and let
CD be joined.

Then, since *AC* is equal to *AD*,

the angle *ACD* is also equal to the angle *ADC*;   [I. 5]

20 therefore the angle *ADC* is greater than the angle *DCB*;
therefore the angle *CDB* is much greater than the angle
DCB.

Again, since *CB* is equal to *DB*,

the angle *CDB* is also equal to the angle *DCB*.

25 But it was also proved much greater than it:

which is impossible.

Therefore etc.   Q. E. D.

1—6. In an English translation of the enunciation of this proposition it is absolutely
necessary, in order to make it intelligible, to insert some words which are not in the Greek.
The reason is partly that the Greek enunciation is itself very elliptical, and partly that some
words used in it conveyed more meaning than the corresponding words in English do.
Particularly is this the case with οὐ συσταθήσονται ἐπί "there shall not be constructed upon,"
since συνίστασθαι is the regular word for constructing a *triangle* in particular. Thus a Greek
would easily understand συσταθήσονται ἐπί as meaning the construction of two lines *forming
a triangle on* a given straight line as base; whereas to "construct two straight lines on a
straight line" is not in English sufficiently definite unless we explain that they are drawn
from the *ends* of the straight line to *meet* at a point. I have had the less hesitation in putting
in the words "from its extremities" because they are actually used by Euclid in the somewhat
similar enunciation of I. 21.

How impossible a literal translation into English is, if it is to convey the meaning of the
enunciation intelligibly, will be clear from the following attempt to render literally: "On the
same straight line there shall not be constructed two other straight lines equal, each to each,
to the same two straight lines, (terminating) at different points on the same side, having the
same extremities as the original straight lines" (ἐπὶ τῆς αὐτῆς εὐθείας δύο ταῖς αὐταῖς εὐθείαις
ἄλλαι δύο εὐθεῖαι ἴσαι ἑκατέρα ἑκατέρᾳ οὐ συσταθήσονται πρὸς ἄλλῳ καὶ ἄλλῳ σημείῳ ἐπὶ τὰ αὐτὰ
μέρη τὰ αὐτὰ πέρατα ἔχουσαι ταῖς ἐξ ἀρχῆς εὐθείαις).

The reason why Euclid allowed himself to use, in this enunciation, language apparently
so obscure is no doubt that the phraseology was traditional and therefore, vague as it was,
had a conventional meaning which the contemporary geometer well understood. This is
proved, I think, by the occurrence in Aristotle (*Meteorologica* III. 5, 376 a 2 sqq.) of the very
same, evidently technical, expressions. Aristotle is there alluding to the theorem given by
Eutocius from Apollonius' *Plane Loci* to the effect that, if *H*, *K* be two fixed points and *M*
such a variable point that the ratio of *MH* to *MK* is a given ratio (not one of equality), the
locus of *M* is a circle. (For an account of this theorem see note on VI. 3 below.) Now
Aristotle says "The lines drawn up from *H*, *K* in this ratio cannot be constructed to two
different points of the semicircle *A*" (αἱ οὖν ἀπὸ τῶν ΗΚ ἀναγόμεναι γραμμαὶ ἐν τούτῳ τῷ
λόγῳ οὐ συσταθήσονται τοῦ ἐφ᾽ ᾧ Α ἡμικυκλίου πρὸς ἄλλο καὶ ἄλλο σημεῖον).

If a paraphrase is allowed instead of a translation adhering as closely as possible to the
original, Simson's is the best that could be found, since the fact that the straight lines form
triangles on the same base is really conveyed in the Greek. Simson's enunciation is, *Upon
the same base, and on the same side of it, there cannot be two triangles that have their sides
which are terminated in one extremity of the base equal to one another, and likewise those
which are terminated at the other extremity.* Th. Taylor (the translator of Proclus) attacks
Simson's alteration as "indiscreet" and as detracting from the beauty and accuracy of
Euclid's enunciation which are enlarged upon by Proclus in his commentary. Yet, when
Taylor says "Whatever difficulty learners may find in conceiving this proposition abstractedly
is easily removed by its exposition in the figure," he really gives his case away. The fact is
that Taylor, always enthusiastic over his author, was nettled by Simson's slighting remarks
on Proclus' comments on the proposition. Simson had said, with reference to Proclus'
explanation of the bearing of the second part of I. 5 on I. 7, that it was not "worth while

to relate his trifles at full length," to which Taylor retorts "But Mr Simson was no philosopher; and therefore the greatest part of these Commentaries must be considered by him as trifles, from the want of a philosophic genius to comprehend their meaning, and a taste superior to that of a *mere mathematician*, to discover their beauty and elegance."

20. It would be natural to insert here the step "but the angle *ACD* is greater than the angle *BCD*. [*C. N.* 5.]"

21. **much greater,** literally "greater by much" (πολλῷ μείζων). Simson and those who follow him translate: "*much more then* is the angle *BDC greater* than the angle *BCD*," but the Greek for this would have to be πολλῷ (or πολὺ) μᾶλλόν ἐστι...μείζων. πολλῷ μᾶλλον, however, though used by Apollonius, is not, apparently, found in Euclid or Archimedes.

Just as in I. 6 we need a Postulate to justify theoretically the statement that *CD* falls within the angle *ACB*, so that the triangle *DBC* is less than the triangle *ABC*, so here we need Postulates which shall satisfy us as to the relative positions of *CA*, *CB*, *CD* on the one hand and of *DC*, *DA*, *DB* on the other, in order that we may be able to infer that the angle *BDC* is greater than the angle *ADC*, and the angle *ACD* greater than the angle *BCD*.

De Morgan (*op. cit.* p. 7) observes that I. 7 would be made easy to beginners if they were first familiarised, as a common notion, with "if two magnitudes be equal, any magnitude greater than the one is greater than any magnitude less than the other." I doubt however whether a beginner would follow this easily; perhaps it would be more easily apprehended in the form "if any magnitude *A* is greater than a magnitude *B*, the magnitude *A* is greater than any magnitude equal to *B*, and (*a fortiori*) greater than any magnitude less than *B*."

It has been mentioned already (note on I. 5) that the second case of I. 7 given by Simson and in our text-books generally is not in the original text (the omission being in accordance with Euclid's general practice of giving only one case, and that the most difficult, and leaving the others to be worked out by the reader for himself). The second case is given by Proclus as the answer to a possible *objection* to Euclid's proposition, which should assert that the proposition is not proved to be universally true, since the proof given does not cover all possible cases. Here the objector is supposed to contend that what Euclid declares to be impossible may still be possible if one pair of lines lie wholly within the other pair of lines; and the second part of I. 5 enables the objection to be refuted.

If possible, let *AD*, *DB* be entirely within the triangle formed by *AC*, *CB* with *AB*, and let *AC* be equal to *AD* and *BC* to *BD*.

Join *CD*, and produce *AC*, *AD* to *E* and *F*.

Then, since *AC* is equal to *AD*,

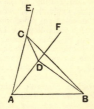

the triangle *ACD* is isosceles,

and the angles *ECD*, *FDC* under the base are equal.

But the angle *ECD* is greater than the angle *BCD*,

therefore the angle *FDC* is also greater than the angle *BCD*.

Therefore the angle *BDC* is greater by far than the angle *BCD*.

Again, since *DB* is equal to *CB*,

the angles at the base of the triangle *BDC* are equal,   [I. 5]

that is, the angle *BDC* is equal to the angle *BCD*.

Therefore the same angle *BDC* is both greater than and equal to the angle *BCD*: which is impossible.

The case in which *D* falls on *AC* or *BC* does not require proof.

I have already referred (note on I. 1) to the mistake made by those editors who regard I. 7 as being of no use except to prove I. 8. What I. 7 proves is that if, in addition to the base of a triangle, the length of the side terminating at each extremity of the base is given, only one triangle satisfying these conditions can be constructed on one and the same side of the given base. Hence not only does I. 7 enable us to prove I. 8, but it supplements I. 1 and I. 22 by showing that the constructions of those propositions give one triangle only on one and the same side of the base. But for I. 7 this could not be proved except by anticipating III. 10, of which therefore I. 7 is the equivalent for Book I. purposes. Dodgson (*Euclid and his modern Rivals*, pp. 194—5) puts it in another way. "It [I. 7] shows that, of all plane figures that can be made by hingeing rods together, the *three*-sided ones (and these only) are *rigid* (which is another way of stating the fact that there cannot be *two* such figures on the same base). This is analogous to the fact, in relation to solids contained by plane surfaces hinged together, that *any* such solid is rigid, there being no maximum number of sides. And there is a close analogy between I. 7, 8 and III. 23, 24. These analogies give to geometry much of its beauty, and I think that they ought not to be lost sight of." It will therefore be apparent how ill-advised are those editors who eliminate I. 7 altogether and rely on Philo's proof for I. 8.

Proclus, it may be added, gives (pp. 268, 19—269, 10) another explanation of the retention of I. 7, notwithstanding that it was apparently only required for I. 8. It was said that astronomers used it to prove that three successive eclipses could not occur at equal intervals of time, i.e. that the third could not follow the second at the same interval as the second followed the first; and it was argued that Euclid had an eye to this astronomical application of the proposition. But, as we have seen, there are other grounds for retaining the proposition which are quite sufficient of themselves.

PROPOSITION 8.

If two triangles have the two sides equal to two sides respectively, and have also the base equal to the base, they will also have the angles equal which are contained by the equal straight lines.

5　　　Let *ABC*, *DEF* be two triangles having the two sides *AB*, *AC* equal to the two sides *DE*, *DF* respectively, namely *AB* to *DE*, and *AC* to *DF*; and let them have the base *BC* equal
10 to the base *EF*;

I say that the angle *BAC* is also equal to the angle *EDF*.

For, if the triangle *ABC* be applied to the triangle *DEF*, and if the point *B* be placed on
15 the point *E* and the straight line *BC* on *EF*,

the point *C* will also coincide with *F*,

because *BC* is equal to *EF*.

Then, *BC* coinciding with *EF*,

    BA, AC will also coincide with *ED, DF*;

20 for, if the base *BC* coincides with the base *EF*, and the sides
BA, AC do not coincide with *ED, DF* but fall beside them
as *EG, GF*,

    then, given two straight lines constructed on a straight
line (from its extremities) and meeting in a point, there will
25 have been constructed on the same straight line (from its
extremities), and on the same side of it, two other straight
lines meeting in another point and equal to the former
two respectively, namely each to that which has the same
extremity with it.

30   But they cannot be so constructed.   [I. 7]

    Therefore it is not possible that, if the base *BC* be applied
to the base *EF*, the sides *BA, AC* should not coincide with
ED, DF;

    they will therefore coincide,

35 so that the angle *BAC* will also coincide with the angle
EDF, and will be equal to it.   •

    If therefore etc.   Q. E. D.

19. **BA, AC.** The text has here "*BA, CA*."
21. **fall beside them.** The Greek has the future, παραλλάξουσι. παραλλάττω means
"to pass by without touching," "to miss" or "to deviate."

As pointed out above (p. 257) I. 8 is a *partial* converse of I. 4.

It is to be observed that in I. 8 Euclid is satisfied with proving the equality
of the vertical angles and does not, as in I. 4, add that the triangles are equal,
and the remaining angles are equal respectively. The reason is no doubt (as
pointed out by Proclus and by Savile after him) that, when once the vertical
angles are proved equal, the rest follows from I. 4, and there is no object in
proving again what has been proved already.

Aristotle has an allusion to the theorem of this proposition in *Meteorologica*
III. 3, 373 a 5—16. He is speaking of the rainbow and observes that, if equal
rays be reflected from one and the same point to one and the same point, the
points at which reflection takes place are on the circumference of a circle.
"For let the broken lines *ACB, AFB, ADB* be all reflected from the point
A to the point *B* (in such a way that) *AC, AF, AD* are all equal to one
another, and the lines (terminating) at *B*, i.e. *CB, FB, DB*, are likewise all
equal; and let *AEB* be joined. It follows that *the triangles are equal*; for
they are upon the equal (base) *AEB*."

Heiberg (*Mathematisches zu Aristoteles*, p. 18) thinks that the form of the
conclusion quoted is an indication that in the corresponding proposition to
Eucl. I. 8, as it lay before Aristotle, it was maintained that the *triangles* were
equal, and not only the angles, and "we see here therefore, in a clear example,
how the stones of the ancient fabric were recut for the rigid structure of his

Elements." I do not, however, think that this inference from Aristotle's language as to the form of the pre-Euclidean proposition is safe. Thus if we, nowadays, were arguing from the data in the passage of Aristotle, we should doubtless infer directly that the triangles are equal in all respects, quoting I. 8 alone. Besides, Aristotle's language is rather careless, as the next sentences of the same passage show. " Let perpendiculars,"
he says, " be drawn to *AEB* from the angles, *CE* from *C*, *FE* from *F* and *DE* from *D*. These, then, are equal; for they are all in equal triangles, and in one plane; for all of them are perpendicular to *AEB*, and they meet at one point *E*. There-fore the (line) drawn (through *C*, *F*, *D*) will be a circle, and its centre (will be) *E*." Aristotle should

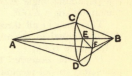

obviously have proved that the three perpendiculars *will* meet at one point *E* on *AEB* before he spoke of drawing the perpendiculars *CE*, *FE*, *DE*. This of course follows from their being "in equal triangles" (by means of Eucl. I. 26); and then, from the fact that the perpendiculars meet at one point on *AB*, it can be inferred that all three are in one plane.

Philo's proof of I. 8.

This alternative proof avoids the use of I. 7, and it is elegant; but it is inconvenient in one respect, since three cases have to be distinguished. Proclus gives the proof in the following order (pp. 266, 15—268, 14).

Let *ABC*, *DEF* be two triangles having the sides *AB*, *AC* equal to the sides *DE*, *DF* respectively, and the base *BC* equal to the base *EF*.

Let the triangle *ABC* be applied to the triangle *DEF*, so that *B* is placed on *E* and *BC* on *EF*, but so that *A* falls on the opposite side of *EF* from *D*, taking the position *G*. Then *C* will coincide with *F*, since *BC* is equal to *EF*.

Now *FG* will either be in a straight line with *DF*, or make an angle with it, and in the latter case the angle will either be *interior* (κατὰ τὸ ἐντός) to the figure or *exterior* (κατὰ τὸ ἐκτός).

I. Let *FG* be in a straight line with *DF*.

Then, since *DE* is equal to *EG*, and *DFG* is a straight line,

DEG is an isosceles triangle, and the angle at *D* is equal to the angle at *G*.
[I. 5].

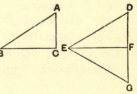

II. Let *DF*, *FG* form an angle *interior* to the figure.
Let *DG* be joined.
Then, since *DE*, *EG* are equal,
the angle *EDG* is equal to the angle *EGD*.
Again, since *DF* is equal to *FG*,
the angle *FDG* is equal to the angle *FGD*.
Therefore, by addition,
the whole angle *EDF* is equal to the whole angle *EGF*.

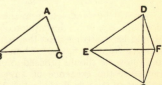

III. Let *DF, FG* form an angle *exterior* to the figure.
Let *DG* be joined.

The proof proceeds as in the last case,
except that subtraction takes the place of
addition, and

the remaining angle *EDF* is equal to the
remaining angle *EGF.*

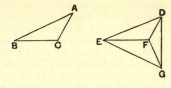

Therefore in all three cases the angle
EDF is equal to the angle *EGF,* that is,
to the angle *BAC.*

It will be observed that, in accordance with the practice of the Greek
geometers in not recognising as an "angle" any angle not less than two right
angles, the re-entrant angle of the quadrilateral *DEGF* is ignored and the angle
DFG is said to be *outside* the figure.

PROPOSITION 9.

To bisect a given rectilineal angle.

Let the angle *BAC* be the given rectilineal angle.
Thus it is required to bisect it.

Let a point *D* be taken at random on *AB*;
let *AE* be cut off from *AC* equal to *AD*;   [I. 3]
let *DE* be joined, and on *DE* let the equilateral
triangle *DEF* be constructed;
let *AF* be joined.

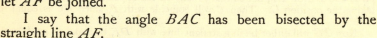

I say that the angle *BAC* has been bisected by the
straight line *AF.*

For, since *AD* is equal to *AE,*
and *AF* is common,

the two sides *DA, AF* are equal to the two sides
EA, AF respectively.

And the base *DF* is equal to the base *EF*;

therefore the angle *DAF* is equal to the angle *EAF.*

[I. 8]

Therefore the given rectilineal angle *BAC* has been
bisected by the straight line *AF.*   Q. E. F.

It will be observed from the translation of this proposition that Euclid
does not say, in his description of the construction, that the equilateral triangle
should be constructed on the side of *DE* opposite to *A*; he leaves this to be
inferred from his figure. There is no particular value in Proclus' explanation
as to how we should proceed in case any one should assert that he could not
recognise the existence of any space below *DE*. He supposes, then, the
equilateral triangle described on the side of *DE* towards *A*, and hence has to
consider three cases according as the vertex of the equilateral triangle falls
on *A*, above *A* or below it. The second and third cases do not differ

substantially from Euclid's. In the first case, where ADE is the equilateral triangle constructed on DE, take any point F on AD, and from AE cut off AG equal to AF. Join DG, EF meeting in H; and join AH. Then AH is the bisector required.

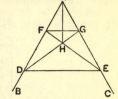

Proclus also answers the possible *objection* that might be raised to Euclid's proof on the ground that it assumes that, if the equilateral triangle be described on the side of DE opposite to A, its vertex F will lie within the angle BAC. The objector is supposed to argue that this is not necessary, but that F might fall either on one of the lines forming the angle or outside it altogether. The two cases are disposed of thus.

Suppose F to fall as shown in the two figures below respectively.

Then, since FD is equal to FE,
the angle FDE is equal to the angle FED.

Therefore the angle CED is greater than the angle FDE; and, in the second figure, *a fortiori*, the angle CED is greater than the angle BDE.

But, since ADE is an isosceles triangle, and the equal sides are produced,

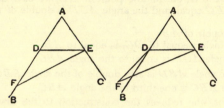

the angles under the base are equal,
i.e., the angle CED is equal to the angle BDE.

But the angle CED was proved greater: which is impossible.

Here then is the second case in which, in Proclus' view, the second part of I. 5 is useful for refuting objections.

On this proposition Proclus takes occasion (p. 271, 15—19) to emphasize the fact that the given angle must be *rectilineal*, since the bisection of any sort of angle (including angles made by curves with one another or with straight lines) is not matter for an elementary treatise, besides which it is questionable whether such bisection is always possible. "Thus it is difficult to say whether it is possible to bisect the so-called *horn-like* angle" (formed by the circumference of a circle and a tangent to it).

Trisection of an angle.

Further it is here that Proclus gives us his valuable historical note about the *trisection* of any acute angle, which (as well as the division of an angle in any given ratio) requires resort to other curves than circles, i.e. curves of the species which, after Geminus, he calls "mixed." "This," he says (p. 272, 1—12), "is shown by those who have set themselves the task of trisecting such a given rectilineal angle. For Nicomedes trisected any rectilineal angle by means of the *conchoidal* lines, the origin, order, and properties of which he has handed down to us, being himself the discoverer of their peculiarity. Others have done the same thing by means of the *quadratrices* of Hippias and Nicomedes, thereby again using 'mixed' curves. But others, starting from the Archimedean spirals, cut a given rectilineal angle in a given ratio."

(*a*)   Trisection by means of the *conchoid*.

I have already spoken of the *conchoid* of Nicomedes (note on Def. 2, pp. 160—1); it remains to show how it could be used for trisecting an angle. Pappus explains this (IV. pp. 274—5) as follows.

Let *ABC* be the given acute angle, and from any point *A* in *AB* draw *AC* perpendicular to *BC*.

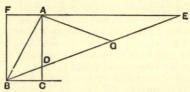

Complete the parallelogram *FBCA* and produce *FA* to a point *E* such that, if *BE* be joined, *BE intercepts between AC and AE a length DE equal to twice AB*.

I say that the angle *EBC* is one-third of the angle *ABC*.

For, joining *A* to *G*, the middle point of *DE*, we have the three straight lines *AG*, *DG*, *EG* equal, and the angle *AGD* is double of the angle *AED* or *EBC*.

But *DE* is double of *AB*; therefore *AG*, which is equal to *DG*, is equal to *AB*.

Hence the angle *AGD* is equal to the angle *ABG*.

Therefore the angle *ABD* is also double of the angle *EBC*; so that the angle *EBC* is one-third of the angle *ABC*.

So far Pappus, who reduces the construction to the drawing of *BE* so that *DE* shall be equal to twice *AB*.

This is what the conchoid constructed with *B* as *pole*, *AC* as *directrix*, and *distance* equal to twice *AB* enables us to do; for that conchoid cuts *AE* in the required point *E*.

(*b*)   Use of the *quadratrix*.

The plural *quadratrices* in the above passage is a Hellenism for the singular *quadratrix*, which was a curve discovered by Hippias of Elis about 420 B.C. According to Proclus (p. 356, 11) Hippias proved its properties; and we are told (1) in the passage quoted above that Nicomedes also investigated it and that it was used for trisecting an angle, and (2) by Pappus (IV. pp. 250, 33—252, 4) that it was used by Dinostratus and Nicomedes and some more recent writers for squaring the circle, whence its name. It is described thus (Pappus IV. p. 252).

Suppose that *ABCD* is a square and *BED* a quadrant of a circle with centre *A*.

Suppose (1) that a radius of the circle moves uniformly about *A* from the position *AB* to the position *AD*, and (2) that *in the same time* the line *BC* moves uniformly, always parallel to itself, and with its extremity *B* moving along *BA*, from the position *BC* to the position *AD*.

Then the radius *AE* and the moving line *BC* determine at any instant by their intersection a point *F*.

The locus of *F* is the *quadratrix*.

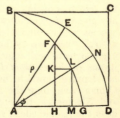

The property of the curve is that, if F is any point, the arc BED is to the arc ED as AB is to FH.

In other words, if ϕ is the angle FAD, ρ the radius vector AF and a the side of the square,

$$(\rho \sin \phi)/a = \phi/\tfrac{1}{2}\pi.$$

Now the angle EAD can not only be *trisected* but *divided in any given ratio* by means of the quadratrix (Pappus IV. p. 286).

For let FH be divided at K in the given ratio.

Draw KL parallel to AD, meeting the curve in L; join AL and produce it to meet the circle in N.

Then the angles EAN, NAD are in the ratio of FK to KH, as is easily proved.

(c) Use of the *spiral of Archimedes*.

The trisection of an angle, or the division of an angle in any ratio, by means of the *spiral of Archimedes* is of course an equally simple matter. Suppose any angle included between the two radii vectores OA and OB of the spiral, and let it be required to cut the angle AOB in a given ratio. Since the radius vector increases proportionally with the angle described by the vector which generates the curve (reckoned from the original position of the vector coinciding with the initial line to the particular position assumed), we have only to take the radius vector OB (the greater of the two OA, OB), mark off OC along it equal to OA, cut CB in the given ratio (at D say), and then draw the circle with centre O and radius OD cutting the spiral in E. Then OE will divide the angle AOB in the required manner.

PROPOSITION 10.

To bisect a given finite straight line.

Let AB be the given finite straight line.

Thus it is required to bisect the finite straight line AB.

Let the equilateral triangle ABC be constructed on it,　　　　　　　　　　　[I. 1]
and let the angle ACB be bisected by the straight line CD;　　　　　　　　　　　[I. 9]

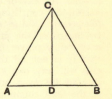

I say that the straight line AB has been bisected at the point D.

For, since AC is equal to CB, and CD is common,

the two sides AC, CD are equal to the two sides BC, CD respectively;

and the angle ACD is equal to the angle BCD;

therefore the base AD is equal to the base BD.　　　　　[I. 4]

Therefore the given finite straight line AB has been bisected at D.　　　　　　　　　　　Q. E. F.

Apollonius, we are told (Proclus, pp. 279, 16—280, 4), bisected a straight
line *AB* by a construction like that of I. I.
With centres *A*, *B*, and radii *AB*, *BA* respectively, two circles are described, intersecting in
C, *D*. Joining *CD*, *AC*, *CB*, *AD*, *DB*, Apollonius proves in two steps that *CD* bisects *AB*.

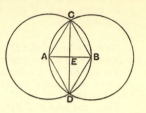

(1) Since, in the triangles *ACD*, *BCD*,

two sides *AC*, *CD* are equal to two sides
BC, *CD*,

and the bases *AD*, *BD* are equal,

the angle *ACD* is equal to the angle
BCD.   [I. 8]

(2) The latter angles being equal, and *AC* being equal to *CB*, while *CE*
is common,

the equality of *AE*, *EB* follows by I. 4.

The objection to this proof is that, instead of *assuming* the bisection of
the angle *ACB*, as already effected by I. 9, Apollonius goes a step further
back and embodies a construction for bisecting the angle. That is, he
unnecessarily does over again what has been done before, which is open to
objection from a theoretical point of view.

Proclus (pp. 277, 25—279, 4) warns us against being moved by this
proposition to conclude that geometers assumed, as a preliminary hypothesis,
that a line is not made up of indivisible parts (ἐξ ἀμερῶν). This might be
argued thus. If a line is made up of indivisibles, there must be in a finite
line either an odd or an even number of them. If the number were odd,
it would be necessary in order to bisect the line to bisect an indivisible (the
odd one). In that case therefore it would not be possible to bisect a straight
line, if it is a magnitude made up of indivisibles. But, if it is not so made
up, the straight line can be divided *ad infinitum* or without limit (ἐπ᾽ ἄπειρον
διαιρεῖται). Hence it was argued (φασίν), says Proclus, that the divisibility
of magnitudes without limit was admitted and assumed as a geometrical
principle. To this he replies, following Geminus, that geometers did indeed
assume, by way of a common notion, that a continuous magnitude, i.e. a
magnitude consisting of parts connected together (συνημμένων), is divisible
(διαιρετόν). But *infinite* divisibility was not assumed by them; it was *proved*
by means of the first principles applicable to the case. "For when," he
says, "they prove that the incommensurable exists among magnitudes, and
that it is not all things that are commensurable with one another, what
else will any one say that they prove but that every magnitude can be
divided for ever, and that we shall never arrive at the indivisible, that
is, the least common measure of the magnitudes? This then is matter of
demonstration, whereas it is an *axiom* that everything continuous is divisible,
so that a finite continuous line is divisible. The writer of the Elements
bisects a finite straight line, starting from the latter notion, and not from any
assumption that it is divisible without limit." Proclus adds that the proposition
may also serve to refute Xenocrates' theory of indivisible lines (ἄτομοι γραμμαί).
The argument given by Proclus to disprove the existence of indivisible lines
is substantially that used by Aristotle as regards magnitudes generally (cf.
Physics VI. I, 231 a 21 sqq. and especially VI. 2, 233 b 15—32).

PROPOSITION 11.

To draw a straight line at right angles to a given straight line from a given point on it.

Let AB be the given straight line, and C the given point on it.

5   Thus it is required to draw from the point C a straight line at right angles to the straight line AB.

Let a point D be taken at random on AC;

10 let CE be made equal to CD; [I. 3]

on DE let the equilateral triangle FDE be constructed, [I. 1]

and let FC be joined;

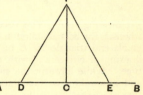

I say that the straight line FC has been drawn at right 15 angles to the given straight line AB from C the given point on it.

For, since DC is equal to CE, and CF is common,

    the two sides DC, CF are equal to the two sides EC, 20 CF respectively;

and the base DF is equal to the base FE;

    therefore the angle DCF is equal to the angle ECF;

                         [I. 8]

and they are adjacent angles.

But, when a straight line set up on a straight line makes 25 the adjacent angles equal to one another, each of the equal angles is right; [Def. 10]

    therefore each of the angles DCF, FCE is right.

Therefore the straight line CF has been drawn at right angles to the given straight line AB from the given point 30 C on it.

                        Q. E. F.

10. let CE be made equal to CD. The verb is κείσθω which, as well as the other parts of κεῖμαι, is constantly used for the passive of τίθημι "to *place*"; and the latter word is constantly used in the sense of *making*, e.g., one straight line equal to another straight line.

De Morgan remarks that this proposition, which is "to bisect the angle made by a straight line and its continuation" [i.e. a *flat* angle], should be a particular case of I. 9, the constructions being the same. This is certainly worth noting, though I doubt the advantage of rearranging the propositions in consequence.

Apollonius gave a construction for this proposition (see Proclus, p. 282, 8) differing from Euclid's in much the same way as his construction for bisecting a straight line differed from that of I. 10. Instead of assuming an equilateral triangle drawn without repeating the process of I. 1, Apollonius takes D and E equidistant from C as in Euclid, and then draws circles in the manner of

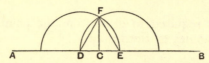

I. 1 meeting at F. This necessitates proving again that DF is equal to FE; whereas Euclid's assumption of the construction of I. 1 in the words "let the equilateral triangle FDE be constructed" enables him to dispense with the drawing of circles and with the proof that DF is equal to FE at the same time. While however the substitution of Apollonius' constructions for I. 10 and 11 would show faulty arrangement in a theoretical treatise like Euclid's, they are entirely suitable for what we call *practical* geometry, and such may have been Apollonius' object in these constructions and in his alternative for I. 23.

Proclus gives a construction for drawing a straight line at right angles to another straight line but from *one end* of it, instead of from an intermediate point on it, it being supposed (for the sake of argument) that we are not permitted to *produce* the straight line. In the commentary of an-Nairīzī (ed. Besthorn-Heiberg, pp. 73—4; ed. Curtze, pp. 54—5) this construction is attributed to Heron.

Let it be required to draw from A a straight line at right angles to AB.

On AB take any point C, and in the manner of the proposition draw CE at right angles to AB.

From CE cut off CD equal to AC, bisect the angle ACE by the straight line CF,   [I. 9] and draw DF at right angles to CE meeting CF in F. Join FA.

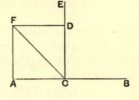

Then the angle FAC will be a right angle.

For, since, in the triangles ACF, DCF, the two sides AC, CF are equal to the two sides DC, CF respectively, and the included angles ACF, DCF are equal,

the triangles are equal in all respects.   [I. 4]

Therefore the angle at A is equal to the angle at D, and is accordingly a right angle.

PROPOSITION 12.

To a given infinite straight line, from a given point which is not on it, to draw a perpendicular straight line.

Let AB be the given infinite straight line, and C the given point which is not on it ;

5 thus it is required to draw to the given infinite straight
line *AB*, from the given point
C which is not on it, a perpendicular straight line.

For let a point *D* be taken
10 at random on the other side of
the straight line *AB*, and with
centre *C* and distance *CD* let
the circle *EFG* be described;
            [Post. 3]
let the straight line *EG*
15   be bisected at *H*,   [I. 10]
and let the straight lines *CG*, *CH*, *CE* be joined.

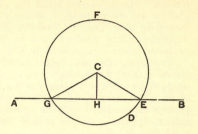

                                                    [Post. 1]
I say that *CH* has been drawn perpendicular to the given
infinite straight line *AB* from the given point *C* which is
not on it.

20 For, since *GH* is equal to *HE*,
and *HC* is common,

      the two sides *GH*, *HC* are equal to the two sides
      EH, *HC* respectively;

and the base *CG* is equal to the base *CE*;

25   therefore the angle *CHG* is equal to the angle *EHC*.
                                                    [I. 8]
      And they are adjacent angles.

But, when a straight line set up on a straight line makes
the adjacent angles equal to one another, each of the equal
angles is right, and the straight line standing on the other is
30 called a perpendicular to that on which it stands.   [Def. 10]

Therefore *CH* has been drawn perpendicular to the given
infinite straight line *AB* from the given point *C* which is
not on it.

                                    Q. E. F.

2. a perpendicular straight line, κάθετον εὐθεῖαν γραμμήν. This is the full expression
for a *perpendicular*, κάθετος meaning *let fall* or *let down*, so that the expression corresponds
to our *plumb-line*. ἡ κάθετος is however constantly used alone for a perpendicular, γραμμή
being understood.

10. on the other side of the straight line AB, literally "towards the other parts of
the straight line *AB*," ἐπὶ τὰ ἕτερα μέρη τῆς AB. Cf. "on the same side" (ἐπὶ τὰ αὐτὰ
μέρη) in Post. 5 and "in both directions" (ἐφ' ἑκάτερα τὰ μέρη) in Def. 23.

"This problem," says Proclus (p. 283, 7—10), "was first investigated
by Oenopides [5th cent. B.C.], who thought it useful for astronomy. He
however calls the perpendicular, in the archaic manner, (a line drawn)

gnomon-wise (κατὰ γνώμονα), because the gnomon is also at right angles to the horizon." In this earlier sense the *gnomon* was a staff placed in a vertical position for the purpose of casting shadows and so serving as a means of measuring time (Cantor, *Geschichte der Mathematik*, I₃, p. 161). The later meanings of the word as used in Eucl. Book II. and elsewhere will be explained in the note on Book II. Def. 2.

Proclus says that two kinds of perpendicular were distinguished, the "plane" (ἐπίπεδος) and the "solid" (στερεά), the former being the perpendicular dropped on a line in a plane and the latter the perpendicular dropped on a plane. The term "solid perpendicular" is sufficiently curious, but it may perhaps be compared with the Greek term "solid locus" applied to a conic section, apparently on the ground that it has its origin in the section of a solid, namely a cone.

Attention is called by most editors to the assumption in this proposition that, if only D be taken on the side of AB remote from C, the circle described with CD as radius must necessarily cut AB in two points. To satisfy us of this we need, as in I. I, some postulate of continuity, e.g. something like that suggested by Killing (see note on the Principle of Continuity above, p. 235): " If a point [here the point describing the circle] moves in a figure which is divided into two parts [by the straight line], and if it belongs at the beginning of the motion to one part and at another stage of the motion to the other part, it must during the motion cut the boundary between the two parts," and this of course applies to the motion in *two* directions from D.

But the editors have not, as a rule, noticed a possible *objection* to the Euclidean statement of this problem which is much more difficult to dispose of at this stage, i.e. without employing any proposition later than this in Euclid's order. How do we know, says the supposed critic, that the circle does not cut AB in *three* or more points, in which case there would be not *one* perpendicular but *three* or more? Proclus (pp. 286, 12—289, 6) tries to refute this objection, and it is interesting to follow his argument, though it will easily be seen to be inconclusive. He takes in order three possible suppositions.

  1. May not the circle meet AB in a third point K between the middle point of GE and either extremity of it, taking the form drawn in the figure appended?

Suppose this possible. Bisect GE in H. Join CH, and produce it to meet the circle in L. Join CG, CK, CE.

Then, since CG is equal to CE, and CH is common, while the base GH is equal to the base HE,
  the angles CHG, CHE are equal and, since they are adjacent, they are both right.

Again, since CG is equal to CE,
  the angles at G and E are equal.

Lastly, since CK is equal to CG and also to CE, the angles CGK, CKG are equal, as also are the angles CKE, CEK.

Since the angles CGK, CEK are equal, it follows that

  the angles CKG, CKE are equal and therefore both right.

Therefore the angle CKH is equal to the angle CHK,

  and CH is equal to CK.

But *CK* is equal to *CL*, by the definition of the circle; therefore *CH* is equal to *CL*: which is impossible.

Thus Proclus; but why should not the circle meet *AB* in *H* as well as *K*?

2. May not the circle meet *AB* in *H* the middle point of *GE* and take the form shown in the second figure?

In that case, says Proclus, join *CG*, *CH*, *CE* as before. Then bisect *HE* at *K*, join *CK* and produce it to meet the circumference at *L*.

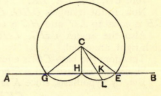

Now, since *HK* is equal to *KE*, *CK* is common, and the base *CH* is equal to the base *CE*,

the angles at *K* are equal and therefore both right angles.

Therefore the angle *CHK* is equal to the angle *CKH*, whence *CK* is equal to *CH* and therefore to *CL*: which is impossible.

So Proclus; but why should not the circle meet *AB* in *K* as well as *H*?

3. May not the circle meet *AB* in *two* points besides *G*, *E* and pass, between those two points, to the side of *AB* towards *C*, as in the next figure?

Here again, by the same method, Proclus proves that, *K*, *L* being the other two points in which the circle cuts *AB*,

CK is equal to *CH*,

and, since the circle cuts *CH* in *M*,

CM is equal to *CK* and therefore to

CH: which is impossible.

But, again, why should the circle not cut *AB* in the point *H* as well?

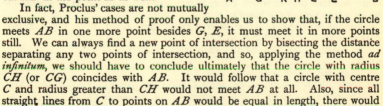

In fact, Proclus' cases are not mutually exclusive, and his method of proof only enables us to show that, if the circle meets *AB* in one more point besides *G*, *E*, it must meet it in more points still. We can always find a new point of intersection by bisecting the distance separating any two points of intersection, and so, applying the method *ad infinitum*, we should have to conclude ultimately that the circle with radius *CH* (or *CG*) coincides with *AB*. It would follow that a circle with centre *C* and radius greater than *CH* would not meet *AB* at all. Also, since all straight lines from *C* to points on *AB* would be equal in length, there would be an infinite number of perpendiculars from *C* on *AB*.

Is this under any circumstances possible? It is not possible in Euclidean space, but it is possible, under the Riemann hypothesis (where a straight line is a "closed series" and returns on itself), in the case where *C* is the pole of the straight line *AB*.

It is natural therefore that, for a proof that in Euclidean space there is only one perpendicular from a point to a straight line, we have to wait until I. 16, the precise proposition which under the Riemann hypothesis is only valid with a certain restriction and not universally. There is no difficulty involved by waiting until I. 16, since I. 12 is not used before that proposition is reached; and we are only in the same position as when, in order to satisfy ourselves of the number of possible solutions of I. 1, we have to wait till I. 7.

But if we wish, after all, to prove the truth of the assumption *without* recourse to any later proposition than I. 12, we can do so by means of this same invaluable I. 7.

If the circle intersects AB as before in G, E, let H be the middle point of GE, and suppose, if possible, that the circle also intersects AB in any other point K on AH.

From H, on the side of AB opposite to C, draw HL at right angles to AB, and make HL equal to HC.

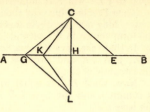

Join CG, LG, CK, LK.

Now, in the triangles CHG, LHG, CH is equal to LH, and HG is common.

Also the angles CHG, LHG, being both right, are equal.

Therefore the base CG is equal to the base LG.

Similarly we prove that CK is equal to LK.

But, by hypothesis, since K is on the circle,

$$CK \text{ is equal to } CG.$$

Therefore CG, CK, LG, LK are all equal.

Now the next proposition, I. 13, will tell us that CH, HL are in a straight line; but we will not assume this. Join CL.

Then on the same base CL and on the same side of it we have two pairs of straight lines drawn from C, L to G and K such that CG is equal to CK and LG to LK.

But this is impossible [I. 7].

Therefore the circle cannot cut BA or BA produced in any point other than G on that side of CL on which G is.

Similarly it cannot cut AB or AB produced at any point other than E on the other side of CL.

The only possibility left therefore is that the circle might cut AB in the same point as that in which CL cuts it. But this is shown to be impossible by an adaptation of the proof of I. 7.

For the assumption is that there may be some point M on CL such that CM is equal to CG and LM to LG.

If possible, let this be the case, and produce CG to N.

Then, since CM is equal to CG, the angle NGM is equal to the angle GML [I. 5, part 2].

Therefore the angle GML is greater than the angle MGL.

Again, since LG is equal to LM, the angle GML is equal to the angle MGL.

But it was also greater: which is impossible.

Hence the circle in the original figure cannot cut AB in the point in which CL cuts it.

Therefore the circle cannot cut AB in any point whatever except G and E.

[This proof of course does not prove that CK is *less* than CG, but only that it is not equal to it. The proposition that, of the obliques drawn from C to AB, that is less the foot of which is nearer to H can only be proved later. The proof by I. 7 also fails, under the Riemann hypothesis, if C, L are the poles of the straight line AB, since the broken lines CGL, CKL etc. become equal straight lines, all perpendicular to AB.]

Proclus rightly adds (p. 289, 18 sqq.) that it is not *necessary* to take D on the side of AB away from A if an objector "says that there is no space on that side." If it is not desired to trespass on that side of *AB*, we can take *D* anywhere on *AB* and describe the *arc* of a circle between *D* and the point where it meets *AB* again, drawing the arc on the side of *AB* on which *C* is. If it should happen that the selected point *D* is such that the circle only meets *AB* in *one* point (*D* itself), we have only to describe the circle with *CD* as radius, then, if *E* be a point on this circle, take *F* a point further from *C* than *E* is, and describe with *CF* as radius the circular arc meeting *AB* in two points.

PROPOSITION 13.

If a straight line set up on a straight line make angles, it will make either two right angles or angles equal to two right angles.

For let any straight line *AB* set up on the straight line
5 *CD* make the angles *CBA*, *ABD*;
I say that the angles *CBA*, *ABD* are either two right angles or equal to two right angles.

Now, if the angle *CBA* is equal to
10 the angle *ABD*,

they are two right angles.   [Def. 10]

But, if not, let *BE* be drawn from the point *B* at right angles to *CD*;   [I. 11]

therefore the angles *CBE*, *EBD* are two right angles.
15   Then, since the angle *CBE* is equal to the two angles *CBA*, *ABE*,

let the angle *EBD* be added to each;

therefore the angles *CBE*, *EBD* are equal to the three
angles *CBA*, *ABE*, *EBD*.   [C. N. 2]
20   Again, since the angle *DBA* is equal to the two angles *DBE*, *EBA*,

let the angle *ABC* be added to each;

therefore the angles *DBA*, *ABC* are equal to the three
angles *DBE*, *EBA*, *ABC*.   [C. N. 2]
25   But the angles *CBE*, *EBD* were also proved equal to the same three angles;

and things which are equal to the same thing are also equal to one another;   [C. N. 1]

therefore the angles *CBE*, *EBD* are also equal to the
30   angles *DBA*, *ABC*.

But the angles *CBE*, *EBD* are two right angles ;

therefore the angles *DBA*, *ABC* are also equal to two right angles.

Therefore etc.

Q. E. D.

17. **let the angle EBD be added to each,** literally "let the angle *EBD* be added (so as to be) common," κοινὴ προσκείσθω ἡ ὑπὸ ΕΒΔ. Similarly κοινὴ ἀφῃρήσθω is used of subtracting a straight line or angle from each of two others. "Let the common angle *EBD* be added" is clearly an inaccurate translation, for the angle is not common before it is added, i.e. the κοινή is proleptic. "Let the common angle be *subtracted*" as a translation of κοινὴ ἀφῃρήσθω would be less unsatisfactory, it is true, but, as it is desirable to use corresponding words when translating the two expressions, it seems hopeless to attempt to keep the word "common," and I have therefore said "to each" and "from each" simply.

PROPOSITION 14.

If with any straight line, and at a point on it, two straight lines not lying on the same side make the adjacent angles equal to two right angles, the two straight lines will be in a straight line with one another.

5   For with any straight line *AB*, and at the point *B* on it, let the two straight lines *BC*, *BD* not lying on the same side make the adjacent angles *ABC*, *ABD* equal to two right angles ;

I say that *BD* is in a straight line with *CB*.

10   For, if *BD* is not in a straight line with *BC*, let *BE* be in a straight line with *CB*.

Then, since the straight line *AB* stands on the straight line *CBE*,

15   the angles *ABC*, *ABE* are equal to two right angles.

[I. 13]

But the angles *ABC*, *ABD* are also equal to two right angles ;

therefore the angles *CBA*, *ABE* are equal to the angles *CBA*, *ABD*.

[Post. 4 and *C. N.* 1]

Let the angle *CBA* be subtracted from each ;

20 therefore the remaining angle *ABE* is equal to the remaining angle *ABD*,

[*C. N.* 3]

the less to the greater : which is impossible.

Therefore *BE* is not in a straight line with *CB*.

Similarly we can prove that neither is any other straight

25 line except *BD*.

Therefore *CB* is in a straight line with *BD*.
Therefore etc.

Q. E. D.

1. **If with any straight line....** There is no greater difficulty in translating the works of the Greek geometers than that of accurately giving the force of prepositions. πρός, for instance, is used in all sorts of expressions with various shades of meaning. The present enunciation begins Ἐὰν πρός τινι εὐθείᾳ καὶ τῷ πρὸς αὐτῇ σημείῳ, and it is really necessary in this one sentence to translate πρός by three different words, *with*, *at*, and *on*. The first πρός must be translated by *with* because two straight lines "make" an angle *with* one another. On the other hand, where the similar expression πρὸς τῇ δοθείσῃ εὐθείᾳ occurs in I. 23, but it is a question of "constructing" an angle (συστήσασθαι), we have to say "to construct *on* a given straight line." *Against* would perhaps be the English word coming nearest to expressing all these meanings of πρός, but it would be intolerable as a translation.

17. Todhunter points out that for the inference in this line Post. 4, that all right angles are equal, is necessary as well as the Common Notion that things which are equal to the same thing (or rather, here, to *equal things*) are equal. A similar remark applies to steps in the proofs of I. 15 and I. 28.

24. **we can prove.** The Greek expresses this by the future of the verb, δείξομεν, "we shall prove," which however would perhaps be misleading in English.

Proclus observes (p. 297) that two straight lines on the *same* side of another straight line and meeting it in one and the same point may make with one and the same portion of the straight line terminated at the point two angles which are together equal to two right angles, in which case however the two straight lines would not be in a straight line with one another. And he quotes from Porphyry a construction for two such straight lines in the particular case where they form with the given straight line angles equal respectively to half a right angle and one and a half right angles. There is no particular value in

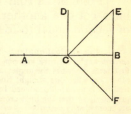

the construction, which will be gathered from the annexed figure where *CE*, *CF* are drawn at the prescribed inclinations to *CD*.

PROPOSITION 15.

If two straight lines cut one another, they make the vertical angles equal to one another.

For let the straight lines *AB*, *CD* cut one another at the point *E* ;

5   I say that the angle *AEC* is equal to the angle *DEB*,

and the angle *CEB* to the angle *AED*.

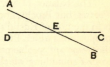

For, since the straight line *AE* stands
10 on the straight line *CD*, making the angles *CEA*, *AED*, the angles *CEA*, *AED* are equal to two right angles

Again, since the straight line *DE* stands on the straight line *AB*, making the angles *AED*, *DEB*,

the angles *AED*, *DEB* are equal to two right angles.

[I. 13]

15 But the angles *CEA*, *AED* were also proved equal to two right angles;

therefore the angles *CEA*, *AED* are equal to the angles *AED DEB*. [Post. 4 and *C. N.* 1]

Let the angle *AED* be subtracted from each;

20 therefore the remaining angle *CEA* is equal to the remaining angle *BED*. [*C. N.* 3]

Similarly it can be proved that the angles *CEB*, *DEA* are also equal.

Therefore etc. Q. E. D.

25 [PORISM. From this it is manifest that, if two straight lines cut one another, they will make the angles at the point of section equal to four right angles.]

1. **the vertical angles.** The difference between *adjacent* angles (αἱ ἐφεξῆς γωνίαι) and *vertical* angles (αἱ κατὰ κορυφὴν γωνίαι) is thus explained by Proclus (p. 298, 14—24). The first term describes the angles made by two straight lines when one only is divided by the other, i.e. when one straight line meets another at a point which is not either of its extremities, but is not itself produced beyond the point of meeting. When the first straight line *is* produced, so that the lines cross at the point, they make two pairs of *vertical* angles (which are more clearly described as *vertically opposite* angles), and which are so called because their convergence is from opposite directions to one point (the intersection of the lines) as vertex (κορυφή).

26. **at the point of section,** literally "at the section," πρὸς τῇ τομῇ.

This theorem, according to Eudemus, was first discovered by Thales, but found its scientific demonstration in Euclid (Proclus, p. 299, 3—6).

Proclus gives a converse theorem which may be stated thus. *If a straight line is met at one and the same point intermediate in its length by two other straight lines on different sides of it and such as to make the vertical angles equal, the latter straight lines are in a straight line with one another.* The proof need not be given, since it is almost self-evident, whether (1) it is direct, by means of I. 13, 14, or (2) indirect, by *reductio ad absurdum* depending on I. 15.

The balance of MS. authority seems to be against the genuineness of this *Porism*, but Proclus and Psellus both have it. The word is not here used, as it is in the title of Euclid's lost *Porisms*, to signify a particular class of independent propositions which Proclus describes as being in some sort intermediate between theorems and problems (requiring us, not to bring a thing into existence, but to *find* something which we know to exist). *Porism* has here (and wherever the term is used in the *Elements*) its second meaning; it is what we call a corollary, i.e. an incidental result springing from the proof of a theorem or the solution of a problem, a result not directly sought but appearing as it were by chance without any additional labour, and constituting, as Proclus says, a sort of *windfall* (ἕρμαιον) and *bonus* (κέρδος). These Porisms appear in both the geometrical and arithmetical Books of the *Elements*, and may either result from theorems or problems. Here the Porism is geometrical, and springs out of a theorem; VII. 2 affords an instance of an arithmetical Porism. As an instance of a Porism to a problem Proclus cites "that which is found in the second Book" (τὸ ἐν τῷ δευτέρῳ βιβλίῳ κείμενον); but as to this see notes on II. 4 and IV. 15.

The present Porism, says Proclus, formed the basis of "that paradoxical theorem which proves that only the following three (regular) polygons can fill up the whole space surrounding one point, the equilateral triangle, the square, and the equilateral and equiangular hexagon." We can in fact place round a point in this manner six equilateral triangles, three regular hexagons, or four squares. "But only the angles of these regular figures, to the number specified, can make up four right angles: a theorem due to the Pythagoreans."

Proclus further adds that it results from the Porism that, if any number of straight lines intersect one another at one point, the sum of all the angles so formed will still be equal to four right angles. This is of course what is generally given in the text-books as Corollary 2.

PROPOSITION 16.

In any triangle, if one of the sides be produced, the exterior angle is greater than either of the interior and opposite angles.

Let ABC be a triangle, and let one side of it BC be produced to D;

5   I say that the exterior angle ACD is greater than either of the interior and opposite angles CBA, BAC.

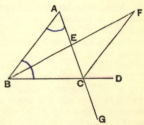

Let AC be bisected at E [I. 10], and let BE be joined and produced
10 in a straight line to F;

let EF be made equal to BE [I. 3], let FC be joined [Post. 1], and let AC be drawn through to G [Post. 2].

Then, since AE is equal to EC,
15 and BE to EF,

the two sides AE, EB are equal to the two sides CE, EF respectively;

and the angle AEB is equal to the angle FEC,

for they are vertical angles.   [I. 15]

20   Therefore the base AB is equal to the base FC,

and the triangle ABE is equal to the triangle CFE,

and the remaining angles are equal to the remaining angles respectively, namely those which the equal sides subtend;   [I. 4]

therefore the angle BAE is equal to the angle ECF.

25   But the angle *ECD* is greater than the angle *ECF*;

[C. N. 5]

therefore the angle *ACD* is greater than the angle *BAE*.

Similarly also, if *BC* be bisected, the angle *BCG*, that is, the angle *ACD* [I. 15], can be proved greater than the angle *ABC* as well.

Therefore etc.   Q. E. D.

1. **the exterior angle**, literally "the outside angle," ἡ ἐκτὸς γωνία.
2. **the interior and opposite angles**, τῶν ἐντὸς καὶ ἀπεναντίον γωνιῶν.
12. **let AC be drawn through to G.** The word is διήχθω, a variation on the more usual ἐκβεβλήσθω, "let it be *produced*."
21. **CFE**, in the text "*FEC*."

As is well known, this proposition is not universally true under the Riemann hypothesis of a space endless in extent but not infinite in size. On this hypothesis a straight line is a "closed series" and returns on itself; and two straight lines which have one point of intersection have another point of intersection also, which bisects the whole length of the straight line measured from the first point on it to the same point again; thus the axiom of Euclidean geometry that two straight lines do not enclose a space does not hold. If 4Δ denotes the finite length of a straight line measured from any point once round to the same point again, 2Δ is the distance between the two intersections of two straight lines which meet. Two points *A*, *B* do not determine one sole straight line unless the distance between them is different from 2Δ. In order that there may only be one perpendicular from a point *C* to a straight line *AB*, *C* must not be one of the two "poles" of the straight line.

Now, in order that the proof of the present proposition may be universally valid, it is necessary that *CF* should always fall within the angle *ACD* so that the angle *ACF* may be less than the angle *ACD*. But this will not always be so on the Riemann hypothesis. For, (1) if *BE* is equal to Δ, so that *BF* is equal to 2Δ, *F* will be the second point in which *BE and BD* intersect; i.e. *F* will lie on *CD*, and the angle *ACF* will be *equal* to the angle *ACD*. In this case the exterior angle *ACD* will be *equal* to the interior angle *BAC*. (2) If *BE* is greater than Δ and less than 2Δ, so that *BF* is greater than 2Δ and less than 4Δ, the angle *ACF* will be *greater* than the angle *ACD*, and therefore the angle *ACD* will be *less* than the interior angle *BAC*. Thus, e.g., in the particular case of a right-angled triangle, the angles other than the right angle may be (1) both acute, (2) one acute and one obtuse, or (3) both obtuse according as the perpendicular sides are (1) both less than Δ, (2) one less and the other greater than Δ, (3) both greater than Δ.

Proclus tells us (p. 307, 1—12) that some combined this theorem with the next in one enunciation thus: *In any triangle, if one side be produced, the exterior angle of the triangle is greater than either of the interior and opposite angles, and any two of the interior angles are less than two right angles*, the combination having been suggested by the similar enunciation of Euclid I. 32, *In any triangle, if one of the sides be produced, the exterior angle is equal to the two interior and opposite angles, and the three interior angles of the triangle are equal to two right angles*.

The present proposition enables Proclus to prove what he did not succeed in establishing conclusively in his note on I. 12, namely that *from one point there cannot be drawn to the same straight line three straight lines equal in length*.

For, if possible, let *AB*, *AC*, *AD* be all equal, *B*, *C*, *D* being in a straight line.

Then, since *AB*, *AC* are equal, the angles *ABC*, *ACB* are equal.

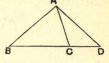

Similarly, since *AB*, *AD* are equal, the angles *ABD*, *ADB* are equal.

Therefore the angle *ACB* is equal to the angle *ADC*, i.e. the exterior angle to the interior and opposite angle: which is impossible.

Proclus next (p. 308, 14 sqq.) undertakes to prove by means of 1. 16 that, *if a straight line falling on two straight lines make the exterior angle equal to the interior and opposite angle, the two straight lines will not form a triangle or meet,* for in that case the same angle would be both greater and equal.

The proof is really equivalent to that of Eucl. 1. 27.   If *BE* falls on the two straight lines *AB*, *CD* in such a way that the angle *CDE* is equal to the interior and opposite angle *ABD*, *AB* and *CD* cannot form a triangle or meet.   For, if they did, then (by 1. 16) the angle *CDE* would be *greater* than the angle *ABD*, while by the hypothesis it is at the same time *equal* to it.

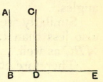

Hence, says Proclus, in order that *BA*, *DC* may form a triangle it is necessary for them to *approach* one another in the sense of being turned round one pair of corresponding extremities, e.g. *B*, *D*, so that the other extremities *A*, *C* come nearer.   This may be brought about in one of three ways: (1) *AB* may remain fixed and *CD* be turned about *D* so that the angle *CDE* increases; (2) *CD* may remain fixed and *AB* be turned about *B* so that the angle *ABD* becomes smaller; (3) both *AB* and *CD* may move so as to make the angle *ABD* smaller and the angle *CDE* larger at the same time.   The *reason*, then, of the straight lines *AB*, *CD* coming to form a triangle or to meet is (says Proclus) *the movement of the straight lines.*

Though he does not mention it here, Proclus does in another passage (p. 371, 2—10, quoted on p. 207 above) hint at the possibility that, while 1. 16 may remain universally true, either of the straight lines *BA*, *DC* (or both together) may be turned through any angle not greater than a certain finite angle and yet may not meet (the Bolyai-Lobachewsky hypothesis).

PROPOSITION 17.

In any triangle two angles taken together in any manner are less than two right angles.

Let *ABC* be a triangle;

I say that two angles of the triangle *ABC* taken together in any manner are less than two right angles.

For let *BC* be produced to *D*.   [Post. 2]

Then, since the angle *ACD* is an exterior angle of the triangle *ABC*,

it is greater than the interior and opposite angle *ABC*.

Let the angle ACB be added to each;
therefore the angles ACD, ACB are greater than the angles
ABC, BCA.

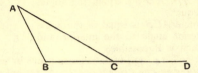

But the angles ACD, ACB are equal to two right angles.

[I. 13]

Therefore the angles ABC, BCA are less than two right
angles.

Similarly we can prove that the angles BAC, ACB are
also less than two right angles, and so are the angles CAB,
ABC as well.

Therefore etc.

Q. E. D.

1. taken together in any manner, πάντη μεταλαμβανόμεναι, i.e. any pair added
together.

As in his note on the previous proposition, Proclus tries to state the *cause*
of the property.   He takes the case of two straight lines forming right angles
with a transversal and observes that it is the *convergence of the straight lines
towards one another* (σύνευσις τῶν εὐθειῶν), the *lessening* of the two right angles,
which produces the triangle.   He will not have it that the fact of the exterior
angle being greater than the interior and opposite angle is the *cause* of the
property, for the odd reason that "it is not necessary that a side should be
produced, or that there should be any exterior angle constructed...and how can
what is not necessary be the cause of what is necessary?" (p. 311, 17—21).

Agreeably to this view, Proclus then sets himself to prove the theorem
without producing a side of the triangle.

Let ABC be a triangle.   Take any point D on
BC, and join AD.

Then the exterior angle ADC of the triangle ABD
is greater than the interior and opposite angle ABD.

Similarly the exterior angle ADB of the triangle
ADC is greater than the interior and opposite angle
ACD.

Therefore, by addition, the angles ADB, ADC are together greater than
the angles ABC, ACB.

But the angles ADB, ADC are equal to two right angles; therefore the
angles ABC, ACB are less than two right angles.

Lastly, Proclus proves (what is obvious from this proposition) that *there
cannot be more than one perpendicular to a straight line from a point without
it*.   For, if this were possible, two of such perpendiculars would form a triangle
in which two angles would be right angles: which is impossible, since any two
angles of a triangle are together less than two right angles.

PROPOSITION 18.

In any triangle the greater side subtends the greater angle.

For let ABC be a triangle having the side AC greater than AB;

I say that the angle ABC is also greater than the angle BCA.

For, since AC is greater than AB, let AD be made equal to AB [I. 3], and let BD be joined.

Then, since the angle ADB is an exterior angle of the triangle BCD,

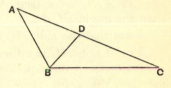

it is greater than the interior and opposite angle DCB.   [I. 16]

But the angle ADB is equal to the angle ABD,

since the side AB is equal to AD;

therefore the angle ABD is also greater than the angle ACB;

therefore the angle ABC is much greater than the angle ACB.

Therefore etc.

Q. E. D.

In the enunciation of this proposition we have ὑποτείνειν ("subtend") used with the simple accusative instead of the more usual ὑπό with accusative. The latter construction is used in the enunciation of I. 19, which otherwise only differs from that of I. 18 in the order of the words. The point to remember in order to distinguish the two is that the *datum* comes first and the *quaesitum* second, the *datum* being in this proposition the greater *side* and in the next the greater *angle*. Thus the enunciations are (I. 18) παντὸς τριγώνου ἡ μείζων πλευρὰ τὴν μείζονα γωνίαν ὑποτείνει and (I. 19) παντὸς τριγώνου ὑπὸ τὴν μείζονα γωνίαν ἡ μείζων πλευρὰ ὑποτείνει. In order to keep the proper order in English we must use the passive of the verb in I. 19. Aristotle quotes the result of I. 19, using the exact wording, ὑπὸ γὰρ τὴν μείζω γωνίαν ὑποτείνει (*Meteorologica* III. 5, 376 a 12).

"In order to assist the student in remembering which of these two propositions [I. 18, 19] is demonstrated directly and which indirectly, it may be observed that the order is similar to that in I. 5 and I. 6" (Todhunter).

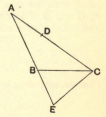

An alternative proof of I. 18 given by Porphyry (see Proclus, pp. 315, 11—316, 13) is interesting. It starts by supposing a length equal to AB cut off from the other end of AC; that is, CD and not AD is made equal to AB.

Produce AB to E so that BE is equal to AD, and join EC.

Then, since AB is equal to CD, and BE to AD,

AE is equal to AC.

Therefore the angle *AEC* is equal to the angle *ACE*.

Now the angle *ABC* is greater than the angle *AEC*,   [I. 16]

and therefore greater than the angle *ACE*.

Hence, *a fortiori*, the angle *ABC* is greater than the angle *ACB*.

PROPOSITION 19.

In any triangle the greater angle is subtended by the greater side.

Let *ABC* be a triangle having the angle *ABC* greater than the angle *BCA* ;

I say that the side *AC* is also greater than the side *AB*.

For, if not, *AC* is either equal to *AB* or less.

Now *AC* is not equal to *AB* ;

for then the angle *ABC* would also have been equal to the angle *ACB* ;   [I. 5]

but it is not ;

therefore *AC* is not equal to *AB*.

Neither is *AC* less than *AB*,

for then the angle *ABC* would also have been less than the angle *ACB* ;   [I. 18]

but it is not ;

therefore *AC* is not less than *AB*.

And it was proved that it is not equal either.

Therefore *AC* is greater than *AB*.

Therefore etc.   Q. E. D.

This proposition, like I. 6, can be proved by merely *logical* deduction from I. 5 and I. 18 taken together, as pointed out by De Morgan. The general form of the argument used by De Morgan is given in his *Formal Logic* (1847), p. 25, thus :

"*Hypothesis.* Let there be any number of propositions or assertions— three for instance, *X*, *Y* and *Z*—of which it is the property that one or the other must be true, *and one only*. Let there be three other propositions *P*, *Q* and *R* of which it is also the property that one, and one only, must be true. Let it be a connexion of those assertions that :

when *X* is true, *P* is true,
when *Y* is true, *Q* is true,
when *Z* is true, *R* is true.

Consequence : then it follows that,

when *P* is true, *X* is true,
when *Q* is true, *Y* is true,
when *R* is true, *Z* is true."

To apply this to the case before us, let us denote the sides of the triangle *ABC* by *a*, *b*, *c*, and the angles opposite to these sides by *A*, *B*, *C* respectively, and suppose that *a* is the base.

Then we have the three propositions,

when b is equal to c, B is equal to C,   [I. 5]

when b is greater than c, B is greater than C,   ⎫
when b is less than c, B is less than C,   ⎬ [I. 18]

and it follows *logically* that,

when B is equal to C, b is equal to c,   [I. 6]

when B is greater than C, b is greater than c,   ⎫
when B is less than C, b is less than c.   ⎬ [I. 19]

Reductio ad absurdum by exhaustion.

Here, says Proclus (p. 318, 16—23), Euclid proves the impossibility "by means of *division*" (ἐκ διαιρέσεως). This means simply the separation of different hypotheses, each of which is inconsistent with the truth of the theorem to be proved, and which therefore must be successively shown to be impossible. If a straight line is not greater than a straight line, it must be either equal to it or less; thus in a *reductio ad absurdum* intended to prove such a theorem as I. 19 it is necessary to dispose successively of *two* hypotheses inconsistent with the truth of the theorem.

Alternative (direct) proof.

Proclus gives a direct proof (pp. 319—321) which an-Nairīzī also has and attributes to Heron. It requires a lemma and is consequently open to the slight objection of separating a theorem from its converse. But the lemma and proof are worth giving.

Lemma.

If an angle of a triangle be bisected and the straight line bisecting it meet the base and divide it into unequal parts, the sides containing the angle will be unequal, and the greater will be that which meets the greater segment of the base, and the less that which meets the less.

Let *AD*, the bisector of the angle *A* of the triangle *ABC*, meet *BC* in *D*, making *CD* greater than *BD*.

I say that *AC* is greater than *AB*.

Produce *AD* to *E* so that *DE* is equal to *AD*. And, since *DC* is greater than *BD*, cut off *DF* equal to *BD*.

Join *EF* and produce it to *G*.

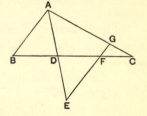

Then, since the two sides *AD*, *DB* are equal to the two sides *ED*, *DF*, and the vertical angles at *D* are equal,

AB is equal to EF,

and the angle *DEF* to the angle *BAD*,

i.e. to the angle DAG (by hypothesis).

Therefore *AG* is equal to *EG*,

and therefore greater than EF, or AB.

Hence, *a fortiori*, *AC* is greater than *AB*.

Proof of I. 19.

Let *ABC* be a triangle in which the angle *ABC* is greater than the angle *ACB*.

Bisect *BC* at *D*, join *AD*, and produce it to *E* so that *DE* is equal to *AD*. Join *BE*.

Then the two sides *BD*, *DE* are equal to the two sides *CD*, *DA*, and the vertical angles at *D* are equal;

therefore *BE* is equal to *AC*,

and the angle *DBE* to the angle at *C*.

But the angle at *C* is less than the angle *ABC*;

therefore the angle *DBE* is less than the angle *ABD*.

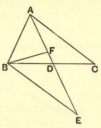

Hence, if *BF* bisect the angle *ABE*, *BF* meets *AE* between *A* and *D*. Therefore *EF* is greater than *FA*.

It follows, by the lemma, that *BE* is greater than *BA*,

that is, *AC* is greater than *AB*.

PROPOSITION 20.

In any triangle two sides taken together in any manner are greater than the remaining one.

For let *ABC* be a triangle;

I say that in the triangle *ABC* two sides taken together in any manner are greater than the remaining one, namely

BA, *AC* greater than *BC*,

AB, *BC* greater than *AC*,

BC, *CA* greater than *AB*.

For let *BA* be drawn through to the point *D*, let *DA* be made equal to *CA*, and let *DC* be joined.

Then, since *DA* is equal to *AC*,

the angle *ADC* is also equal to the angle *ACD*;   [I. 5]

therefore the angle *BCD* is greater than the angle *ADC*.   [C. N. 5]

And, since *DCB* is a triangle having the angle *BCD* greater than the angle *BDC*,

and the greater angle is subtended by the greater side,

[I. 19]

therefore *DB* is greater than *BC*.

But *DA* is equal to *AC*;

therefore *BA*, *AC* are greater than *BC*.

Similarly we can prove that *AB*, *BC* are also greater than *CA*, and *BC*, *CA* than *AB*.

Therefore etc.

Q. E. D.

It was the habit of the Epicureans, says Proclus (p. 322), to ridicule this theorem as being evident even to an ass and requiring no proof, and their allegation that the theorem was "known" (γνώριμον) even to an ass was based on the fact that, if fodder is placed at one angular point and the ass at another, he does not, in order to get to his food, traverse the two sides of the triangle but only the one side separating them (an argument which makes Savile exclaim that its authors were "digni ipsi, qui cum Asino foenum essent," p. 78). Proclus replies truly that a mere perception of the truth of the theorem is a different thing from a scientific proof of it and a knowledge of the reason *why* it is true. Moreover, as Simson says, the number of axioms should not be increased without necessity.

Alternative Proofs.

Heron and Porphyry, we are told (Proclus, pp. 323—6), proved this theorem in different ways as follows, without producing one of the sides.

First proof.

Let *ABC* be the triangle, and let it be required to prove that the sides *BA*, *AC* are greater than *BC*.

Bisect the angle *BAC* by *AD* meeting *BC* in *D*.

Then, in the triangle *ABD*,

the exterior angle *ADC* is greater than the interior and opposite angle *BAD*,　　　[I. 16]

that is, greater than the angle *DAC*.

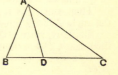

Therefore the side *AC* is greater than the side *CD*.　　　　　　　　　　　[I. 19]

Similarly we can prove that *AB* is greater than *BD*.

Hence, by addition, *BA*, *AC* are greater than *BC*.

Second proof.

This, like the first proof, is direct. There are several cases to be considered.

(1) If the triangle is *equilateral*, the truth of the proposition is obvious.

(2) If the triangle is *isosceles*, the proposition needs no proof in the case (*a*) where each of the equal sides is greater than the base.

(*b*) If the base is greater than either of the other sides, we have to prove that the sum of the two equal sides is greater than the base. Let *BC* be the base in such a triangle.

Cut off from *BC* a length *BD* equal to *AB*, and join *AD*.

Then, in the triangle *ADB*, the exterior angle *ADC* is greater than the interior and opposite angle *BAD*.　　　　　　　　[I. 16]

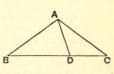

Similarly, in the triangle *ADC*, the exterior angle *ADB* is greater than the interior and opposite angle *CAD*.

By addition, the two angles BDA, ADC are together greater than the two angles BAD, DAC (or the whole angle BAC).

Subtracting the equal angles BDA, BAD, we have the angle ADC greater than the angle CAD.

It follows that AC is greater than CD;   [I. 19]
and, adding the equals AB, BD respectively, we have BA, AC together greater than BC.

(3) If the triangle be *scalene*, we can arrange the sides in order of length. Suppose BC is the greatest, AB the intermediate and AC the least side. Then it is obvious that AB, BC are together greater than AC, and BC, CA together greater than AB.

It only remains therefore to prove that CA, AB are together greater than BC.

We cut off from BC a length BD equal to the adjacent side, join AD, and proceed exactly as in the above case of the isosceles triangle.

Third proof.

This proof is by *reductio ad absurdum*.

Suppose that BC is the greatest side and, as before, we have to prove that BA, AC are greater than BC.

If they are not, they must be either equal to or less than BC.

(1) Suppose BA, AC are together equal to BC.

From BC cut off BD equal to BA, and join AD.

It follows from the hypothesis that DC is equal to AC.

Then, since BA is equal to BD, the angle BDA is equal to the angle BAD.

Similarly, since AC is equal to CD, the angle CDA is equal to the angle CAD.

By addition, the angles BDA, ADC are together equal to the whole angle BAC.

That is, the angle BAC is equal to two right angles: which is impossible.

(2) Suppose BA, AC are together less than BC.

From BC cut off BD equal to BA, and from CB cut off CE equal to CA. Join AD, AE.

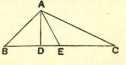

In this case, we prove in the same way that the angle BDA is equal to the angle BAD, and the angle CEA to the angle CAE.

By addition, the sum of the angles BDA, AEC is equal to the sum of the angles BAD, CAE.

Now, by I. 16, the angle BDA is greater than the angle DAC, and therefore, *a fortiori*, greater than the angle EAC.

Similarly the angle AEC is greater than the angle BAD.

Hence the sum of the angles BDA, AEC is greater than the sum of the angles BAD, EAC.

But the former sum was also equal to the latter: which is impossible.

PROPOSITION 21.

If on one of the sides of a triangle, from its extremities, there be constructed two straight lines meeting within the triangle, the straight lines so constructed will be less than the remaining two sides of the triangle, but will contain a greater angle.

On *BC*, one of the sides of the triangle *ABC*, from its extremities *B*, *C*, let the two straight lines *BD*, *DC* be constructed meeting within the triangle;

I say that *BD*, *DC* are less than the remaining two sides of the triangle *BA*, *AC*, but contain an angle *BDC* greater than the angle *BAC*.

For let *BD* be drawn through to *E*.

Then, since in any triangle two sides are greater than the remaining one,   [I. 20]

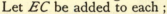

therefore, in the triangle *ABE*, the two sides *AB*, *AE* are greater than *BE*.

Let *EC* be added to each;

therefore *BA*, *AC* are greater than *BE*, *EC*.

Again, since, in the triangle *CED*,

the two sides *CE*, *ED* are greater than *CD*,

let *DB* be added to each;

therefore *CE*, *EB* are greater than *CD*, *DB*.

But *BA*, *AC* were proved greater than *BE*, *EC*;

therefore *BA*, *AC* are much greater than *BD*, *DC*.

Again, since in any triangle the exterior angle is greater than the interior and opposite angle,   [I. 16]

therefore, in the triangle *CDE*,

the exterior angle *BDC* is greater than the angle *CED*.

For the same reason, moreover, in the triangle *ABE* also,

the exterior angle *CEB* is greater than the angle *BAC*.

But the angle *BDC* was proved greater than the angle *CEB*;

therefore the angle *BDC* is much greater than the angle *BAC*.

Therefore etc.   Q. E. D.

2. **be constructed...meeting within the triangle.** The word "meeting" is not in the Greek, where the words are ἐντὸς συσταθῶσιν. συνίστασθαι is the word used of constructing two straight lines *to a point* (cf. I. 7) or so as to form a triangle; but it is necessary in English to indicate that they *meet*.

3. **the straight lines so constructed.** Observe the elegant brevity of the Greek αἱ συσταθεῖσαι.

The editors generally call attention to the fact that the lines drawn within
the triangle in this proposition must be drawn,
as the enunciation says, from the *ends* of the
side; otherwise it is not necessary that their
sum should be less than that of the remaining
sides of the triangle. Proclus (p. 327, 12 sqq.)
gives a simple illustration.

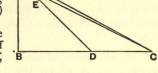

Let *ABC* be a right-angled triangle. Take
any point *D* on *BC*, join *DA*, and cut off
from it *DE* equal to *AB*. Bisect *AE* at *F*,
and join *FC*.

Then shall *CF, FD* be together greater than *CA, AB*.

For *CF, FE* are equal to *CF, FA*,

and therefore greater than *CA*.

Add the equals *ED, AB* respectively;

therefore *CF, FD* are together greater than *CA, AB*.

Pappus gives the same proposition as that just proved, but follows it up
by a number of others more elaborate in character, selected apparently from
"the so-called paradoxes" of one Erycinus (Pappus, III. p. 106 sqq.). Thus
he proves the following:

1. In any triangle, except an equilateral triangle or an isosceles triangle
with base less than one of the other sides, it is possible to construct on the
base and within the triangle two straight lines the sum of which is equal to
the sum of the other two sides of the triangle.

2. In any triangle in which it is possible to construct two straight lines on
the base which are equal to the sum of the other two sides of the triangle it is
also possible to construct two others the sum of which is *greater* than that sum.

3. Under the same conditions, if the base is greater than either of the
other two sides, two straight lines can be constructed in the manner described
which are *respectively* greater than the other two sides of the triangle; and the
lines may be constructed so as to be respectively *equal* to the two sides, if one
of those two sides is less than the other and each of them less than the base.

4. The lines may be so constructed that their sum will bear to the sum
of the two sides of the triangle any ratio less than 2 : 1.

As a specimen of the proofs we will give that of the proposition which has

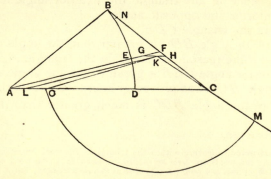

been numbered (1) for the case where the triangle is isosceles (Pappus, III.
pp. 108—110).

Let ABC be an isosceles triangle in which the base AC is greater than either of the equal sides AB, BC.

With centre A and radius AB describe a circle meeting AC in D.

Draw any radius AEF such that it meets BC in a point F outside the circle.

Take any point G on EF, and through it draw GH parallel to AC. Take any point K on GH, and draw KL parallel to FA meeting AC in L.

From BC cut off BN equal to EG.

Thus AG, or LK, is equal to the sum of AB, BN, and CN is less than LK.

Now GF, FH are together greater than GH,

and CH, HK together greater than CK.

Therefore, by addition,

CF, FG, HK are together greater than CK, HG.

Subtracting HK from each side, we see that

CF, FG are together greater than CK, KG;

therefore, if we add AG to each,

AF, FC are together greater than AG, GK, KC.

And AB, BC are together greater than AF, FC.   [I. 21]

Therefore AB, BC are together greater than AG, GK, KC.

But, by construction, AB, BN are together equal to AG;

therefore, by subtraction, NC is greater than GK, KC,

and *a fortiori* greater than KC.

Take on KC produced a point M such that KM is equal to NC;

with centre K and radius KM describe a circle meeting CL in O, and join KO.

Then shall LK, KO be equal to AB, BC.

For, by construction, LK is equal to the sum of AB, BN, and KO is equal to NC;

therefore LK, KO are together equal to AB, BC.

It is after I. 21 that (as remarked by De Morgan) the important proposition about the perpendicular and obliques drawn from a point to a straight line of unlimited length is best introduced:

Of all straight lines that can be drawn to a given straight line of unlimited length from a given point without it:

(a)   the perpendicular is the shortest;

(b)   of the obliques, that is the greater the foot of which is further from the perpendicular;

(c)   given one oblique, only one other can be found of the same length, namely that the foot of which is equally distant with the foot of the given one from the perpendicular, but on the other side of it.

Let A be the given point, BC the given straight line; let AD be the perpendicular from A on BC, and AE, AF any two obliques of which AF makes the greater angle with AD.

Produce AD to A', making $A'D$ equal to AD, and join $A'E$, $A'F$.

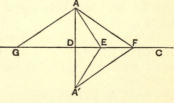

Then the triangles ADE, $A'DE$ are equal in all respects; and so are the triangles ADF, $A'DF$.

Now (1) in the triangle AEA' the two sides AE, EA' are greater than AA' [I. 20], that is, twice AE is greater than twice AD.

Therefore AE is greater than AD.

(2)   Since AE, $A'E$ are drawn to E, a point within the triangle AFA',
AF, FA' are together greater than AE, EA',   [I. 21]
or twice AF is greater than twice AE.

Therefore AF is greater than AE.

(3)   Along DB measure off DG equal to DF, and join AG.

The triangles AGD, AFD are then equal in all respects, so that the angles GAD, FAD are equal, and AG is equal to AF.

PROPOSITION 22.

Out of three straight lines, which are equal to three given straight lines, to construct a triangle : thus it is necessary that two of the straight lines taken together in any manner should be greater than the remaining one.   [I. 20]

Let the three given straight lines be A, B, C, and of these let two taken together in any manner be greater than the remaining one,

namely A, B greater than C,

A, C greater than B,

and B, C greater than A ;

thus it is required to construct a triangle out of straight lines equal to A, B, C.

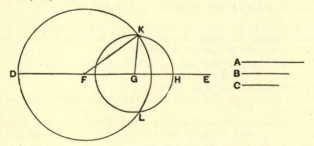

Let there be set out a straight line DE, terminated at D but of infinite length in the direction of E,

and let DF be made equal to A, FG equal to B, and GH equal to C.   [I. 3]

With centre F and distance FD let the circle DKL be described ;

again, with centre G and distance GH let the circle KLH be described ;

and let KF, KG be joined ;

I say that the triangle KFG has been constructed out of three straight lines equal to A, B, C.

For, since the point F is the centre of the circle DKL,
FD is equal to FK.

But FD is equal to A ;
therefore KF is also equal to A.

Again, since the point G is the centre of the circle LKH,
GH is equal to GK.

But GH is equal to C ;
therefore KG is also equal to C.

And FG is also equal to B ;

therefore the three straight lines KF, FG, GK are equal to the three straight lines A, B, C.

Therefore out of the three straight lines KF, FG, GK, which are equal to the three given straight lines A, B, C, the triangle KFG has been constructed.

Q. E. F.

2—4. This is the first case in the *Elements* of a διορισμός to a problem in the sense of a statement of the conditions or limits of the possibility of a solution. The criterion is of course supplied by the preceding proposition.

2. **thus it is necessary.** This is usually translated (e.g. by Williamson and Simson) "*But* it is necessary," which is however inaccurate, since the Greek is not δεῖ δέ but δεῖ δή. The words are the same as those used to introduce the διορισμός in the other sense of the "definition" or "particular statement" of a construction to be effected. Hence, as in the latter case we say "thus it is required" (e.g. to bisect the finite straight line AB, I. 10), we should here translate "*thus* it is necessary."

4. To this enunciation all the MSS. and Boethius add, after the διορισμός, the words "because in any triangle two sides taken together in any manner are greater than the remaining one." But this explanation has the appearance of a gloss, and it is omitted by Proclus and Campanus. Moreover there is no corresponding addition to the διορισμός of VI. 28.

It was early observed that Euclid assumes, without giving any reason, that the circles drawn as described will meet if the condition that any two of the straight lines A, B, C are together greater than the third be fulfilled. Proclus (p. 331, 8 sqq.) argues the matter by means of *reductio ad absurdum*, but does not exhaust the possible hypotheses inconsistent with the contention. He says the circles must do one of three things, (1) cut one another, (2) touch one another, (3) stand apart (διεστάναι) from one another. He then considers the hypotheses (*a*) of their touching *externally*, (*b*) of their being separated from one another by a space. He should have considered also the hypothesis (*c*) of one circle touching the other *internally* or lying entirely within the other without touching. These three hypotheses being successively disproved, it follows that the circles must meet (this is the line taken by Camerer and Todhunter).

Simson says: "Some authors blame Euclid because he does not demonstrate that the two circles made use of in the construction of this problem must cut one another: but this is very plain from the determination he has given, namely, that any two of the straight lines DF, FG, GH must be greater than the third. For who is so dull, though only beginning to learn the Elements, as not to perceive that the circle described from the centre F, at the distance FD, must meet FH betwixt F and H, because FD is less than FH; and that, for the like reason, the circle described from the

centre G at the distance GH must meet DG betwixt D and G; and that
these circles must meet one another, because FD and GH are together
greater than FG."

We have in fact only to satisfy ourselves that one of the circles, e.g. that
with centre G, has at least one point of its circumference outside the other
circle and also at least one point of its circumference inside the same circle;
and this is best shown with reference to the points in which the first circle
cuts the straight line DE.   For (1) FH, being equal to the sum of B and C,
is greater than A, i.e. than the radius of the circle with centre F, and therefore
H is outside that circle.   (2) If GM be measured along GF equal to GH
or C, then, since GM is either (*a*) less or (*b*) greater than GF, M will fall
either (*a*) between G and F or (*b*) beyond F towards D; in the first case
(*a*) the sum of FM and C is equal to FG and therefore less than the sum
of A and C, so that FM is less than A or FD; in the second case (*b*) the
sum of MF and FG, i.e. the sum of MF and B, is equal to GM or C, and
therefore less than the sum of A and B, so that MF is less than A or FD;
hence in either case M falls within the circle with centre F.

It being now proved that the circumference of the circle with centre G
has at least one point outside, and at least one point inside, the circle with
centre F, we have only to invoke the Principle of Continuity, as we have to
do in I. 1 (cf. the note on that proposition, p. 242, where the necessary
postulate is stated in the form suggested by Killing).

That the construction of the proposition gives only *two* points of
intersection between the circles, and therefore only two triangles satisfying
the condition, one on each side of FG, is clear from I. 7, which, as before
pointed out, takes the place, in Book I., of III. 10 proving that two circles
cannot intersect in more points than two.

PROPOSITION 23.

*On a given straight line and at a point on it to construct a
rectilineal angle equal to a given rectilineal angle.*

Let AB be the given straight line, A the point on it, and
the angle DCE the given rectilineal angle;

thus it is required to construct on the given straight line
AB, and at the point A on it, a rectilineal angle equal to the
given rectilineal angle DCE.

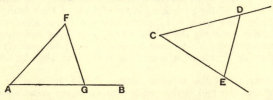

On the straight lines CD, CE respectively let the points
D, E be taken at random;
let DE be joined,
and out of three straight lines which are equal to the three

straight lines CD, DE, CE let the triangle AFG be constructed in such a way that CD is equal to AF, CE to AG,
and further DE to FG.   [I. 22]

Then, since the two sides DC, CE are equal to the two
sides FA, AG respectively,

and the base DE is equal to the base FG,

the angle DCE is equal to the angle FAG.   [I. 8]

Therefore on the given straight line AB, and at the point
A on it, the rectilineal angle FAG has been constructed equal
to the given rectilineal angle DCE.
                                              Q. E. F.

This problem was, according to Eudemus (see Proclus, p. 333, 5), "rather
the discovery of Oenopides," from which we must apparently infer, not that
Oenopides was the first to find any solution of it, but that it was he who discovered the particular solution given by Euclid. (Cf. Bretschneider, p. 65.)

The editors do not seem to have noticed the fact that the construction of
the triangle assumed in this proposition is not exactly the construction given
in I. 22. We have here to construct a triangle on a certain finite straight line
AG as base; in I. 22 we have only to construct a triangle with sides of given
length without any restriction as to how it is to be placed. Thus in I. 22 we
set out any line whatever and measure successively three lengths along it
beginning from the given extremity, and what we must regard as the base is the
intermediate length, not the length beginning at the given extremity of the
straight line arbitrarily set out. Here the base is a given straight line abutting
at a given point. Thus the construction has to be modified somewhat from

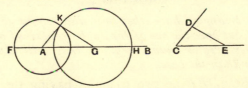

that of the preceding proposition. We must measure AG along AB so that
AG is equal to CE (or CD), and GH along GB equal to DE; and then we
must produce BA, in the opposite direction, to F, so that AF is equal to CD
(or CE, if AG has been made equal to CD).

Then, by drawing circles (1) with centre A and radius AF, (2) with centre
G and radius GH, we determine K, one of their points of intersection, and we
prove that the triangle KAG is equal in all respects to the triangle DCE, and
then that the angle at A is equal to the angle DCE.

I think that Proclus must (though he does not say so) have felt the same
difficulty with regard to the use in I. 23 of the result of I. 22, and that this is
probably the reason why he gives over again the construction which I have
given above, with the remark (p. 334, 6) that "you may obtain the construction
of the triangle in a more instructive manner (διδασκαλικώτερον) as follows."

Proclus objects to the procedure of Apollonius in constructing an angle
under the same conditions, and certainly, if he quotes Apollonius correctly, the
latter's exposition must have been somewhat slipshod.

"He takes an angle *CDE* at random," says Proclus (p. 335, 19 sqq.), "and a straight line *AB*, and with centre *D* and distance *CD* describes the circumference *CE*, and in the same way with centre *A* and distance *AB* the circumference *FB*. Then, cutting off *FB* equal to *CE*, he joins *AF*. And he declares that the angles *A*, *D* standing on equal circumferences are equal."

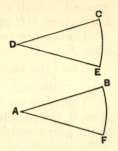

In the first place, as Proclus remarks, it should be premised that *AB* is equal to *CD* in order that the circles may be equal; and the use of Book III. for such an elementary construction is objectionable. The omission to state that *AB* must be taken equal to *CD* was no doubt a slip, if it occurred. And, as regards the equal angles "standing on equal circumferences," it would seem possible that Apollonius said this in *explanation*, for the sake of brevity, rather than by way of proof. It seems to me probable that his construction was only given from the point of view of *practical*, not theoretical, geometry. It really comes to the same thing as Euclid's except that *DC* is taken equal to *DE*. For cutting off the arc *BF* equal to the arc *CE* can only be meant in the sense of measuring the *chord CE*, say, with a pair of compasses, and then drawing a circle with centre *B* and radius equal to the chord *CE*. Apollonius' direction was therefore probably intended as a practical short cut, avoiding the actual drawing of the chords *CE*, *BF*, which, as well as a proof of the equality in all respects of the triangles *CDE*, *BAF*, would be required to establish *theoretically* the correctness of the construction.

PROPOSITION 24.

If two triangles have the two sides equal to two sides respectively, but have the one of the angles contained by the equal straight lines greater than the other, they will also have the base greater than the base.

5   Let *ABC*, *DEF* be two triangles having the two sides *AB*, *AC* equal to the two sides *DE*, *DF* respectively, namely *AB* to *DE*, and *AC* to *DF*, and let the angle at *A* be greater than the angle at *D* ;

I say that the base *BC* is also greater than the base *EF*.

10   For, since the angle *BAC* is greater than the angle *EDF*, let there be constructed, on the straight line *DE*, and at the point *D* on it, the angle *EDG* equal to the angle *BAC*; [I. 23]

15 let *DG* be made equal to either of the two straight lines *AC*, *DF*, and let *EG*, *FG* be joined.

Then, since *AB* is equal to *DE*, and *AC* to *DG*,
20 the two sides *BA*, *AC* are equal to the two sides *ED*, *DG*,
respectively ;

and the angle *BAC* is equal to the angle *EDG* ;

therefore the base *BC* is equal to the base *EG*.   [I. 4]

Again, since *DF* is equal to *DG*,
25 the angle *DGF* is also equal to the angle *DFG* ;   [I. 5]

therefore the angle *DFG* is greater than the angle *EGF*.

Therefore the angle *EFG* is much greater than the angle
EGF.

And, since *EFG* is a triangle having the angle *EFG*
30 greater than the angle *EGF*,

and the greater angle is subtended by the greater side,

[I. 19]

the side *EG* is also greater than *EF*.

But *EG* is equal to *BC*.

Therefore *BC* is also greater than *EF*.

35 Therefore etc.

Q. E. D.

10.   I have naturally left out the well-known words added by Simson in
order to avoid the necessity of considering three cases : "Of the two sides
DE, *DF* let *DE* be the side which is not greater than the other." I doubt
whether Euclid could have been induced to insert the words himself, even if
it had been represented to him that their omission meant leaving two possible
cases out of consideration.   His habit and that of the great Greek geometers
was, not to set out all possible cases, but to give as a rule one case, generally
the most difficult, as here, and to leave the others to the reader to work out for
himself.   We have already seen one instance in I. 7.

Proclus of course gives the other
two cases which arise if we do not
first provide that *DE* is not greater
than *DF*.

(1)   In the first case *G* may fall
on *EF* produced, and it is then
obvious that *EG* is greater than *EF*.

(2)   In the second case *EG* may
fall below *EF*.

If so, by I. 21, *DF*, *FE* are
together less than *DG*, *GE*.

But *DF* is equal to *DG* ; therefore *EF* is less than *EG*, i.e. than
BC.

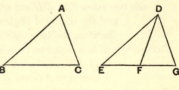

These two cases are therefore
decidedly simpler than the case taken
by Euclid as typical, and could well be left to the ingenuity of the learner.

If however after all we prefer to insert Simson's words and avoid the latter two cases, the proof is not complete unless we show that, with his assumption, *F* must, in the figure of the proposition, fall *below EG*.

De Morgan would make the following proposition precede: *Every straight line drawn from the vertex of a triangle to the base is less than the greater of the two sides, or than either if they are equal,* and he would prove it by means of the proposition relating to perpendicular and obliques given above, p. 291.

But it is easy to prove directly that *F* falls below *EG,* if *DE* is not greater than *DG,* by the method employed by Pfleiderer, Lardner, and Todhunter.

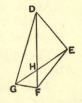

Let *DF,* produced if necessary, meet *EG* in *H.*

Then the angle *DHG* is greater than the angle *DEG;*
[I. 16]

and the angle *DEG* is not less than the angle *DGE;*
[I. 18]

therefore the angle *DHG* is greater than the angle *DGH.*

Hence *DH* is less than *DG,*   [I. 19]

and therefore *DH* is less than *DF.*

Alternative proof.

Lastly, the modern alternative proof is worth giving.

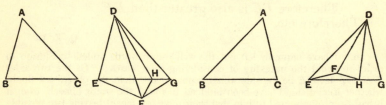

Let *DH* bisect the angle *FDG* (after the triangle *DEG* has been made equal in all respects to the triangle *ABC,* as in the proposition), and let *DH* meet *EG* in *H.*   Join *HF.*

Then, in the triangles *FDH, GDH,*

the two sides *FD, DH* are equal to the two sides *GD, DH,*

and the included angles *FDH, GDH* are equal;

therefore the base *HF* is equal to the base *HG.*

Accordingly *EG* is equal to the sum of *EH, HF;*

and *EH, HF* are together greater than *EF;*   [I. 20]

therefore *EG,* or *BC,* is greater than *EF.*

Proclus (p. 339, 11 sqq.) answers by anticipation the possible question that might occur to any one on this proposition, viz. why does Euclid not compare the areas of the triangles as he does in I. 4?   He observes that inequality of the areas does not follow from the inequality of the angles contained by the equal sides, and that Euclid leaves out all reference to the question both for this reason and because the areas cannot be compared without the help of the theory of parallels.   "But if," says Proclus, "we must anticipate what is to come and make our comparison of the areas at once, we assert that (1) *if the angles A, D*—supposing that our argument proceeds with reference to the figure in the proposition—*are (together) equal to two right angles, the triangles are proved equal,* (2) *if greater than two right angles, that triangle which has the greater angle is less, and* (3) *if they are less, greater."* Proclus then gives the proof, but without any reference to the source from which he quoted the proposition. Now an-Nairīzī adds a similar proposition to I. 38, but definitely attributes it to Heron. I shall accordingly give it in the place where Heron put it

Proposition 25.

If two triangles have the two sides equal to two sides respectively, but have the base greater than the base, they will also have the one of the angles contained by the equal straight lines greater than the other.

Let *ABC, DEF* be two triangles having the two sides *AB, AC* equal to the two sides *DE, DF* respectively, namely *AB* to *DE*, and *AC* to *DF*; and let the base *BC* be greater than the base *EF*;

I say that the angle *BAC* is also greater than the angle *EDF*.

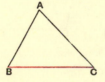

For, if not, it is either equal to it or less.
Now the angle *BAC* is not equal to the angle *EDF*;
for then the base *BC* would also have been equal to the base *EF*,   [I. 4]
                   but it is not ;
therefore the angle *BAC* is not equal to the angle *EDF*.
Neither again is the angle *BAC* less than the angle *EDF*;
for then the base *BC* would also have been less than the base *EF*,   [I. 24]
                   but it is not ;
therefore the angle *BAC* is not less than the angle *EDF*.
But it was proved that it is not equal either ;
therefore the angle *BAC* is greater than the angle *EDF*.
Therefore etc.

Q. E. D.

De Morgan points out that this proposition (as also I. 8) is a purely *logical* consequence of I. 4 and I. 24 in the same way as I. 19 and I. 6 are purely *logical* consequences of I. 18 and I. 5.   If *a*, *b*, *c* denote the sides, *A*, *B*, *C* the angles opposite to them in a triangle *ABC*, and *a'*, *b'*, *c'*, *A'*, *B'*, *C'* the sides and opposite angles respectively in a triangle *A'B'C'*, I. 4 and I. 24 tell us that, *b*, *c* being respectively equal to *b'*, *c'*,

(1) if *A* is equal to *A'*, then *a* is equal to *a'*,

(2) if *A* is less than *A'*, then *a* is less than *a'*,

(3) if *A* is greater than *A'*, then *a* is greater than *a'* ;

and it follows *logically* that,

(1) if *a* is equal to *a'*, the angle *A* is equal to the angle *A'*,   [I. 8]

(2) if *a* is less than *a'*, *A* is less than *A'*,

(3) if *a* is greater than *a'*, *A* is greater than *A'*. }   [I. 25]

Two alternative proofs of this theorem are given by Proclus (pp. 345—7), and they are both interesting.   Moreover both are *direct*.

I.   Proof by Menelaus of Alexandria.

Let *ABC*, *DEF* be two triangles having the two sides *BA*, *AC* equal to the two sides *ED*, *DF*, but the base *BC* greater than the base *EF*.

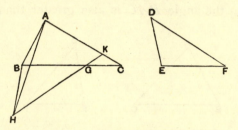

Then shall the angle at *A* be greater than the angle at *D*.

From *BC* cut off *BG* equal to *EF*.   At *B*, on the straight line *BC*, make the angle *GBH* (on the side of *BG* remote from *A*) equal to the angle *FED*.

Make *BH* equal to *DE*; join *HG*, and produce it to meet *AC* in *K*. Join *AH*.

Then, since the two sides *GB*, *BH* are equal to the two sides *FE*, *ED* respectively,

and the angles contained by them are equal,

HG is equal to *DF* or *AC*,

and the angle *BHG* is equal to the angle *EDF*.

Now *HK* is greater than *HG* or *AC*,

and *a fortiori* greater than *AK* ;

therefore the angle *KAH* is greater than the angle *KHA*.

And, since *AB* is equal to *BH*,

the angle *BAH* is equal to the angle *BHA*.

Therefore, by addition,

the whole angle *BAC* is greater than the whole angle *BHG*,

that is, greater than the angle *EDF*.

II. Heron's proof.

Let the triangles be given as before.

Since BC is greater than EF, produce EF to G so that EG is equal to BC.

Produce ED to H so that DH is equal to DF. The circle with centre D and radius DF will then pass through H. Let it be described, as FKH.

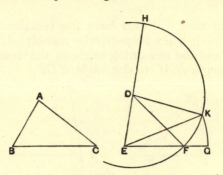

Now, since BA, AC are together greater than BC,

and BA, AC are equal to ED, DH respectively,

while BC is equal to EG,

EH is greater than EG.

Therefore the circle with centre E and radius EG will cut EH, and therefore will cut the circle already drawn. Let it cut that circle in K, and join DK, KE.

Then, since D is the centre of the circle FKH,

DK is equal to DF or AC.

Similarly, since E is the centre of the circle KG,

EK is equal to EG or BC,

And DE is equal to AB.

Therefore the two sides BA, AC are equal to the two sides ED, DK respectively;

and the base BC is equal to the base EK;

therefore the angle BAC is equal to the angle EDK.

Therefore the angle BAC is greater than the angle EDF.

PROPOSITION 26.

If two triangles have the two angles equal to two angles respectively, and one side equal to one side, namely, either the side adjoining the equal angles, or that subtending one of the equal angles, they will also have the remaining sides equal to
5 *the remaining sides and the remaining angle to the remaining angle.*

Let *ABC*, *DEF* be two triangles having the two angles *ABC*, *BCA* equal to the two angles *DEF*, *EFD* respectively, namely the angle *ABC* to the angle *DEF*, and the angle
10 *BCA* to the angle *EFD*; and let them also have one side equal to one side, first that adjoining the equal angles, namely *BC* to *EF*;

I say that they will also have the remaining sides equal to the remaining sides respectively, namely *AB* to *DE* and
15 *AC* to *DF*, and the remaining angle to the remaining angle, namely the angle *BAC* to the angle *EDF*.

For, if *AB* is unequal to *DE*, one of them is greater.

Let *AB* be greater, and let *BG* be made equal to *DE*; and let *GC* be joined.

20   Then, since *BG* is equal to *DE*, and *BC* to *EF*, the two sides *GB*, *BC* are equal to the two sides *DE*, *EF* respectively;
and the angle *GBC* is equal to the angle *DEF*;

    therefore the base *GC* is equal to the base *DF*,
25 and the triangle *GBC* is equal to the triangle *DEF*,
and the remaining angles will be equal to the remaining angles, namely those which the equal sides subtend;   [I. 4]

    therefore the angle *GCB* is equal to the angle *DFE*.

But the angle *DFE* is by hypothesis equal to the angle *BCA*;
30   therefore the angle *BCG* is equal to the angle *BCA*,

    the less to the greater: which is impossible.

    Therefore *AB* is not unequal to *DE*,

        and is therefore equal to it.

    But *BC* is also equal to *EF*;
35   therefore the two sides *AB*, *BC* are equal to the two sides *DE*, *EF* respectively,
and the angle *ABC* is equal to the angle *DEF*;

    therefore the base *AC* is equal to the base *DF*,
and the remaining angle *BAC* is equal to the remaining
40 angle *EDF*.   [I. 4]

Again, let sides subtending equal angles be equal, as AB to DE;

I say again that the remaining sides will be equal to the remaining sides, namely AC to DF and BC to EF, and
45 further the remaining angle BAC is equal to the remaining angle EDF.

For, if BC is unequal to EF, one of them is greater.

Let BC be greater, if possible, and let BH be made equal to EF; let AH be joined.

50   Then, since BH is equal to EF, and AB to DE, the two sides AB, BH are equal to the two sides DE, EF respectively, and they contain equal angles;

therefore the base AH is equal to the base DF, and the triangle ABH is equal to the triangle DEF,
55 and the remaining angles will be equal to the remaining angles, namely those which the equal sides subtend;   [I. 4]

therefore the angle BHA is equal to the angle EFD.

But the angle EFD is equal to the angle BCA;

therefore, in the triangle AHC, the exterior angle BHA is
60 equal to the interior and opposite angle BCA:

which is impossible.   [I. 16]

Therefore BC is not unequal to EF, and is therefore equal to it.

But AB is also equal to DE;
65 therefore the two sides AB, BC are equal to the two sides DE, EF respectively, and they contain equal angles;

therefore the base AC is equal to the base DF, the triangle ABC equal to the triangle DEF,

and the remaining angle BAC equal to the remaining angle
70 EDF.   [I. 4]

Therefore etc.

Q. E. D.

2—3.   the side adjoining the equal angles, πλευρὰν τὴν πρὸς ταῖς ἴσαις γωνίαις.
29.   is by hypothesis equal. ὑπόκειται ἴση, according to the elegant Greek idiom. ὑπόκειμαι is used for the passive of ὑποτίθημι, as κεῖμαι is used for the passive of τίθημι, and so with the other compounds. Cf. προσκεῖσθαι, "to be added."

The alternative method of proving this proposition, viz. by applying one triangle to the other, was very early discovered, at least so far as regards the case where the equal sides are adjacent to the equal angles in each. An-Nairīzī gives it for this case, observing that the proof is one which he had found, but of which he did not know the author.

Proclus has the following interesting note (p. 352, 13—18): "Eudemus in his geometrical history refers this theorem to Thales. For he says that, in the method by which they say that Thales proved the distance of ships in the sea, it was necessary to make use øf this theorem." As, unfortunately, this information is not sufficient of itself to enable us to determine how Thales solved this problem, there is considerable room for conjecture as to his method.

The suggestions of Bretschneider and Cantor agree in the assumption that the necessary observations were probably made from the top of some tower or structure of known height, and that a right-angled triangle was used in which the tower was the perpendicular, and the line connecting the bottom of the tower and the ship was the base, as in the annexed figure, where AB is the tower and C the ship. Bretschneider (*Die Geometrie und die Geometer vor Eukleides*, § 30) says that it was only necessary for the observer to observe the angle CAB, and then the triangle would be completely determined by means of this angle and the known length AB. As Bretschneider says that the result would be obtained "in a moment" by this method, it is not clear in what sense he supposes Thales to have "observed" the angle BAC. Cantor is more definite (*Gesch. d. Math.* I_3, p. 145), for he says that the problem was nearly related to that of finding the *Seqt* from given sides. By the *Seqt* in the Papyrus Rhind is meant the ratio to one another of certain lines in pyramids or obelisks. Eisenlohr and Cantor took the one word to be equivalent, sometimes to the *cosine* of the angle made by the *edge* of the pyramid with the co-terminous diagonal of the base, sometimes to the *tangent* of the angle of slope of the *faces* of the pyramid. It is now certain that it meant one thing, viz. the ratio of half the side of the base to the height of the pyramid, i.e. the *cotangent* of the angle of slope. The calculation of the *Seqt* thus implying a sort of theory of similarity, or even of trigonometry, the suggestion of Cantor is apparently that the *Seqt* in this case would be found from a *small* right-angled triangle ADE having a common angle A with ABC as shown in the figure, and that the ascertained value of the *Seqt* with the length AB would determine BC. This amounts to the use of the property of similar triangles; and Bretschneider's suggestion must apparently come to the same thing, since, even if Thales measured the *angle* in our sense (e.g. by its ratio to a right angle), he would, in the absence of something corresponding to a table of trigonometrical ratios, have gained nothing and would have had to work out the proportions all the same.

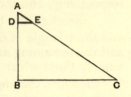

Max C. P. Schmidt also (*Kulturhistorische Beiträge zur Kenntnis des griechischen und römischen Altertums*, 1906, p. 32) similarly supposes Thales to have had a right angle made of wood or bronze with the legs graduated, and to have placed it in the position ADE (A being the position of his eye), and then to have read off the lengths AD, DE respectively, and worked out the length of BC by the rule of three.

How then does the supposed use of similar triangles and their property square with Eudemus' remark about I. 26? As it stands, it asserts the *equality* of *two* triangles which have two angles and one side respectively equal, and the theorem can only be brought into relation with the above explanations by taking it as asserting that, if two angles and one side of *one* triangle are given, the triangle is completely determined. But, if Thales practically used *proportions*, as supposed, I. 26 is surely not at all the theorem which this procedure would naturally suggest as underlying it and being "necessarily used"; the use of proportions or of similar but not equal triangles would surely have taken attention altogether away from I. 26 and fixed it on VI. 4.

For this reason I think Tannery is on the right road when he tries to find a solution using I. 26 as it stands, and withal as primitive as any recorded solution of such a problem. His suggestion (*La Géométrie grecque*, pp. 90—1) is based on the *fluminis varatio* of the Roman agrimensor Marcus Junius Nipsus and is as follows.

To find the distance from a point A to an inaccessible point B. From A measure along a straight line at right angles to AB a
length AC and bisect it at D. From C draw $C\acute{E}$ at right angles to CA on the side of it remote from B, and let E be the point on it which is in a straight line with B and D.

Then, by I. 26, CE is obviously equal to AB.

As regards the equality of angles, it is to be observed that those at D are equal because they are vertically opposite, and, curiously enough, Thales is expressly credited with the discovery of the equality of such angles.

The only objection which I can see to Tannery's solution is that it would require, in the case of the ship, a certain extent of free and level ground for the construction and measurements.

I suggest therefore that the following may have been Thales' method. Assuming that he was on the top of a
tower, he had only to use a rough instrument made of a straight stick and a cross-piece fastened to it so as to be capable of turning about the fastening (say a nail) so that it could form any angle with the stick and would remain where it was put. Then the natural thing would be to fix the stick upright (by means of a plumb-line) and direct the cross-piece towards the ship. Next, leaving the cross-piece at the angle so found, the stick could be turned round, still remaining vertical, until the cross-piece pointed to some visible object on the shore, when the object could be mentally noted and the distance from the bottom of the tower to it could be subsequently measured. This would, by I. 26, give the distance from the bottom of the tower to the ship. This solution has the advantage of corresponding better to the simpler and more probable version of Thales' method of measuring the height of the pyramids; Diogenes Laertius says namely (I. 27, p. 6, ed. Cobet) on the authority of Hieronymus of Rhodes (B.C. 290—230), that he waited for this purpose until the moment when *our shadows are of the same length as ourselves*.

Recapitulation of congruence theorems.

Proclus, like other commentators, gives at this point (p. 347, 20 sqq.) a summary of the cases in which the equality of two triangles in all respects can be established. We may, he says, seek the conditions of such equality by successively considering as hypotheses the equality (1) of sides alone, (2) of angles alone, (3) of sides and angles combined. Taking (1) first, we can only establish the equality of the triangles in all respects if all three sides are respectively equal; we cannot establish the equality of the triangles by any hypothesis of class (2), not even the hypothesis that all the three angles are respectively equal; among the hypotheses of class (3), the equality of one side and one angle in each triangle is not enough, the equality (*a*) of one side and all three angles is more than enough, as is also the equality (*b*) of two sides and two or three angles, and (*c*) of three sides and one or two angles.

The only hypotheses therefore to be examined from this point of view are the equality of

(α)　three sides [Eucl. I. 8].

(β)　two sides and one angle [I. 4 proves one case of this, where the angle is that contained by the sides which are by hypothesis equal].

(γ)　one side and two angles [I. 26 covers all cases].

It is curious that Proclus makes no allusion to what we call the *ambiguous case*, that case namely of (β) in which it is an angle opposite to one of the two specified sides in one triangle which is equal to the angle opposite to the equal side in the other triangle. Camerer indeed attributes to Proclus the observation that in this case the equality of the triangles cannot, unless some other condition is added, be asserted generally; but it would appear that Camerer was probably misled by a figure (Proclus, p. 351) which looks like a figure of the ambiguous case but is really only used to show that in I. 26 the equal sides must be *corresponding* sides, i.e., they must be either adjacent to the equal angles in each triangle, or opposite to corresponding equal angles, and that, e.g., one of the equal sides must not be adjacent to the two angles in one triangle, while the side equal to it in the other triangle is opposite to one of the two corresponding angles in that triangle.

The ambiguous case.

If two triangles have two sides equal to two sides respectively, and if the angles opposite to one pair of equal sides be also equal, then will the angles opposite the other pair of equal sides be either equal or supplementary; and, in the former case, the triangles will be equal in all respects.

Let *ABC*, *DEF* be two triangles such that *AB* is equal to *DE*, and *AC* to *DF*, while the angle *ABC* is equal to the angle *DEF*;
it is required to prove that the angles *ACB*, *DFE* are either equal or supplementary.

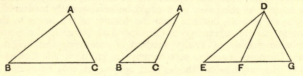

Now (1), if the angle *BAC* be equal to the angle *EDF*, it follows, since the two sides *AB*, *AC* are equal to the two sides *DE*, *DF* respectively, that

the triangles *ABC*, *DEF* are equal in all respects,　　　　　[I. 4]

and the angles *ACB*, *DFE* are equal.

(2)　If the angles *BAC*, *EDF* be not equal, make the angle *EDG* (on the same side of *ED* as the angle *EDF*) equal to the angle *BAC*.

Let *EF*, produced if necessary, meet *DG* in *G*.

Then, in the triangles *ABC*, *DEG*,
the two angles *BAC*, *ABC* are equal to the two angles *EDG*, *DEG* respectively,
and the side *AB* is equal to the side *DE*;

therefore the triangles ABC, DEG are equal in all respects,   [I. 26]
so that the side AC is equal to the side DG,
and the angle ACB is equal to the angle DGE.

Again, since AC is equal to DF as well as to DG,

DF is equal to DG,

and therefore the angles DFG, DGF are equal.

But the angle DFE is supplementary to the angle DFG; and the angle DGF was proved equal to the angle ACB;

therefore the angle DFE is supplementary to the angle ACB.

If it is desired to avoid the ambiguity and secure that the triangles may be congruent, we can introduce the necessary conditions into the enunciation, on the analogy of Eucl. VI. 7.

If two triangles have two sides of the one equal to two sides of the other respectively, and the angles opposite to a pair of equal sides equal, then, if the angles opposite to the other pair of equal sides are both acute, or both obtuse, or if one of them is a right angle, the two triangles are equal in all respects.

The proof of the three cases (by *reductio ad absurdum*) was given by Todhunter.

PROPOSITION 27.

If a straight line falling on two straight lines make the alternate angles equal to one another, the straight lines will be parallel to one another.

For let the straight line EF falling on the two straight
5 lines AB, CD make the alternate angles AEF, EFD equal to one another;

I say that AB is parallel to CD.

For, if not, AB, CD when produced will meet either in the direction
10 of B, D or towards A, C.

Let them be produced and meet, in the direction of B, D, at G.

Then, in the triangle GEF,

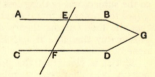

the exterior angle AEF is equal to the interior and opposite
15 angle EFG:

which is impossible.   [I. 16]

Therefore AB, CD when produced will not meet in the direction of B, D.

Similarly it can be proved that neither will they meet
20 towards A, C.

But straight lines which do not meet in either direction
are parallel;   [Def. 23]
    therefore *AB* is parallel to *CD*.

Therefore etc.

Q. E. D.

1. **falling on two straight lines**, εἰς δύο εὐθείας ἐμπίπτουσα, the phrase being the same
as that used in Post. 5, meaning a *transversal*.

2. **the alternate angles**, αἱ ἐναλλὰξ γωνίαι.   Proclus (p. 357, 9) explains that Euclid
uses the word *alternate* (or, more exactly, *alternately*, ἐναλλάξ) in two connexions, (1) of a
certain transformation of a proportion, as in Book V. and the arithmetical Books, (2) as here,
of certain of the angles formed by parallels with a straight line crossing them.   *Alternate*
angles are, according to Euclid as interpreted by Proclus, those which are not on the same
side of the transversal, and are not adjacent, but are separated by the transversal, both being
within the parallels but one "above" and the other "below."   The meaning is natural
enough if we imagine the four internal angles to be taken in cyclic order and *alternate* angles
to be any two of them not successive but separated by one angle of the four.

9. **in the direction of B, D or towards A, C**, literally "towards the *parts B, D* or
towards *A, C*," ἐπὶ τὰ B, Δ μέρη ἢ ἐπὶ τὰ A, Γ.

With this proposition begins the second section of the first Book.   Up
to this point Euclid has dealt mainly with triangles, their construction
and their properties in the sense of the relation of their parts, the sides and
angles, to one another, and the comparison of different triangles in respect of
their parts, and of their area in the particular cases where they are congruent.

The second section leads up to the third, in which we pass to relations
between the areas of triangles, parallelograms and squares, the special feature
being a new conception of *equality* of areas, equality not dependent on
congruence.   This whole subject requires the use of parallels.   Consequently
the second section beginning at I. 27 establishes the theory of parallels,
introduces the cognate matter of the equality of the sum of the angles of a
triangle to two right angles (I. 32), and ends with two propositions forming the
transition to the third section, namely I. 33, 34, which introduce the parallelogram for the first time.

Aristotle on parallels.

We have already seen reason to believe that Euclid's personal contribution
to the subject was nothing less than the formulation of the famous Postulate
5 (see the notes on that Postulate and on Def. 23), since Aristotle indicates
that the then current theory of parallels contained a *petitio principii*, and
presumably therefore it was Euclid who saw the defect and proposed the
remedy.

But it is clear that the propositions I. 27, 28 were contained in earlier
text-books.   They were familiar to Aristotle, as we may judge from two
interesting passages.

(1) In *Anal. Post.* I. 5 he is explaining that a scientific demonstration
must not only prove a fact of every individual of a class (κατὰ παντός) but
must prove it primarily and generally true (πρῶτον καθόλου) of the *whole* of
the class as one; it will not do to prove it first of one part, then of another
part, and so on, until the class is exhausted.   He illustrates this (74 a 13—16)
by a reference to parallels : "If then one were to show that right (angles) do
not meet, the proof of this might be thought to depend on the fact that this
is true of all (pairs of actual) right angles.   But this is not so, inasmuch as
the result does not follow because (the two angles are) equal (to two right

angles) *in the particular way* [i.e. because each is a right angle], but by virtue of their being equal (to two right angles) in any way whatever [i.e. because the *sum* only needs to be equal to two right angles, and the angles themselves may vary as much as we please subject to this]."

(2) The second passage has already been quoted in the note on Def. 23 : "there is nothing surprising in different hypotheses leading to the same false conclusion ; e.g. the conclusion that parallels meet might equally be drawn from either of the assumptions (*a*) that the interior (angle) is greater than the exterior or (*b*) that the sum of the angles of a triangle is greater than two right angles" (*Anal. Prior.* II. 17, 66 a 11—15).

I do not quite concur in the interpretation which Heiberg places upon these passages (*Mathematisches zu Aristoteles*, pp. 18—19). He says, first, that the allusion to the "interior angle" being "greater than the exterior" in the second passage shows that the reference in the first passage must be to Eucl. I. 28 and not to I. 27, and he therefore takes the words ὅτι ὡδὶ ἴσαι in the first passage (which I have translated "because the two angles are equal to two right angles in the particular way") as meaning "because the angles, viz. the *exterior* and the *interior*, are equal in the particular way." He also takes αἱ ὀρθαὶ οὐ συμπίπτουσι (which I have translated "right angles do not meet," an expression quite in Aristotle's manner) to mean "perpendicular *straight lines* do not meet"; this is very awkward, especially as he is obliged to supply *angles* with ἴσαι in the next sentence.

But I think that the first passage certainly refers to I. 28, although I do not think that the alternative (*a*) in the second passage suggests it. This alternative may, I think, equally with the alternative (*b*) refer to I. 27. That proposition is proved by *reductio ad absurdum* based on the fact that, if the straight lines do meet, they must form a *triangle*, in which case the exterior angle must be greater than the interior (while according to the hypothesis these angles are equal). It is true that Aristotle speaks of the hypothesis that the *interior* angle is greater than the *exterior* ; but after all Aristotle had only to state *some* incorrect hypothesis. It is of course only in connexion with straight lines *meeting*, as the hypothesis in I. 27 makes them, that the alternative (*b*) about the sum of the angles of a triangle could come in, and alternative (*a*) implies alternative (*b*).

It seems clear then from Aristotle that I. 27, 28 at least are pre-Euclidean, and that it was only in I. 29 that Euclid made a change by using his Postulate.

De Morgan observes that I. 27 is a *logical* equivalent of I. 16. Thus, if *A* means "straight lines forming a triangle with a transversal," *B* "straight lines making angles with a transversal on the same side which are together less than two right angles," we have

All A is B,

and it follows *logically* that

All not-B is not-A.

PROPOSITION 28.

If a straight line falling on two straight lines make the exterior angle equal to the interior and opposite angle on the same side, or the interior angles on the same side equal to two right angles, the straight lines will be parallel to one another.

For let the straight line *EF* falling on the two straight lines *AB*, *CD* make the exterior angle *EGB* equal to the interior and opposite angle *GHD*, or the interior angles on the same side, namely *BGH*, *GHD*, equal to two right angles;

I say that *AB* is parallel to *CD*.

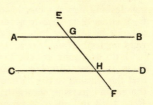

For, since the angle *EGB* is equal to the angle *GHD*, while the angle *EGB* is equal to the angle *AGH*,   [I. 15]
the angle *AGH* is also equal to the angle *GHD*;

and they are alternate;

                    therefore *AB* is parallel to *CD*.   [I. 27]

Again, since the angles *BGH*, *GHD* are equal to two right angles, and the angles *AGH*, *BGH* are also equal to two right angles,   [I. 13]
the angles *AGH*, *BGH* are equal to the angles *BGH*, *GHD*.

Let the angle *BGH* be subtracted from each;

therefore the remaining angle *AGH* is equal to the remaining angle *GHD*;

and they are alternate;

                    therefore *AB* is parallel to *CD*.   [I. 27]

          Therefore etc.

                                        Q. E. D.

One criterion of parallelism, the equality of alternate angles, is given in I. 27; here we have two more, each of which is easily reducible, and is actually reduced, to the other.

Proclus observes (pp. 358—9) that Euclid could have stated six criteria as well as three, by using, in addition, other pairs of angles in the figure (not adjacent) of which it could be predicated that the two angles are equal or that their sum is equal to two right angles. A natural division is to consider, first the pairs which are on the same side of the transversal, and secondly the pairs which are on different sides of it.

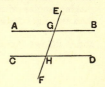

Taking (1) the possible pairs on the *same* side, we may have a pair consisting of

(*a*) two internal angles, viz. the pairs (*BGH*, *GHD*) and (*AGH*, *GHC*);

(*b*) two external angles, viz. the pairs (*EGB*, *DHF*) and (*EGA*, *CHF*);

(*c*) one external and one internal angle, viz. the pairs (*EGB*, *GHD*), (*FHD*, *HGB*), (*EGA*, *GHC*) and (*FHC*, *HGA*).

And (2) the possible pairs on *different* sides of the transversal may consist respectively of

(*a*) two internal angles, viz. the pairs (*AGH*, *GHD*) and (*CHG*, *HGB*);

(*b*) two external angles, viz. the pairs (*AGE*, *DHF*) and (*EGB*, *CHF*);

(*c*) one external and one internal, viz. the pairs (*AGE*, *GHD*), (*EGB*, *GHC*), (*FHC*, *HGB*) and (*FHD*, *HGA*).

The angles are equal in the pairs (1) (*c*), (2) (*a*) and (2) (*b*), and the sum is equal to two right angles in the case of the pairs (1) (*a*), (1) (*b*) and (2) (*c*). For his criteria Euclid selects the cases (2) (*a*) [I. 27] and (1) (*c*), (1) (*a*) [I. 28], leaving out the other three, which are of course equivalent but are not quite so easily expressed.

From Proclus' note on I. 28 (p. 361) we learn that one Aigeias (? Aineias) of Hierapolis wrote an epitome or abridgment of the *Elements*. This seems to be the only mention of this editor and his work; and they are only mentioned as having combined Eucl. I. 27, 28 into one proposition. To do this, or to make the three hypotheses the subject of *three* separate theorems, would, Proclus thinks, have been more natural than to deal with them, as Euclid does, in two propositions. Proclus has no suggestion for explaining Euclid's arrangement unless the ground were that I. 27 deals with angles on different sides, I. 28 with angles on one and the same side, of the transversal. But may not the reason have been one of convenience, namely that the criterion of I. 27 is that actually used to prove parallelism, and is moreover the basis of the construction of parallels in I. 31, while I. 28 only reduces the other two hypotheses to that of I. 27, so that precision of reference, as well as clearness of exposition, is better secured by the arrangement adopted?

PROPOSITION 29.

A straight line falling on parallel straight lines makes the alternate angles equal to one another, the exterior angle equal to the interior and opposite angle, and the interior angles on the same side equal to two right angles.

5   For let the straight line *EF* fall on the parallel straight lines *AB*, *CD* ;

I say that it makes the alternate angles *AGH*, *GHD* equal, the exterior angle *EGB* equal to the interior and opposite angle *GHD*, and the interior angles on the same side, namely *BGH*, *GHD*, equal to two right angles.

For, if the angle *AGH* is unequal to the angle *GHD*, one of them is greater.

Let the angle *AGH* be greater.

15   Let the angle *BGH* be added to each ;

therefore the angles *AGH*, *BGH* are greater than the angles *BGH*, *GHD*.

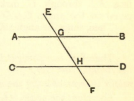

20   But the angles *AGH*, *BGH* are equal to two right angles;
[I. 13]
        therefore the angles *BGH*, *GHD* are less than two right angles.

        But straight lines produced indefinitely from angles less than two right angles meet;   [Post. 5]
25   therefore *AB*, *CD*, if produced indefinitely, will meet; but they do not meet, because they are by hypothesis parallel.

        Therefore the angle *AGH* is not unequal to the angle *GHD*,

                        and is therefore equal to it.

30   Again, the angle *AGH* is equal to the angle *EGB*; [I. 15]
        therefore the angle *EGB* is also equal to the angle *GHD*.   [C. N. 1]

        Let the angle *BGH* be added to each;

        therefore the angles *EGB*, *BGH* are equal to the
35 angles *BGH*, *GHD*.   [C. N. 2]

        But the angles *EGB*, *BGH* are equal to two right angles;
[I. 13]
        therefore the angles *BGH*, *GHD* are also equal to two right angles.

        Therefore etc.   Q. E. D.

23. **straight lines produced indefinitely from angles less than two right angles,**
αἱ δὲ ἀπ᾽ ἐλασσόνων ἢ δύο ὀρθῶν ἐκβαλλόμεναι εἰς ἄπειρον συμπίπτουσιν, a variation from the more explicit language of Postulate 5. A good deal is left to be understood, namely that the straight lines begin from points at which they meet a transversal, and make with it internal angles on the same side the sum of which is less than two right angles.

26. **because they are by hypothesis parallel,** literally "because they are supposed parallel," διὰ τὸ παραλλήλους αὐτὰς ὑποκεῖσθαι.

Proof by "Playfair's" axiom.

        If, instead of Postulate 5, it is preferred to use "Playfair's" axiom in the proof of this proposition, we proceed thus.

        To prove that the alternate angles *AGH*, *GHD* are equal.

        If they are not equal, draw another straight line *KL* through *G* making the angle *KGH* equal to the angle *GHD*.

        Then, since the angles *KGH*, *GHD* are equal,

                KL is parallel to *CD*.   [I. 27]

        Therefore *two straight lines* KL, AB *intersecting at* G *are both parallel to the straight line* CD :
which is impossible (by the axiom).

        Therefore the angle *AGH* cannot but be equal to the angle *GHD*.

        The rest of the proposition follows as in Euclid.

Proof of Euclid's Postulate 5 from "Playfair's" axiom.

Let *AB, CD* make with the transversal *EF* the angles *AEF, EFC*
together less than two right angles.

To prove that *AB, CD* meet towards *A, C.*

Through *E* draw *GH* making with *EF* the angle
GEF equal (and alternate) to the angle *EFD.*

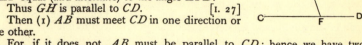

Thus *GH* is parallel to *CD.*   [I. 27]

Then (1) *AB* must meet *CD* in one direction or
the other.

For, if it does not, *AB* must be parallel to *CD;* hence we have two
straight lines *AB, GH* intersecting at *E* and both parallel to *CD*:
which is impossible.

Therefore *AB, CD* must meet.

(2) Since *AB, CD* meet, they must form a triangle with *EF.*

But in any triangle any two angles are together less than two right angles.

Therefore the angles *AEF, EFC* (which are less than two right angles),
and not the angles *BEF, EFD* (which are together greater than two right
angles, by I. 13), are the angles of the triangle ;

that is, *EA, FC* meet in the direction of *A, C,* or on the side of *EF* on
which are the angles together less than two right angles.

The usual course in modern text-books which use "Playfair's" axiom in
lieu of Euclid's Postulate is apparently to prove I. 29 by means of the axiom,
and then Euclid's Postulate by means of I. 29.

De Morgan would introduce the proof of Postulate 5 by means of
"Playfair's" axiom *before* I. 29, and would therefore apparently prove I. 29 as
Euclid does, without any change.

As between Euclid's Postulate 5 and "Playfair's" axiom, it would appear
that the tendency in modern text-books is rather in favour of the latter.
Thus, to take a few noteworthy foreign writers, we find that Rausenberger
stands almost alone in using Euclid's Postulate, while Hilbert, Henrici and
Treutlein, Rouché and De Comberousse, Enriques and Amaldi all use
"Playfair's" axiom.

Yet the case for preferring Euclid's Postulate is argued with some force by
Dodgson (*Euclid and his modern Rivals*, pp. 44—6). He maintains (1) that
"Playfair's" axiom in fact involves Euclid's Postulate, but at the same time
involves *more* than the latter, so that, to that extent, it is a needless strain on
the faith of the learner. This is shown as follows.

Given *AB, CD* making with *EF* the angles *AEF, EFC* together less than
two right angles, draw *GH* through *E* so that the angles *GEF, EFC* are
together *equal* to two right angles.

Then, by I. 28, *GH, CD* are "separational."

We see then that any lines which have the property (*a*) that they make
with a transversal angles less than two right angles have also the property (*β*)
that one of them intersects a straight line which is "separational" from
the other.

Now Playfair's axiom asserts that the lines which have property (*β*) meet
if produced : for, if they did not, we should have two intersecting straight
lines both "separational" from a third, which is impossible.

We then argue that lines having property (*a*) meet because lines having
property (*a*) are lines having property (*β*). But we do not know, until we
have proved I. 29, that all pairs of lines having property (*β*) have also property

(a). For anything we know to the contrary, class (β) *may* be greater than class (a). Hence, if you assert anything of class (β), the logical effect is more extensive than if you assert it of class (a); for you assert it, not only of that portion of class (β) which is known to be included in class (a), but also of the unknown (but possibly existing) portion which is *not* so included.

(2) Euclid's Postulate puts before the beginner clear and *positive* conceptions, a pair of straight lines, a transversal, and two angles together less than two right angles, whereas "Playfair's" axiom requires him to realise a pair of straight lines which never meet though produced to infinity: a *negative* conception which does not convey to the mind any clear notion of the relative position of the lines. And (p. 68) Euclid's Postulate gives a direct criterion for judging that two straight lines meet, a criterion which is constantly required, e.g. in I. 44. It is true that the Postulate can be *deduced* from "Playfair's" axiom, but editors frequently omit to deduce it, and then tacitly assume it afterwards: which is the least justifiable course of all.

PROPOSITION 30.

Straight lines parallel to the same straight line are also parallel to one another.

Let each of the straight lines *AB*, *CD* be parallel to *EF*; I say that *AB* is also parallel to *CD*.

5   For let the straight line *GK* fall upon them.

Then, since the straight line *GK* has fallen on the parallel straight lines *AB*, *EF*,

10   the angle *AGK* is equal to the angle *GHF*.   [I. 29]

Again, since the straight line *GK* has fallen on the parallel straight lines *EF*, *CD*,

        the angle *GHF* is equal to the angle *GKD*.   [I. 29]

15   But the angle *AGK* was also proved equal to the angle *GHF*;

        therefore the angle *AGK* is also equal to the angle *GKD*;   [C. N. 1]

and they are alternate.

20   Therefore *AB* is parallel to *CD*.

                                                        Q. E. D.

20. The usual *conclusion* in general terms ("Therefore etc.") repeating the enunciation is, curiously enough, wanting at the end of this proposition.

The proposition is, as De Morgan points out, the *logical* equivalent of "Playfair's" axiom. Thus, if *X* denote "pairs of straight lines intersecting one

another," Y "pairs of straight lines parallel to one and the same straight line,"
we have

No X is Y,

and it follows logically that

No Y is X.

De Morgan adds that a proposition is much wanted about parallels (or
perpendiculars) to two straight lines respectively making the same angles with
one another as the latter do. The proposition may be enunciated thus :

If the sides of one angle be respectively (1) *parallel or* (2) *perpendicular to
the sides of another angle, the two angles are either
equal or supplementary.*

(1)   Let *DE* be parallel to *AB* and *GEF* parallel
to *BC*.

To prove that the angles *ABC*, *DEG* are equal
and the angles *ABC*, *DEF* supplementary.

Produce *DE* to meet *BC* in *H*.

Then [I. 29] the angle *DEG* is equal to the angle
DHC,

and the angle *ABC* is equal to the angle *DHC*.

Therefore the angle *DEG* is equal to the angle *ABC*; whence also the
angle *DEF* is supplementary to the angle *ABC*.

(2)   Let *ED* be perpendicular to *AB*, and *GEF* perpendicular to *BC*.

To prove that the angles *ABC*, *DEG* are
equal, and the angles *ABC*, *DEF* supplementary.

Draw *ED′* at right angles to *ED* on the side
of it opposite to *B*, and draw *EG′* at right angles
to *EF* on the side of it opposite to *B*.

Then, since the angles *BDE*, *DED′*, being
right angles, are equal,

ED′ is parallel to *BA*.   [I. 27]

Similarly *EG′* is parallel to *BC*.

Therefore [Part (1)] the angle *D′EG′* is equal to the angle *ABC*.

But, the right angle *DED′* being equal to the right angle *GEG′*, if the
common angle *GED′* be subtracted,

the angle *DEG* is equal to the angle *D′EG′*.

Therefore the angle *DEG* is equal to the angle *ABC*; and hence the
angle *DEF* is supplementary to the angle *ABC*.

PROPOSITION 31.

*Through a given point to draw a straight line parallel to a
given straight line.*

Let *A* be the given point, and *BC* the given straight
line ;

thus it is required to draw through the point *A* a straight
line parallel to the straight line *BC*.

Let a point D be taken at random on BC, and let AD be
joined; on the straight line DA,
and at the point A on it, let the
angle DAE be constructed equal
to the angle ADC [I. 23]; and let the
straight line AF be produced in a
straight line with EA.

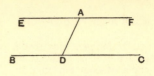

Then, since the straight line AD falling on the two
straight lines BC, EF has made the alternate angles EAD,
ADC equal to one another,

therefore EAF is parallel to BC.   [I. 27]

Therefore through the given point A the straight line
EAF has been drawn parallel to the given straight line BC.

Q. E. F.

Proclus rightly remarks (p. 376, 14—20) that, as it is implied in I. 12
that only one perpendicular can be drawn to a straight line from an external
point, so here it is implied that only one straight line can be drawn through a
point parallel to a given straight line. The construction, be it observed,
depends only upon I. 27, and might therefore have come directly after that
proposition. Why then did Euclid postpone it until after I. 29 and I. 30?
Presumably because he considered it necessary, before giving the construction,
to place beyond all doubt the fact that only one such parallel can be drawn.
Proclus infers this fact from I. 30; for, he says, if two straight lines could be
drawn through one and the same point parallel to the same straight line, the two
straight lines would be *parallel*, though intersecting at the given point: which
is impossible.   I think it is a fair inference that Euclid would have considered
it necessary to justify the assumption that only one parallel can be drawn
by some such argument, and that he deliberately determined that his own
assumption was more appropriate to be made the subject of a Postulate
than the assumption of the uniqueness of the parallel.

PROPOSITION 32.

*In any triangle, if one of the sides be produced, the exterior
angle is equal to the two interior and opposite angles, and the
three interior angles of the triangle are equal to two right
angles.*

Let ABC be a triangle, and let one side of it BC be
produced to D;

I say that the exterior angle ACD is equal to the two
interior and opposite angles CAB, ABC, and the three
interior angles of the triangle ABC, BCA, CAB are equal
to two right angles.

For let *CE* be drawn through the point *C* parallel to the straight line *AB*.   [I. 31]

Then, since *AB* is parallel to *CE*,

and *AC* has fallen upon them,

the alternate angles *BAC*, *ACE* are equal to one another.   [I. 29]

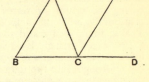

Again, since *AB* is parallel to *CE*,

and the straight line *BD* has fallen upon them,

the exterior angle *ECD* is equal to the interior and opposite angle *ABC*.   [I. 29]

But the angle *ACE* was also proved equal to the angle *BAC*;

therefore the whole angle *ACD* is equal to the two interior and opposite angles *BAC*, *ABC*.

Let the angle *ACB* be added to each;

therefore the angles *ACD*, *ACB* are equal to the three angles *ABC*, *BCA*, *CAB*.

But the angles *ACD*, *ACB* are equal to two right angles;
[I. 13]
therefore the angles *ABC*, *BCA*, *CAB* are also equal to two right angles.

Therefore etc.

Q. E. D.

This theorem was discovered in the very early stages of Greek geometry. What we know of the history of it is gathered from three allusions found in Eutocius, Proclus and Diogenes Laertius respectively.

1. Eutocius at the beginning of his commentary on the *Conics* of Apollonius (ed. Heiberg, Vol. II. p. 170) quotes Geminus as saying that "the ancients (οἱ ἀρχαῖοι) investigated the theorem of the two right angles in each individual species of triangle, first in the equilateral, again in the isosceles, and afterwards in the scalene triangle, and later geometers demonstrated the general theorem to the effect that in *any* triangle the three interior angles are equal to two right angles."

2. Now, according to Proclus (p. 379, 2—5), Eudemus the Peripatetic refers the discovery of this theorem to the Pythagoreans and gives what he affirms to be their demonstration of it. This demonstration will be given below, but it should be remarked that it is general, and therefore that the "later geometers" spoken of by Geminus were presumably the Pythagoreans, whence it appears that the "ancients" contrasted with them must have belonged to the time of Thales, if they were not his Egyptian instructors.

3. That the truth of the theorem was known to Thales might also be inferred from the statement of Pamphile (quoted by Diogenes Laertius, I. 24—5, p. 6, ed. Cobet) that "he, having learnt geometry from the

Egyptians, was the first to inscribe a right-angled triangle in a circle and sacrificed an ox" (on the strength of it); in other words, he discovered that the angle in a semicircle is a right angle.   No doubt, when this fact was once discovered (*empirically*, say), the consideration of the two isosceles triangles having the centre for vertex and the sides of the right angle for bases respectively, with the help of the theorem of Eucl. I. 5, also known to Thales, would easily lead to the conclusion that the sum of the angles of a *right-angled* triangle is equal to two right angles, and it could be readily inferred that the angles of *any* triangle were likewise equal to two right angles (by resolving it into two right-angled triangles).   But it is not easy to see how the property of the angle in a semicircle could be *proved* except (in the reverse order) by means of the equality of the sum of the angles of a *right-angled* triangle to two right angles; and hence it is most natural to suppose, with Cantor, that Thales proved it (if he did prove it) practically as Euclid does in III. 31, i.e. by means of I. 32 as applied to *right-angled* triangles at all events.

If the theorem of I. 32 was proved before Thales' time, or by Thales himself, by the stages indicated in the note of Geminus, we may be satisfied that the reconstruction of the argument of the older proof by Hankel (pp. 96—7) and Cantor (I_3, pp. 143—4) is not far wrong.   First, it must have been observed that six angles equal to an angle of an equilateral triangle would, if placed adjacent to one another round a common vertex, fill up the whole space round that vertex.   It is true that Proclus attributes to the Pythagoreans the general theorem that only three kinds of regular polygons, the equilateral triangle, the square and the regular hexagon, can fill up the entire space round a point, but the practical knowledge that equilateral triangles have this property could hardly have escaped the Egyptians, whether they made floors with tiles in the form of equilateral triangles or regular hexagons (Allman, *Greek Geometry from Thales to Euclid*, p. 12) or joined the ends of adjacent radii of a figure like the six-spoked wheel, which was their common form of wheel from the time of Ramses II. of the nineteenth Dynasty, say 1300 B.C. (Cantor, I_3, p. 109).   It would then be clear that six angles equal to an angle of an equilateral triangle are equal to four right angles, and therefore that the three angles of an equilateral triangle are equal to two right angles.   (It would be as clear or clearer, from observation of a square divided into two triangles by a diagonal, that an isosceles right-angled triangle has each of its equal angles equal to half a right angle, so that an isosceles right-angled triangle must have the sum of its angles equal to two right angles.)   Next, with regard to the equilateral triangle, it could not fail be observed that, if AD were drawn from the vertex A perpendicular to the base BC, each of the two right-angled triangles so formed would have the sum of its angles equal to two right angles; and this would be confirmed by completing the rectangle $ADCE$, when it would be seen that the rectangle (with its angles equal to four right angles) was divided by its diagonal into two equal triangles, each of which had the sum of its angles equal to two right angles.   Next it would be inferred, as the result of drawing the diagonal of *any* rectangle and observing the equality of the triangles forming the two halves, that the sum of the angles of *any* right-angled triangle is equal to two right angles, and hence (the two congruent right-angled triangles being then placed so as to form one isosceles triangle) that the same is true of *any isosceles* triangle.   Only the last step remained, namely that of observing that *any* triangle could be regarded as the half of a rectangle (drawn as indicated in the next figure), or simply that any triangle could be divided into two right-angled triangles, whence it would be inferred that in general the sum of the angles of any triangle is equal to two right angles.

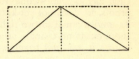

Such would be the probabilities if we could absolutely rely upon the statements attributed to Pamphile and Geminus respectively. But in fact there is considerable ground for doubt in both cases.

1. Pamphile's story of the sacrifice of an ox by Thales for joy at his discovery that the angle in a semicircle is a right angle is too suspiciously like the similar story told with reference to Pythagoras and his discovery of the theorem of Eucl. I. 47 (Proclus, p. 426, 6—9). And, as if this were not enough, Diogenes Laertius immediately adds that "others, among whom is Apollodorus the calculator (ὁ λογιστικός), say it was Pythagoras" (sc. who "inscribed the right-angled triangle in a circle"). Now Pamphile lived in the reign of Nero (A.D. 54—68) and therefore some 700 years after the birth of Thales (about 640 B.C.). I do not know on what Max Schmidt bases his statement (*Kulturhistorische Beiträge zur Kenntnis des griechischen und römischen Altertums*, 1906, p. 31) that "other, *much older*, sources name Pythagoras as the discoverer of the said proposition," because nothing more seems to be known of Apollodorus than what is stated here by Diogenes Laertius. But it would at least appear that Apollodorus was only one of several authorities who attributed the proposition to Pythagoras, while Pamphile is alone mentioned as referring it to Thales. Again, the connexion of Pythagoras with the investigation of the right-angled triangle makes it *a priori* more likely that it would be he who would discover its relation to a semicircle. On the whole, therefore, the attribution to Thales would seem to be more than doubtful.

2. As regards Geminus' account of the three stages through which the proof of the theorem of I. 32 passed, we note, first, that it is certainly not confirmed by Eudemus, who referred to the Pythagoreans the *discovery* of the theorem that the sum of the angles of *any* triangle is equal to two right angles and says nothing about any gradual stages by which it was proved. Secondly, it must be admitted, I think, that in the evolution of the proof as reconstructed by Hankel the middle stage is rather artificial and unnecessary, since, once it is proved that *any right-angled* triangle has the sum of its angles equal to two right angles, it is just as easy to pass at once to any *scalene* triangle (which is decomposable into two *unequal* right-angled triangles) as to the isosceles triangle made up of two congruent right-angled triangles. Thirdly, as Heiberg has recently pointed out (*Mathematisches zu Aristoteles*, p. 20), it is quite possible that the statement of Geminus from beginning to end is simply due to a misapprehension of a passage of Aristotle (*Anal. Post.* I. 5, 74 a 25). Aristotle is illustrating his contention that a property is not scientifically proved to belong to a class of things unless it is proved to belong *primarily* (πρῶτον) and *generally* (καθόλου) to the *whole* of the class. His first illustration relates to parallels making with a transversal angles on the same side together equal to two right angles, and has been quoted above in the note on I. 27 (pp. 308—9). His second illustration refers to the transformation of a proportion *alternando*, which (he says) "used at one time to be proved separately" for numbers, lines, solids, and times, although it admits of being proved of all at once by one demonstration. The third illustration is: "For the same reason, even *if one should prove* (οὐδ᾽ ἄν τις δείξῃ) with reference to

each (sort of) triangle, the equilateral, scalene and isosceles, separately, that each has its angles equal to two right angles, either by one proof or by different proofs, he does not yet know that *the triangle*, i.e. the triangle *in general*, has its angles equal to two right angles, except in a sophistical sense, even though there exists no triangle other than triangles of the kinds mentioned. For he knows it, not *quâ* triangle, nor of *every* triangle, except in a numerical sense (κατ' ἀριθμόν); he does not know it *notionally* (κατ' εἶδος) of every triangle, even though there be actually no triangle which he does not know."

The difference between the phrase "used at one time to be proved" in the second illustration and "if any one should prove" in the third appears to indicate that, while the former referred to a historical fact, the latter does not; the reference to a person who should prove the theorem of I. 32 for the three kinds of triangle separately, and then claim that he had proved it generally, states a purely hypothetical case, a mere illustration. Yet, coming after the historical fact stated in the preceding illustration, it might not unnaturally give the impression, at first sight, that it was historical too.

On the whole, therefore, it would seem that we cannot safely go behind the dictum of Eudemus that the discovery and proof of the theorem of I. 32 in all its generality were Pythagorean. This does not however preclude its having been discovered by stages such as those above set out after Hankel and Cantor. Nor need it be doubted that Thales and even his Egyptian instructors had advanced some way on the same road, so far at all events as to see that in an equilateral triangle, and in an isosceles right-angled triangle, the sum of the angles is equal to two right angles.

The Pythagorean proof.

This proof, handed down by Eudemus (Proclus, p. 379, 2—15), is no less elegant than that given by Euclid, and is a natural development from the last figure in the reconstructed argument of Hankel. It would be seen, after the theory of parallels was added to geometry, that the actual drawing of the perpendicular and the complete rectangle on BC as base was unnecessary, and that the parallel to BC through A was all that was required.

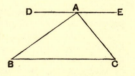

Let ABC be a triangle, and through A draw DE parallel to BC.   [I. 31]
Then, since BC, DE are parallel,
the alternate angles DAB, ABC are equal,   [I. 29]
and so are the alternate angles EAC, ACB also.

Therefore the angles ABC, ACB are together equal to the angles DAB, EAC.

Add to each the angle BAC;

therefore the sum of the angles ABC, ACB, BAC is equal to the sum of the angles DAB, BAC, CAE, that is, to two right angles.

Euclid's proof pre-Euclidean.

The theorem of I. 32 is Aristotle's favourite illustration when he wishes to refer to some truth generally acknowledged, and so often does it occur that it is often indicated by two or three words in themselves hardly intelligible, e.g. τὸ δυσὶν ὀρθαῖς (*Anal. Post.* I. 24, 85 b 5) and ὑπάρχει παντὶ τριγώνῳ τὸ δύο (*ibid.* 85 b 11).

One passage (*Metaph.* 1051 a 24) makes it clear, as Heiberg (*op. cit.*

p. 19) acutely observes, that in the proof as Aristotle knew it Euclid's construction was used. "Why does the triangle make up two right angles? Because *the angles about one point* are equal to two right angles. If then the parallel to the side had been *drawn up* (ἀνῆκτο), the fact would at once have been clear from merely looking at the figure." The words "the angles about one point" would equally fit the Pythagorean construction, but "drawn *upwards*" applied to the parallel to a side can only indicate Euclid's.

Attempts at proof independently of parallels.

The most indefatigable worker on these lines was Legendre, and a sketch of his work has been given in the note on Postulate 5 above.

One other attempted proof needs to be mentioned here because it has found much favour. I allude to

Thibaut's method.

This appeared in Thibaut's *Grundriss der reinen Mathematik*, Göttingen (2 ed. 1809, 3 ed. 1818), and is to the following effect.

Suppose *CB* produced to *D*, and let *BD* (produced to any necessary extent either way) revolve in one direction (say clockwise) first about *B* into the position *BA*, then about *A* into the position of *AC* produced both ways, and lastly about *C* into the position *CB* produced both ways.

The argument then is that the straight line *BD* has revolved through the sum of the three exterior angles of the triangle. But, since it has at the end of the revolution assumed a position in the same straight line with its original position, it must have revolved *through four right angles*.

Therefore the sum of the three exterior angles is equal to four right angles;

from which it follows that the sum of the three angles of the triangle is equal to two right angles.

But it is to be observed that the straight line *BD* revolves about *different points in it*, so that there is *translation* combined with *rotatory* motion, and it is necessary to assume as an axiom that the two motions are independent, and therefore that the *translation* may be neglected.

Schumacher (letter to Gauss of 3 May, 1831) tried to represent the rotatory motion graphically in a second figure as mere motion round a point; but Gauss (letter of 17 May, 1831) pointed out in reply that he really assumed, without proving it, a proposition to the effect that "If two straight lines (1) and (2) which cut one another make angles *A*, *A''* with a straight line (3) cutting both of them, and if a straight line (4) in the same plane is likewise cut by (1) at an angle *A'*, then (4) will be cut by (2) at the angle *A''*. But this proposition not only needs proof, but we may say that it is, in essence, the very proposition to be proved" (see Engel and Stäckel, *Die Theorie der Parallellinien von Euklid bis auf Gauss*, 1895, p. 230).

How easy it is to be deluded in this way is plainly shown by Proclus' attempt on the same lines. He says (p. 384, 13—21) that the truth of the theorem is borne in upon us by the help of "common notions" only. "For, if we conceive a straight line with two perpendiculars drawn to it at its extremities, and if we then suppose the perpendiculars to (revolve about their feet and) approach one another, so as to form a triangle, we see that, to the extent to which they converge, they diminish the right angles which they made with the straight line, so that the amount taken from the right angles is also the amount added to the vertical angle of the triangle, and the three angles are necessarily made equal to two right angles." But a moment's reflection shows that, so far from being founded on mere "common notions," the supposed proof assumes, to begin with, that, if the perpendiculars approach one another ever so little, they will then form a triangle immediately, i.e., it assumes Postulate 5 itself; and the fact about the vertical angle can only be seen by means of the equality of the alternate angles exhibited by drawing a perpendicular from the vertex of the triangle to the base, i.e. a *parallel* to either of the original perpendiculars.

Extension to polygons.

The two important corollaries added to I. 32 in Simson's edition are given by Proclus; but Proclus' proof of the first is different from, and perhaps somewhat simpler than, Simson's.

1. *The sum of the interior angles of a convex rectilineal figure is equal to twice as many right angles as the figure has sides, less four.*

For let one angular point A be joined to all the other angular points with which it is not connected already.

The figure is then divided into triangles, and mere inspection shows

(1)   that the number of triangles is two less than the number of sides in the figure,

(2)   that the sum of the angles of all the triangles is equal to the sum of all the interior angles of the figure.

Since then the sum of the angles of each triangle is equal to two right angles the sum of the interior angles of the figure is equal to $2(n-2)$ right angles, i.e. $(2n-4)$ right angles, where n is the number of sides in the figure.

2. *The exterior angles of any convex rectilineal figure are together equal to four right angles.*

For the interior and exterior angles together are equal to $2n$ right angles, where n is the number of sides.

And the interior angles are together equal to $(2n-4)$ right angles.

Therefore the exterior angles are together equal to four right angles.

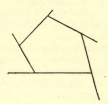

This last property is already quoted by Aristotle as true of all rectilineal figures in two passages (*Anal. Post.* I. 24, 85 b 38 and II. 17, 99 a 19).

PROPOSITION 33.

The straight lines joining equal and parallel straight lines (at the extremities which are) in the same directions (respectively) are themselves also equal and parallel.

Let AB, CD be equal and parallel, and let the straight
5 lines AC, BD join them (at the extremities which are) in the same directions (respectively);

I say that *AC, BD* are also equal and parallel.

Let *BC* be joined.

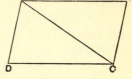

Then, since *AB* is parallel to *CD*,
10 and *BC* has fallen upon them,
　　　the alternate angles *ABC, BCD*
are equal to one another.　　　[I. 29]

And, since *AB* is equal to *CD*,
　　　and *BC* is common,
15　　the two sides *AB, BC* are equal to the two sides *DC, CB* ;
and the angle *ABC* is equal to the angle *BCD* ;
　　　therefore the base *AC* is equal to the base *BD*,
　　and the triangle *ABC* is equal to the triangle *DCB*,
and the remaining angles will be equal to the remaining angles
20 respectively, namely those which the equal sides subtend ;　[I. 4]
　　　therefore the angle *ACB* is equal to the angle *CBD*.

And, since the straight line *BC* falling on the two straight
lines *AC, BD* has made the alternate angles equal to one
another,

25　　　*AC* is parallel to *BD*.　　　　　[I. 27]

And it was also proved equal to it.

Therefore etc.　　　　　　　　　　　　Q. E. D.

1. joining…(at the extremities which are) in the same directions (respectively).
I have for clearness' sake inserted the words in brackets though they are not in the original
Greek, which has "joining…in the same directions" or "on the same sides," ἐπὶ τὰ αὐτὰ μέρη
ἐπιζευγνύουσαι. The expression "towards the same parts," though usage has sanctioned it,
is perhaps not quite satisfactory.

15. DC, CB and 18. DCB. The Greek has " *BC, CD* " and " *BCD* " in these places
respectively. Euclid is not always careful to write in corresponding order the letters denoting
corresponding points in congruent figures. On the contrary, he evidently prefers the alpha-
betical order, and seems to disdain to alter it for the sake of beginners or others who might
be confused by it. In the case of angles alteration is perhaps unnecessary ; but in the case
of triangles and pairs of corresponding sides I have ventured to alter the order to that which
the mathematician of to-day expects.

This proposition is, as Proclus says (p. 385, 5), the connecting link between
the exposition of the theory of parallels and the investigation of parallelograms.
For, while it only speaks of equal and parallel straight lines connecting those
ends of equal and parallel straight lines which are in the same directions, it
gives, without expressing the fact, the construction or origin of the parallelogram,
so that in the next proposition Euclid is able to speak of "parallelogrammic
areas" without any further explanation.

PROPOSITION 34.

In parallelogrammic areas the opposite sides and angles
are equal to one another, and the diameter bisects the areas.

Let *ACDB* be a parallelogrammic area, and *BC* its
diameter ;

5 I say that the opposite sides and angles of the parallelogram
ACDB are equal to one another, and the diameter *BC*
bisects it.

For, since *AB* is parallel to *CD*,
and the straight line *BC* has fallen
10 upon them,
the alternate angles *ABC*, *BCD*
are equal to one another.   [I. 29]

Again, since *AC* is parallel to *BD*,
and *BC* has fallen upon them,
15   the alternate angles *ACB*, *CBD* are equal to one
another.   [I. 29]

Therefore *ABC*, *DCB* are two triangles having the two
angles *ABC*, *BCA* equal to the two angles *DCB*, *CBD*
respectively, and one side equal to one side, namely that
20 adjoining the equal angles and common to both of them, *BC*;
therefore they will also have the remaining sides equal
to the remaining sides respectively, and the remaining angle
to the remaining angle ;   [I. 26]
therefore the side *AB* is equal to *CD*,
25   and *AC* to *BD*,
and further the angle *BAC* is equal to the angle *CDB*.

And, since the angle *ABC* is equal to the angle *BCD*,
and the angle *CBD* to the angle *ACB*,
the whole angle *ABD* is equal to the whole angle *ACD*.
          [C. N. 2]
30 And the angle *BAC* was also proved equal to the angle *CDB*.

Therefore in parallelogrammic areas the opposite sides
and angles are equal to one another.

I say, next, that the diameter also bisects the areas.

For, since *AB* is equal to *CD*,
35 and *BC* is common,
the two sides *AB*, *BC* are equal to the two sides *DC*, *CB*
respectively ;
and the angle *ABC* is equal to the angle *BCD* ;
therefore the base *AC* is also equal to *DB*,
40   and the triangle *ABC* is equal to the triangle *DCB*.   [I. 4]

Therefore the diameter *BC* bisects the parallelogram
ACDB.   Q. E. D.

1. It is to be observed that, when parallelograms have to be mentioned for the first time, Euclid calls them "**parallelogrammic areas**" or, more exactly, "**parallelogram**" areas (παραλληλόγραμμα χωρία). The meaning is simply areas bounded by parallel straight lines with the further limitation placed upon the term by Euclid that only *four-sided* figures are so called, although of course there are certain regular polygons which have opposite sides parallel, and which therefore might be said to be areas bounded by parallel straight lines. We gather from Proclus (p. 393) that the word "parallelogram" was first introduced by Euclid, that its use was suggested by I. 33, and that the formation of the word παραλληλόγραμμος (parallel-lined) was analogous to that of εὐθύγραμμος (straight-lined or rectilineal).

17, 18, 40. **DCB** and 36. **DC, CB**. The Greek has in these places "*BCD*" and "*CD, BC*" respectively. Cf. note on I. 33, lines 15, 18.

After specifying the particular kinds of parallelograms (squares and rhombi) in which the diagonals bisect the angles which they join, as well as the areas, and those (rectangles and rhomboids) in which the diagonals do not bisect the angles, Proclus proceeds (pp. 390 sqq.) to analyse this proposition with reference to the distinction in Aristotle's *Anal. Post.* (I. 4, 5, 73 a 21—74 b 4) between attributes which are only predicable of every individual thing (κατὰ παντός) in a class and those which are true of it *primarily* (τούτου πρώτου) and *generally* (καθόλου). We are apt, says Aristotle, to mistake a proof κατὰ παντός for a proof τούτου πρώτου καθόλου because it is either impossible to find a higher generality to comprehend all the particulars of which the predicate is true, or to find a name for it. (Part of this passage of Aristotle has been quoted above in the note on I. 32, pp. 319—320.)

Now, says Proclus, adapting Aristotle's distinction to *theorems*, the present proposition exhibits the distinction between theorems which are *general* and theorems which are *not general*. According to Proclus, the first part of the proposition stating that the opposite sides and angles of a parallelogram are equal is *general* because the property is only true of parallelograms; but the second part which asserts that the diameter bisects the area is *not general* because it does not include all the figures of which this property is true, e.g. circles and ellipses. Indeed, says Proclus, the first attempts upon problems seem usually to have been of this partial character (μερικώτεραι), and generality was only attained by degrees. Thus "the ancients, after investigating the fact that the diameter bisects an ellipse, a circle, and a parallelogram respectively, proceeded to investigate what was common to these cases," though "it is difficult to show what is common to an ellipse, a circle and a parallelogram."

I doubt whether the supposed distinction between the two parts of the proposition, in point of "generality," can be sustained. Proclus himself admits that it is presupposed that the subject of the proposition is a *quadrilateral*, because there are other figures (e.g. regular polygons of an even number of sides) besides parallelograms which have their opposite sides and angles equal; therefore the second part of the theorem is, in this respect, no more *general* than the other, and, if we are entitled to the tacit limitation of the theorem to quadrilaterals in one part, we are equally entitled to it in the other.

It would almost appear as though Proclus had drawn the distinction for the mere purpose of alluding to investigations by Greek geometers on the general subject of *diameters* of all sorts of figures; and it may have been these which brought the subject to the point at which Apollonius could say in the first definitions at the beginning of his *Conics* that "In *any bent line*, such as is in one plane, I give the name *diameter* to any straight line which, being drawn from the bent line, bisects all the straight lines (chords) drawn in the line parallel to any straight line." The term *bent line* (καμπύλη γραμμή) includes, e.g. in Archimedes, not only curves, but any composite line made up of straight lines and curves joined together in any manner.   It is of course clear that either diagonal of a parallelogram bisects all lines drawn within the parallelogram parallel to the other diagonal.

An-Nairīzī gives after I. 31 a neat construction for dividing a straight line into any number of equal parts (ed. Curtze, p. 74, ed. Besthorn-Heiberg, pp. 141—3) which requires only one measurement repeated, together with the properties of parallel lines including I. 33, 34.   As I. 33, 34 are assumed, I place the problem here.   The particular case taken is the problem of dividing a straight line into *three* equal parts.

Let *AB* be the given straight line.   Draw *AC, BD* at right angles to it on opposite sides.

An-Nairīzī takes *AC, BD* of the same length and then bisects *AC* at *E* and *BD* at *F.*   But of course it is even simpler to measure *AE, EC* along one perpendicular equal and of any length, and *BF, FD* along the other also equal and of the same length.

Join *ED, CF* meeting *AB* in *G, H* respectively.

Then shall *AG, GH, HB* all be equal.

Draw *HK* parallel to *AC,* or at right angles to *AB.*

Since now *EC, FD* are equal and parallel, *ED, CF* are equal and parallel.   [I. 33]

And *HK* was drawn parallel to *AC.*

Therefore *ECHK* is a parallelogram; whence *KH* is equal as well as parallel to *EC,* and therefore to *EA.*

The triangles *EAG, KHG* have now two angles respectively equal and the sides *AE, HK* equal.

Thus the triangles are equal in all respects, and

AG is equal to *GH.*

Similarly the triangles *KHG, FBH* are equal in all respects, and

GH is equal to *HB.*

If now we wish to extend the problem to the case where *AB* is to be divided into *n* parts, we have only to measure (*n*−1) successive equal lengths along *AC* and (*n*−1) successive lengths, each equal to the others, along *BD.* Then join the first point arrived at on *AC* to the last point on *BD,* the second on *AC* to the last but one on *BD,* and so on; and the joining lines cut *AB* in points dividing it into *n* equal parts.

PROPOSITION 35.

Parallelograms which are on the same base and in the same parallels are equal to one another.

Let *ABCD, EBCF* be parallelograms on the same base *BC* and in the same parallels *AF, BC*;

5 I say that *ABCD* is equal to the parallelogram *EBCF.*

For, since *ABCD* is a parallelogram,

AD is equal to *BC.*   [I. 34]

For the same reason also

EF is equal to BC,

so that *AD* is also equal to *EF*;   [*C. N.* 1]

and *DE* is common;

therefore the whole *AE* is equal to the whole *DF*.

[*C. N.* 2]

But *AB* is also equal to *DC*;   [I. 34]

therefore the two sides *EA, AB* are equal to the two sides *FD, DC* respectively,

and the angle *FDC* is equal to the angle *EAB*, the exterior to the interior;   [I. 29]

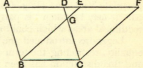

therefore the base *EB* is equal to the base *FC*,

and the triangle *EAB* will be equal to the triangle *FDC*.

[I. 4]

Let *DGE* be subtracted from each;

therefore the trapezium *ABGD* which remains is equal to the trapezium *EGCF* which remains.   [*C. N.* 3]

Let the triangle *GBC* be added to each;

therefore the whole parallelogram *ABCD* is equal to the whole parallelogram *EBCF*.   [*C. N.* 2]

Therefore etc.

Q. E. D.

21. **FDC.** The text has "*DFC*."

22. **Let DGE be subtracted.** Euclid speaks of the triangle *DGE* without any explanation that, in the case which he takes (where *AD, EF* have no point in common), *BE, CD* must meet at a point *G* between the two parallels. He allows this to appear from the figure simply.

Equality in a new sense.

It is important to observe that we are in this proposition introduced for the first time to a new conception of equality between figures. Hitherto we have had equality in the sense of *congruence* only, as applied to straight lines, angles, and even triangles (cf. I. 4). Now, without any explicit reference to any change in the meaning of the term, figures are inferred to be *equal* which are equal in *area* or in *content* but need not be of the same *form*. No *definition* of equality is anywhere given by Euclid; we are left to infer its meaning from the few *axioms* about "equal things." It will be observed that in the above proof the "equality" of two parallelograms on the same base and between the same parallels is inferred by the successive steps (1) of subtracting one and the same area (the triangle *DGE*) from two areas equal in the sense of *congruence* (the triangles *AEB, DFC*), and inferring that the remainders (the trapezia *ABGD, EGCF*) are "equal"; (2) of adding one and the same area (the triangle GBC) to each of the latter "equal" trapezia, and inferring the equality of the respective sums (the two given parallelograms).

As is well known, Simson (after Clairaut) slightly altered the proof in order to make it applicable to all the three possible cases. The alteration substituted *one* step of subtracting congruent areas (the triangles AEB, DFC) from one and the same area (the trapezium $ABCF$) for the *two* steps above shown of first subtracting and then adding a certain area.

While, in either case, nothing more is explicitly used than the axioms that, *if equals be added to equals, the wholes are equal* and that, *if equals be subtracted from equals, the remainders are equal,* there is the further *tacit* assumption that it is indifferent to *what part* or from *what part* of the same or equal areas the same or equal areas are added or subtracted. De Morgan observes that the postulate "an area taken from an area leaves the same area from whatever part it may be taken" is particularly important as the key to equality of non-rectilineal areas which could not be cut into coincidence geometrically.

Legendre introduced the word *equivalent* to express this wider sense of equality, restricting the term *equal* to things equal in the sense of congruent; and this distinction has been found convenient.

I do not think it necessary, nor have I the space, to give any account of the recent developments of the theory of equivalence on new lines represented by the researches of W. Bolyai, Duhamel, De Zolt, Stolz, Schur, Veronese, Hilbert and others, and must refer the reader to Ugo Amaldi's article *Sulla teoria dell' equivalenza* in *Questioni riguardanti le matematiche elementari,* I. (Bologna, 1912), pp. 145—198, and to Max Simon, *Über die Entwicklung der Elementar-geometrie im XIX. Jahrhundert* (Leipzig, 1906), pp. 115—120, with their full references to the literature of the subject. I may however refer to the suggestive distinction of phraseology used by Hilbert (*Grundlagen der Geometrie,* pp. 39, 40):

(1) "Two polygons are called *divisibly-equal* (*zerlegungsgleich*) if they can be divided into a *finite* number of triangles which are congruent two and two."

(2) "Two polygons are called *equal in content* (*inhaltsgleich*) or *of equal content* if it is possible to add *divisibly-equal* polygons to them in such a way that the two combined polygons are *divisibly-equal.*"

(Amaldi suggests as alternatives for the terms in (1) and (2) the expressions *equivalent by sum* and *equivalent by difference* respectively.)

From these definitions it follows that "by combining *divisibly-equal* polygons we again arrive at *divisibly-equal* polygons; and, if we subtract *divisibly-equal* polygons from *divisibly-equal* polygons, the polygons remaining are *equal in content.*"

The proposition also follows without difficulty that, "if two polygons are *divisibly-equal* to a third polygon, they are also *divisibly-equal* to one another ; and, if two polygons are *equal in content* to a third polygon, they are *equal in content* to one another."

The different cases.

As usual, Proclus (pp. 399—400), observing that Euclid has given only the most difficult of the three possible cases, adds the other two with separate proofs. In the case where E in the figure of the proposition falls between A and D, he *adds* the congruent triangles ABE, DCF respectively to the smaller trapezium $EBCD$, instead of subtracting them (as Simson does) from the larger trapezium $ABCF$.

An ancient "Budget of Paradoxes."

Proclus observes (p. 396, 12 sqq.) that the present theorem and the similar one relating to triangles are among the so-called paradoxical theorems of mathematics, since the uninstructed might well regard it as impossible that the area of the parallelograms should remain the same while the length of the sides other than the base and the side opposite to it may increase indefinitely. He adds that mathematicians had made a collection of such paradoxes, the so-called *treasury of paradoxes* (ὁ παράδοξος τόπος)—cf. the similar expressions τόπος ἀναλυόμενος (treasury of analysis) and τόπος ἀστρονομούμενος—in the same way as the Stoics with their *illustrations* (ὥσπερ οἱ ἀπὸ τῆς Στοᾶς ἐπὶ τῶν δειγμάτων). It may be that this *treasury of paradoxes* was the work of Erycinus quoted by Pappus (III. p. 107, 8) and mentioned above (note on I. 21, p. 290).

Locus-theorems and loci in Greek geometry.

The proposition I. 35 is, says Proclus (pp. 394—6), the first *locus-theorem* (τοπικὸν θεώρημα) given by Euclid. Accordingly it is in his note on this proposition that Proclus gives us his view of the nature of a locus-theorem and of the meaning of the word *locus* (τόπος); and great importance attaches to his words because he is one of the three writers (Pappus and Eutocius being the two others) upon whom we have to rely for all that is known of the Greek conception of geometrical loci.

Proclus' explanation (pp. 394, 15—395, 2) is as follows. "I call those (theorems) *locus-theorems* (τοπικά) in which the same property is found to exist on the whole of some locus (πρὸς ὅλῳ τινὶ τόπῳ), and (I call) a locus a position of a line or a surface producing one and the same property (γραμμῆς ἢ ἐπιφανείας θέσιν ποιοῦσαν ἓν καὶ ταὐτὸν σύμπτωμα). For, of locus-theorems, some are constructed on lines and others on surfaces (τῶν γὰρ τοπικῶν τὰ μέν ἐστι πρὸς γραμμαῖς συνιστάμενα, τὰ δὲ πρὸς ἐπιφανείαις). And, since some lines are plane (ἐπίπεδοι) and others solid (στερεαί)—those being plane which are simply conceived of in a plane (ὧν ἐν ἐπιπέδῳ ἁπλῆ ἡ νόησις), and those solid the origin of which is revealed from some section of a solid figure, as the cylindrical helix and the conic lines (ὡς τῆς κυλινδρικῆς ἕλικος καὶ τῶν κωνικῶν γραμμῶν)—I should say (φαίην ἄν) further that, of locus-theorems on lines, some give a plane locus and others a solid locus."

Leaving out of sight for the moment the class of *loci on surfaces*, we find that the distinction between *plane* and *solid loci*, or *plane* and *solid lines*, was similarly understood by Eutocius, who says (Apollonius, ed. Heiberg, II. p. 184) that "*solid loci* have obtained their name from the fact that the lines used in the solution of problems regarding them have their origin in the section of solids, for example the sections of the cone and several others." Similarly we gather from Pappus that *plane loci* were straight lines and circles, and *solid loci* were conics. Thus he tells us (VII. p. 672, 20) that Aristaeus wrote five books of *Solid Loci* "supplementary to (literally, continuous with) the conics"; and, though Hultsch brackets the passage (VII. p. 662, 10—15) which says plainly that *plane loci* are straight lines and circles, while *solid loci* are sections of cones, i.e. parabolas, ellipses and hyperbolas, we have the exactly corresponding distinction drawn by Pappus (III. p. 54, 7—16) between *plane* and *solid problems*, plane problems being those solved by means of straight lines and circumferences of circles, and solid problems those solved by means of one or more of the sections of the cone. But, whereas Proclus

and Eutocius speak of other *solid loci* besides conics, there is nothing in
Pappus to support the wider application of the term. According to Pappus
(III. p. 54, 16—21) problems which could not be solved by means of straight
lines, circles, or conics were *linear* (γραμμικά) because they used for their
construction lines having a more complicated and unnatural origin than those
mentioned, namely such curves as *quadratrices*, conchoids and cissoids.
Similarly, in the passage supposed to be interpolated, *linear loci* are distinguished as those which are neither straight lines nor circles nor any of the
conic sections (VII. p. 662, 13—15). Thus the classification given by Proclus
and Eutocius is less precise than that which we find in Pappus; and the
inclusion by Proclus of the cylindrical helix among solid loci, on the ground
that it arises from a section of a solid figure, would seem to be, in any case,
due to some misapprehension.

Comparing these passages and the hints in Pappus about *loci on surfaces*
(τόποι πρὸς ἐπιφανείᾳ) with special reference to Euclid's two books under that
title, Heiberg concludes that *loci on lines* and *loci on surfaces* in Proclus'
explanation are loci which *are* lines and loci which *are* surfaces respectively.
But some qualification is necessary as regards Proclus' conception of *loci on
lines*, because he goes on to say (p. 395, 5), with reference to this proposition,
that, while the locus is a *locus on lines* and moreover *plane*, it is "the whole
space between the parallels" which is the locus of the various parallelograms
on the same base proved to be equal in area. Similarly, when he quotes
III. 21 about the equality of the angles in the same segment and III. 31 about
the right angle in a semicircle as cases where a circumference of a circle
takes the place of a straight line in a *plane* locus-theorem, he appears to
imply that it is the segment or semicircle as an *area* which is regarded as the
locus of an infinite number of *triangles* with the same base and equal vertical
angles, rather than that it is the *circumference* which is the locus of the angular
points. Likewise he gives the equality of parallelograms inscribed in "the
asymptotes and the hyperbola" as an example of a *solid* locus-theorem, as if
the area included between the curve and its asymptotes was regarded as the
locus of the equal parallelograms. However this may be, it is clear that the
locus in the present proposition can only be either (1) a *line*-locus of a *line*,
not a point, or (2) an *area*-locus of an *area*, not a point or a line; and we
seem to be thus brought to another and different classification of loci
corresponding to that quoted by Pappus (VII. p. 660, 18 sqq.) from the preliminary exposition given by Apollonius in his *Plane Loci*. According to this,
loci in general are of three kinds: (1) ἐφεκτικοί, *holding-in*, in which sense
the locus of a point is a point, of a line a line, of a surface a surface, and of a
solid a solid, (2) διεξοδικοί, *moving along*, a line being in this sense a locus of a
point, a surface of a line and a solid of a surface, (3) ἀναστροφικοί, where a
surface is a locus of a point and a solid of a line. Thus the locus in this
proposition, whether it is the space between the two parallels regarded as the
locus of the equal parallelograms, or the line parallel to the base regarded as
the locus of the sides opposite to the base, would seem to be of the first class
(ἐφεκτικός); and, as Proclus takes the former view of it, a *locus on lines* is
apparently not merely a locus which *is* a line but a locus *bounded by lines*
also, the locus being *plane* in the particular case because it is bounded by
straight lines, or, in the case of III. 21, 31, by straight lines and circles, but
not by any higher curves.

Proclus notes lastly (p. 395, 13—21) that, according to Geminus,
"Chrysippus likened locus-theorems to the *ideas*. For, as the ideas confine

the genesis of unlimited (particulars) within defined limits, so in such theorems the unlimited (particular figures) are confined within defined *places* or *loci* (τόποι). And it is this boundary which is the cause of the equality; for the height of the parallels, which remains the same, while an infinite number of parallelograms are conceived on the same base, is what makes them all equal to one another."

PROPOSITION 36.

Parallelograms which are on equal bases and in the same parallels are equal to one another.

Let *ABCD*, *EFGH* be parallelograms which are on equal bases *BC*, *FG* and in the same parallels *AH*, *BG*;

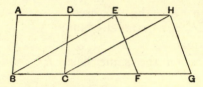

I say that the parallelogram *ABCD* is equal to *EFGH*.

For let *BE*, *CH* be joined.

Then, since *BC* is equal to *FG*

while   *FG* is equal to *EH*,

            BC is also equal to *EH*.   [*C. N.* 1]

But they are also parallel.

And *EB*, *HC* join them;

but straight lines joining equal and parallel straight lines (at the extremities which are) in the same directions (respectively) are equal and parallel.   [I. 33]

Therefore *EBCH* is a parallelogram.   [I. 34]

And it is equal to *ABCD*;

for it has the same base *BC* with it, and is in the same parallels *BC*, *AH* with it.   [I. 35]

For the same reason also *EFGH* is equal to the same *EBCH*;   [I. 35]

so that the parallelogram *ABCD* is also equal to *EFGH*.

            [*C. N.* 1]

Therefore etc.

            Q. E. D.

PROPOSITION 37.

Triangles which are on the same base and in the same parallels are equal to one another.

Let *ABC, DBC* be triangles on the same base *BC* and in the same parallels *AD, BC* ;

5 I say that the triangle *ABC* is equal to the triangle *DBC*.

Let *AD* be produced in both directions to *E, F* ;

through *B* let *BE* be drawn parallel to *CA*,   [I. 31]

10 and through *C* let *CF* be drawn parallel to *BD*.   [I. 31]

Then each of the figures *EBCA, DBCF* is a parallelogram ; and they are equal,

15 for they are on the same base *BC* and in the same parallels *BC, EF*.   [I. 35]

Moreover the triangle *ABC* is half of the parallelogram *EBCA* ; for the diameter *AB* bisects it.   [I. 34]

And the triangle *DBC* is half of the parallelogram *DBCF*;

20 for the diameter *DC* bisects it.   [I. 34]

[But the halves of equal things are equal to one another.]
Therefore the triangle *ABC* is equal to the triangle *DBC*.
Therefore etc.

Q. E. D.

21. Here and in the next proposition Heiberg brackets the words "But the halves of equal things are equal to one another" on the ground that, since the *Common Notion* which asserted this fact was interpolated at a very early date (before the time of Theon), it is probable that the words here were interpolated at the same time. Cf. note above (p. 224) on the interpolated *Common Notion*.

There is a lacuna in the text of Proclus' notes to I. 36 and I. 37. Apparently the end of the former and the beginning of the latter are missing, the MSS. and the *editio princeps* showing no separate note for I. 37 and no lacuna, but going straight on without regard to sense. Proclus had evidently remarked again in the missing passage that, in the case of both parallelograms and triangles between the same parallels, the two sides which stretch from one parallel to the other may increase in length to any extent, while the area remains the same. Thus the *perimeter* in parallelograms or triangles is of itself no criterion as to their area. Misconception on this subject was rife among non-mathematicians ; and Proclus (p. 403, 5 sqq.) tells us (1) of describers of countries (χωρογράφοι) who drew conclusions regarding the size of cities from their perimeters, and (2) of certain members of communistic societies in his own time who cheated their fellow members by giving them land of greater perimeter but less area than they took themselves, so that, on the one hand, they got a reputation for greater honesty while, on the other, they took more than their share of produce. Cantor (*Gesch. d. Math.* i_3, p. 172) quotes several remarks of ancient authors which show the prevalence of the same misconception. Thus Thucydides estimates the size of Sicily according to the time required for circumnavigating it. About 130 B.C. Polybius said that there were people who could not understand that camps of the same periphery might have different capacities. Quintilian has a similar remark, and Cantor thinks he may have had in his mind the calculations of Pliny, who compares the size of different parts of the earth by adding their length to their breadth.

The comparison however of the areas of different figures of equal contour had not been neglected by mathematicians. Theon of Alexandria, in his commentary on Book I. of Ptolemy's *Syntaxis*, has preserved a number of propositions on the subject taken from a treatise by Zenodorus περὶ ἰσομέτρων σχημάτων (reproduced in Latin on pp. 1190—1211 of Hultsch's edition of Pappus) which was written at some date between, say, 200 B.C. ...d 90 A.D., and probably not long after the former date. Pappus too has at the beginning of Book V. of his *Collection* (pp. 308 sqq.) the same propositions, in which he appears to have followed Zenodorus pretty closely while making some changes in detail. The propositions proved by Zenodorus and Pappus include the following: (1) that, *of all polygons of the same number of sides and equal perimeter, the equilateral and equiangular polygon is the greatest in area*, (2) that, *of regular polygons of equal perimeter, that is the greatest in area which has the most angles*, (3) that *a circle is greater than any regular polygon of equal contour*, (4) that, *of all circular segments in which the arcs are equal in length, the semicircle is the greatest*. The treatise of Zenodorus was not confined to propositions about plane figures, but gave also the theorem that, *of all solid figures the surfaces of which are equal, the sphere is the greatest in volume*.

PROPOSITION 38.

Triangles which are on equal bases and in the same parallels are equal to one another.

Let ABC, DEF be triangles on equal bases BC, EF and in the same parallels BF, AD ;
I say that the triangle ABC is equal to the triangle DEF.

For let AD be produced in both directions to G, H ;
through B let BG be drawn parallel to CA,   [I. 31]

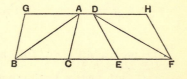

and through F let FH be drawn parallel to DE.

Then each of the figures $GBCA$, $DEFH$ is a parallelogram ;
and $GBCA$ is equal to $DEFH$;

for they are on equal bases *BC*, *EF* and in the same parallels *BF*, *GH*.   [I. 36]

Moreover the triangle *ABC* is half of the parallelogram *GBCA* ; for the diameter *AB* bisects it.   [I. 34]

And the triangle *FED* is half of the parallelogram *DEFH*; for the diameter *DF* bisects it.   [I. 34]

[But the halves of equal things are equal to one another.]
Therefore the triangle *ABC* is equal to the triangle *DEF*.
Therefore etc.

                                                    Q. E. D.

On this proposition Proclus remarks (pp. 405—6) that Euclid seems to him to have given in VI. I one proof including all the four theorems from I. 35 to I. 38, and that most people had failed to notice this. When Euclid, he says, proves that triangles and parallelograms of the same altitude have to one another the same ratio as their bases, he simply proves all these propositions more generally by the use of proportion; for of course to be of the same altitude is equivalent to being in the same parallels. It is true that VI. I generalises these propositions, but it must be observed that it does not prove the propositions themselves, as Proclus seems to imply; they are in fact assumed in order to prove VI. I.

Comparison of areas of triangles of I. 24.

The theorem already mentioned as given by Proclus on I. 24 (pp. 340—4) is placed here by Heron, who also enunciates it more clearly (an-Nairīzī, ed. Besthorn-Heiberg, pp. 155—161, ed. Curtze, pp. 75—8).

If in two triangles two sides of the one be equal to two sides of the other respectively, and the angle of the one be greater than the angle of the other, namely the angles contained by the equal sides, then, (1) *if the sum of the two angles contained by the equal sides is equal to two right angles, the two triangles are equal to one another ;* (2) *if less than two right angles, the triangle which has the greater angle is also itself greater than the other ;* (3) *if greater than two right angles, the triangle which has the less angle is greater than the other triangle.*

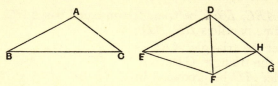

Let two triangles *ABC*, *DEF* have the sides *AB*, *AC* respectively equal to *DE*, *DF*.

(1) First, suppose that the angles at *A* and *D* in the triangles *ABC*, *DEF* are together equal to two right angles.

Heron's construction is now as follows.

Make the angle *EDG* equal to the angle *BAC*.
Draw *FH* parallel to *ED* meeting *DG* in *H*.
Join *EH*.

Then, since the angles *BAC*, *EDF* are equal to two right angles, the angles *EDH*, *EDF* are equal to two right angles.

But so are the angles *EDH*, *DHF*.

Therefore the angles *EDF*, *DHF* are equal.

And the alternate angles *EDF*, *DFH* are equal.   [I. 29]

Therefore the angles *DHF*, *DFH* are equal,

and *DF* is equal to *DH*.   [I. 6]

Hence the two sides *ED*, *DH* are equal to the two sides *BA*, *AC*; and the included angles are equal.

Therefore the triangles *ABC*, *DEH* are equal in all respects.

And the triangles *DEF*, *DEH* between the same parallels are equal.

[I. 37]

Therefore the triangles *ABC*, *DEF* are equal.

[Proclus takes the construction of Eucl. I. 24, i.e., he makes *DH* equal to *DF* and then proves that *ED*, *FH* are parallel.]

(2) Suppose the angles *BAC*, *EDF* together *less* than two right angles.

As before, make the angle *EDG* equal to the angle *BAC*, draw *FH* parallel to *ED*, and join *EH*.

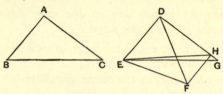

In this case the angles *EDH*, *EDF* are together less than two right angles, while the angles *EDH*, *DHF* are equal to two right angles.   [I. 29]

Hence the angle *EDF*, and therefore the angle *DFH*, is less than the angle *DHF*.

Therefore *DH* is less than *DF*.   [I. 19]

Produce *DH* to *G* so that *DG* is equal to *DF* or *AC*, and join *EG*.

Then the triangle *DEG*, which is equal to the triangle *ABC*, is greater than the triangle *DEH*, and therefore greater than the triangle *DEF*.

(3) Suppose the angles *BAC*, *EDF* together greater than two right angles.

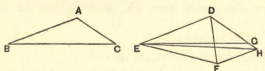

We make the same construction in this case, and we prove in like manner that the angle *DHF* is less than the angle *DFH*,

whence *DH* is greater than *DF* or *AC*.

Make *DG* equal to *AC*, and join *EG*.

It then follows that the triangle *DEF* is greater than the triangle *ABC*.

[In the second and third cases again Proclus starts from the construction in I. 24, and proves, in the second case, that the parallel, *FH*, to *ED* cuts *DG* and, in the third case, that it cuts *DG* produced.]

There is no necessity for Heron to take account of the position of *F* in relation to the side opposite *D*.   For in the first and third cases *F must* fall

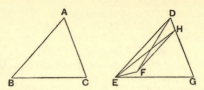

in the position in which Euclid draws it in I. 24, whatever be the relative lengths of *AB*, *AC*. In the second case the figure *may* be as annexed, but the proof is the same, or rather the case needs no proof at all.

PROPOSITION 39.

Equal triangles which are on the same base and on the same side are also in the same parallels.

Let *ABC*, *DBC* be equal triangles which are on the same base *BC* and on the same side of it ;

5 [I say that they are also in the same parallels.]

And [For] let *AD* be joined ;

I say that *AD* is parallel to *BC*.

For, if not, let *AE* be drawn through the point *A* parallel to the straight line 10 *BC*,   [I. 31]

and let *EC* be joined.

Therefore the triangle *ABC* is equal to the triangle *EBC* ;

for it is on the same base *BC* with it and in the same 15 parallels.   [I. 37]

But *ABC* is equal to *DBC* ;

therefore *DBC* is also equal to *EBC*,   [C. N. 1]

the greater to the less : which is impossible.

Therefore *AE* is not parallel to *BC*.

20   Similarly we can prove that neither is any other straight line except *AD* ;

therefore *AD* is parallel to *BC*.

Therefore etc.

5. [**I say that they are also in the same parallels.**] Heiberg has proved (*Hermes*, XXXVIII., 1903, p. 50) from a recently discovered papyrus-fragment (*Fayūm towns and their papyri*, p. 96, No. IX.) that these words are an interpolation by some one who did not observe that the words "And let *AD* be joined" are part of the *setting-out* (ἔκθεσις), but took them as belonging to the *construction* (κατασκευή) and consequently thought that a διορισμός or "definition" (of the thing to be proved) should precede. The interpolator then altered "And" into "For" in the next sentence.

This theorem is of course the *partial* converse of I. 37. In I. 37 we have triangles which are (1) on the same base, (2) in the same parallels, and the theorem proves (3) that the triangles are equal. Here the hypothesis (1) and the conclusion (3) are combined as hypotheses, and the conclusion is the hypothesis (2) of I. 37, that the triangles are in the same parallels. The additional qualification in this proposition that the triangles must be *on the same side* of the base is necessary because it is not, as in I. 37, involved in the other hypotheses.

Proclus (p. 407, 4—17) remarks that Euclid only converts I. 37 and I. 38 relative to triangles, and omits the converses of I. 35, 36 about parallelograms as unnecessary because it is easy to see that the method would be the same, and therefore the reader may properly be left to prove them for himself.

The proof is, as Proclus points out (p. 408, 5—21), equally easy on the supposition that the assumed parallel *AE* meets *BD* or *CD* produced beyond *D*.

[PROPOSITION 40.

Equal triangles which are on equal bases and on the same side are also in the same parallels.

Let *ABC*, *CDE* be equal triangles on equal bases *BC*, *CE* and on the same side.

I say that they are also in the same parallels.

For let *AD* be joined;

I say that *AD* is parallel to *BE*.

For, if not, let *AF* be drawn through *A* parallel to *BE* [I. 31], and let *FE* be joined.

Therefore the triangle *ABC* is equal to the triangle *FCE*;

for they are on equal bases *BC*, *CE* and in the same parallels *BE*, *AF*.   [I. 38]

But the triangle *ABC* is equal to the triangle *DCE*;

therefore the triangle *DCE* is also equal to the triangle *FCE*,   [*C. N.* 1]

the greater to the less: which is impossible.

Therefore *AF* is not parallel to *BE*.

Similarly we can prove that neither is any other straight line except *AD*;

therefore *AD* is parallel to *BE*.

Therefore etc.

Q. E. D.]

Heiberg has proved by means of the papyrus-fragment mentioned in the last note that this proposition is an interpolation by some one who thought that there should be a proposition following I. 39 and related to it in the same way as I. 38 is related to I. 37, and I. 36 to I. 35.

PROPOSITION 41.

If a parallelogram have the same base with a triangle and be in the same parallels, the parallelogram is double of the triangle.

For let the parallelogram *ABCD* have the same base *BC* with the triangle *EBC*, and let it be in the same parallels *BC, AE*;

I say that the parallelogram *ABCD* is double of the triangle *BEC*.

For let *AC* be joined.

Then the triangle *ABC* is equal to the triangle *EBC*;

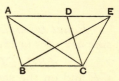

for it is on the same base *BC* with it and in the same parallels *BC, AE*.

[I. 37]

But the parallelogram *ABCD* is double of the triangle *ABC*;

for the diameter *AC* bisects it ;   [I. 34]

so that the parallelogram *ABCD* is also double of the triangle *EBC*.

Therefore etc.

Q. E. D.

On this proposition Proclus (pp. 414, 15—415, 16), "by way of practice" (γυμνασίας ἕνεκα), considers the area of a *trapezium* (a quadrilateral with only one pair of opposite sides parallel) in comparison with that of the triangles in the same parallels and having the greater and less of the parallel sides of the trapezium for bases respectively, and proves that the trapezium is less than double of the former triangle and more than double of the latter.

He next (pp. 415, 22—416, 14) proves the proposition that,

If a triangle be formed by joining the middle point of either of the non-parallel sides to the extremities of the opposite side, the area of the trapezium is always double of that of the triangle.

Let *ABCD* be a trapezium in which *AD, BC* are the parallel sides, and
E the middle point of one of the non-parallel sides,
say *DC*.

Join *EA, EB* and produce *BE* to meet *AD*
produced in *F*.

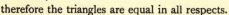

Then the triangles *BEC, FED* have two angles
equal respectively, and one side *CE* equal to one
side *DE*;

therefore the triangles are equal in all respects.   [I. 26]

Add to each the quadrilateral *ABED*;

therefore the trapezium *ABCD* is equal to the triangle *ABF*,

that is, to twice the triangle *AEB*, since *BE* is equal to *EF*.   [I. 38]

The three properties proved by Proclus may be combined in one enunciation thus:

*If a triangle be formed by joining the middle point of one side of a trapezium
to the extremities of the opposite side, the area of the trapezium is* (1) *greater
than,* (2) *equal to, or* (3) *less than, double the area of the triangle according as
the side the middle point of which is taken is* (1) *the greater of the parallel sides,*
(2) *either of the non-parallel sides, or* (3) *the lesser of the parallel sides.*

PROPOSITION 42.

*To construct, in a given rectilineal angle, a parallelogram
equal to a given triangle.*

Let *ABC* be the given triangle, and *D* the given rectilineal angle;

thus it is required to construct in the rectilineal angle *D* a
parallelogram equal to the
triangle *ABC*.

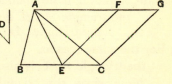

Let *BC* be bisected at *E*,
and let *AE* be joined;

on the straight line *EC*, and
at the point *E* on it, let the
angle *CEF* be constructed
equal to the angle *D*;   [I. 23]

through *A* let *AG* be drawn parallel to *EC*, and   [I. 31]
through *C* let *CG* be drawn parallel to *EF*.

Then *FECG* is a parallelogram.

And, since *BE* is equal to *EC*,

the triangle *ABE* is also equal to the triangle *AEC*,

for they are on equal bases *BE, EC* and in the same parallels
BC, AG;   [I. 38]

therefore the triangle *ABC* is double of the triangle
AEC.

But the parallelogram *FECG* is also double of the triangle *AEC*, for it has the same base with it and is in the same parallels with it;   [I. 41]

therefore the parallelogram *FECG* is equal to the triangle *ABC*.

And it has the angle *CEF* equal to the given angle *D*.

Therefore the parallelogram *FECG* has been constructed equal to the given triangle *ABC*, in the angle *CEF* which is equal to *D*.   Q. E. F.

Proposition 43.

In any parallelogram the complements of the parallelograms about the diameter are equal to one another.

Let *ABCD* be a parallelogram, and *AC* its diameter;
and about *AC* let *EH*, *FG* be parallelograms, and *BK*, *KD*
5 the so-called complements;

I say that the complement *BK* is equal to the complement *KD*.

For, since *ABCD* is a parallelogram, and *AC* its diameter,
the triangle *ABC* is equal to
10   the triangle *ACD*. [I. 34]

Again, since *EH* is a parallelogram, and *AK* is its diameter,
the triangle *AEK* is equal to
the triangle *AHK*.

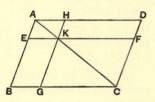

15 For the same reason
the triangle *KFC* is also equal to *KGC*.

Now, since the triangle *AEK* is equal to the triangle *AHK*,

and *KFC* to *KGC*,
20 the triangle *AEK* together with *KGC* is equal to the triangle *AHK* together with *KFC*.   [C. N. 2]

And the whole triangle *ABC* is also equal to the whole *ADC*;

therefore the complement *BK* which remains is equal to the
25 complement *KD* which remains.   [C. N. 3]

Therefore etc.

                                    Q. E. D.

1. complements, παραπληρώματα, the figures put in to fill up (interstices).

4. and about AC.... Euclid's phraseology here and in the next proposition implies that the complements as well as the other parallelograms are "about" the diagonal. The words are here περὶ δὲ τὴν ΑΓ παραλληλόγραμμα μὲν ἔστω τὰ ΕΘ, ΖΗ, τὰ δὲ λεγόμενα παραπληρώματα τὰ ΒΚ, ΚΔ. The expression "the so-called complements" indicates that this technical use of παραπληρώματα was not new, though it might not be universally known.

In the text of Proclus' commentary as we have it, the end of the note on I. 41, the whole of that on I. 42, and the beginning of that on I. 43 are missing.

Proclus remarks (p. 418, 15—20) that Euclid did not need to give a formal definition of *complement* because the name was simply suggested by the facts; when once we have the two "parallelograms about the diameter," the *complements* are necessarily the areas remaining over on each side of the diameter, which fill up the complete parallelogram. Thus (p. 417, 1 sqq.) the complements need not be parallelograms. They are so if the two "parallelograms about the diameter" are formed by straight lines drawn through *one point* of the diameter parallel to the sides of the original parallelogram, but not 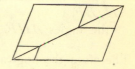 otherwise. If, as in the first of the accompanying figures, the parallelograms have no common point, the complements are five-sided figures as shown. When the parallelograms overlap, as in the second figure, Proclus regards the complements as being the small parallelograms *FG*, *EH*. But, if complements are strictly the areas required to fill up the original parallelogram, Proclus is inaccurate in describing *FG*, *EH* as the complements. The complements are really (1) the parallelogram *FG minus* the triangle *LMN*, and (2) the parallelogram *EH minus* the triangle *KMN*, respectively; the possibility that the re- 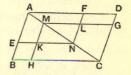 spective differences may be negative merely means the possibility that the sum of the two parallelograms about the diameter may be together greater than the original parallelogram.

In all the cases it is easy to show, as Proclus does, that the complements are still equal.

PROPOSITION 44.

To a given straight line to apply, in a given rectilineal angle, a parallelogram equal to a given triangle.

Let *AB* be the given straight line, *C* the given triangle and *D* the given rectilineal angle;

5 thus it is required to apply to the given straight line *AB*, in an angle equal to the angle *D*, a parallelogram equal to the given triangle *C*.

Let the parallelogram *BEFG* be constructed equal to the triangle *C*, in the angle *EBG* which is equal to *D* [I. 42];

10 let it be placed so that *BE* is in a straight line with *AB*; let

FG be drawn through to *H*, and let *AH* be drawn through
A parallel to either *BG* or *EF*.   [I. 31]
Let *HB* be joined.

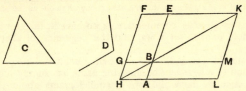

Then, since the straight line *HF* falls upon the parallels
15 *AH*, *EF*,
the angles *AHF*, *HFE* are equal to two right angles.

                                                    [I. 29]
Therefore the angles *BHG*, *GFE* are less than two right
angles.;
and straight lines produced indefinitely from angles less than
20 two right angles meet ;   [Post. 5]
therefore *HB*, *FE*, when produced, will meet.

Let them be produced and meet at *K*; through the point
K let *KL* be drawn parallel to either *EA* or *FH*,   [I. 31]
and let *HA*, *GB* be produced to the points *L*, *M*.
25   Then *HLKF* is a parallelogram,
HK is its diameter, and *AG*, *ME* are parallelograms. and
LB, *BF* the so-called complements, about *HK* ;
therefore *LB* is equal to *BF*.   [I. 43]
But *BF* is equal to the triangle *C* ;
30   therefore *LB* is also equal to *C*.   [C. N. 1]
And, since the angle *GBE* is equal to the angle *ABM*,

                                                    [I. 15]
while the angle *GBE* is equal to *D*,
the angle *ABM* is also equal to the angle *D*.

Therefore the parallelogram *LB* equal to the given triangle
35 *C* has been applied to the given straight line *AB*, in the angle
ABM which is equal to *D*.

                                        Q. E. F.

14. since the straight line HF falls.... The verb is in the aorist (ἐνέπεσεν) here and
in similar expressions in the following propositions.

This proposition will always remain one of the most impressive in all
geometry when account is taken (1) of the great importance of the result

obtained, the transformation of a parallelogram of any shape into another with the same angle and of equal area but with one side of any given length, e.g. a *unit* length, and (2) of the simplicity of the means employed, namely the mere application of the property that the complements of the "parallelograms about the diameter" of a parallelogram are equal. The marvellous ingenuity of the solution is indeed worthy of the "godlike men of old," as Proclus calls the discoverers of the method of "application of areas"; and there would seem to be no reason to doubt that the particular solution, like the whole theory, was Pythagorean, and not a new solution due to Euclid himself.

Application of areas.

On this proposition Proclus gives (pp. 419, 15—420, 23) a valuable note on the method of "application of areas" here introduced, which was one of the most powerful methods on which Greek geometry relied. The note runs as follows :

"These things, says Eudemus (οἱ περὶ τὸν Εὔδημον), are ancient and are discoveries of the Muse of the Pythagoreans, I mean the *application of areas* (παραβολὴ τῶν χωρίων), their *exceeding* (ὑπερβολή) and their *falling-short* (ἔλλειψις). It was from the Pythagoreans that later geometers [i.e. Apollonius] took the names, which they again transferred to the so-called *conic* lines, designating one of these a *parabola* (application), another a *hyperbola* (exceeding) and another an *ellipse* (falling-short), whereas those godlike men of old saw the things signified by these names in the construction, in a plane, of areas upon a finite straight line. For, when you have a straight line set out and lay the given area exactly alongside the whole of the straight line, then they say that you *apply* (παραβάλλειν) the said area ; when however you make the length of the area greater than the straight line itself, it is said to *exceed* (ὑπερβάλλειν), and when you make it less, in which case, after the area has been drawn, there is some part of the straight line extending beyond it, it is said to *fall short* (ἐλλείπειν). Euclid too, in the sixth book, speaks in this way both of *exceeding* and *falling-short*; but in this place he needed the *application* simply, as he sought to apply to a given straight line an area equal to a given triangle in order that we might have in our power, not only the *construction* (σύστασις) of a parallelogram equal to a given triangle, but also the *application* of it to a finite straight line. For example, given a triangle with an area of 12 feet, and a straight line set out the length of which is 4 feet, we apply to the straight line the area equal to the triangle if we take the whole length of 4 feet and find how many feet the breadth must be in order that the parallelogram may be equal to the triangle. In the particular case, if we find a breadth of 3 feet and multiply the length into the breadth, supposing that the angle set out is a right angle, we shall have the area. Such then is the *application* handed down from early times by the Pythagoreans."

Other passages to a similar effect are quoted from Plutarch. (1) "Pythagoras sacrificed an ox on the strength of his proposition (διάγραμμα) as Apollodotus (?-rus) says...whether it was the theorem of the hypotenuse, viz. that the square on it is equal to the squares on the sides containing the right angle, or the problem about the *application of an area.*" (*Non posse suaviter vivi secundum Epicurum*, c. 11.) (2) "Among the most geometrical theorems, or rather problems, is the following : given two figures, to *apply* a third equal to the one and similar to the other, on the strength of which discovery they say moreover that Pythagoras sacrificed. This is indeed unquestionably more subtle and more scientific than the theorem which demonstrated that the square on the hypotenuse is equal to the squares on the sides about the right angle" (*Symp.* VIII. 2, 4).

The story of the sacrifice must (as noted by Bretschneider and Hankel) be given up as inconsistent with Pythagorean ritual, which forbade such sacrifices; but there is no reason to doubt that the first distinct formulation and introduction into Greek geometry of the method of *application of areas* was due to the Pythagoreans. The complete exposition of the *application* of areas, their *exceeding* and their *falling-short,* and of the construction of a rectilineal figure equal to one given figure and similar to another, takes us into the sixth Book of Euclid; but it will be convenient to note here the general features of the theory of *application, exceeding* and *falling-short.*

The simple *application* of a parallelogram of given area to a given straight line as one of its sides is what we have in I. 44 and 45; the general form of the problem with regard to *exceeding* and *falling-short* may be stated thus:

"To apply to a given straight line a rectangle (or, more generally, a parallelogram) equal to a given rectilineal figure and (1) *exceeding* or (2) *falling-short* by a square (or, in the more general case, a parallelogram similar to a given parallelogram)."

What is meant by saying that the applied parallelogram (1) *exceeds* or (2) *falls short* is that, while its base coincides and is coterminous *at one end* with the straight line, the said base (1) overlaps or (2) falls short of the straight line *at the other end,* and the portion by which the applied parallelogram exceeds a parallelogram of the same angle and height on the given straight line (exactly) as base is a parallelogram similar to a given parallelogram (or, in particular cases, a square). In the case where the parallelogram is to *fall short,* a διορισμός is necessary to express the condition of possibility of solution.

We shall have occasion to see, when we come to the relative propositions in the second and sixth Books, that the general problem here stated is equivalent to that of solving geometrically a mixed quadratic equation. We shall see that, even by means of II. 5 and 6, we can solve geometrically the equations

$$ax \pm x^2 = b^2,$$
$$x^2 - ax = b^2;$$

but in VI. 28, 29 Euclid gives the equivalent of the solution of the general equations

$$ax \pm \frac{b}{c} x^2 = \frac{C}{m}.$$

We are now in a position to understand the application of the terms *parabola* (application), *hyperbola* (exceeding) and *ellipse* (falling-short) to conic sections. These names were first so applied by Apollonius as expressing in each case the fundamental property of the curves as stated by him. This fundamental property is the geometrical equivalent of the Cartesian equation referred to any diameter of the conic and the tangent at its extremity as (in general, oblique) axes. If the *parameter* of the ordinates from the several points of the conic drawn to the given diameter be denoted by p (p being accordingly, in the case of the hyperbola and ellipse, equal to $\dfrac{d'^2}{d}$, where d is the length of the given diameter and d' that of its conjugate), Apollonius gives the properties of the three conics in the following form.

(1)   For the *parabola*, the square on the ordinate at any point is equal to a rectangle applied to *p* as base with altitude equal to the corresponding abscissa.   That is to say, with the usual notation,

$$y^2 = px.$$

(2)   For the *hyperbola* and *ellipse*, the square on the ordinate is equal to the rectangle applied to *p* having as its width the abscissa and *exceeding* (for the hyperbola) or *falling-short* (for the ellipse) by a figure similar and similarly situated to the rectangle contained by the given diameter and *p*.

That is, in the *hyperbola* $y^2 = px + \dfrac{x^2}{d^2}\, pd,$

or $y^2 = px + \dfrac{p}{d}\, x^2\,;$

and in the *ellipse* $y^2 = px - \dfrac{p}{d}\, x^2.$

The form of these equations will be seen to be exactly the same as that of the general equations above given, and thus Apollonius' nomenclature followed exactly the traditional theory of *application*, *exceeding*, and *falling-short*.

PROPOSITION 45.

To construct, in a given rectilineal angle, a parallelogram equal to a given rectilineal figure.

Let *ABCD* be the given rectilineal figure and *E* the given rectilineal angle ;

5 thus it is required to construct, in the given angle *E*, a parallelogram equal to the rectilineal figure *ABCD*.

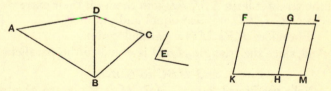

Let *DB* be joined, and let the parallelogram *FH* be constructed equal to the triangle *ABD*, in the angle *HKF* which is equal to *E* ;   [I. 42]

10 let the parallelogram *GM* equal to the triangle *DBC* be applied to the straight line *GH*, in the angle *GHM* which is equal to *E*.   [I. 44]

Then, since the angle *E* is equal to each of the angles *HKF*, *GHM*,

15   the angle *HKF* is also equal to the angle *GHM*. [C. N. 1]

Let the angle *KHG* be added to each ;
therefore the angles *FKH*, *KHG* are equal to the angles
KHG, *GHM*.

But the angles *FKH*, *KHG* are equal to two right angles;
[I. 29]
20 therefore the angles *KHG*, *GHM* are also equal to two right
angles.

Thus, with a straight line *GH*, and at the point *H* on it,
two straight lines *KH*, *HM* not lying on the same side make
the adjacent angles equal to two right angles ;
25   therefore *KH* is in a straight line with *HM*.   [I. 14]

And, since the straight line *HG* falls upon the parallels
KM, *FG*, the alternate angles *MHG*, *HGF* are equal to one
another.   [I. 29]

Let the angle *HGL* be added to each ;
30 therefore the angles *MHG*, *HGL* are equal to the angles
HGF, *HGL*.   [C. N. 2]

But the angles *MHG*, *HGL* are equal to two right angles;
[I. 29]
therefore the angles *HGF*, *HGL* are also equal to two right
angles.   [C. N. 1]
35   Therefore *FG* is in a straight line with *GL*.   [I. 14]

And, since *FK* is equal and parallel to *HG*,   [I. 34]
and *HG* to *ML* also,

KF is also equal and parallel to *ML* ;   [C. N. 1 ; I. 30]
and the straight lines *KM*, *FL* join them (at their extremities);
40 therefore *KM*, *FL* are also equal and parallel.   [I. 33]

Therefore *KFLM* is a parallelogram.

And, since the triangle *ABD* is equal to the parallelogram
FH,
and *DBC* to *GM*,
45 the whole rectilineal figure *ABCD* is equal to the whole
parallelogram *KFLM*.

Therefore the parallelogram *KFLM* has been constructed
equal to the given rectilineal figure *ABCD*, in the angle *FKM*
which is equal to the given angle *E*.   Q. E. F.

2, 3, 6, 45, 48.   **rectilineal figure,** in the Greek "rectilineal" simply, without "figure,"
εὐθύγραμμον being here used as a substantive, like the similarly formed παραλληλόγραμμον.

Transformation of areas.

We can now take stock of how far the propositions I. 43—45 bring us in
the matter of *transformation of areas*, which constitutes so important a part of

what has been fitly called the *geometrical algebra* of the Greeks. We have now learnt how to represent any rectilineal area, which can of course be resolved into triangles, by a single parallelogram having one side equal to any given straight line and one angle equal to any given rectilineal angle. Most important of all such parallelograms is the rectangle, which is one of the simplest forms in which an area can be shown. Since a rectangle corresponds to the product of two magnitudes in algebra, we see that *application* to a given straight line of a rectangle equal to a given area is the geometrical equivalent of algebraical *division* of the product of two quantities by a third. Further than this, it enables us to *add* or *subtract* any rectilineal areas and to represent the sum or difference by *one* rectangle with one side of any given length, the process being the equivalent of obtaining a common factor. But one step still remains, the finding of a *square* equal to a given rectangle, i.e. to a given rectilineal figure; and this step is not taken till II. 14. In general, the transformation of combinations of rectangles and squares into other combinations of rectangles and squares is the subject-matter of Book II., with the exception of the expression of the sum of two squares as a single square which appears earlier in the other Pythagorean theorem I. 47. Thus the transformation of rectilineal areas is made complete *in one direction*, i.e. in the direction of their simplest expression in terms of rectangles and squares, by the end of Book II. The reverse process of transforming the simpler rectangular area into an equal area which shall be similar to any rectilineal figure requires, of course, the use of proportions, and therefore does not appear till VI. 25.

Proclus adds to his note on this proposition the remark (pp. 422, 24—423, 6): "I conceive that it was in consequence of this problem that the ancient geometers were led to investigate the squaring of the circle as well. For, if a parallelogram can be found equal to any rectilineal figure, it is worth inquiring whether it be not also possible to prove rectilineal figures equal to circular. And Archimedes actually proved that any circle is equal to the right-angled triangle which has one of its sides about the right angle [the perpendicular] equal to the radius of the circle and its base equal to the perimeter of the circle. But of this elsewhere."

PROPOSITION 46.

On a given straight line to describe a square.

Let *AB* be the given straight line; thus it is required to describe a square on the straight line *AB*.

5   Let *AC* be drawn at right angles to the straight line *AB* from the point *A* on it [I. 11], and let *AD* be made equal to *AB*;

through the point *D* let *DE* be drawn
10 parallel to *AB*,

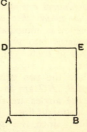

and through the point *B* let *BE* be drawn parallel to *AD*.

Therefore *ADEB* is a parallelogram ;
therefore *AB* is equal to *DE*, and *AD* to *BE*.   [I. 34]

But *AB* is equal to *AD* ;

15   therefore the four straight lines *BA*, *AD*, *DE*, *EB*
are equal to one another ;

therefore the parallelogram *ADEB* is equilateral.

I say next that it is also right-angled.

For, since the straight line *AD* falls upon the parallels
20 *AB*, *DE*,

the angles *BAD*, *ADE* are equal to two right angles.

[I. 29]

But the angle *BAD* is right ;
therefore the angle *ADE* is also right.

And in parallelogrammic areas the opposite sides and
25 angles are equal to one another ;   [I. 34]

therefore each of the opposite angles *ABE*, *BED* is also
right.

Therefore *ADEB* is right-angled.

And it was also proved equilateral.

30   Therefore it is a square ; and it is described on the straight
line *AB*.

Q. E. F.

1, 3, 30. Proclus (p. 423, 18 sqq.) notes the difference between the word *construct*
(συστήσασθαι) applied by Euclid to the construction of a *triangle* (and, he might have added,
of an *angle*) and the words *describe on* (ἀναγράφειν ἀπό) used of drawing a square on a given
straight line as one side. The *triangle* (or *angle*) is, so to say, pieced together, while the
describing of a square on a given straight line is the making of a figure " from " *one* side,
and corresponds to the multiplication of the number representing the side by itself.

Proclus (pp. 424—5) proves that, *if squares are described on equal straight
lines, the squares are equal*; and, *conversely*, that,
*if two squares are equal, the straight lines are
equal on which they are described*. The first
proposition is immediately obvious if we divide
the squares into two triangles by drawing a
diagonal in each. The converse is proved as
follows.

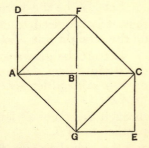

Place the two equal squares *AF*, *CG* so
that *AB*, *BC* are in a straight line. Then,
since the angles are right, *FB*, *BG* will also
be in a straight line. Join *AF*, *FC*, *CG*, *GA*.

Now, since the squares are equal, the
triangles *ABF*, *CBG* are equal.

Add to each the triangle *FBC*; therefore the triangles *AFC*, *GFC* are
equal, and hence they must be in the same parallels.

Therefore *AG*, *CF* are parallel.

Also, since each of the alternate angles *AFG*, *FGC* is half a right angle,
AF, *CG* are parallel.

Hence *AFCG* is a parallelogram; and *AF*, *CG* are equal.

Thus the triangles *ABF*, *CBG* have two angles and one side respectively equal;

therefore *AB* is equal to *BC*, and *BF* to *BG*.

PROPOSITION 47.

In right-angled triangles the square on the side subtending the right angle is equal to the squares on the sides containing the right angle.

Let *ABC* be a right-angled triangle having the angle
5 *BAC* right;

I say that the square on *BC* is equal to the squares on
BA, *AC*.

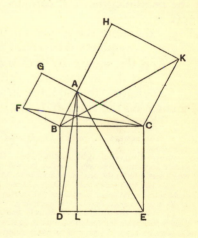

For let there be described
on *BC* the square *BDEC*,
10 and on *BA*, *AC* the squares
GB, *HC*;   [I. 46]
through *A* let *AL* be drawn
parallel to either *BD* or *CE*,
and let *AD*, *FC* be joined.

15   Then, since each of the
angles *BAC*, *BAG* is right,
it follows that with a straight
line *BA*, and at the point *A*
on it, the two straight lines
20 *AC*, *AG* not lying on the
same side make the adjacent
angles equal to two right
angles;

therefore *CA* is in a straight line with *AG*.   [I. 14]

25   For the same reason

BA is also in a straight line with *AH*.

And, since the angle *DBC* is equal to the angle *FBA*: for
each is right:

let the angle *ABC* be added to each;

30   therefore the whole angle *DBA* is equal to the whole
angle *FBC*.   [C. N. 2]

And, since *DB* is equal to *BC*, and *FB* to *BA*,
the two sides *AB*, *BD* are equal to the two sides *FB*, *BC*
respectively,
35   and the angle *ABD* is equal to the angle *FBC*;
       therefore the base *AD* is equal to the base *FC*,
and the triangle *ABD* is equal to the triangle *FBC*. [I. 4]
Now the parallelogram *BL* is double of the triangle *ABD*,
for they have the same base *BD* and are in the same parallels
40 *BD*, *AL*.   [I. 41]
And the square *GB* is double of the triangle *FBC*,
for they again have the same base *FB* and are in the same
parallels *FB*, *GC*.   [I. 41]
[But the doubles of equals are equal to one another.]
45   Therefore the parallelogram *BL* is also equal to the
       square *GB*.
Similarly, if *AE*, *BK* be joined,
the parallelogram *CL* can also be proved equal to the square
HC;
50   therefore the whole square *BDEC* is equal to the two
       squares *GB*, *HC*.   [C. N. 2]
And the square *BDEC* is described on *BC*,
       and the squares *GB*, *HC* on *BA*, *AC*.
Therefore the square on the side *BC* is equal to the
55 squares on the sides *BA*, *AC*.
Therefore etc.   Q. E. D.

1. **the square on,** τὸ ἀπὸ...τετράγωνον, the word ἀναγραφέν or ἀναγεγραμμένον being
understood.
    subtending the right angle. Here ὑποτεινούσης, "subtending," is used with the
simple accusative (τὴν ὀρθὴν γωνίαν) instead of being followed by ὑπό and the accusative,
which seems to be the original and more orthodox construction. Cf. I. 18, note.
33. **the two sides AB, BD**.... Euclid actually writes "*DB, BA*," and therefore the
equal sides in the two triangles are not mentioned in corresponding order, though he adheres
to the words ἑκατέρα ἑκατέρᾳ "respectively." Here *DB* is equal to *BC* and *BA* to *FB*.
44. **[But the doubles of equals are equal to one another.]** Heiberg brackets
these words as an interpolation, since it quotes a *Common Notion* which is itself interpolated.
Cf. notes on I. 37, p. 332, and on interpolated *Common Notions*, pp. 223—4.

"If we listen," says Proclus (p. 426, 6 sqq.), "to those who wish to
recount ancient history, we may find some of them referring this theorem to
Pythagoras and saying that he sacrificed an ox in honour of his discovery.
But for my part, while I admire those who first observed the truth of this
theorem, I marvel more at the writer of the Elements, not only because he
made it fast (κατεδήσατο) by a most lucid demonstration, but because he
compelled assent to the still more general theorem by the irrefragable
arguments of science in the sixth Book. For in that Book he proves
generally that, in right-angled triangles, the figure on the side subtending
the right angle is equal to the similar and similarly situated figures described
on the sides about the right angle."

In addition, Plutarch (in the passages quoted above in the note on I. 44), Diogenes Laertius (VIII. 12) and Athenaeus (x. 13) agree in attributing this proposition to Pythagoras. It is easy to point out, as does G. Junge ("Wann haben die Griechen das Irrationale entdeckt?" in *Novae Symbolae Joachimicae*, Halle a. S., 1907, pp. 221—264), that these are late witnesses, and that the Greek literature which we possess belonging to the first five centuries after Pythagoras contains no statement specifying this or any other particular great geometrical discovery as due to him.   Yet the distich of Apollodorus the "calculator," whose date (though it cannot be fixed) is at least earlier than that of Plutarch and presumably of Cicero, is quite definite as to the existence of *one* "famous proposition" discovered by Pythagoras, whatever it was.   Nor does Cicero, in commenting apparently on the verses (*De nat. deor.* III. c. 36, § 88), seem to dispute the fact of the geometrical discovery, but only the story of the sacrifice.   Junge naturally emphasises the apparent uncertainty in the statements of Plutarch and Proclus.   But, as I read the passages of Plutarch, I see nothing in them inconsistent with the supposition that Plutarch un-hesitatingly accepted as discoveries of Pythagoras *both* the theorem of the square of the hypotenuse and the problem of the application of an area, and the only doubt he felt was as to which of the two discoveries was the more appropriate occasion for the supposed sacrifice.   There is also other evidence not without bearing on the question.   The theorem is closely connected with the whole of the matter of Eucl. Book II., in which one of the most prominent features is the use of the *gnomon*.   Now the gnomon was a well-understood term with the Pythagoreans (cf. the fragment of Philolaus quoted on p. 141 of Boeckh's *Philolaos des Pythagoreers Lehren*, 1819).   Aristotle also (*Physics* III. 4, 203 a 10—15) clearly attributes to the Pythagoreans the placing of odd numbers as *gnomons* round successive squares beginning with 1, thereby forming new squares, while in another place (*Categ.* 14, 15 a 30) the word *gnomon* occurs in the same (obviously familiar) sense : "e.g. a square, when a gnomon is placed round it, is increased in size but is not altered in form." The inference must therefore be that practically the whole doctrine of Book II. is Pythagorean.   Again Heron (? 3rd cent. A.D.), like Proclus, credits Pythagoras with a general rule for forming right-angled triangles with rational whole numbers for sides.   Lastly, the "summary" of Proclus appears to credit Pythagoras with the discovery of the theory, or study, of irrationals (τὴν τῶν ἀλόγων πραγματείαν).   But it is now more or less agreed that the reading here should be, not τῶν ἀλόγων, but τῶν ἀναλόγων, or rather τῶν ἀνὰ λόγον ("of proportionals"), and that the author intended to attribute to Pythagoras a theory of *proportion*, i.e. the (arithmetical) theory of proportion applicable only to commensurable magnitudes, as distinct from the theory of Eucl. Book V., which was due to Eudoxus.   It is not however disputed that the *Pythagoreans* discovered the irrational (cf. the scholium No. 1 to Book X.). Now everything goes to show that this discovery of the irrational was made with reference to $\sqrt{2}$, the ratio of the diagonal of a square to its side.   It is clear that this presupposes the knowledge that I. 47 is true of an isosceles right-angled triangle ; and the fact that some triangles of which it had been discovered to be true were *rational* right-angled triangles was doubtless what suggested the inquiry whether the ratio between the lengths of the diagonal and the side of a square could also be expressed in whole numbers. On the whole, therefore, I see no sufficient reason to question the tradition that, *so far as Greek geometry is concerned* (the possible priority of the discovery of the same proposition in India will be considered later), Pythagoras was the first to introduce the theorem of I. 47 and to give a general proof of it.

On this assumption, how was Pythagoras led to this discovery? It has been suggested and commonly assumed that the Egyptians were aware that a triangle with its sides in the ratio 3, 4, 5 was right-angled. Cantor inferred this from the fact that this was precisely the triangle with which Pythagoras began, if we may accept the testimony of Vitruvius (IX. 2) that Pythagoras taught how to make a right angle by means of three lengths measured by the numbers 3, 4, 5. If then he took from the Egyptians the triangle 3, 4, 5, he presumably learnt its property from them also. Now the Egyptians must certainly be credited from a period at least as far back as 2000 B.C. with the knowledge that $4^2 + 3^2 = 5^2$. Cantor finds proof of this in a fragment of papyrus belonging to the time of the 12th Dynasty newly discovered at Kahun. In this papyrus we have extractions of square roots: e.g. that of 16 is 4, that of $1\frac{9}{16}$ is $1\frac{1}{4}$, that of $6\frac{1}{4}$ is $2\frac{1}{2}$, and the following equations can be traced :

$$1^2 + (\tfrac{3}{4})^2 = (1\tfrac{1}{4})^2$$
$$8^2 + 6^2 = 10^2$$
$$2^2 + (1\tfrac{1}{2})^2 = (2\tfrac{1}{2})^2$$
$$16^2 + 12^2 = 20^2.$$

It will be seen that $4^2 + 3^2 = 5^2$ can be derived from each of these by multiplying, or dividing out, by one and the same factor. We may therefore admit that the Egyptians knew that $3^2 + 4^2 = 5^2$. But there seems to be no evidence that they knew that the triangle (3, 4, 5) is *right-angled*; indeed, according to the latest authority (T. Eric Peet, *The Rhind Mathematical Papyrus*, 1923), nothing in Egyptian mathematics suggests that the Egyptians were acquainted with this or any special cases of the Pythagorean theorem.

How then did Pythagoras discover the general theorem? Observing that 3, 4, 5 was a right-angled triangle, while $3^2 + 4^2 = 5^2$, he was probably led to

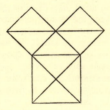

consider whether a similar relation was true of the sides of right-angled triangles other than the particular one. The simplest case (geometrically) to investigate was that of the *isosceles* right-angled triangle; and the truth of the theorem in this particular case would easily appear from the mere construction of a figure. Cantor (I₃, p. 185) and Allman (*Greek Geometry from Thales to Euclid*, p. 29) illustrate by a figure in which the squares are drawn outwards, as in I. 47, and divided by diagonals into equal triangles; but I think that the truth was more likely to be first observed from a figure of the kind suggested by Bürk (*Das Āpastamba-Śulba-Sūtra* in *Zeitschrift der deutshen morgenländ. Gesellschaft*, LV., 1901, p. 557) to explain how the Indians arrived at the same thing. The two figures are as shown above. When the geometrical consideration of the figure had shown that the isosceles right-angled triangle had the property in question, the investigation of the same fact from the arithmetical point of view would ultimately lead to the other momentous discovery of the irrationality of the length of the diagonal of a square expressed in terms of its side.

The *irrational* will come up for discussion later; and our next question is: Assuming that Pythagoras had observed the geometrical truth of the theorem in the case of the two particular triangles, and doubtless of other rational right-angled triangles, how did he establish it generally?

There is no positive evidence on this point. Two possible lines are however marked out. (1) Tannery says (*La Géométrie grecque*, p. 105) that the geometry of Pythagoras was sufficiently advanced to make it possible for him to prove the theorem by *similar triangles*. He does not say in what particular manner similar triangles would be used, but their use must apparently have involved the use of *proportions*, and, in order that the proof should be conclusive, of the theory of proportions in its complete form applicable to incommensurable as well as commensurable magnitudes. Now Eudoxus was the first to make the theory of proportion independent of the hypothesis of commensurability; and as, before Eudoxus' time, this had not been done, any proof of the general theorem by means of proportions given by Pythagoras must at least have been inconclusive. But this does not constitute any objection to the supposition that the truth of the general theorem may have been discovered in such a manner; on the contrary, the supposition that Pythagoras proved it by means of an imperfect theory of proportions would better than anything else account for the fact that Euclid had to devise an entirely new proof, as Proclus says he did in I. 47. This proof had to be independent of the theory of proportion even in its rigorous form, because the plan of the *Elements* postponed that theory to Books V. and VI., while the Pythagorean theorem was required as early as Book II. On the other hand, if the Pythagorean proof had been based on the doctrine of Books I. and II. only, it would scarcely have been necessary for Euclid to supply a new proof.

The possible proofs by means of proportion would seem to be practically limited to two.

(*a*) One method is to prove, from the similarity of the triangles ABC, DBA, that the rectangle CB, BD is equal to the square on BA, and, from the similarity of the triangles ABC, DAC, that the rectangle BC, CD is equal to the square on CA; whence the result follows by addition.

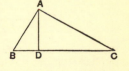

It will be observed that this proof is *in substance* identical with that of Euclid, the only difference being that the equality of the two smaller squares to the respective rectangles is inferred by the method of Book VI. instead of from the relation between the areas of parallelograms and triangles on the same base and between the same parallels established in Book I. It occurred to me whether, if Pythagoras' proof had come, even in substance, so near to Euclid's, Proclus would have emphasised so much as he does the originality of Euclid's, or would have gone so far as to say that he marvelled more at that proof than at the original discovery of the theorem. But on the whole I see no difficulty; for there can be little doubt that the proof by proportion is what suggested to Euclid the method of I. 47, and the transformation of

the method of proportions into one based on Book I. only, effected by a
construction and proof so extraordinarily ingenious, is a veritable *tour de
force* which compels admiration, notwithstanding the ignorant strictures of
Schopenhauer, who wanted something as obvious as the second figure in
the case of the isosceles right-angled triangle (p. 352), and accordingly
(*Sämmtliche Werke*, III. § 39 and I. § 15) calls Euclid's proof "a mouse-trap
proof" and "a proof walking on stilts, nay, a mean, underhand, proof" ("Des
Eukleides stelzbeiniger, ja, hinterlistiger Beweis").

(*b*) The other possible method is this. As it would be seen that the
triangles into which the original triangle is divided by the perpendicular from
the right angle on the hypotenuse are similar to one another and to the whole
triangle, while in these three triangles the two sides about the right angle in the
original triangle, and the hypotenuse of the original triangle, are corresponding
sides, and that the sum of the two former similar triangles is identically equal
to the similar triangle on the hypotenuse, it might be inferred that the same
would also be true of *squares* described on the corresponding three sides
respectively, because squares as well as similar triangles are to one another in
the duplicate ratio of corresponding sides. But the same thing is equally true
of any similar rectilineal figures, so that this proof would practically establish
the extended theorem of Eucl. VI. 31, which theorem, however, Proclus
appears to regard as being entirely Euclid's discovery.

On the whole, the most probable supposition seems to me to be that
Pythagoras used the first method (*a*) of proof by means of the theory of
proportion as he knew it, i.e. in the defective form which was in use up to the
date of Eudoxus.

(2) I have pointed out the difficulty in the way of the supposition that
Pythagoras' proof depended upon the principles of Eucl. Books I. and II. only.

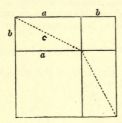 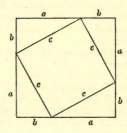

Were it not for this difficulty, the conjecture of Bretschneider (p. 82), followed
by Hankel (p. 98), would be the most tempting hypothesis. According to this
suggestion, we are to suppose a figure like that of Eucl. II. 4 in which *a*, *b* are
the sides of the two inner squares respectively, and *a* + *b* is the side of the
complete square. Then, if the two complements, which are equal, are divided
by their two diagonals into four equal triangles of sides *a*, *b*, *c*, we can place
these triangles round another square of the same size as the whole square, in the
manner shown in the second figure, so that the sides *a*, *b* of *successive* triangles
make up one of the sides of the square and are arranged in cyclic order. It
readily follows that the remainder of the square when the four triangles are
deducted is, in the one case, a square whose side is *c*, and in the other the sum of
two squares whose sides are *a*, *b* respectively. Therefore the square on *c* is equal

to the sum of the squares on a, b. All that can be said against this conjectural proof is that it has no specifically Greek colouring but rather recalls the Indian method. Thus Bhāskara (born 1114 A.D.; see Cantor, I_3, p. 656) simply draws four right-angled triangles equal to the original one inwards, one on each side of the square on the hypotenuse, and says "see!", without even adding that inspection shows that

$$c^2 = 4\frac{ab}{2} + (a - b)^2 = a^2 + b^2.$$

Though, for the reason given, there is difficulty in supposing that Pythagoras used a general proof of this kind, which applies of course to right-angled triangles with sides incommensurable as well as commensurable, there is no objection, I think, to supposing that the truth of the proposition in the case of the first *rational* right-angled triangles discovered, e.g. 3, 4, 5, was proved by a method of this sort. Where the sides are commensurable in this way, the squares can be divided up into small (unit) squares, which would much facilitate the comparison between them. That this subdivision was in fact resorted to in adding and subtracting squares is made probable by Aristotle's allusion to odd numbers as *gnomons* placed round unity to form successive squares in *Physics* III. 4; this must mean that the squares were represented by dots arranged in the form of a square and a gnomon formed of dots put round, or that (if the given square was drawn in the usual way) the gnomon was divided up into unit squares. Zeuthen has shown ("*Théorème de Pythagore,*" *Origine de la Géométrie scientifique* in *Comptes rendus du IIme Congrès international de Philosophie,* Genève, 1904), how easily the proposition could be proved by a method of this kind for the triangle 3, 4, 5. To admit of the two smaller squares being shown side by side, take a square on a line containing 7 units of length $(4 + 3)$, and divide it up into 49 small squares. It would be obvious that the whole square could be exhibited as containing four rectangles of sides 4, 3 cyclically arranged round the figure with one unit square in the middle. (This same figure is given by Cantor, I_3, p. 680. to illustrate the method given in the Chinese "Chóu-peï".) It would be seen that

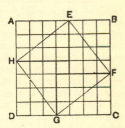

(i) the whole square (7^2) is made up of two squares 3^2 and 4^2, and two rectangles 3, 4;

(ii) the same square is made up of the square *EFGH* and the halves of four of the same rectangles 3, 4, whence the square *EFGH*, being equal to the sum of the squares 3^2 and 4^2, must contain 25 unit squares and its side, or the diagonal of one of the rectangles, must contain 5 units of length.

Or the result might equally be seen by observing that

(i) the square *EFGH* on the diagonal of one of the rectangles is made up of the halves of four rectangles and the unit square in the middle, while

(ii) the squares 3^2 and 4^2 placed at adjacent corners of the large square make up two rectangles 3, 4 with the unit square in the middle.

The procedure would be equally easy for any *rational* right-angled triangle, and would be a natural method of trying to *prove* the property when it had once been *empirically* observed that triangles like 3, 4, 5 did in fact contain a right angle.

Zeuthen has, in the same paper, shown in a most ingenious way how the property of the triangle 3, 4, 5 could be verified by a sort of combination of the second possible method by similar triangles, (*b*) on p. 354 above, with subdivision of *rectangles* into similar small *rectangles*. I give the method on account of its interest, although it is no doubt too advanced to have been used by those who first proved the property of the particular triangle.

Let *ABC* be a triangle right-angled at *A*, and such that the lengths of the sides *AB*, *AC* are 4 and 3 units respectively.

Draw the perpendicular *AD*, divide up *AB*, *AC* into unit lengths, complete the rectangle on *BC* as base and with *AD* as altitude, and subdivide this rectangle into small rectangles by drawing parallels to *BC*, *AD* through the points of division of *AB*, *AC*.

Now, since the diagonals of the small rectangles are all equal, each being of unit length, it follows by similar triangles that the small rectangles are all equal. And the rectangle with *AB* for diagonal contains 16 of the small rectangles, while the rectangle with diagonal *AC* contains 9 of them.

But the sum of the triangles *ABD*, *ADC* is equal to the triangle *ABC*.

Hence the rectangle with *BC* as diagonal contains 9 + 16 or 25 of the small rectangles;

and therefore *BC* = 5.

Rational right-angled triangles from the arithmetical standpoint.

Pythagoras investigated the *arithmetical* problem of finding rational numbers which could be made the sides of right-angled triangles, or of finding square numbers which are the sum of two squares; and herein we find the beginning of the *indeterminate analysis* which reached so high a stage of development in Diophantus. Fortunately Proclus has preserved Pythagoras' method of solution in the following passage (pp. 428, 7—429, 8). "Certain methods for the discovery of triangles of this kind are handed down, one of which they refer to Plato, and another to Pythagoras. [The latter] starts from odd numbers. For it makes the odd number the smaller of the sides about the right angle; then it takes the square of it, subtracts unity, and makes half the difference the greater of the sides about the right angle; lastly it adds unity to this and so forms the remaining side, the hypotenuse. For example, taking 3, squaring it, and subtracting unity from the 9, the method takes half of the 8, namely 4; then, adding unity to it again, it makes 5, and a right-angled triangle has been found with one side 3, another 4 and another 5. But the method of Plato argues from even numbers. For it takes the given even number and makes it one of the sides about the right angle; then, bisecting this number and squaring the half, it adds unity to the square to form the hypotenuse, and subtracts unity from the square to form the other side about the right angle. For example, taking 4, the method squares half of this, or 2, and so makes 4; then, subtracting unity, it produces 3, and adding unity it produces 5. Thus it has formed the same triangle as that which was obtained by the other method."

The formula of Pythagoras amounts, if m be an odd number, to

$$m^2 + \left(\frac{m^2 - 1}{2}\right)^2 = \left(\frac{m^2 + 1}{2}\right)^2,$$

the sides of the right-angled triangle being m, $\dfrac{m^2 - 1}{2}$, $\dfrac{m^2 + 1}{2}$. Cantor (I₃, pp. 185—6), taking up an idea of Röth (*Geschichte der abendländischen Philosophie*, II. 527), gives the following as a possible explanation of the way in which Pythagoras arrived at his formula. If $c^2 = a^2 + b^2$, it follows that

$$a^2 = c^2 - b^2 = (c + b)(c - b).$$

Numbers can be found satisfying the first equation if (1) $c + b$ and $c - b$ are either both even or both odd, and if further (2) $c + b$ and $c - b$ are such numbers as, when multiplied together, produce a square number. The first condition is necessary because, in order that c and b may both be whole numbers, the sum and difference of $c + b$ and $c - b$ must both be even. The second condition is satisfied if $c + b$ and $c - b$ are what were called *similar numbers* (ὅμοιοι ἀριθμοί); and that such numbers were most probably known in the time before Plato may be inferred from their appearing in Theon of Smyrna (*Expositio rerum mathematicarum ad legendum Platonem utilium*, ed. Hiller, p. 36, 12), who says that similar plane numbers are, first, all square numbers and, secondly, such oblong numbers as have the sides which contain them proportional. Thus 6 is an oblong number with length 3 and breadth 2; 24 is another with length 6 and breadth 4. Since therefore 6 is to 3 as 4 is to 2, the numbers 6 and 24 are similar.

Now the simplest case of two similar numbers is that of 1 and a^2, and, since 1 is odd, the condition (1) requires that a^2, and therefore a, is also odd. That is, we may take 1 and $(2n + 1)^2$ and equate them respectively to $c - b$ and $c + b$, whence we have

$$b = \frac{(2n + 1)^2 - 1}{2},$$

$$c = \frac{(2n + 1)^2 - 1}{2} + 1,$$

while $a = 2n + 1.$

As Cantor remarks, the form in which c and b appear correspond sufficiently closely to the description in the text of Proclus.

Another obvious possibility would be, instead of equating $c - b$ to unity, to put $c - b = 2$, in which case the similar number $c + b$ must be equated to double of some square, i.e. to a number of the form $2n^2$, or to the half of an even square number, say $\dfrac{(2n)^2}{2}$. This would give

$$a = 2n,$$
$$b = n^2 - 1,$$
$$c = n^2 + 1,$$

which is Plato's solution, as given by Proclus.

The two solutions supplement each other. It is interesting to observe that the method suggested by Röth and Cantor is very like that of Eucl. x. (Lemma 1 following Prop. 28). We shall come to this later, but it may be mentioned here that the problem is *to find two square numbers such that their sum is also a square.* Euclid there uses the property of II. 6 to the effect that, if *AB* is bisected at *C* and produced to *D*,

$$AD . DB + BC^2 = CD^2.$$

We may write this $\qquad uv = c^2 - b^2,$

where $\qquad u = c + b, \quad v = c - b.$

In order that uv may be a square, Euclid points out that u and v must be similar numbers, and further that u and v must be either both odd or both even in order that b may be a whole number. We may then put for the similar numbers, say, $a\beta^2$ and $a\gamma^2$, whence (if $a\beta^2$, $a\gamma^2$ are either both odd or both even) we obtain the solution

$$a\beta^2 . a\gamma^2 + \left(\frac{a\beta^2 - a\gamma^2}{2}\right)^2 = \left(\frac{a\beta^2 + a\gamma^2}{2}\right)^2.$$

But I think a serious, and even fatal, objection to the conjecture of Cantor and Röth is the very fact that the method enables both the Pythagorean and the Platonic series of triangles to be deduced with equal ease. If this had been the case with the method used by Pythagoras, it would not, I think, have been left to Plato to discover the second series of such triangles. It seems to me therefore that Pythagoras must have used some method which would produce his rule *only*; and further it would be some less recondite method, suggested by direct *observation* rather than by argument from general principles.

One solution satisfying these conditions is that of Bretschneider (p. 83), who suggests the following simple method. Pythagoras was certainly aware that the successive odd numbers are *gnomons*, or the differences between successive square numbers. It was then a simple matter to write down in three rows (*a*) the natural numbers, (*b*) their squares, (*c*) the successive odd numbers constituting the differences between the successive squares in (*b*), thus:

1	2	3	4	5	6	7	8	9	10	11	12	13	14
1	4	9	16	25	36	49	64	81	100	121	144	169	196
1	3	5	7	9	11	13	15	17	19	21	23	25	27

Pythagoras had then only to pick out the numbers in the third row which are squares, and his rule would be obtained by finding the formula connecting the square in the third line with the two adjacent squares in the second line. But even this would require some little argument; and I think a still better suggestion, because making pure observation play a greater part, is that of P. Treutlein (*Zeitschrift für Mathematik und Physik*, XXVIII., 1883, Hist.-litt. Abtheilung, pp. 209 sqq.).

We have the best evidence (e.g. in Theon of Smyrna) of the practice of representing square numbers and other figured numbers, e.g. oblong, triangular, hexagonal, by dots or signs arranged in the shape of the particular figure. (Cf. Aristotle, *Metaph.* 1092 b 12). Thus, says Treutlein, it would be easily seen that any square number can be turned into the next higher square by putting a single row of dots round two adjacent sides, in the form of a gnomon (see figures on next page).

If a is the side of a particular square, the gnomon round it is shown by simple inspection to contain $2a + 1$ dots or units. Now, in order that $2a + 1$ may itself be a square, let us suppose

$$2a + 1 = n^2,$$

whence $\qquad a = \tfrac{1}{2}(n^2 - 1),$

and $\qquad a + 1 = \tfrac{1}{2}(n^2 + 1).$

In order that a and $a + 1$ may be integral, n must be odd, and we have at once the Pythagorean formula

$$n^2 + \left(\frac{n^2 - 1}{2}\right)^2 = \left(\frac{n^2 + 1}{2}\right)^2.$$

I think Treutlein's hypothesis is shown to be the correct one by the passage in Aristotle's *Physics* already quoted, where the reference is undoubtedly to the Pythagoreans, and odd numbers are clearly identified with *gnomons* "placed round 1." But the ancient commentaries on the passage make the matter clearer still. Philoponus says: "As a proof...the Pythagoreans refer to what

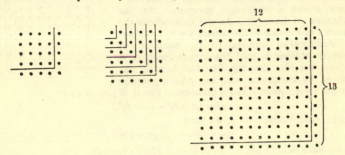

happens with the addition of numbers; for when the odd numbers are successively added to a square number they keep it square and equilateral.... Odd numbers are accordingly called *gnomons* because, when added to what are already squares, they preserve the square form....Alexander has excellently said in explanation that the phrase 'when gnomons are placed round' means *making a figure* with the odd numbers (τὴν κατὰ τοὺς περιττοὺς ἀριθμοὺς σχηματογραφίαν)...for it is the practice with the Pythagoreans to *represent things in figures* (σχηματογραφεῖν)."

The next question is: assuming this explanation of the Pythagorean formula, what are we to say of the origin of Plato's? It could of course be obtained as a particular case of the general formula of Eucl. x. already referred to; but there are two simple alternative explanations in this case also. (1) Bretschneider observes that, to obtain Plato's formula, we have only to double the sides of the squares in the Pythagorean formula,

for $$(2n)^2 + (n^2 - 1)^2 = (n^2 + 1)^2,$$

where however n is not necessarily odd.

(2) Treutlein would explain by means of an extension of the gnomon idea. As, he says, the Pythagorean formula was obtained by placing a gnomon consisting of a single row of dots round two adjacent sides of a square, it would be natural to try whether another solution could not be found by placing round the square a gnomon consisting of a *double* row of dots. Such a gnomon would equally turn the square into a larger square; and the question would be whether the double-row gnomon itself could be a square. If the side of the original square was a, it would easily be seen that the number of units in the double-row gnomon would be $4a + 4$, and we have only to put

$$4a + 4 = 4n^2,$$

whence
$$a = n^2 - 1,$$
$$a + 2 = n^2 + 1,$$
and we have the Platonic formula
$$(2n)^2 + (n^2 - 1)^2 = (n^2 + 1)^2.$$

I think this is, in substance, the right explanation, but, in form, not quite correct. The Greeks would not, I think, have treated the *double* row as a gnomon. Their comparison would have been between (1) a certain square *plus* a single-row gnomon and (2) the same square *minus* a single-row gnomon. As the application of Eucl. II. 4 to the case where the segments of the side of the square are a, 1 enables the Pythagorean formula to be obtained as Treutlein obtains it, so I think that Eucl. II. 8 confirms the idea that the Platonic formula was obtained by comparing a square *plus* a gnomon with the same square *minus* a gnomon. For II. 8 proves that

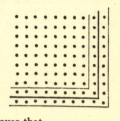

$$4ab + (a - b)^2 = (a + b)^2,$$
whence, substituting 1 for b, we have
$$4a + (a - 1)^2 = (a + 1)^2,$$
and we have only to put $a = n^2$ to obtain Plato's formula.

The "theorem of Pythagoras" in India.

This question has been discussed anew in the last few years as the result of the publication of two important papers by Albert Bürk on *Das Āpastamba-Sulba-Sūtra* in the *Zeitschrift der deutschen morgenländischen Gesellschaft* (LV., 1901, pp. 543—591, and LVI., 1902, pp. 327—391). The first of the two papers contains the introduction and the text, the second the translation with notes. A selection of the most important parts of the material was made and issued by G. Thibaut in the *Journal of the Asiatic Society of Bengal*, XLIV., 1875, Part I. (reprinted also at Calcutta, 1875, as *The Śulvasūtras*, by G. Thibaut). Thibaut in this work gave a most valuable comparison of extracts from the three Śulvasūtras by Baudhāyana, Āpastamba and Kātyāyana respectively, with a running commentary and an estimate of the date and originality of the geometry of the Indians. Bürk has however done good service by making the Āpastamba-S.-S. accessible in its entirety and investigating the whole subject afresh. With the natural enthusiasm of an editor for the work he is editing, he roundly maintains, not only that the Pythagorean theorem was known and proved in all its generality by the Indians long before the date of Pythagoras (about 580—500 B.C.), but that they had also discovered the irrational; and further that, so far from Indian geometry being indebted to the Greek, the much-travelled Pythagoras probably obtained his theory from India (*loc. cit.* LV., p. 575 note). Three important notices and criticisms of Bürk's work have followed, by H. G. Zeuthen ("*Théorème de Pythagore*," *Origine de la Géométrie scientifique*, 1904, already quoted), by Moritz Cantor (*Über die älteste indische Mathematik* in the *Archiv der Mathematik und Physik*, VIII., 1905, pp. 63—72) and by Heinrich Vogt (*Haben die alten Inder den Pythagoreischen Lehrsatz und das Irrationale gekannt?* in the *Bibliotheca Mathematica*, VII₃, 1906, pp. 6—23. See also Cantor's *Geschichte der Mathematik*, I₃, pp. 635—645.

The general effect of the criticisms is, I think, to show the necessity for the greatest caution, to say the least, in accepting Bürk's conclusions.

I proceed to give a short summary of the portions of the contents of the Āpastamba-Ś.-S. which are important in the present connexion. It may be premised that the general object of the book is to show how to construct altars of certain shapes, and to vary the dimensions of altars without altering the form. It is a collection of *rules* for carrying out certain constructions. There are no proofs, the nearest approach to a proof being in the rule for obtaining the area of an isosceles trapezium, which is done by drawing a perpendicular from one extremity of the smaller of the two parallel sides to the greater, and then taking away the triangle so cut off and placing it, the other side up, adjacent to the other equal side of the trapezium, thereby transforming the trapezium into a rectangle. It should also be observed that Āpastamba does not speak of *right-angled triangles*, but of two adjacent sides and the diagonal of a *rectangle*. For brevity, I shall use the expression "rational rectangle" to denote a rectangle the two sides and the diagonal of which can be expressed in terms of rational numbers. The references in brackets are to the chapters and numbers of Āpastamba's work.

(1) Constructions of right angles by means of cords of the following relative lengths respectively:

$$\begin{cases} 3, \; 4, \; 5 \quad (\text{I. 3, v. 3}) \\ 12, 16, 20 \quad (\text{v. 3}) \\ 15, 20, 25 \quad (\text{v. 3}) \end{cases}$$

$$\begin{cases} 5, \; 12, \; 13 \quad (\text{v. 4}) \\ 15, 36, 39 \quad (\text{I. 2, v. 2, 4}) \end{cases}$$

$$8, 15, 17 \quad (\text{v. 5})$$

$$12, 35, 37 \quad (\text{v. 5})$$

(2) A general enunciation of the Pythagorean theorem thus: "The diagonal of a rectangle produces [i.e. the square on the diagonal is equal to] the sum of what the longer and shorter sides separately produce [i.e. the squares on the two sides]." (I. 4)

(3) The application of the Pythagorean theorem to a *square* instead of a rectangle [i.e. to an *isosceles* right-angled triangle]: "The diagonal of a square produces an area double [of the original square]." (I. 5)

(4) An approximation to the value of $\sqrt{2}$; the diagonal of a square is $\left(1 + \dfrac{1}{3} + \dfrac{1}{3 \cdot 4} - \dfrac{1}{3 \cdot 4 \cdot 34}\right)$ times the side. (I. 6)

(5) Application of this approximate value to the construction of a square with side of any length. (II. 1)

(6) The construction of $a \sqrt{3}$, by means of the Pythagorean theorem, as the diagonal of a rectangle with sides a and $a \sqrt{2}$. (II. 2)

(7) Remarks equivalent to the following:

(a) $a \sqrt{\frac{1}{3}}$ is the side of $\frac{1}{9} (a \sqrt{3})^2$, or $a \sqrt{\frac{1}{3}} = \frac{1}{3} a \sqrt{3}$. (II. 3)

(b) A square on length of 1 unit gives 1 unit square (III. 4)

„	„	2	units gives 4	unit squares	(III. 6)
„	„	3	„	9 „	(III. 6)
„	„	$1\frac{1}{2}$	„	$2\frac{1}{4}$ „	(III. 8)

A square on length of $2\frac{1}{2}$ units gives $6\frac{1}{4}$ unit squares   (III. 8)

„   „ $\frac{1}{2}$ unit gives $\frac{1}{4}$ unit square   (III. 10)

„   „ $\frac{1}{3}$ „ $\frac{1}{9}$ „   (III. 10)

(c)   Generally, the square on any length contains as many rows (of small, unit, squares) as the length contains units.   (III. 7)

(8)   Constructions, by means of the Pythagorean theorem, of

    (a)   the *sum* of two squares as one square,   (II. 4)

    (b)   the *difference* of two squares as one square.   (II. 5)

(9)   A transformation of a rectangle into a square.   (II. 7)

[This is not directly done as by Euclid in II. 14, but the rectangle is first transformed into a gnomon, i.e. into the difference between two squares, which difference is then transformed into one square by the preceding rule. If *ABCD* be the given rectangle of which *BC* is the longer side, cut off the square *ABEF*, bisect the rectangle *DE* left over by *HG* parallel to *FE*, move the upper half *DG* and place it on *AF* as base in the position *AK*. Then the rectangle *ABCD* is equal to the gnomon which is the difference between the square *LB* and the square *LF*. In other words, Āpastamba transforms the rectangle *ab* into the difference between the squares $\left(\dfrac{a+b}{2}\right)^2$ and $\left(\dfrac{a-b}{2}\right)^2$.]

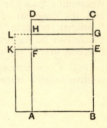

(10)   An attempt at a transformation of a square (a^2) into a rectangle which shall have one side of given length (*b*).   (III. 1)

[This shows no sign of such a procedure as that of Eucl. I. 44, and indeed does no more than say that we must subtract *ab* from a^2 and then adapt the remainder $a^2 - ab$ so that it may "fit on" to the rectangle *ab*. The problem is therefore only reduced to another of the same kind, and presumably it was only solved *arithmetically* in the case where *a*, *b* are given numerically. The Indian was therefore far from the general, geometrical, solution.]

(11)   Increase of a given square into a larger square.   (III. 9)

[This amounts to saying that you must add two rectangles (*a*, *b*) and another square (b^2) in order to transform a square a^2 into a square $(a+b)^2$. The formula is therefore that of Eucl. II. 4, $a^2 + 2ab + b^2 = (a+b)^2$.]

The first important question in relation to the above is that of date. Bürk assigns to the *Āpastamba-Śulba-Sūtra* a date at least as early as the 5th or 4th century B.C. He observes however (what is likely enough) that the matter of it must have been much older than the book itself. Further, as regards one of the constructions for right angles, that by means of cords of lengths 15, 36, 39, he shows that it was known at the time of the *Taittirīya-Saṃhitā* and the *Śatapatha-Brāhmaṇa*, still older works belonging to the 8th century B.C. at latest. It may be that (as Bürk maintains) the discovery that triangles with sides (*a*, *b*, *c*) in rational numbers such that $a^2 + b^2 = c^2$ are right-angled was nowhere made so early as in India. We find however in two ancient Chinese treatises (1) a statement that the diagonal of the rectangle (3, 4) is 5 and (2) a rule for finding the hypotenuse of a "right triangle" from the sides, while tradition connects both works with the name of Chou Kung who died 1105 B.C. (D. E. Smith, *History of Mathematics*, I. pp. 30—33, II. p. 288).

As regards the various "rational rectangles" used by Apastamba, it is to be observed that two of the seven, viz. 8, 15, 17 and 12, 35, 37, do not belong to the Pythagorean series, the others consist of two which belong to it, viz. 3, 4, 5 and 5, 12, 13, and multiples of these. It is true, as remarked by Zeuthen (*op. cit.* p. 842), that the rules of II. 7 and III. 9, numbered (9) and (11) above respectively, would furnish the means of finding any number of "rational rectangles." But it would not appear that the Indians had been able to formulate any general rule; otherwise their list of such rectangles would hardly have been so meagre. Apastamba mentions seven only, really reducible to four (though one other, 7, 24, 25, appears in the Baudhāyana-Ś.-S., supposed to be older than Apastamba). These are all that Apastamba knew of, for he adds (v. 6): "So many *recognisable* (erkennbare) constructions are there," implying that he knew of no other "rational rectangles" that could be employed. But the words also imply that the theorem of the square on the diagonal is also true of other rectangles not of the "recognisable" kind, i.e. rectangles in which the sides and the diagonal are not in the ratio of integers; this is indeed implied by the constructions for $\sqrt{2}$, $\sqrt{3}$ etc. up to $\sqrt{6}$ (cf. II. 2, VIII. 5). This is all that can be said. The theorem is, it is true, enunciated as a general proposition, but there is no sign of anything like a general proof; there is nothing to show that the assumption of its universal truth was founded on anything better than an imperfect induction from a certain number of cases, discovered empirically, of triangles with sides in the ratio of whole numbers in which the property (1) that the square on the longest side is equal to the sum of the squares on the other two sides was found to be always accompanied by the property (2) that the latter two sides include a right angle.

It remains to consider Bürk's claim that the Indians had discovered the *irrational*. This is based upon the approximate value of $\sqrt{2}$ given by Apastamba in his rule I. 6 numbered (4) above. There is nothing to show how this was arrived at, but Thibaut's suggestion certainly seems the best and most natural. The Indians may have observed that $17^2 = 289$ is nearly double of $12^2 = 144$. If so, the next question which would naturally occur to them would be, by how much the side 17 must be diminished in order that the square on it may be 288 *exactly*. If, in accordance with the Indian fashion, a gnomon with unit area were to be subtracted from a square with 17 as side, this would approximately be secured by giving the gnomon the breadth $\frac{1}{34}$, for $2 \times 17 \times \frac{1}{34} = 1$. The side of the smaller square thus arrived at would be $17 - \frac{1}{34} = 12 + 4 + 1 - \frac{1}{34}$, whence, dividing out by 12, we have

$$\sqrt{2} = 1 + \frac{1}{3} + \frac{1}{3 \cdot 4} - \frac{1}{3 \cdot 4 \cdot 34}, \text{ approximately.}$$

But it is a far cry from this calculation of an approximate value to the discovery of the *irrational*. First, we ask, is there any sign that this value was known to be inexact? It comes directly after the statement (I. 6) that the square on the diagonal of a square is double of that square, and the rule is quite boldly stated without any qualification : "lengthen the unit by one-third and the latter by one-quarter of itself less one-thirty-fourth of this part." Further, the approximate value is actually used for the purpose of constructing a square when the side is given (II. 1). So familiar was the formula that it was apparently made the basis of a sub-division of measures of length.

Thibaut observes (*Journal of the Asiatic Society of Bengal*, XLIX., p. 241) that, according to Bāudhāyana, the unit of length was divided into 12 *fingerbreadths*, and that one of two divisions of the *fingerbreadth* was into 34 *sesame-corns*, and he adds that he has no doubt that this division, which he has not elsewhere met, owes its origin to the formula for $\sqrt{2}$. The result of using this sub-division would be that, in a square with side equal to 12 *fingerbreadths*, the diagonal would be 17 *fingerbreadths* less 1 *sesame-corn*. Is it conceivable that a sub-division of a measure of length would be based on an evaluation known to be inexact? No doubt the first discoverer would be aware that the area of a gnomon with breadth $\frac{1}{34}$ and outer side 17 is not exactly equal to 1 but less than it by the square of $\frac{1}{34}$ or by $\frac{1}{1156}$, and therefore that, in taking that gnomon as the proper area to be subtracted from 17^2, he was leaving out of account the small fraction $\frac{1}{1156}$; as, however, the object of the whole proceeding was purely practical, he would, without hesitation, ignore this as being of no practical importance, and, thereafter, the formula would be handed down and taken as a matter of course without arousing suspicion as to its accuracy. This supposition is confirmed by reference to the sort of rules which the Indians allowed themselves to regard as accurate. Thus Āpastamba himself gives a construction for a circle equal in area to a given square, which is equivalent to taking $\pi = 3·09$, and yet observes that it gives the required circle "*exactly*" (III. 2), while his construction of a square equal to a circle, which he equally calls "exact," makes the side of the square equal to $\frac{13}{15}$ths of the diameter of the circle (III. 3), and is equivalent to taking $\pi = 3·004$. But, even if some who used the approximation for $\sqrt{2}$ were conscious that it was not quite accurate (of which there is no evidence), there is an immeasurable difference between arrival at this consciousness and the discovery of the irrational. As Vogt says, three stages had to be passed through before the irrationality of the diagonal of a square was discovered in any real sense. (1) All values found by direct measurement or calculations based thereon have to be recognised as being inaccurate. Next (2) must supervene the conviction that it is *impossible* to arrive at an accurate arithmetical expression of the value. And lastly (3) the impossibility must be proved. Now there is no real evidence that the Indians, at the date in question, had even reached the first stage, still less the second or third.

The net results then of Bürk's papers and of the criticisms to which they have given rise appear to be these. (1) It must be admitted that Indian geometry had reached the stage at which we find it in Āpastamba quite independently of Greek influence. But (2) the old Indian geometry was purely empirical and practical, far removed from abstractions such as the irrational. The Indians had indeed, by·trial in particular cases, persuaded themselves of the truth of the Pythagorean theorem and enunciated it in all its generality; but they had not established it by scientific proof.

Alternative proofs.

I. The well-known proof of I. 47 obtained by putting two squares side by side, with their bases continuous, and cutting off right-angled triangles which can then be put on again in different positions, is attributed by an-Nairīzī to Thābit b. Qurra (826—901 A.D.).

His actual construction proceeds thus.

Let ABC be the given triangle right-angled at A.

Construct on AB the square AD;

produce AC to F so that EF may be equal to AC.

Construct on *EF* the square *EG*, and produce *DH* to *K* so that *DK* may be equal to *AC*.

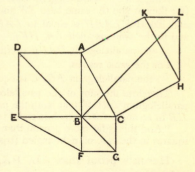

It is then proved that, in the triangles
BAC, *CFG*, *KHG*, *BDK*,
the sides *BA*, *CF*, *KH*, *BD* are all equal,
and
the sides *AC*, *FG*, *HG*, *DK* are all equal.

The angles included by the equal sides are all right angles; hence the four triangles are equal in all respects.   [I. 4]

Hence *BC*, *CG*, *GK*, *KB* are all equal.
Further the angles *DBK*, *ABC* are equal;
hence, if we add to each the angle *DBC*,
the angle *KBC* is equal to the angle *ABD* and is therefore a right angle.

In the same way the angle *CGK* is right;
therefore *BCGK* is a square, i.e. the square on *BC*.

Now the sum of the quadrilateral *GCLH* and the triangle *LDB* together with two of the equal triangles make the squares on *AB*, *AC*, and together with the other two make the square on *BC*.

Therefore etc.

II.   Another proof is easily arrived at by taking the particular case of Pappus' more general proposition given below in which the given triangle is right-angled and the parallelograms on the sides containing the right angles are squares.   If the figure is drawn, it will be seen that, with no more than one additional line inserted, it contains Thābit's figure, so that Thābit's proof may have been practically derived from that of Pappus.

III.   The most interesting of the remaining proofs seems to be that shown in the accompanying figure.

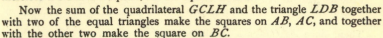

It is given by J. W. Müller, *Systema-tische Zusammenstellung der wichtigsten bisher bekannten Beweise des Pythag. Lehrsatzes* (Nürnberg, 1819), and in the second edition (Mainz, 1821) of Ign. Hoffmann, *Der Pythag. Lehrsatz mit 32 theils bekannten theils neuen Beweisen* [3 more in second edition]. It appears to come from one of the scientific papers of Lionardo da Vinci (1452—1519).

The triangle *HKL* is constructed on the base *KH* with the side *KL* equal to *BC* and the side *LH* equal to *AB*.

Then the triangle *HLK* is equal in all respects to the triangle *ABC*, and to the triangle *EBF*.

Now *DB*, *BG*, which bisect the angles *ABE*, *CBF* respectively, are in a straight line.   Join *BL*.

It is easily proved that the four quadrilaterals *ADGC*, *EDGF*, *ABLK*, *HLBC* are all equal.

Hence the hexagons *ADEFGC, ABCHLK* are equal.

Subtracting from the former the two triangles *ABC, EBF*, and from the · latter the two equal triangles *ABC, HLK*, we prove that

the square *CK* is equal to the sum of the squares *AE, CF*.

Pappus' extension of I. 47.

In this elegant extension the triangle may be *any* triangle (not necessarily right-angled), and *any* parallelograms take the place of squares on two of the sides.

Pappus (IV. p. 177) enunciates the theorem as follows :

If ABC *be a triangle, and any parallelograms whatever* ABED, BCFG *be described on* AB, BC, *and if* DE, FG *be produced to* H, *and* HB *be joined, the parallelograms* ABED, BCFG *are equal to the parallelogram contained by* AC, HB *in an angle which is equal to the sum of the angles* BAC, DHB.

Produce *HB* to *K*; through *A, C* draw *AL, CM* parallel to *HK*, and join *LM.*

Then, since *ALHB* is a parallelogram, *AL, HB* are equal and parallel. Similarly *MC, HB* are equal and parallel.

Therefore *AL, MC* are equal and parallel;

whence *LM, AC* are also equal and parallel,

and *ALMC* is a parallelogram.

Further, the angle *LAC* of this parallelogram is equal to the sum of the angles *BAC, DHB*, since the angle *DHB* is equal to the angle *LAB*.

Now, since the parallelogram *DABE* is equal to the parallelogram *LABH* (for they are on the same base *AB* and in the same parallels *AB, DH*),

and likewise *LABH* is equal to *LAKN* (for they are on the same base *LA* and in the same parallels *LA, HK*),

the parallelogram *DABE* is equal to the parallelogram *LAKN*.

For the same reason,

the parallelogram *BGFC* is equal to the parallelogram *NKCM*.

Therefore the sum of the parallelograms *DABE, BGFC* is equal to the parallelogram *LACM*, that is, to the parallelogram which is contained by *AC, HB* in an angle *LAC* which is equal to the sum of the angles *BAC, BHD*.

"And this is far more general than what is proved in the Elements about squares in the case of right-angled (triangles)."

Heron's proof that AL, BK, CF in Euclid's figure meet in a point.

The final words of Proclus' note on I. 47 (p. 429, 9—15) are historically interesting. He says: "The demonstration by the writer of the Elements being clear, I consider that it is unnecessary to add anything further, and that we may be satisfied with what has been written, since, in fact those who have added anything more, like Pappus and Heron, were obliged to draw upon what is proved in the sixth Book, for no really useful object." These words cannot of course refer to the extension of I. 47 given by Pappus; but the key to them, so far as Heron is concerned, is to be found in the commentary of an-Nairīzī (pp. 175—185, ed. Besthorn-Heiberg; pp. 78—84, ed. Curtze) on I. 47, wherein he gives Heron's proof that the lines *AL*, *FC*, *BK* in Euclid's figure meet in a point. Heron proved this by means of three lemmas which would most naturally be proved from the principle of similitude as laid down in Book VI., but which Heron, as a *tour de force*, proved on the principles of Book I. only. The *first* lemma is to the following effect.

If, in a triangle ABC, DE be drawn parallel to the base BC, and if AF be drawn from the vertex A to the middle point F of BC, then AF will also bisect DE.

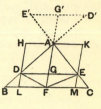

This is proved by drawing *HK* through *A* parallel to *DE* or *BC*, and *HDL*, *KEM* through *D*, *E* respectively parallel to *AGF*, and lastly joining *DF*, *EF*.

Then the triangles *ABF*, *AFC* are equal (being on equal bases), and the triangles *DBF*, *EFC* are also equal (being on equal bases and between the same parallels).

Therefore, by subtraction, the triangles *ADF*, *AEF* are equal, and hence the parallelograms *AL*, *AM* are equal.

These parallelograms are between the same parallels *LM*, *HK*; therefore *LF*, *FM* are equal, whence *DG*, *GE* are also equal.

The *second* lemma is an extension of this to the case where *DE* meets *BA*, *CA* produced beyond *A*.

The *third* lemma proves the converse of Euclid I. 43, that, *If a parallelogram AB is cut into four others ADGE, DF, FGCB, CE, so that DF, CE are equal, the common vertex G will be on the diagonal AB.*

Heron produces *AG* till it meets *CF* in *H*. Then, if we join *HB*, we have to prove that *AHB* is one straight line. The proof is as follows. Since the areas *DF*, *EC* are equal, the triangles *DGF*, *ECG* are equal.

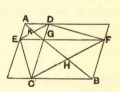

If we add to each the triangle *GCF*,

the triangles *ECF*, *DCF* are equal;

therefore *ED*, *CF* are parallel.

Now it follows from I. 34, 29 and 26 that the triangles *AKE*, *GKD* are equal in all respects;

therefore *EK* is equal to *KD*.

Hence, by the second lemma,

CH is equal to *HF*.

Therefore, in the triangles *FHB*, *CHG*,
the two sides *BF*, *FH* are equal to the two sides *GC*, *CH*,

and the angle *BFH* is equal to the angle *GCH*;

hence the triangles are equal in all respects,
and the angle *BHF* is equal to the angle *GHC*.

Adding to each the angle *GHF*, we find that the angles *BHF*, *FHG* are equal to the angles *CHG*, *GHF*,

and therefore to two right angles.

Therefore *AHB* is a straight line.

Heron now proceeds to prove the proposition that, in the accompanying figure, if *AKL* perpendicular to *BC* meet *EC* in *M*, and if *BM*, *MG* be joined,

BM, *MG* are in one straight line.

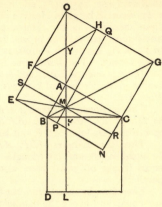

Parallelograms are completed as shown in the figure, and the diagonals *OA*, *FH* of the parallelogram *FH* are drawn.

Then the triangles *FAH*, *BAC* are clearly equal in all respects;

therefore the angle *HFA* is equal to the angle *ABC*, and therefore to the angle *CAK* (since *AK* is perpendicular to *BC*).

But, the diagonals of the rectangle *FH* cutting one another in *Y*,

FY is equal to *YA*,

and the angle *HFA* is equal to the angle *OAF*.

Therefore the angles *OAF*, *CAK* are equal, and accordingly

OA, *AK* are in a straight line.

Hence *OM* is the diagonal of *SQ*;

therefore *AS* is equal to *AQ*,

and, if we add *AM* to each,

FM is equal to *MH*.

But, since *EC* is the diagonal of the parallelogram *FN*,

FM is equal to *MN*.

Therefore *MH* is equal to *MN*;

and, by the third lemma, *BM*, *MG* are in a straight line.

PROPOSITION 48.

If in a triangle the square on one of the sides be equal to the squares on the remaining two sides of the triangle, the angle contained by the remaining two sides of the triangle is right.

For in the triangle *ABC* let the square on one side *BC* be equal to the squares on the sides *BA*, *AC*;

I say that the angle *BAC* is right.

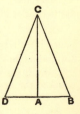

For let *AD* be drawn from the point *A* at right angles to the straight line *AC*, let *AD* be made equal to *BA*, and let *DC* be joined.

Since *DA* is equal to *AB*, the square on *DA* is also equal to the square on *AB*.

Let the square on *AC* be added to each;

therefore the squares on DA, AC are equal to the squares on BA, AC.

But the square on DC is equal to the squares on DA, AC, for the angle DAC is right ;   [I. 47]

and the square on BC is equal to the squares on BA, AC, for this is the hypothesis ;

therefore the square on DC is equal to the square on BC,

so that the side DC is also equal to BC.

And, since DA is equal to AB,

and AC is common,

the two sides DA, AC are equal to the two sides BA, AC ;

and the base DC is equal to the base BC ;

therefore the angle DAC is equal to the angle BAC. [I. 8]

But the angle DAC is right ;

therefore the angle BAC is also right.

Therefore etc.   Q. E. D

Proclus' note (p. 430) on this proposition, though it does not mention Heron's name, gives an alternative proof, which is the same as that definitely attributed by an-Nairīzī to Heron, the only difference being that Proclus demonstrates two cases in full, while Heron dismisses the second with a "similarly." The alternative proof is another instance of the use of I. 7 as a means of answering objections. If, says Proclus, it be not admitted that the perpendicular AD may be drawn on the opposite side of AC from B, we may draw it on the same side as AB, in which case it is impossible that it should not coincide with AB. Proclus takes two cases, first supposing that the perpendicular falls, as AD, within the angle CAB, and secondly that it falls, as AE, outside that angle. In either case the absurdity results that, on the same straight line AC and on the same side of it, AD, DC must be respectively equal to AB, BC, which contradicts I. 7.

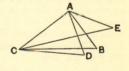

Much to the same effect is the note of De Morgan that there is here "an appearance of avoiding indirect demonstration by drawing the triangles on different sides of the base and appealing to I. 8, because drawing them on the same side would make the appeal to I. 7 (on which, however, I. 8 is founded)."

BOOK II.

DEFINITIONS.

1. Any rectangular parallelogram is said to be **contained** by the two straight lines containing the right angle.

2. And in any parallelogrammic area let any one whatever of the parallelograms about its diameter with the two complements be called a **gnomon**.

DEFINITION 1.

Πᾶν παραλληλόγραμμον ὀρθογώνιον περιέχεσθαι λέγεται ὑπὸ δύο τῶν τὴν ὀρθὴν γωνίαν περιεχουσῶν εὐθειῶν.

As the full expression in Greek for "the angle BAC" is "the angle contained by the (straight lines) BA, AC," ἡ ὑπὸ τῶν ΒΑ, ΑΓ περιεχομένη γωνία, so the full expression for "the rectangle contained by BA, AC" is τὸ ὑπὸ τῶν ΒΑ, ΑΓ περιεχόμενον ὀρθογώνιον. In this case too ΒΑ, ΑΓ is commonly abbreviated by the Greek geometers into ΒΑΓ. Thus in Archimedes and Apollonius τὸ ὑπὸ ΒΑΓ or τὸ ὑπὸ τῶν ΒΑΓ means *the rectangle* BA, AC, just as ἡ ὑπὸ ΒΑΓ means *the angle* BAC; the gender of the article shows which is meant in each case. In the early Books Euclid uses the full expression τὸ ὑπὸ τῶν ΒΑ, ΑΓ; but the shorter form τὸ ὑπὸ τῶν ΒΑΓ is found from Book x. onwards. Cf. XII. 11, where τὰ (τμήματα) ἐπὶ τῶν ΘΟΕ, ΕΠΖ, ΖΡΗ, ΗΣΘ means the segments on the eight straight lines ΘΟ, ΟΕ, ΕΠ, ΠΖ, ΖΡ, ΡΗ, ΗΣ, ΣΘ.

DEFINITION 2.

Παντὸς δὲ παραλληλογράμμου χωρίου τῶν περὶ τὴν διάμετρον αὐτοῦ παραλληλογράμμων ἓν ὁποιονοῦν σὺν τοῖς δυσὶ παραπληρώμασι γνώμων καλείσθω.

Meaning literally a thing enabling something to be *known*, *observed* or *verified*, a *teller* or *marker*, as we might say, the word *gnomon* (γνώμων) was first used in the sense (1) in which it appears in a passage of Herodotus (II. 109) stating that "the Greeks learnt the πόλος, the *gnomon* and the twelve parts of the day from the Babylonians." According to Suidas, it was Anaximander (611—545 B.C.) who introduced the *gnomon* into Greece. Whatever may be the details of the construction of the two instruments called the πόλος and the *gnomon*, so much is certain, that the gnomon had to do with the measurement of time by shadows thrown by the sun, and that the word signified the placing of a staff perpendicular to the horizon. This is borne out by the statement of Proclus that Oenopides of Chios, who first investigated the problem (Eucl. I. 12) of drawing a perpendicular from an external point to a given straight line, called the perpendicular a straight line drawn "*gnomon-wise*" (κατὰ γνώμονα). Then (2) we find the term used of a mechanical instrument for drawing right angles, as shown in the figure annexed. This seems to be the meaning in Theognis 805, where it is said that the envoy sent to consult the oracle at Delphi should be "straighter (ἰθύτερος) than the τόρνος, the στάθμη and the *gnomon*," and all three words evidently denote appliances, the τόρνος being an instrument for drawing a circle

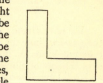

(probably a string stretched between a fixed and a moving point), and the στάθμη a plumb-line. Next (3) it was natural that the *gnomon*, owing to its shape, should become the figure which remained of a square when a smaller square was cut out of one corner (or the figure, as Aristotle says, which when *added* to a square increases its size but does not alter its form). We have seen (note on I. 47, p. 351) that the Pythagoreans used the term in this sense, and further applied it, by analogy, to the series of odd numbers as having the same property in relation to square *numbers*. The earliest evidence for this is the fragment of Philolaus (*c.* 460 B.C.) already mentioned (see Boeckh, *Philolaos des Pythagoreers Lehren*, p. 141) where he says that "number makes all things knowable and mutually agreeing (ποτάγορα ἀλλάλοις) in the way characteristic of the gnomon" (κατὰ γνώμονος φύσιν). As Boeckh says (p. 144), it would appear from the fragment that the connexion between the gnomon and the square to which it is added was regarded as symbolical of union and agreement, and that Philolaus used the idea to explain the knowledge of things, making the *knowing* embrace and grasp the *known* as the gnomon does the square. Cf. Scholium II. No. 11 (Euclid, ed. Heiberg, Vol. v. p. 225), which says "It is to be noted that the gnomon was discovered by geometers with a view to brevity, while the name came from its incidental property, namely that from it the whole is known, whether of the whole area or of the remainder, when it is either placed round or taken away. In sundials too its sole function is to make the actual time of day known."

The geometrical meaning of the word is extended in the definition of *gnomon* given by Euclid, where (4) the gnomon has the same relation to *any parallelogram* as it before had to a *square*. From the fact that Euclid says "*let*" the figure described "*be called* a gnomon" we may infer that he was using the word in the wider sense for the first time. Later still (5) we find Heron of Alexandria defining a *gnomon in general* as any figure which, when added to any figure

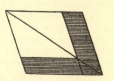

whatever, makes the whole figure similar to that to which it is added. In this definition of Heron (Def. 58) Hultsch brackets the words which make it apply to any *number* as well; but Theon of Smyrna, who explains that plane, triangular, square, solid and other kinds of numbers are so called after the likeness of the areas which they measure, does make the term in its most general sense apply to numbers. "All the successive numbers which [by being successively added] produce triangles or squares or polygons are called gnomons" (p. 37, 11—13, ed. Hiller). Thus the successive odd numbers added together make square numbers; the gnomons in the case of triangular numbers are the successive numbers 1, 2, 3, 4...; those for pentagonal numbers are the series 1, 4, 7, 10...(the common difference being 3), and so on. In general, the successive *gnomonic* numbers for any polygonal number, say of *n* sides, have $n - 2$ for their common difference (Theon of Smyrna, p. 34, 13—15).

Geometrical Algebra.

We have already seen (cf. part of the note on I. 47 and the above note on the *gnomon*) how the Pythagoreans and later Greek mathematicians exhibited different kinds of numbers as forming different geometrical figures. Thus, says Theon of Smyrna (p. 36, 6—11), "plane numbers, triangular, square and solid numbers, and the rest, are not so called independently (κυρίως) but in virtue of their similarity to the areas which they measure; for 4, since it measures a square area, is called square by adaptation from it, and 6 is called oblong for the same reason." A "plane number" is similarly described as a number obtained by multiplying two numbers together, which two numbers are sometimes spoken of as "sides," sometimes as the "length" and "breadth" respectively, of the number which is their product.

The *product* of two numbers was thus represented geometrically by the *rectangle* contained by the straight lines representing the two numbers respectively. It only needed the discovery of incommensurable or irrational straight lines in order to represent geometrically by a rectangle the product of any two quantities whatever, rational or irrational; and it was possible to advance from a geometrical arithmetic to a geometrical *algebra*, which indeed by Euclid's time (and probably long before) had reached such a stage of development that it could solve the same problems as our algebra so far as they do not involve the manipulation of expressions of a degree higher than the second. In order to make the geometrical algebra so generally effective, the theory of proportions was essential. Thus, suppose that *x, y, z* etc. are quantities which can be represented by straight lines, while α, β, γ etc. are coefficients which can be expressed by ratios between straight lines. We can then by means of Book VI. find a single straight line *d* such that

$$\alpha x + \beta y + \gamma z + \ldots = d.$$

To solve the simple equation in its general form

$$\alpha x + a = b,$$

where α represents any ratio between straight lines also requires recourse to the sixth Book, though, e.g., if α is $\frac{1}{2}$ or $\frac{1}{3}$ or any submultiple of unity, or if α is 2, 4 or any power of 2, we should not require anything beyond Book I. for solving the equation. Similarly the general form of a quadratic equation requires Book VI. for its geometrical solution, though particular quadratic equations may be so solved by means of Book II. alone.

Besides enabling us to solve geometrically these particular quadratic equations, Book II. gives the geometrical proofs of a number of algebraical formulae. Thus the first ten propositions give the equivalent of the several identities

1. $a (b + c + d + \ldots) = ab + ac + ad + \ldots,$

2. $(a + b) a + (a + b) b = (a + b)^2,$

3. $(a + b) a = ab + a^2,$

4. $(a + b)^2 = a^2 + b^2 + 2ab.$

5. $ab + \left(\dfrac{a+b}{2} - b\right)^2 = \left(\dfrac{a+b}{2}\right)^2$,

or $(\alpha + \beta)(\alpha - \beta) + \beta^2 = \alpha^2$,

6. $(2a + b)b + a^2 = (a + b)^2$,

or $(\alpha + \beta)(\beta - \alpha) + \alpha^2 = \beta^2$,

7. $(a + b)^2 + a^2 = 2(a + b)a + b^2$,

or $\alpha^2 + \beta^2 = 2\alpha\beta + (\alpha - \beta)^2$,

8. $4(a + b)a + b^2 = \{(a + b) + a\}^2$,

or $4\alpha\beta + (\alpha - \beta)^2 = (\alpha + \beta)^2$,

9. $a^2 + b^2 = 2\left\{\left(\dfrac{a+b}{2}\right)^2 + \left(\dfrac{a+b}{2} - b\right)^2\right\}$,

or $(\alpha + \beta)^2 + (\alpha - \beta)^2 = 2(\alpha^2 + \beta^2)$,

10. $(2a + b)^2 + b^2 = 2\{a^2 + (a + b)^2\}$,

or $(\alpha + \beta)^2 + (\beta - \alpha)^2 = 2(\alpha^2 + \beta^2)$.

The form of these identities may of course be varied according to the different symbols which we may use to denote particular portions of the lines given in Euclid's figures. They are, for the most part, simple identities, but there is no reason to suppose that these were the only applications of the geometrical algebra that Euclid and his predecessors had been able to make. We may infer the very contrary from the fact that Apollonius in his *Conics* frequently states without proof much more complicated propositions of the kind.

It is important however to bear in mind that the whole procedure of Book II. is *geometrical*; rectangles and squares are shown in the figures, and the equality of certain combinations to other combinations is proved by those figures. We gather that this was the classical or standard method of proving such propositions, and that the *algebraical* method of proving them, with no figure except a line with points marked thereon, was a later introduction. Accordingly Eutocius' method of proving certain lemmas assumed by Apollonius (*Conics*, II. 23 and III. 29) probably represents more nearly than Pappus' proof of the same the point of view from which Apollonius regarded them.

It would appear that Heron was the first to adopt the *algebraical* method of demonstrating the propositions of Book II., beginning from the second, without figures, as consequences of the first proposition corresponding to

$$a(b + c + d) = ab + ac + ad.$$

According to an-Nairīzī (ed. Curtze, p. 89), Heron explains that it is not possible to prove II. 1 without drawing a number of lines (i.e. without actually drawing the rectangles), but that the following propositions up to II. 10 inclusive can be proved by merely drawing one line. He distinguishes two varieties of the method, one by *dissolutio*, the other by *compositio*, by which he seems to mean *splitting-up* of rectangles and squares, and *combination* of them into others. But in his proofs he sometimes combines the two varieties.

When he comes to II. 11, he says that it is not possible to do without a figure because the proposition is a problem, which accordingly requires an *operation* and therefore the drawing of a figure.

The algebraical method has been preferred to Euclid's by some English editors; but it should not find favour with those who wish to preserve the essential features of Greek geometry as presented by its greatest exponents, or to appreciate their point of view.

It may not be out of place to add a word with reference to the geometrical equivalent of the algebraical operations. The addition and subtraction of quantities represented in the geometrical algebra by lines is of course effected by producing the line to the required extent or cutting off a portion of it. The equivalent of multiplication is the construction of the rectangle of which the given lines are adjacent sides. The equivalent of the division of one quantity represented by a line by another quantity represented by a line is simply the statement of a *ratio* between lines on the principles of Books v. and vi. The division of a product of two quantities by a third is represented in the geometrical algebra by the finding of a rectangle with one side of a given length and equal to a given rectangle or square. This is the problem of *application of areas* solved in I. 44, 45. The addition and subtraction of products is, in the geometrical algebra, the addition and subtraction of rectangles or squares; the sum or difference can be transformed into a single rectangle by means of the *application of areas* to any line of given length, corresponding to the algebraical process of finding a common measure. Lastly, the extraction of the square root is, in the geometrical algebra, the finding of a square equal to a given rectangle, which is done in II. 14 with the help of I. 47.

BOOK II. PROPOSITIONS.

PROPOSITION 1.

*If there be two straight lines, and one of them be cut into
any number of segments whatever, the rectangle contained by
the two straight lines is equal to the rectangles contained by the
uncut straight line and each of the segments.*

5   Let A, BC be two straight lines, and let BC be cut at
random at the points D, E;

I say that the rectangle contained by A, BC is equal to the
rectangle contained by A, BD,
that contained by A, DE and
10 that contained by A, EC.

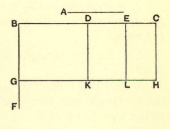

For let BF be drawn from B
at right angles to BC;   [I. 11]
let BG be made equal to A, [I. 3]
through G let GH be drawn
15 parallel to BC,   [I. 31]
and through D, E, C let DK,
EL, CH be drawn parallel to
BG.

Then BH is equal to BK, DL, EH.

20   Now BH is the rectangle A, BC, for it is contained by
GB, BC, and BG is equal to A;

BK is the rectangle A, BD, for it is contained by GB,
BD, and BG is equal to A;

and DL is the rectangle A, DE, for DK, that is BG [I. 34],
25 is equal to A.

Similarly also EH is the rectangle A, EC.

Therefore the rectangle A, BC is equal to the rectangle
A, BD, the rectangle A, DE and the rectangle A, EC.

Therefore etc.

Q. E. D.

20. **the rectangle A, BC.** From this point onward I shall translate thus in cases where Euclid leaves out the word *contained* (περιεχόμενον). Though the word "rectangle" is also omitted in the Greek (the neuter article being sufficient to show that the rectangle is meant), it cannot be dispensed with in English. De Morgan advises the use of the expression "the rectangle *under* two lines." This does not seem to me a very good expression, and, if used in a translation from the Greek, it might suggest that ὑπό in τὸ ὑπό meant *under*, which it does not.

This proposition, the geometrical equivalent of the algebraical formula

$$a (b + c + d + ...) = ab + ac + ad + ...,$$

can, of course, easily be extended so as to correspond to the more general algebraical proposition that the product of an expression consisting of any number of terms added together and another expression also consisting of any number of terms added together is equal to the sum of all the products obtained by multiplying each term of one expression by all the terms of the other expression, one after another. The geometrical proof of the more general proposition would be effected by means of a figure showing all the rectangles corresponding to the partial products, in the same way as they are shown in the simpler case of II. 1; the difference would be that a series of parallels to *BC* would have to be drawn as well as the series of parallels to *BF*.

PROPOSITION 2.

If a straight line be cut at random, the rectangle contained by the whole and both of the segments is equal to the square on the whole.

For let the straight line *AB* be cut at random at the point *C*;

I say that the rectangle contained by *AB*, *BC* together with the rectangle contained by *BA*, *AC* is equal to the square on *AB*.

For let the square *ADEB* be described on *AB* [I. 46], and let *CF* be drawn through *C* parallel to either *AD* or *BE*.   [I. 31]

Then *AE* is equal to *AF*, *CE*.

Now *AE* is the square on *AB*;

AF is the rectangle contained by *BA*, *AC*, for it is contained by *DA*, *AC*, and *AD* is equal to *AB*;

and *CE* is the rectangle *AB*, *BC*, for *BE* is equal to *AB*.

Therefore the rectangle *BA*, *AC* together with the rectangle *AB*, *BC* is equal to the square on *AB*.

Therefore etc.

Q. E. D.

The *fact* asserted in the enunciation of this proposition has already been used in the proof of I. 47 ; but there was no occasion in that proof to observe that the two rectangles *BL*, *CL* making up the square on *BC* are the rectangles contained by *BC* and the two parts, respectively, into which it is divided by the perpendicular from *A* on *BC*. It is this fact which it is necessary to state in this proposition, in accordance with the plan of Book II.

The second and third propositions are of course particular cases of the first. They were no doubt separately enunciated by Euclid in order that they might be immediately available for use hereafter, instead of having to be deduced for the particular occasion from II. I. For, if they had not been thus separately stated, it would scarcely have been practicable to quote them later without explaining at the same time that they are included in II. I as particular cases. And, though the propositions are not used by Euclid in the later propositions of Book II., they are used afterwards in XIII. 10 and IX. 15 respectively; and they are of extreme importance for geometry generally, being constantly used by Pappus, for example, who frequently quotes the third proposition by the Book and number.

Attention has been called to the fact that II. I is never used by Euclid; and this may seem no less remarkable than the fact that II. 2, 3 are not again used in Book II. But it is important, I think, to observe that the proofs of all the first ten propositions of Book II. are practically independent of each other, though the results are really so interwoven that they can often be deduced from each other in a variety of ways. What then was Euclid's intention, first in inserting some propositions not immediately required, and secondly in making the proofs of the first ten practically independent of each other? Surely the object was to show the power of the *method* of geometrical algebra as much as to arrive at results. From the point of view of illustrating the *method*, there can be no doubt that Euclid's procedure is far more instructive than the semi-algebraical substitutes which seem to find a good deal of favour; practically it means that, instead of relying on our memory of a few standard formulae, we can use the machinery given us by Euclid's method to prove immediately *ab initio* any of the propositions taken at random.

Let us contrast with Euclid's plan the semi-algebraical alternative. One editor, for example, thinks that, as II. I is not used by Euclid afterwards, it seems more logical to deduce from it those of the subsequent propositions which can be readily so deduced. Putting this idea into practice, he proves II. 2 and 3 by quoting II. I, then proves II. 4 by means of II. I and 3, II. 5 and 6 by means of II. I, 3 and 4, and so on. The result is ultimately to deduce the whole of the first ten propositions from II. I, which Euclid does not use at all; and this is to give an importance to II. I which is altogether disproportionate and, by starting with such a narrow foundation, to make the whole structure of Book II. top-heavy.

Editors have of course been much influenced by a desire to make the proofs of the propositions of Book II. easier, as they think, for schoolboys. But, even from this point of view, is it an improvement to deduce II. 2 and 3 from II. I as corollaries? I doubt it. For, in the first place, Euclid's figures *visualise* the results and so make it easier to grasp their meaning ; the truth of the propositions is made clear even to the eye. Then, in the matter of brevity, to which such an exaggerated importance is attached, Euclid's proof positively has the advantage. Counting a capital letter or a collocation of such as one word, I find, e.g., that Mr H. M. Taylor's proof of II. 2 contains

120 words, of which 8 represent the construction. Euclid's as above translated has 126 words, of which 22 are descriptive of the construction; therefore the actual *proof* by Euclid has 8 words fewer than Mr Taylor's, and the extra words due to the construction in Euclid are much more than atoned for by the advantage of picturing the result in the figure.

The advantages then which Euclid's method may claim are, I think, these: in the case of II. 2, 3 it produces the result more easily and clearly than does the alternative proof by means of II. 1, and, in its general application, it is more powerful in that it makes us independent of any recollection of results.

PROPOSITION 3.

If a straight line be cut at random, the rectangle contained by the whole and one of the segments is equal to the rectangle contained by the segments and the square on the aforesaid segment.

For let the straight line *AB* be cut at random at *C*;

I say that the rectangle contained by *AB*, *BC* is equal to the rectangle contained by *AC*, *CB* together with the square on *BC*.

For let the square *CDEB* be described on *CB*;   [I. 46]
let *ED* be drawn through to *F*,
and through *A* let *AF* be drawn parallel to either *CD* or *BE*.   [I. 31]

Then *AE* is equal to *AD*, *CE*.

Now *AE* is the rectangle contained by *AB*, *BC*, for it is contained by *AB*, *BE*, and *BE* is equal to *BC*;

AD is the rectangle *AC*, *CB*, for *DC* is equal to *CB*;

and *DB* is the square on *CB*.

Therefore the rectangle contained by *AB*, *BC* is equal to the rectangle contained by *AC*, *CB* together with the square on *BC*.

Therefore etc.
                                        Q. E. D.

If we leave out of account the contents of Book II. itself and merely look to the applicability of propositions to general use, this proposition and the preceding are, as already indicated, of great importance, and particularly so to the semi-algebraical method just described, which seems to have found its first exponents in Heron and Pappus. Thus the proposition that *the difference of the squares on two straight lines is equal to the rectangle contained by the sum*

and the difference of the straight lines, which is generally given as equivalent to
II. 5, 6, can be proved by means of II. 1, 2, 3, as shown
by Lardner.   For suppose the given straight lines are   A   C   B
AB, *BC*, the latter being measured along *BA*.

. Then, by II. 2, the square on *AB* is equal to the sum of the rectangles
AB, *BC* and *AB*, *AC*.

By II. 3, the rectangle *AB*, *BC* is equal to the sum of the square on *BC*
and the rectangle *AC*, *CB*.

Therefore the square on *AB* is equal to the square *BC* together with the
sum of the rectangles *AC*, *AB* and *AC*, *CB*.

But, by II. 1, the sum of the latter rectangles is equal to the rectangle
contained by *AC* and the sum of *AB*, *BC*, i.e. the rectangle contained by the
sum and difference of *AB*, *BC*.

Hence the square on *AB* is equal to the square on *BC* and the rectangle
contained by the sum and difference of *AB*, *BC*;

that is, the difference of the squares on *AB*, *BC* is equal to the rectangle
contained by the sum and difference of *AB*, *BC*.

PROPOSITION 4.

*If a straight line be cut at random, the square on the whole
is equal to the squares on the segments and twice the rectangle
contained by the segments.*

For let the straight line *AB* be cut at random at *C* ;

5   I say that the square on *AB* is equal to the squares on *AC*,
CB and twice the rectangle contained
by *AC*, *CB*.

For let the square *ADEB* be described on *AB*,   [I. 46]

10 let *BD* be joined ;

through *C* let *CF* be drawn parallel to
either *AD* or *EB*,

and through *G* let *HK* be drawn parallel
to either *AB* or *DE*.   [I. 31]

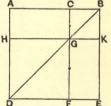

15   Then, since *CF* is parallel to *AD*,
and *BD* has fallen on them,

the exterior angle *CGB* is equal to the interior and opposite
angle *ADB*.   [I. 29]

But the angle *ADB* is equal to the angle *ABD*,

20   since the side *BA* is also equal to *AD* ;   [I. 5]

therefore the angle *CGB* is also equal to the angle *GBC*,

so that the side *BC* is also equal to the side *CG*.   [I. 6]

But CB is equal to GK, and CG to KB;   [I. 34]
    therefore GK is also equal to KB;
25   therefore $CGKB$ is equilateral.
I say next that it is also right-angled.
For, since CG is parallel to BK,
    the angles KBC, GCB are equal to two right angles.
                        [I. 29]
But the angle KBC is right;
30   therefore the angle BCG is also right,
so that the opposite angles CGK, GKB are also right.
                        [I. 34]
Therefore $CGKB$ is right-angled;
and it was also proved equilateral;
        therefore it is a square;
35 and it is described on CB.
For the same reason
        HF is also a square;
and it is described on HG, that is AC.   [I. 34]
    Therefore the squares HF, KC are the squares on AC, CB.
40   Now, since AG is equal to GE,
and AG is the rectangle AC, CB, for GC is equal to CB,
    therefore GE is also equal to the rectangle AC, CB.
    Therefore AG, GE are equal to twice the rectangle AC, CB.
45   But the squares HF, CK are also the squares on AC, CB;
therefore the four areas HF, CK, AG, GE are equal to
the squares on AC, CB and twice the rectangle contained by
AC, CB.
    But HF, CK, AG, GE are the whole $ADEB$,
50 which is the square on AB.
    Therefore the square on AB is equal to the squares on
AC, CB and twice the rectangle contained by AC, CB.
    Therefore etc.   Q. E. D.

2. **twice the rectangle contained by the segments.** By a curious idiom this is in
Greek "the rectangle *twice contained* by the segments." Similarly "twice the rectangle
contained by AC, CB" is expressed as "the rectangle *twice contained* by AC, CB" (τὸ δὶς
ὑπὸ τῶν ΑΓ, ΓΒ περιεχόμενον ὀρθογώνιον).

35, 38. **described.** 39, 45. **the squares** (before "on"). These words are not in the
Greek, which simply says that the squares "are on" (εἰσὶν ἀπό) their respective sides.

46. **areas.** It is necessary to supply some substantive (the Greek leaves it to be understood); and I prefer "areas" to "figures."

The editions of the Greek text which preceded that of E. F. August (Berlin, 1826—9) give a second proof of this proposition introduced by the usual word ἄλλως or "otherwise thus." Heiberg follows August in omitting this proof, which is attributed to Theon, and which is indeed not worth reproducing, since it only differs from the genuine proof in that portion of it which proves that *CGKB* is a square. The proof that *CGKB* is equilateral is rather longer than Euclid's, and the only interesting point to notice is that, whereas Euclid still, as in I. 46, seems to regard it as necessary to prove that *all* the angles of *CGKB* are right angles before he concludes that it is *right-angled*, Theon says simply "And it also has the angle *CBK* right; therefore *CK* is a square." The shorter form indicates a legitimate abbreviation of the genuine proof; because there can be no need to repeat exactly that part of the proof of I. 46 which shows that *all* the angles of the figure there constructed are right when one is.

There is also in the Greek text a Porism which is undoubtedly interpolated: "From this it is manifest that in square areas the parallelograms about the diameter are squares." Heiberg doubted its genuineness when preparing his edition, and conjectured that it too may have been added by Theon; but the matter is placed beyond doubt by a papyrus-fragment referred to already (see Heiberg, *Paralipomena zu Euklid*, in *Hermes* XXXVIII., 1903, p. 48) in which the Porism was evidently wanting. It is the only Porism in Book II., but does not correspond to Proclus' remark (p. 304, 2) that "the Porism found in the second book belongs to a *problem*." Heiberg regards these words as referring to the Porism to IV. 15, the correct reading having probably been not δευτέρῳ but δ´, i.e. τετάρτῳ.

The semi-algebraical proof of this proposition is very easy, and is of course old enough, being found in Clavius and in most later editions. It proceeds thus.

By II. 2, the square on *AB* is equal to the sum of the rectangles *AB*, *AC* and *AB*, *CB*.

But, by II. 3, the rectangle *AB*, *AC* is equal to the sum of the square on *AC* and the rectangle *AC*, *CB*;

while, by II. 3, the rectangle *AB*, *CB* is equal to the sum of the square on *BC* and the rectangle *AC*, *CB*.

Therefore the square on *AB* is equal to the sum of the squares on *AC*, *CB* and twice the rectangle *AC*, *CB*.

The figure of the proposition also helps to visualise, in the orthodox manner, the proof of the theorem deduced above from II. 1—3, viz. that *the difference of the squares on two given straight lines is equal to the rectangle contained by the sum and the difference of the lines.*

For, if the lines be *AB*, *BC* respectively, the shorter of the lines being measured along *BA*, the figure shows that

the square *AE* is equal to the sum of the square *CK* and the rectangles *AF*, *FK*;

that is, the square on *AB* is equal to the sum of the square on *BC* and the rectangles *AB*, *AC* and *AC*, *BC*.

But the rectangles *AB*, *AC* and *BC*, *AC* are, by II. 1, together equal to the rectangle contained by *AC* and the sum of *AB*, *BC*,

i.e to the rectangle contained by the sum and difference of *AB*, *BC*.

Whence the result follows as before.

The proposition II. 4 can also be extended to the case where a straight line is divided into any number of segments; for the figure will show in like manner that the square on the whole line is equal to the sum of the squares on all the parts together with twice the rectangles contained by every pair of the parts.

PROPOSITION 5.

If a straight line be cut into equal and unequal segments, the rectangle contained by the unequal segments of the whole together with the square on the straight line between the points of section is equal to the square on the half.

For let a straight line *AB* be cut into equal segments at *C* and into unequal segments at *D* ;

I say that the rectangle contained by *AD, DB* together with the square on *CD* is equal to the square on *CB*.

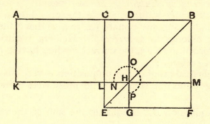

For let the square *CEFB* be described on *CB*,   [I. 46]
and let *BE* be joined ;
through *D* let *DG* be drawn parallel to either *CE* or *BF*,
through *H* again let *KM* be drawn parallel to either *AB* or *EF*,
and again through *A* let *AK* be drawn parallel to either *CL* or *BM*.   [I. 31]

Then, since the complement *CH* is equal to the complement *HF*,   [I. 43]
let *DM* be added to each ;
therefore the whole *CM* is equal to the whole *DF*.
But *CM* is equal to *AL*,
since *AC* is also equal to *CB* ;·   [I. 36]
therefore *AL* is also equal to *DF*.
Let *CH* be added to each ;
therefore the whole *AH* is equal to the gnomon *NOP*.

But *AH* is the rectangle *AD*, *DB*, for *DH* is equal to *DB*,

therefore the gnomon *NOP* is also equal to the rectangle *AD*, *DB*.

Let *LG*, which is equal to the square on *CD*, be added to each ;

therefore the gnomon *NOP* and *LG* are equal to the rectangle contained by *AD*, *DB* and the square on *CD*.

But the gnomon *NOP* and *LG* are the whole square *CEFB*, which is described on *CB* ;

therefore the rectangle contained by *AD*, *DB* together with the square on *CD* is equal to the square on *CB*.

Therefore etc.   Q. E. D.

3. **between the points of section**, literally "between the *sections*," the word being the same (τομή) as that used of a conic *section*.

It will be observed that the gnomon is indicated in the figure by three separate letters and a dotted curve. This is no doubt a clearer way of showing what exactly the gnomon is than the method usual in our text-books. In this particular case the figure of the MSS. has *two* M's in it, the gnomon being MNΞ. I have corrected the lettering to avoid confusion.

It is easily seen that this proposition and the next give exactly the theorem already alluded to under the last propositions, namely that *the difference of the squares on two straight lines is equal to the rectangle contained by their sum and difference.* The two given lines are, in II. 5, the lines *CB* and *CD*, and their sum and difference are respectively equal to *AD* and *DB*. To show that II. 6 gives the same theorem we have only to make *CD* the greater line and *CB* the less, i.e. to draw *C'B'* equal to *CB*, measure *C'B'* along it equal to *CD*, and then produce *B'C'* to *A'*, making *A'C'* equal to *B'C'*, whence it is immediately clear that *A'D'* on the second line is equal

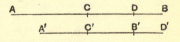

to *AD* on the first, while *D'B'* is also equal to *DB*, so that the rectangles *AD*, *DB* and *A'D'*, *D'B'* are equal, while the difference of the squares on *CB*, *CD* is equal to the difference of the squares on *C'D'*, *C'B'*.

Perhaps the most important fact about II. 5, 6 is however their bearing on the

Geometrical solution of a quadratic equation.

Suppose, in the figure of II. 5, that $AB = a$, $DB = x$;

then $ax - x^2 =$ the rectangle *AH*

                               $=$ the gnomon *NOP*.

Thus, if the area of the gnomon is given ($= b^2$, say), and if a is given ($= AB$), the problem of solving the equation

$$ax - x^2 = b^2$$

is, in the language of geometry, *To a given straight line (a) to apply a rectangle which shall be equal to a given square (b^2) and shall fall short by a square figure*, i.e. to construct the rectangle *AH* or the gnomon *NOP*.

Now we are told by Proclus (on I. 44) that "these propositions are ancient

and the discoveries of the Muse of the Pythagoreans, the application of
areas, their exceeding and their falling-short." We can therefore hardly
avoid crediting the Pythagoreans with the geometrical solution, based upon
II. 5, 6, of the problems corresponding to the quadratic equations which
are directly obtainable from them. It is certain that the Pythagoreans solved
the problem in II. 11, which corresponds to the quadratic equation

$$a(a-x) = x^2,$$

and Simson has suggested the following easy solution of the equation now in
question,

$$ax - x^2 = b^2,$$

on exactly similar lines.

Draw CO perpendicular to AB and equal to b; produce OC to N so
that $ON = CB$ (or $\frac{1}{2}a$); and with O as centre
and radius ON describe a circle cutting CB
in D.

Then DB (or x) is found, and therefore
the required rectangle AH.

For the rectangle AD, DB together with
the square on CD is equal to the square on
CB,   [II. 5]

　i.e. to the square on OD,

　i.e. to the squares on OC, CD;   [I. 47]

whence the rectangle AD, DB is equal to the square on OC,

or $ax - x^2 = b^2$.

It is of course a necessary condition of the possibility of a real solution
that b^2 must not be greater that $(\frac{1}{2}a)^2$. This condition itself can easily be
obtained from Euclid's proposition; for, since the sum of the rectangle AD,
DB and the square on CD is equal to the square on CB, which is constant,
it follows that, as CD diminishes, i.e. as D moves nearer to C, the rectangle
AD, DB increases and, when D actually coincides with C, so that CD
vanishes, the rectangle AD, DB becomes the rectangle AC, CB, i.e. the
square on CB, and is a maximum. It will be seen also that the geometrical
solution of the quadratic equation derived from Euclid does not differ from
our practice of solving a quadratic by completing the square on the side
containing the terms in x^2 and x.

But, while in this case there are two geometrically real solutions (because
the circle described with ON as radius will not only cut CB in D but will
also cut AC in another point E), Euclid's figure corresponds to one only of
the two solutions. Not that there is any doubt that Euclid was aware that the
method of solving the quadratic gives two solutions; he could not fail to see
that $x = BE$ satisfies the equation as well as $x = BD$. If however he had
actually given us the solution of the equation, he would probably have
omitted to specify the solution $x = BE$ because the rectangle found by means
of it, which would be a rectangle on the base AE (equal to BD) and with
altitude EB (equal to AD), is really an equal rectangle to that corresponding
to the other solution $x = BD$; there is therefore no real object in distinguishing
two solutions. This is easily understood when we regard the equation as a
statement of the problem of finding two magnitudes when their sum (a) and
product (b^2) are given, i.e. as equivalent to the simultaneous equations

$$x + y = a,$$
$$xy = b^2.$$

These symmetrical equations have really only one solution, as the two apparent solutions are simply the result of interchanging the values of x and y.   This form of the problem was known to Euclid, as appears from the *Data*, Prop. 85, which states that, *If two straight lines contain a parallelogram given in magnitude in a given angle, and if the sum of them be given, then shall each of them be given.*

This proposition then enables us to solve the problem of finding a rectangle the area and perimeter of which are both given ; and it also enables us to infer that, of all rectangles of given perimeter, the square has the greatest area, while, the more unequal the sides are, the less is the area.

If in the figure of II. 5 we suppose that $AD = a$, $BD = b$, we find that $CB = (a + b)/2$ and $CD = (a - b)/2$, and we may state the result of the proposition in the following algebraical form

$$\left(\frac{a + b}{2}\right)^2 - \left(\frac{a - b}{2}\right)^2 = ab.$$

This way of stating it (which could hardly have escaped the Pythagoreans) gives a ready means of obtaining the two rules, respectively attributed to the Pythagoreans and Plato, for finding integral square numbers which are the sum of two other integral square numbers. We have only to make ab a perfect square in the above formula.   The simplest way in which this can be done is to put $a = n^2$, $b = 1$, whence we have

$$\left(\frac{n^2 + 1}{2}\right)^2 - \left(\frac{n^2 - 1}{2}\right)^2 = n^2,$$

and in order that the first two squares may be integral n^2, and therefore n, must be odd.   Hence the Pythagorean rule.

Suppose next that $a = 2n^2$, $b = 2$, and we have

$$(n^2 + 1)^2 - (n^2 - 1)^2 = 4n^2,$$

whence Plato's rule starting from an *even* number $2n$.

PROPOSITION 6.

If a straight line be bisected and a straight line be added to it in a straight line, the rectangle contained by the whole with the added straight line and the added straight line together with the square on the half is equal to the square on the straight line made up of the half and the added straight line.

For let a straight line AB be bisected at the point C, and let a straight line BD be added to it in a straight line ;

I say that the rectangle contained by AD, DB together with the square on CB is equal to the square on CD.

For let the square $CEFD$ be described on CD,   [I. 46] and let DE be joined ;

through the point B let BG be drawn parallel to either EC or DF, through the point H let KM be drawn parallel to either AB or EF,

and further through A let AK be drawn parallel to either CL or DM.   [I. 31]

Then, since AC is equal to CB,

AL is also equal to CH. [I. 36]

But CH is equal to HF.   [I. 43]

Therefore AL is also equal to HF.

Let CM be added to each;

therefore the whole AM is equal to the gnomon NOP.

But AM is the rectangle AD, DB,

for DM is equal to DB;

therefore the gnomon NOP is also equal to the rectangle AD, DB.

Let LG, which is equal to the square on BC, be added to each;

therefore the rectangle contained by AD, DB together with the square on CB is equal to the gnomon NOP and LG.

But the gnomon NOP and LG are the whole square $CEFD$, which is described on CD;

therefore the rectangle contained by AD, DB together with the square on CB is equal to the square on CD.

Therefore etc.

Q. E. D.

In this case the rectangle AD, DB is "a rectangle applied to a given straight line (AB) but *exceeding* by a square (the side of which is equal to BD)"; and the problem suggested by II. 6 is to find a rectangle of this description equal to a given area, which we will, for convenience, suppose to be a square; i.e., in the language of geometry, *to apply to a given straight line a rectangle which shall be equal to a given square and shall exceed by a square figure.*

We suppose that in Euclid's figure $AB = a$, $BD = x$; then, if the given square be b^2, the problem is to solve geometrically the equation

$$ax + x^2 = b^2.$$

The solution of a problem theoretically equivalent to the solution of a quadratic equation of this kind is presupposed in the fragment of Hippocrates' *Quadrature of lunes* preserved in a quotation by Simplicius (*Comment. in*

Aristot. Phys. pp. 61—68, ed. Diels) from Eudemus' *History of Geometry*. In this fragment Hippocrates (5th cent. B.C.) assumes the following construction.

AB being the diameter and O the centre of a semicircle, and C being the middle point of OB and CD at right angles to AB, a straight line of length such that its square is $1\frac{1}{2}$ times the square on the radius (i.e. of length $a\sqrt{\frac{3}{2}}$, where a is the radius) is to be so placed, as EF, between CD and the circumference AD that it "verges towards B," that is, EF when produced passes through B.

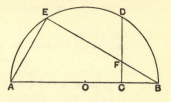

Now the right-angled triangles BFC, BAE are similar, so that

$$BF : BC = BA : BE,$$

and therefore the rectangle $BE, BF = $ rect. BA, BC

$$= \text{sq. on } BO.$$

In other words, $EF\ (= a\sqrt{\frac{3}{2}})$ being given in length, $BF\ (= x,$ say) has to be found such that

$$(\sqrt{\tfrac{3}{2}}\,a + x)\,x = a^2;$$

or the quadratic equation

$$\sqrt{\tfrac{3}{2}}\,ax + x^2 = a^2$$

has to be solved.

A straight line of length $a\sqrt{\frac{3}{2}}$ would easily be constructed, for, in the figure, $CD^2 = AC \cdot CB = \frac{3}{4}a^2$, or $CD = \frac{1}{2}a\sqrt{3}$, and $a\sqrt{\frac{3}{2}}$ is the diagonal of a square of which CD is the side.

There is no doubt that Hippocrates could have solved the equation by the geometrical construction given below, but he may have contemplated, on this occasion, the merely *mechanical* process of *placing* the straight line of the length required between CD and the circumference AD and *moving* it until E, F, B were in a straight line. Zeuthen (*Die Lehre von den Kegelschnitten im Altertum*, pp. 270, 271) thinks this probable because, curiously enough, the fragment speaks immediately afterwards of "joining B to F."

To solve the equation

$$ax + x^2 = b^2,$$

we have to find the rectangle AH, or the gnomon NOP, which is equal in area to b^2 and has one of the sides containing the inner right angle equal to CB or $\frac{1}{2}a$. Thus we know $(\frac{1}{2}a)^2$ and b^2, and we have to find, by I. 47, a square equal to the sum of two given squares.

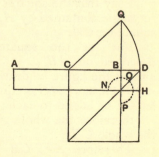

To do this Simson draws BQ at right angles to AB and equal to b, joins CQ and, with centre C and radius CQ, describes a circle cutting AB produced in D. Thus BD, or x, is found.

Now the rectangle AD, DB together with the square on CB

is equal to the square on CD,

i.e. to the square on CQ,

i.e. to the squares on CB, BQ.

Therefore the rectangle AD, DB is equal to the square on BQ, that is,

$$ax + x^2 = b^2.$$

From Euclid's point of view there would only be one solution in this case.
This proposition enables us also to solve the equation

$$x^2 - ax = b^2$$

in a similar manner.

We have only to suppose that $AB = a$, and AD (instead of BD) $= x$; then

$$x^2 - ax = \text{the gnomon}.$$

To find the gnomon we have its area (b^2) and the area, CB^2 or $(\tfrac{1}{2}a)^2$, by which the gnomon differs from CD^2. Thus we can find D (and therefore AD or x) by the same construction as that just given.

Converse propositions to II. 5, 6 are given by Pappus (VII. pp. 948—950) among his lemmas to the *Conics* of Apollonius to the effect that,

(1)   if D be a point dividing AB unequally, and C another point on AB such that the rectangle AD, DB together with the square on CD is equal to the square on AC, then

$$AC \text{ is equal to } CB;$$

(2)   if D be a point on AB produced, and C a point on AB such that the rectangle AD, DB together with the square on CB is equal to the square on CD, then

$$AC \text{ is equal to } CB.$$

PROPOSITION 7.

If a straight line be cut at random, the square on the whole and that on one of the segments both together are equal to twice the rectangle contained by the whole and the said segment and the square on the remaining segment.

For let a straight line AB be cut at random at the point C;

I say that the squares on AB, BC are equal to twice the rectangle contained by AB, BC and the square on CA.

For let the square $ADEB$ be described on AB,   [I. 46] and let the figure be drawn.

Then, since AG is equal to GE, [I. 43] let CF be added to each;

therefore the whole AF is equal to the whole CE.

Therefore AF, CE are double of AF.

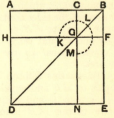

But AF, CE are the gnomon KLM and the square CF; therefore the gnomon KLM and the square CF are double of AF.

But twice the rectangle *AB*, *BC* is also double of *AF*;
for *BF* is equal to *BC*;
therefore the gnomon *KLM* and the square *CF* are equal to twice the rectangle *AB*, *BC*.

Let *DG*, which is the square on *AC*, be added to each;
therefore the gnomon *KLM* and the squares *BG*, *GD* are equal to twice the rectangle contained by *AB*, *BC* and the square on *AC*.

But the gnomon *KLM* and the squares *BG*, *GD* are the whole *ADEB* and *CF*,
　　　which are squares described on *AB*, *BC*;
therefore the squares on *AB*, *BC* are equal to twice the rectangle contained by *AB*, *BC* together with the square on *AC*.

Therefore etc.

Q. E. D.

An interesting variation of the form of this proposition may be obtained by regarding *AB*, *BC* as two given straight lines of which *AB* is the greater, and *AC* as the *difference* between the two straight lines. Thus the proposition shows that the squares on two straight lines are together equal to twice the rectangle contained by them and the square on their difference. That is, *the square on the difference of two straight lines is equal to the sum of the squares on the straight lines diminished by twice the rectangle contained by them.* In other words, just as II. 4 is the geometrical equivalent of the identity

$$(a + b)^2 = a^2 + b^2 + 2ab,$$

so II. 7 proves that

$$(a - b)^2 = a^2 + b^2 - 2ab.$$

The addition and subtraction of these formulae give the algebraical equivalent of the propositions II. 9, 10 and II. 8 respectively; and we have accordingly a suggestion of alternative methods of proving those propositions.

PROPOSITION 8.

If a straight line be cut at random, four times the rectangle contained by the whole and one of the segments together with the square on the remaining segment is equal to the square described on the whole and the aforesaid segment as on one straight line.

For let a straight line *AB* be cut at random at the point *C*;

I say that four times the rectangle contained by *AB*, *BC* together with the square on *AC* is equal to the square described on *AB*, *BC* as on one straight line.

For let [the straight line] BD be produced in a straight line [with AB], and let BD be made equal to CB;

let the square $AEFD$ be described on AD, and let the figure be drawn double.

Then, since CB is equal to BD, while CB is equal to GK, and BD to KN,

therefore GK is also equal to KN.

For the same reason

 QR is also equal to RP.

And, since BC is equal to BD, and GK to KN, therefore CK is also equal to KD, and GR to RN. [I. 36]

But CK is equal to RN, for they are complements of the parallelogram CP; [I. 43]

therefore KD is also equal to GR;

therefore the four areas DK, CK, GR, RN are equal to one another.

Therefore the four are quadruple of CK.

Again, since CB is equal to BD,

while BD is equal to BK, that is CG,

and CB is equal to GK, that is GQ,

therefore CG is also equal to GQ.

And, since CG is equal to GQ, and QR to RP,

 AG is also equal to MQ, and QL to RF. [I. 36]

But MQ is equal to QL, for they are complements of the parallelogram ML; [I. 43]

therefore AG is also equal to RF;

therefore the four areas AG, MQ, QL, RF are equal to one another.

Therefore the four are quadruple of AG.

But the four areas CK, KD, GR, RN were proved to be quadruple of CK;

therefore the eight areas, which contain the gnomon STU, are quadruple of AK.

Now, since AK is the rectangle AB, BD, for BK is equal to BD, therefore four times the rectangle AB, BD is quadruple of AK.

But the gnomon STU was also proved to be quadruple of AK;

therefore four times the rectangle AB, BD is equal to the gnomon STU.

Let OH, which is equal to the square on AC, be added to each;

therefore four times the rectangle AB, BD together with the square on AC is equal to the gnomon STU and OH.

But the gnomon STU and OH are the whole square $AEFD$,

which is described on AD·

therefore four times the rectangle AB, BD together with the square on AC is equal to the square on AD

But BD is equal to BC;

therefore four times the rectangle contained by AB, BC together with the square on AC is equal to the square on AD, that is to the square described on AB and BC as on one straight line.

Therefore etc.

Q. E. D.

This proposition is quoted by Pappus (p. 428, ed. Hultsch) and is used also by Euclid himself in the *Data*, Prop. 86. Further, it is of decided use in proving the fundamental property of a parabola.

Two alternative proofs are worth giving.

The first is that suggested by the consideration mentioned in the last note, though the proof is old enough, being given by Clavius and others. It is of the semi-algebraical type.

Produce AB to D (in the figure of the proposition), so that BD is equal to BC.

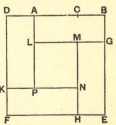

By II. 4, the square on AD is equal to the squares on AB, BD and twice the rectangle AB, BD, i.e. to the squares on AB, BC and twice the rectangle AB, BC.

By II. 7, the squares on AB, BC are equal to twice the rectangle AB, BC together with the square on AC.

Therefore the square on AD is equal to four times the rectangle AB, BC together with the square on AC.

The second proof is after the manner of Euclid but with a difference. Produce BA to D so that AD is equal to BC. On BD construct the square $BEFD$.

Take BG, EH, FK each equal to BC or AD, and draw ALP, HNM parallel to BE and GML, KPN parallel to BD.

Then it can be shown that each of the rectangles BL, AK, FN, EM is equal to the rectangle AB, BC, and that PM is equal to the square on AC.

Therefore the square on BD is equal to four times the rectangle AB, BC together with the square on AC.

PROPOSITION 9.

If a straight line be cut into equal and unequal segments, the squares on the unequal segments of the whole are double of the square on the half and of the square on the straight line between the points of section.

For let a straight line AB be cut into equal segments at C, and into unequal segments at D;

I say that the squares on AD, DB are double of the squares on AC, CD.

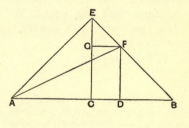

For let CE be drawn from C at right angles to AB, and let it be made equal to either AC or CB;

let EA, EB be joined,

let DF be drawn through D parallel to EC,

and FG through F parallel to AB,

and let AF be joined.

Then, since AC is equal to CE,

the angle EAC is also equal to the angle AEC.

And, since the angle at C is right,

the remaining angles EAC, AEC are equal to one right angle. 　　　　　　　　　　　　　　　　　　　　　[I. 32]

And they are equal;

therefore each of the angles CEA, CAE is half a right angle.

For the same reason

each of the angles CEB, EBC is also half a right angle;

therefore the whole angle AEB is right.

And, since the angle GEF is half a right angle.

and the angle EGF is right, for it is equal to the interior and opposite angle ECB,   [I. 29]

the remaining angle EFG is half a right angle;   [I. 32]

therefore the angle GEF is equal to the angle EFG,

so that the side EG is also equal to GF.   [I. 6]

Again, since the angle at B is half a right angle,

and the angle FDB is right, for it is again equal to the interior and opposite angle ECB,   [I. 29]

the remaining angle BFD is half a right angle;   [I. 32]

therefore the angle at B is equal to the angle DFB,

so that the side FD is also equal to the side DB.   [I. 6]

Now, since AC is equal to CE,

the square on AC is also equal to the square on CE;

therefore the squares on AC, CE are double of the square on AC.

But the square on EA is equal to the squares on AC, CE, for the angle ACE is right;   [I. 47]

therefore the square on EA is double of the square on AC.

Again, since EG is equal to GF,

the square on EG is also equal to the square on GF;

therefore the squares on EG, GF are double of the square on GF.

But the square on EF is equal to the squares on EG, GF;

therefore the square on EF is double of the square on GF.

But GF is equal to CD;   [I. 34]

therefore the square on EF is double of the square on CD.

But the square on EA is also double of the square on AC;

therefore the squares on AE, EF are double of the squares on AC, CD.

And the square on AF is equal to the squares on AE, EF, for the angle AEF is right;   [I. 47]

therefore the square on AF is double of the squares on AC, CD.

But the squares on AD, DF are equal to the square on AF, for the angle at D is right;   [I. 47]

therefore the squares on AD, DF are double of the squares on AC, CD.

And DF is equal to DB;

therefore the squares on AD, DB are double of the squares on AC, CD.

Therefore etc.

Q. E. D.

It is noteworthy that, while the first eight propositions of Book II. are proved independently of the Pythagorean theorem I. 47, all the remaining propositions beginning with the 9th are proved by means of it. Also the 9th and 10th propositions mark a new departure in another respect; the method of demonstration by showing in the figures the various rectangles and squares to which the theorems relate is here abandoned.

The 9th and 10th propositions are related to one another in the same way as the 5th and 6th; they really prove the same result which can, as in the earlier case, be comprised in a single enunciation thus : *The sum of the squares on the sum and difference of two given straight lines is equal to twice the sum of the squares on the lines.*

The semi-algebraical proof of Prop. 9 is that suggested by the remark on the algebraical formulae given at the end of the note on II. 7. It applies with a very slight modification to both II. 9 and II. 10. We will put in brackets the variations belonging to II. 10.

The first of the annexed lines is the figure for II. 9 and the second for II. 10.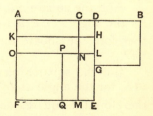

By II. 4, the square on AD is equal to the squares on AC, CD and twice the rectangle AC, CD.

By II. 7, the squares on CB, CD (CD, CB) are equal to

twice the rectangle CB, CD together with the square on BD.

By addition of these equals crosswise,

the squares on AD, DB together with twice the rectangle CB, CD are equal to the squares on AC, CD, CB, CD together with twice the rectangle AC, CD.

But AC, CB are equal, and therefore the rectangles AC, CD and CB, CD are equal.

Taking away the equals, we see that

the squares on AD, DB are equal to the squares on AC, CD, CB, CD,

i.e. to twice the squares on AC, CD.

To show also that the method of geometrical algebra illustrated by II. 1—8 is still effective for the purpose of proving II. 9, 10, we will now prove II. 9 in that manner.

Draw squares on AD, DB respectively as shown in the figure. Measure DH along DE equal to CD, and HL along HE also equal to CD.

Draw HK, LNO parallel to EF, and CNM parallel to DE.

Measure NP along NO equal to CD, and draw PQ parallel to DE.

Now, since *AD*, *CD* are respectively equal to *DE*, *DH*,

HE is equal to *AC* or *CB*;

and, since *HL* is equal to *CD*, *LE* is equal to *DB*.

Similarly, since each of the segments *EM*, *MQ* is equal to *CD*,

FQ is equal to *EL* or *BD*.

Therefore *OQ* is equal to the square on *DB*.

We have to prove that the squares on *AD*, *DB* are equal to twice the squares on *AC*, *CD*.

Now the square on *AD* includes *KM* (the square on *AC*) and *CH*, *HN* (that is, twice the square on *CD*).

Therefore we have to prove that what is left over of the square on *AD* together with the square on *DB* is equal to the square on *AC*.

The parts left over are the rectangles *CK* and *NE*, which are equal to *KN*, *PM* respectively.

But the latter with the square on *DB* are equal to the rectangles *KN*, *PM* and the square *OQ*,

i.e. to the square *KM*, or the square on *AC*.

Hence the required result follows.

PROPOSITION 10.

If a straight line be bisected, and a straight line be added to it in a straight line, the square on the whole with the added straight line and the square on the added straight line both together are double of the square on the half and of the square described on the straight line made up of the half and the added straight line as on one straight line.

For let a straight line *AB* be bisected at *C*, and let a straight line *BD* be added to it in a straight line;

I say that the squares on *AD*, *DB* are double of the squares on *AC*, *CD*.

For let *CE* be drawn from the point *C* at right angles to *AB* [I. 11], and let it be made equal to either *AC* or *CB* [I. 3];

let *EA*, *EB* be joined;

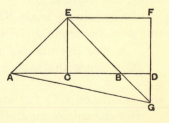

through *E* let *EF* be drawn parallel to *AD*,

and through *D* let *FD* be drawn parallel to *CE*.   [I. 31]

Then, since a straight line *EF* falls on the parallel straight lines *EC*, *FD*,

the angles *CEF, EFD* are equal to two right angles; [I. 29]
therefore the angles *FEB, EFD* are less than two right
angles.

But straight lines produced from angles less than two
right angles meet ;   [I. Post. 5]

therefore *EB, FD*, if produced in the direction *B, D*, will
meet.

Let them be produced and meet at *G*,
and let *AG* be joined.

Then, since *AC* is equal to *CE*,
the angle *EAC* is also equal to the angle *AEC* ;   [I. 5]
and the angle at *C* is right ;

therefore each of the angles *EAC, AEC* is half a right
angle.   [I. 32]

For the same reason
each of the angles *CEB, EBC* is also half a right angle ;
              therefore the angle *AEB* is right.

And, since the angle *EBC* is half a right angle,
the angle *DBG* is also half a right angle.   [I. 15]

But the angle *BDG* is also right,
for it is equal to the angle *DCE*, they being alternate; [I. 29]

therefore the remaining angle *DGB* is half a right angle ;
                                                       [I. 32]
therefore the angle *DGB* is equal to the angle *DBG*,
              so that the side *BD* is also equal to the side *GD*. [I. 6]

Again, since the angle *EGF* is half a right angle,
and the angle at *F* is right, for it is equal to the opposite
angle, the angle at *C*,   [I. 34]

the remaining angle *FEG* is half a right angle ;   [I. 32]
therefore the angle *EGF* is equal to the angle *FEG*,
              so that the side *GF* is also equal to the side *EF*. [I. 6]

Now, since the square on *EC* is equal to the square on
CA,
the squares on *EC, CA* are double of the square on *CA*.

But the square on *EA* is equal to the squares on *EC, CA*;
                                                       [I. 47]
therefore the square on *EA* is double of the square on *AC*.
                                                       [C. N. 1]

Again, since FG is equal to EF,
the square on FG is also equal to the square on FE;
therefore the squares on GF, FE are double of the square on EF.

But the square on EG is equal to the squares on GF, FE;   [I. 47]
therefore the square on EG is double of the square on EF.

And EF is equal to CD;   [I. 34]
therefore the square on EG is double of the square on CD.
But the square on EA was also proved double of the square on AC;
therefore the squares on AE, EG are double of the squares on AC, CD.

And the square on AG is equal to the squares on AE, EG;   [I. 47]
therefore the square on AG is double of the squares on AC, CD.

But the squares on AD, DG are equal to the square on AG;   [I. 47]
therefore the squares on AD, DG are double of the squares on AC, CD.

And DG is equal to DB;
therefore the squares on AD, DB are double of the squares on AC, CD.

Therefore etc.

                              Q. E. D.

The alternative proof of this proposition by means of the principles exhibited in II. 1—8 follows the lines of that which I have given for the preceding proposition.

It is at once obvious from the figure that the square on AD includes within it twice the square on AC together with once the square on CD. What is left over is the sum of the rectangles AH, KE. These, which are equivalent to BH, GK, make up the square on CD *less* the square on BD. Adding therefore the square BG to each side, we have the required result.

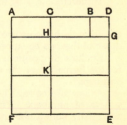

Another alternative proof of the theorem which includes both II. 9 and 10 is worth giving. The theorem states that *the sum of the squares on the sum and difference of two given straight lines is equal to twice the sum of the squares on the lines.*

Let AD, DB be the two given straight lines (of which AD is the greater), placed so as to be in one straight line. Make AC equal to DB and complete the figure as shown, each of the segments CG and DH being equal to AC or DB.

Now, AD, DB being the given straight lines, AB is their sum and CD is equal to their difference.

Also AD is equal to BC.

And AE is the square on AB, GK is equal to the square on CD, AK or FH is the square on AD, and BL the square on CB, while each of the small squares AG, BH, EK, FL is equal to the square on AC or DB.

We have to prove that twice the squares on AD, DB are equal to the squares on AB, CD.

Now twice the square on AD is the sum of the squares on AD, CB, which is equal to the sum of the squares BL, FH; and the figure shows these to be equal to twice the inner square GK and once the remainder of the large square AE excluding the two squares AG, KE, which latter squares are equal to twice the square on AC or DB.

Therefore twice the squares on AD, DB are equal to twice the inner square GK together with once the remainder of the large square AE, that is, to the sum of the squares AE, GK, which are the squares on AB, CD.

" Side " and " diagonal " numbers giving successive approximations to $\sqrt{2}$.

Zeuthen pointed out (*Die Lehre von den Kegelschnitten im Altertum*, 1886, pp. 27, 28) that II. 9, 10 have great interest in connexion with a problem of indeterminate analysis which received much attention from the ancient Greeks.   If we take the straight line AB divided at C and D as in II. 9, and if we put $CD = x$, $DB = y$, the result obtained by Euclid, namely :

$$AD^2 + DB^2 = 2AC^2 + 2CD^2,$$

or $$AD^2 - 2AC^2 = 2CD^2 - DB^2,$$

becomes the formula

$$(2x + y)^2 - 2(x + y)^2 = 2x^2 - y^2.$$

If therefore x, y be numbers which satisfy one of the two equations

$$2x^2 - y^2 = \pm 1,$$

the formula gives us two higher numbers, $x + y$ and $2x + y$, which satisfy the other of the two equations.

Euclid's propositions thus give a general proof of the very formula used for the formation of the succession of what were called "*side*" and "*diagonal*" numbers."

As is well known, Theon of Smyrna (pp. 43, 44, ed. Hiller) describes this system of numbers. The unit, being the beginning of all things, must be potentially both a *side* and a *diameter*. Consequently we begin with two units, the one being the first *side* and the other the first *diameter*, and (*a*) from the sum of them, (*b*) from the sum of twice the first unit and once the second, we form two new numbers

$$1 . 1 + 1 = 2, \quad 2 . 1 + 1 = 3.$$

Of these new numbers the first is a *side-* and the second a *diagonal-*number, or (as we may say)

$$a_2 = 2, \qquad d_2 = 3.$$

In the same way as these numbers were formed from $a_1 = 1$, $d_1 = 1$, successive pairs of numbers are formed from a_2, d_2, and so on, according to the formula

$$a_{n+1} = a_n + d_n, \quad d_{n+1} = 2a_n + d_n.$$

Thus　　　　　$a_3 = 2 + 3 = 5$, 　　$d_3 = 2 \cdot 2 + 3 = 7$,

$$a_4 = 5 + 7 = 12, \quad d_4 = 2 \cdot 5 + 7 = 17,$$

and so on.

Theon states, with reference to these numbers, the general proposition that

$$d_n^2 = 2a_n^2 \pm 1,$$

and he observes (1) that the signs alternate as successive d's and a's are taken, $d_1^2 - 2a_1^2$ being equal to -1, $d_2^2 - 2a_2^2$ equal to $+1$, $d_3^2 - 2a_3^2$ equal to -1, and so on, (2) that the sum of the squares of *all* the d's will be double of the sum of the squares of *all* the a's. [If the number of successive terms in each series is *finite*, it is of course necessary that the number should be even.] The proof, no doubt omitted because it was well known, may be put algebraically thus

$$\begin{aligned} d_n^2 - 2a_n^2 &= (2a_{n-1} + d_{n-1})^2 - 2(a_{n-1} + d_{n-1})^2 \\ &= 2a_{n-1}^2 - d_{n-1}^2 \\ &= -(d_{n-1}^2 - 2a_{n-1}^2) \\ &= +(d_{n-2}^2 - 2a_{n-2}^2), \text{ in like manner,} \end{aligned}$$

and so on, while $d_1^2 - 2a_1^2 = -1$. Thus the theorem is established.

Euclid's propositions enable us to establish the theorem geometrically; and this fact might well be thought to confirm the conjecture that the investigation of the indeterminate equation $2x^2 - y^2 = \pm 1$ in the manner explained by Theon was no new thing but began at a period long before Euclid's time. No one familiar with the truth of the proposition stated by Theon could have failed to observe that, as the corresponding *side-* and *diagonal-*numbers were successively formed, the value of d_n^2/a_n^2 would approach more and more nearly to 2, and consequently that the successive fractions d_n/a_n would give nearer and nearer approximations to the value of $\sqrt{2}$, viz. $\frac{1}{1}, \frac{3}{2}, \frac{7}{5}, \frac{17}{12}, \frac{41}{29}, \ldots$.

It is fairly clear that in the famous passage of Plato's *Republic* (546 c) about the "geometrical number" some such system of approximations is hinted at. Plato there contrasts the "*rational diameter* of five" (ῥητὴ διάμετρος τῆς πεμπάδος) with the "irrational" (diameter). This was certainly taken from the Pythagorean theory of numbers (cf. the expression immediately preceding, 546 B, C πάντα προσήγορα καὶ ῥητὰ πρὸς ἄλληλα ἀπέφηναν, with the phrase πάντα γνωστὰ καὶ ποτάγορα ἀλλάλοις ἀπεργάζεται in the fragment of Philolaus). The reference of Plato is to the following consideration. If the square of side 5 be taken, the diagonal is $\sqrt{2 \cdot 25}$ or $\sqrt{50}$. This is the Pythagorean "*irrational* diameter" of 5; and the "rational diameter" was the approximation $\sqrt{50 - 1}$, or 7.

But the conjecture of Zeuthen, and the attribution of the whole theory of *side-* and *diagonal-*numbers to the Pythagoreans, have now been fully confirmed by the publication of Kroll's edition of *Procli Diadochi in Platonis rempublicam commentarii* (Teubner), Vol. II., 1901. The passages (cc. 23 and 27, pp. 24, 25 and 27—29) which there saw the light for the first time describe the same system of forming *side-* and *diagonal*-numbers and definitely attribute it, as well as the distinction between the "rational" and "irrational diameter," to the Pythagoreans. Proclus further says (p. 27, 16—22) that the property of the *side-* and *diagonal*-numbers " is proved graphically (γραμμικῶς) in the second book of the Elements by 'him' (ἀπ' ἐκείνου). For, *if a straight line be bisected and a straight line be added to it, the square on the whole line including the added straight line and the square on the latter alone are double of the square on the half of the original straight line and of the square on the straight line made up of the half and of the added straight line.*" And this is simply Eucl. II. 10. Proclus then goes on to show specifically how this proposition was used to prove that, with the notation above used, the *diameter* corresponding to the side $a + d$ is $2a + d$.   Let AB be a *side* and BC equal to it, while CD is the *diameter* corresponding to AB, i.e. a straight line such that the square on it is double of the square on AB.   (I use the figure supplied by Hultsch on p. 397 of Kroll's Vol. II.)

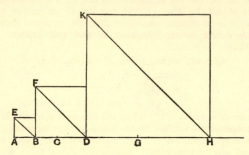

Then, by the theorem of Eucl. II. 10, the squares on AD, DC are double of the squares on AB, BD.

But the square on DC (i.e. BE) is double of the square on AB; therefore, by subtraction, the square on AD is double of the square on BD.

And the square on DF, the *diagonal* corresponding to the *side* BD, is double the square of BD.

Therefore the square on DF is equal to the square on AD, so that DF is equal to AD.

That is, while the *side* BD is, with our notation, $a + d$, the corresponding *diagonal*, being equal to AD, is $2a + d$.

In the above reference by Proclus to II. 10 ἀπ' ἐκείνου " by *him*" must apparently mean ὑπ' Εὐκλείδου, " by Euclid," although Euclid's name has not been mentioned in the chapter; the phrase would be equivalent to saying "in the second Book of the famous Elements."   But, when Proclus says "this *is proved* in the second Book of the Elements," he does not imply that it had not been proved before; on the contrary, it is clear that the theorem had been proved by the Pythagoreans, and we have therefore here a confirmation of the inference from the part played by the *gnomon* and by I. 47 in Book II. that the whole of the substance of that Book was Pythagorean.   For further detailed explanation of the passages of Proclus reference should be made to Hultsch's note in Kroll's Vol. II. pp. 393—400, and to the separate article, also by Hultsch, in the *Bibliotheca Mathematica* I_3, 1900, pp. 8—12.

P. Bergh has an ingenious suggestion (see *Zeitschrift für Math. u. Physik*

XXXI. Hist.-litt. Abt. p. 135, and Cantor, *Geschichte der Mathematik*, I₃, p. 437)
as to the way in which the formation of the successive
side- and *diagonal*-numbers may have been discovered,
namely by observation from a very simple geometrical
figure. Let ABC be an isosceles triangle, right-angled at
A, with sides a_{n-1}, a_{n-1}, d_{n-1} respectively. If now the
two sides AB, AC about the right angle be lengthened
by adding d_{n-1} to each, and the extremities D, E be
joined, it is easily seen by means of the figure (in which
BF, CG are perpendicular to DE) that the new *diagonal*

d_n is equal to $2a_{n-1} + d_{n-1}$, while the equal *sides* a_n are, by construction, equal
to $a_{n-1} + d_{n-1}$.

Important deductions from II. 9, 10.

I.   Pappus (VII. pp. 856—8) uses II. 9, 10 for proving the well-known
theorem that

*The sum of the squares on two sides of a triangle is equal to twice the square
on half the base together with twice the square on the straight line joining the
middle point of the base to the opposite vertex.*

Let ABC be the given triangle and D the middle point of the base BC.
Join AD, and draw AE perpendicular to BC (produced if necessary).

　　　　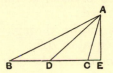

Now, by II. 9, 10,
the squares on BE, EC are equal to twice the squares on BD, DE.

Add to each twice the square on AE.

Then, remembering that

the squares on BE, EA are equal to the square on BA,

the squares on AE, EC are equal to the square on AC,

and the squares on AE, ED are equal to the square on AD,

we find that

the squares on BA, AC are equal to twice the squares on AD, BD.

The proposition is generally proved by means of II. 12, 13, but not, I
think, so conveniently as by the method of Pappus.

II.   The inference was early made by Gregory of St. Vincent (1584–1667)
and Viviani (1622–1703) that *In any parallelogram the squares on the diagonals
are together equal to the squares on the sides, or to twice the squares on adjacent
sides.*

III.   It appears that Leonhard Euler (1707–83) was the first to discover
the corresponding theorem with reference to any quadrilateral, namely that
*In any quadrilateral the sum of the squares on the sides is equal to the sum of the
squares on the diagonals and four times the square on the line joining the middle*

points of the diagonals. Euler seems however to have proved the property from the corresponding theorem for parallelograms just quoted (cf. Camerer's Euclid, Vol. I. pp. 468, 469) and not from the property of the triangle, though the latter brings out the result more easily.

PROPOSITION 11.

To cut a given straight line so that the rectangle contained by the whole and one of the segments is equal to the square on the remaining segment.

Let AB be the given straight line;

thus it is required to cut AB so that the rectangle contained by the whole and one of the segments is equal to the square on the remaining segment.

For let the square $ABDC$ be described on AB;   [I. 46]

let AC be bisected at the point E, and let BE be joined;

let CA be drawn through to F, and let EF be made equal to BE;

let the square FH be described on AF, and let GH be drawn through to K.

I say that AB has been cut at H so as to make the rectangle contained by AB, BH equal to the square on AH.

For, since the straight line AC has been bisected at E, and FA is added to it,

the rectangle contained by CF, FA together with the square on AE is equal to the square on EF.   [II. 6]

But EF is equal to EB;

therefore the rectangle CF, FA together with the square on AE is equal to the square on EB.

But the squares on BA, AE are equal to the square on EB, for the angle at A is right: [I. 47]

therefore the rectangle CF, FA together with the square on AE is equal to the squares on BA, AE.

Let the square on AE be subtracted from each;

therefore the rectangle CF, FA which remains is equal to the square on AB.

Now the rectangle *CF*, *FA* is *FK*, for *AF* is equal to
FG;

and the square on *AB* is *AD*;

therefore *FK* is equal to *AD*.

Let *AK* be subtracted from each;

therefore *FH* which remains is equal to *HD*.

And *HD* is the rectangle *AB*, *BH*, for *AB* is equal to
BD;

and *FH* is the square on *AH*;

therefore the rectangle contained by *AB*, *BH* is equal
to the square on *HA*.

therefore the given straight line *AB* has been cut at *H*
so as to make the rectangle contained by *AB*, *BH* equal to
the square on *HA*.

Q. E. F.

As the solution of this problem is necessary to that of inscribing a regular
pentagon in a circle (Eucl. IV. 10, 11), we must necessarily conclude that it
was solved by the Pythagoreans, or, in other words, that they discovered the
geometrical solution of the quadratic equation

$$a\,(a-x)=x^2,$$
$$\text{or} \qquad x^2 + ax = a^2.$$

The solution in II. 11, too, exactly corresponds to the solution of the more
general equation

$$x^2 + ax = b^2,$$

which, as shown above (pp. 387—8), Simson based upon II. 6. Only Simson's
solution, if applied here, gives us the point *F* on *CA* produced and does not
directly find the point *H*. It takes *E* the middle point of *CA*, draws *AB* at
right angles to *CA* and of length equal to *CA*, and then describes a circle
with *EB* as radius cutting *EA* produced in *F*. The only difference between
the solution in this case and in the more general case is that *AB* is here equal
to *CA* instead of being equal to another given straight line *b*.

As in the more general case, there is, from Euclid's point of view, only one
solution.

The construction shows that *CF* is also divided at *A* in the manner
described in the enunciation, since the rectangle *CF*, *FA* is equal to the
square on *CA*.

The problem in II. 11 reappears in VI. 30 in the form of *cutting a given
straight line in extreme and mean ratio.*

PROPOSITION 12.

*In obtuse-angled triangles the square on the side subtending
the obtuse angle is greater than the squares on the sides containing the obtuse angle by twice the rectangle contained by one
of the sides about the obtuse angle, namely that on which the*

*perpendicular falls, and the straight line cut off outside by the
perpendicular towards the obtuse angle.*

Let ABC be an obtuse-angled triangle having the angle
BAC obtuse, and let BD be drawn from the point B perpendicular to CA produced ;

I say that the square on BC is greater than the squares
on BA, AC by twice the rectangle contained by CA, AD.

For, since the straight line CD has
been cut at random at the point A,
the square on DC is equal to the
squares on CA, AD and twice the rectangle contained by CA, AD.   [II. 4]

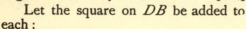

Let the square on DB be added to
each ;

therefore the squares on CD, DB are equal to the squares on
CA, AD, DB and twice the rectangle CA, AD.

But the square on CB is equal to the squares on CD, DB,
for the angle at D is right ;   [I. 47]

and the square on AB is equal to the squares on AD,
DB ;   [I. 47]

therefore the square on CB is equal to the squares on CA, AB
and twice the rectangle contained by CA, AD ;

so that the square on CB is greater than the squares on
CA, AB by twice the rectangle contained by CA, AD.

Therefore etc.   Q. E. D.

Since in this proposition and the next we have to do with the squares on
the sides of triangles, the particular form of graphic representation of areas
which we have had in Book II. up to this point does not help us to visualise
the results of the propositions in the same way, and only two lines of proof
are possible, (1) by means of the *results* of certain earlier propositions in
Book II. combined with the *result* of I. 47 and (2) by means of the procedure
in Euclid's proof of I. 47 itself.   The alternative proofs of II. 12, 13 after the
manner of Euclid's proof of I. 47 are therefore alone worth giving.

These proofs appear in certain modern text-books (e.g. Mehler, Henrici and
Treutlein, H. M. Taylor, Smith and Bryant).   Smith and Bryant are not
correct in saying (p. 142) that they cannot be traced further back than
Lardner's Euclid (1828) ; they are to be found in Gregory of St. Vincent's
work (published in 1647) *Opus geometricum quadraturae circuli et sectionum
coni*, Book I. Pt. 2, Props. 44, 45 (pp. 31, 32).

To prove II. 12, take an obtuse-angled triangle ABC in which the angle at
A is the obtuse angle

Describe squares on BC, CA, AB, as $BCED$, $CAGF$, $ABKH$.

Draw AL, BM, CN, perpendicular to BC, CA, AB (produced if necessary), and produce them to meet the further sides of the squares on them in P, Q, R respectively.

Join AD, CK.

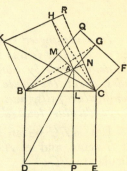

Then, as in I. 47, the triangles KBC, ABD are equal in all respects;

therefore their doubles, the parallelograms in the same parallels respectively, are equal;

that is, the rectangle BP is equal to the rectangle BR.

Similarly the rectangle CP is equal to the rectangle CQ.

Also, if BG, CH be joined, we see that

the triangles BAG, HAC are equal in all respects;

therefore their doubles, the rectangles AQ, AR, are equal.

Now the square on BC is equal to the sum of the rectangles BP, CP,

i.e. to the sum of the rectangles BR, CQ,

i.e. to the sum of the squares BH, CG and

the rectangles AR, AQ.

But the rectangles AR, AQ are equal, and they are respectively the rectangle contained by BA, AN and the rectangle contained by CA, AM.

Therefore the square on BC is equal to the squares on BA, AC together with twice the rectangle BA, AN or CA, AM.

Incidentally this proof shows that the rectangle BA, AN is equal to the rectangle CA, AM: a result which will be seen later on to be a particular case of the theorem in III. 35.

Heron (in an-Nairīzī, ed. Curtze, p. 109) gives a "converse" of II. 12 related to it as I. 48 is related to I. 47.

In any triangle, if the square on one of the sides is greater than the squares on the other two sides, the angle contained by the latter is obtuse.

Let ABC be a triangle such that the square on BC is greater than the squares on BA, AC.

Draw AD at right angles to AC and of length equal to AB.

Join DC.

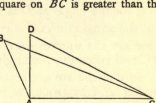

Then, since DAC is a right angle, the square on DC is equal to the squares on DA, AC,   [I. 47]

i.e. to the squares on BA, AC.

But the square on BC is greater than the squares on BA, AC; therefore the square on BC is greater than the square on DC.

Therefore BC is greater than DC.

Thus, in the triangles BAC, DAC,

the two sides BA, AC are equal to the two sides DA, AC respectively, but the base BC is greater than the base DC.

Therefore the angle *BAC* is greater than the angle *DAC*;   [I. 25]
that is, the angle *BAC* is *obtuse*.

PROPOSITION 13.

*In acute-angled triangles the square on the side subtending
the acute angle is less than the squares on the sides containing
the acute angle by twice the rectangle contained by one of the
sides about the acute angle, namely that on which the perpendicular falls, and the straight line cut off within by the
perpendicular towards the acute angle.*

Let *ABC* be an acute-angled triangle having the angle
at *B* acute, and let *AD* be drawn from the point *A* perpendicular to *BC*;

I say that the square on *AC* is less than the squares on
CB, *BA* by twice the rectangle contained
by *CB*, *BD*.

For, since the straight line *CB* has
been cut at random at *D*,

the squares on *CB*, *BD* are equal to
twice the rectangle contained by *CB*, *BD*
and the square on *DC*.   [II. 7]

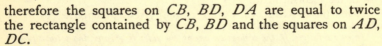

Let the square on *DA* be added to
each;

therefore the squares on *CB*, *BD*, *DA* are equal to twice
the rectangle contained by *CB*, *BD* and the squares on *AD*,
DC.

But the square on *AB* is equal to the squares on *BD*,
DA, for the angle at *D* is right;   [I. 47]
and the square on *AC* is equal to the squares on *AD*, *DC*;
therefore the squares on *CB*, *BA* are equal to the square on
AC and twice the rectangle *CB*, *BD*,

so that the square on *AC* alone is less than the squares
on *CB*, *BA* by twice the rectangle contained by *CB*, *BD*.

Therefore etc.

Q. E. D.

As the text stands, this proposition is unequivocally enunciated of *acute-
angled* triangles; and, as if to obviate any doubt as to whether the restriction
was fully intended, the enunciation speaks of the rectangle contained by one
of the sides containing the acute angle and the straight line intercepted
within by the perpendicular towards the acute angle. On the other hand, it

is curious that it speaks of the square on the side subtending *the* acute angle; and again the *setting-out* begins "let *ABC* be an acute-angled triangle *having the angle at* B *acute*," though the last words have no point if all the angles of the triangle are necessarily acute.

It was however very early noticed, not only by Isaacus Monachus, Campanus, Peletarius, Clavius, Commandinus and the rest, but by the Greek scholiast (Heiberg, Vol. v. p. 253), that the relation between the sides of a triangle established by this theorem is true of the side opposite to, and the sides about, an acute angle respectively in any sort of triangle whether acute-angled, right-angled or obtuse-angled. The scholiast tries to explain away the word "acute-angled" in the enunciation: "Since in the definitions he calls acute-angled the triangle which has three acute angles, you must know that he does not mean that here, but calls all triangles acute-angled because all have an acute angle, one at least, if not all. The enunciation therefore is: 'In any triangle the square on the side subtending the acute angle is less than the squares on the sides containing the acute angle by twice the rectangle, etc.'"

We may judge too by Heron's enunciation of his "converse" of the proposition that he would have left the word "acute-angled" out of the enunciation. His converse is: *In any triangle in which the square on one of the sides is less than the squares on the other two sides, the angle contained by the latter sides is acute.*

If the triangle that we take is a right-angled triangle, and the perpendicular is drawn, not from the right angle, but from the acute angle *not* referred to in the enunciation, the proposition reduces to I. 47, and this case need not detain us.

The other cases can be proved, like II. 12, after the manner of I. 47.

Let us take first the case where all the angles of the triangle are acute.

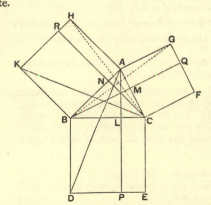

As before, if we draw *ALP, BMQ, CNR* perpendicular to *BC, CA, AB* and meeting the further sides of the squares on *BC, CA, AB* in *P, Q, R,* and if we join *KC, AD,* we have

the triangles *KBC, ABD* equal in all respects,

and consequently the rectangles *BP, BR* equal to one another.

Similarly the rectangles *CP, CQ* are equal to one another.

Next, by joining BG, CH, we prove in like manner that the rectangles AR, AQ are equal.

Now the square on BC is equal to the sum of the rectangles BP, CP,

> i.e. to the sum of the rectangles BR, CQ,

> i.e. to the sum of the squares BH, CG diminished by the rectangles AR, AQ.

But the rectangles AR, AQ are equal, and they are respectively the rectangles contained by BA, AN and by CA, AM.

Therefore the square on BC is less than the squares on BA, AC by twice the rectangle BA, AN or CA, AM.

Next suppose that we have to prove the theorem in the case where the triangle has an obtuse angle at A.

Take B as the acute angle under consideration, so that AC is the side opposite to it.

Now the square on CA is equal to the difference of the rectangles CQ, AQ,

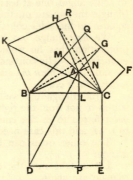

> i.e. to the difference between CP and AQ,

> i.e. to the difference between the square BE and the sum of the rectangles BP, AQ,

> i.e. to the difference between the square BE and the sum of the rectangles BP, AR,

> i.e. to the difference between the sum of the squares BE, BH and the sum of the rectangles BP, BR

(since AR is the difference between BR and BH).

But BP, BR are equal, and they are respectively the rectangles CB, BL and AB, BN.

Therefore the square on CA is less than the squares on AB, BC by twice the rectangle CB, BL or AB, BN.

Heron's proof of his converse proposition (an-Nairīzī, ed. Curtze, p. 110), which is also given by the Greek scholiast above quoted, is of course simple. For let ABC be a triangle in which the square on AC is less than the squares on AB, BC.

Draw BD at right angles to BC and of length equal to BA.

Join DC.

Then, since the angle CBD is right, the square on DC is equal to the squares on DB, BC, i.e. to the squares on AB, BC.   [I. 47]

But the square on AC is less than the squares on AB, BC.

Therefore the square on AC is less than the square on DC.

Therefore AC is less than DC.

Hence in the two triangles DBC, ABC the sides about the angles DBC, ABC are respectively equal, but the base DC is greater than the base AC.

Therefore the angle DBC (a right angle) is greater than the angle ABC [I. 25], which latter is therefore *acute*.

It may be noted, lastly, that II. 12, 13 are supplementary to I. 47 and complete the theory of the relations between the squares on the sides of *any* triangle, whether right-angled or not.

PROPOSITION 14.

To construct a square equal to a given rectilineal figure.

Let A be the given rectilineal figure;
thus it is required to construct a square equal to the rectilineal figure A.

 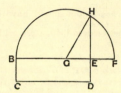

5　For let there be constructed the rectangular parallelogram BD equal to the rectilineal figure A.　　　　　　　　　　[I. 45]

Then, if BE is equal to ED, that which was enjoined will have been done; for a square BD has been constructed equal to the rectilineal figure A.

10　But, if not, one of the straight lines BE, ED is greater. Let BE be greater, and let it be produced to F;
let EF be made equal to ED, and let BF be bisected at G.

With centre G and distance one of the straight lines GB, GF let the semicircle BHF be described; let DE be produced 15 to H, and let GH be joined.

Then, since the straight line BF has been cut into equal segments at G, and into unequal segments at E,

the rectangle contained by BE, EF together with the square on EG is equal to the square on GF.　　　　　　[II. 5]

20　But GF is equal to GH;
therefore the rectangle BE, EF together with the square on GE is equal to the square on GH.

But the squares on HE, EG are equal to the square on GH;　　　　　　　　　　　　　　　　　　　　　　[I. 47]

25 therefore the rectangle BE, EF together with the square on GE is equal to the squares on HE, EG.

Let the square on GE be subtracted from each;

therefore the rectangle contained by BE, EF which remains is equal to the square on EH.

30   But the rectangle BE, EF is BD, for EF is equal to ED; therefore the parallelogram BD is equal to the square on HE.

And BD is equal to the rectilineal figure A.

Therefore the rectilineal figure A is also equal to the square
35 which can be described on EH.

Therefore a square, namely that which can be described on EH, has been constructed equal to the given rectilineal figure A.   Q. E. F.

7.   that which was enjoined will have been done, literally "would have been done," γεγονὸς ἂν εἴη τὸ ἐπιταχθέν.

35, 36.   which can be described, expressed by the future passive participle, ἀναγραφησομένῳ, ἀναγραφησόμενον.

Heiberg (*Mathematisches zu Aristoteles*, p. 20) quotes as bearing on this proposition Aristotle's remark (*De anima* II. 2, 413 a 19: cf. *Metaph.* 996 b 21) that "squaring" (τετραγωνισμός) is better defined as the "finding of the mean (proportional)" than as "the making of an equilateral rectangle equal to a given oblong," because the former definition states the *cause*, the latter the *conclusion* only. This, Heiberg thinks, implies that in the text-books which were in Aristotle's hands the problem of II. 14 was solved by means of proportions. As a matter of fact, the actual *construction* is the same in II. 14 as in VI. 13; and the change made by Euclid must have been confined to substituting in the *proof* of the correctness of the construction an argument based on the principles of Books I. and II. instead of Book VI.

As II. 12, 13 are supplementary to I. 47, so II. 14 completes the theory of *transformation of areas* so far as it can be carried without the use of proportions. As we have seen, the propositions I. 42, 44, 45 enable us to construct a parallelogram having a given side and angle, and equal to any given rectilineal figure. The parallelogram can also be transformed into an equal triangle with the same given side and angle by making the other side about the angle twice the length. Thus we can, as a particular case, construct a rectangle on a given base (or a right-angled triangle with one of the sides about the right angle of given length) equal to a given square. Further, I. 47 enables us to make a square equal to the sum of any number of squares or to the difference between any two squares. The problem still remaining unsolved is to transform any rectangle (as representing an area equal to that of any rectilineal figure) into a square of equal area. The solution of this problem, given in II. 14, is of course the equivalent of the extraction of the square root, or of the solution of the pure quadratic equation

$$x^2 = ab.$$

Simson pointed out that, in the construction given by Euclid in this case, it was not necessary to put in the words "*Let* BE *be greater*," since the construction is not affected by the question whether BE or ED is the greater. This is true, but after all the words do little harm, and perhaps Euclid may have regarded it as conducive to clearness to have the points B, G, E, F in the same relative positions as the corresponding points A, C, D, B in the figure of II. 5 which he quotes in the proof.

EXCURSUS I.

PYTHAGORAS AND THE PYTHAGOREANS.

The problem of determining how much of the Pythagorean discoveries in mathematics can be attributed to Pythagoras himself is not only difficult; it may be said to be insoluble. Tradition on the subject is very meagre and uncertain, and further doubt is thrown upon it by the well-known tendency of the later Pythagoreans to ascribe everything to the Master himself (αὐτὸς ἔφα, *Ipse dixit*). Pythagoras himself left no written exposition of his doctrines, nor did any of his immediate successors, not even Hippasus, about whom the different stories ran (1) that he was expelled from the school because he published doctrines of Pythagoras, and (2) that he was drowned at sea for revealing the construction of the dodecahedron in the sphere and claiming it as his own, or (as others have it) for making known the discovery of the irrational or incommensurable. Nor is the absence of any written record of Pythagorean doctrines down to the time of Philolaus to be put down to a pledge of secrecy binding the school; at all events this did not apply to their mathematics or their physics; and it may be that the supposed secrecy was invented to account for the absence of documents. The fact seems to be that oral communication was the tradition of the school, while their doctrines would in the main be too abstruse to be understood by the generality of people outside. Even Aristotle felt the difficulty; he evidently knew nothing for certain about any ethical or physical doctrines going back to Pythagoras himself; when he speaks of the Pythagorean system, he always refers it to "the Pythagoreans," sometimes even to "the so-called Pythagoreans."

Since my note on Eucl. I. 47 was originally written the part of Pythagoras in the Pythagorean mathematical discoveries has been further discussed and every scrap of evidence closely, and even meticulously, examined in two long articles by Heinrich Vogt, "Die Geometrie des Pythagoras" (*Bibliotheca Mathematica* ix_3, 1908/9, pp. 15—54) and "Die Entstehungsgeschichte des Irrationalen nach Plato und anderen Quellen des 4. Jahrhunderts" (*Bibliotheca Mathematica* x_3, 1910, pp. 97—155). These papers would not indeed have enabled me to modify greatly what I have written regarding the supposed discoveries of Pythagoras and the early Pythagoreans, because I have throughout been careful to give the traditions on the subject for what they are worth and no more, and not to build too much upon them. It is right however to give, in a separate note, a few details of Vogt's arguments.

G. Junge had, in his paper "Wann haben die Griechen das Irrationale entdeckt?" mentioned above (p. 351), tried to prove that Pythagoras himself could not have discovered the irrational; and the object of Vogt's papers is to go further on the same lines and to show (1) that it was only the later Pythagoreans who (before 410 B.C.) recognised the incommensurability of the diagonal with the side of a square, (2) that the *theory* of the irrational was first discovered by Theodorus, to whom Plato refers (*Theaetetus* 147 D), and (3) that Pythagoras himself could not have been the discoverer of any one of the things specifically attributed to him, namely (a) the theorem of Eucl. I. 47, (b) the construction of the five regular solids in the sense in which they are respectively constructed in Eucl. XIII, (c) the application of an area in its widest sense, equivalent to the solution of a quadratic equation in its most general form.

Vogt's main argument as regards (a) the theorem of I. 47 is based on a new translation which he gives of the well-known passage of Proclus' note on the proposition (p. 426, 6—9), Τῶν μὲν ἱστορεῖν τὰ ἀρχαῖα βουλομένων ἀκούοντας τὸ θεώρημα τοῦτο εἰς Πυθαγόραν ἀναπεμπόντων ἐστὶν εὑρεῖν καὶ βουθύτην λεγόντων αὐτὸν ἐπὶ τῇ εὑρέσει. Vogt translates this as follows: "Unter denen, welche das Altertum erforschen wollen, kann man einige finden, welche denen Gehör geben, die dieses Theorem auf Pythagoras zurückführen und ihn als Stieropferer bei dieser Gelegenheit bezeichnen," "Among those who have a taste for research into antiquity, we can find some who give ear to those who refer this theorem to Pythagoras and describe him as sacrificing an ox on the strength of the discovery." According to this version the words τῶν... βουλομένων and the words ἀναπεμπόντων...καὶ...λεγόντων refer respectively to two different sets of persons, in fact two different generations; the latter are older authorities who are supposed to be cited by the former; the former are a later generation, perhaps contemporaries of Proclus, some of whom accepted the view of the older authorities while others did not. But this would have required the article τῶν before ἀναπεμπόντων, or some such expression as ἄλλων τινῶν οἳ ἀναπέμπουσι instead of ἀναπεμπόντων. Vogt's interpretation is therefore quite inadmissible. The persons denoted by ἀναπεμπόντων are some of the persons denoted by τῶν βουλομένων; hence Tannery's translation, to which mine (p. 350 above) is equivalent, is the only possible one, namely "Si l'on écoute ceux qui veulent raconter l'histoire des anciens temps, on peut en trouver qui attribuent ce théorème à Pythagore et lui font sacrifier un bœuf après sa découverte" (La Géométrie grecque, p. 103). ἀκούοντας agrees with the assumed subject of εὑρεῖν; ἀναπεμπόντων and λεγόντων should, strictly speaking, have been ἀναπέμποντας and λέγοντας agreeing with τινὰς (the direct object of εὑρεῖν) understood, but are simply attracted into the case of βουλομένων; the construction is quite intelligible. I agree with Vogt that Eudemus' history contained nothing attributing the theorem to Pythagoras. The words of Proclus imply this; but I do not think that they imply (as Vogt maintains) any pronouncement by Proclus himself against such attribution. In my opinion, Proclus is simply determined not to commit himself to any view; his way of evading a decision is the sentence following, ἐγὼ δὲ θαυμάζω μὲν καὶ τοὺς πρώτους ἐπιστάντας τῇ τοῦδε τοῦ θεωρήματος ἀληθείᾳ, μειζόνως δὲ ἄγαμαι τὸν στοιχειωτήν...; the plural τοὺς πρώτους ἐπιστάντας is, I hold, used for the very purpose of making the statement as vague as possible; he will not even allow it to be inferred that he attributed the discovery to any single person. Returning to ἡ τῶν ἀλόγων πραγματεία (Proclus, p. 65, 19), we may now concede (following Diels) that we should read τῶν ἀνὰ λόγον ("proportionals") instead of τῶν ἀλόγων ("irrationals") and that the author intended to attribute to Pythagoras a theory of proportion (the arithmetical theory applicable to commensurable magnitudes only) rather than the theory of irrationals. But I do not agree in Vogt's contention that the theory of the irrational was first discovered by Theodorus. It seems to me that we have evidence to the contrary in the very passage of Plato referred to. Plato (Theaetetus 147 D) mentions $\sqrt{3}$, $\sqrt{5}$, ... up to $\sqrt{17}$ as dealt with by Theodorus, but omits $\sqrt{2}$. This fact, along with Plato's allusions elsewhere to the irrationality of $\sqrt{2}$, and to approximations to it, in the expressions ἄρρητος and ῥητὴ διάμετρος τῆς πεμπάδος, as if those expressions had a well-known signification, implies that the discovery of the irrationality of √2 had been made before the time of Theodorus. The words ἡ τῶν ἀλόγων πραγματεία might well be used even if the reference is only to √2, because the first step would be the most difficult, and πραγματεία need not mean the establishment of a complete theory or anything more than "investigation" of a subject.

Junge and Vogt hold that the theory of the irrational was not discovered by the early Pythagoreans any more than Pythagoras because, if it had been so discovered, an impossibly long period would intervene between the investigation of the particular case of √2 and the extension of the theory by Theodorus to the cases of √3, √5 etc. But might not this well be due to the fact that in the meantime the minds of geometers were engrossed by other problems of importance, namely the quadrature of the circle (Hippocrates of Chios and his quadratures of lunes), the trisection of any angle (Hippias of Elis and his curve, afterwards known as the *quadratrix*), and the doubling of the cube (reduced by Hippocrates to the problem of finding two mean proportionals in continued proportion between two given straight lines), the last of which problems, which meant finding geometrically the equivalent of ∛2, would naturally follow the investigation of √2 ? Now Hippias was probably born about 460 B.C., while Hippocrates seems to have been in Athens during a considerable portion of the second half of the fifth century, perhaps from about 450 to 430 B.C. Moreover Vogt has to get over the fact that Democritus (born 470/469 B.C.) wrote a book περὶ ἀλόγων γραμμῶν καὶ ναστῶν, *On irrational lines and solids* (or *atoms*). This difficulty he seeks to overcome by maintaining that ἀλόγων does not here mean "irrational" at all, but "without ratio" ("verhältnislos"), in the sense that any two straight lines are "without ratio" because they both contain an infinite number of the indivisible (or atomic) lines, and therefore their ratio, being of the form ∞/∞, is indeterminate. But, if these were so, *all* lines (including commensurable lines) would be "without a ratio" to one another, whereas the title of Democritus' work clearly implies that ἄλογοι γραμμαί are a class or classes of lines distinguished from other lines. The fact is that Democritus was too good a mathematician to have anything to do with "indivisible lines." This is confirmed by a scholium to Aristotle's *De caelo* (p. 469 b 14, Brandis) which implicitly denies to Democritus any theory of indivisible lines: "of those who have maintained the existence of indivisibles, some, as for example Leucippus and Democritus, believe in indivisible bodies, others, like Xenocrates, in indivisible lines." Moreover Simplicius tells us that, according to Democritus himself, even the atoms were, in a mathematical sense, divisible further and in fact *ad infinitum*.

Coming now to (*b*) the construction of the cosmic figures, ἡ τῶν κοσμικῶν σχημάτων σύστασις (Proclus, p. 65, 20), I agree with Vogt to the following extent. It is unlikely that Pythagoras or even the early Pythagoreans "constructed" the five regular solids in the sense of a complete theoretical construction such as we find, say, in Eucl. XIII.; and it is possible that Theaetetus was the first to give these constructions, whether ἔγραψε in Suidas' notice, πρῶτος δὲ τὰ πέντε καλούμενα στερεὰ ἔγραψε, means "constructed" or "wrote upon." But σύστασις in the above phrase of Proclus may well mean something less than the theoretical constructions and proofs of Eucl. XIII.; it may mean, as Vogt says, simply the "putting together" of the figures in the same way as Plato puts them together in the *Timaeus*, i.e. by bringing a certain number of angles of equilateral triangles and of regular pentagons together at one point. There is no reason why the early Pythagoreans should not have "constructed" the five regular solids in this sense; in fact the supposition that they did so agrees well with what we know of their having put angles of certain regular figures together round a point (in connexion with the theorem of Eucl. I. 32) and shown that only three kinds of such angles would fill up the space *in one plane* round the point. But I do not agree in the apparent refusal of Vogt to credit the Pythagoreans with the knowledge of the theoretical construction of the regular pentagon as we find it in Eucl. IV. 10, 11. I do not know of any reason for rejecting the evidence of the Scholia IV. Nos. 2 and 4 which say categorically that "this Book" (Book IV.) and "the whole of the theorems" in it (including therefore Props. 10, 11) are discoveries of the Pythagoreans. And the division of a straight line in extreme and mean ratio, on which the construction of the regular pentagon depends, comes in Eucl. Book II. (Prop. 11), while we have sufficient grounds for regarding the whole of the substance of Book II. as Pythagorean.

I will permit myself one more criticism on Vogt's first paper. I think he bases too much on the fact that it was left for Oenopides (in the period from, say, 470 to 450 B.C.) to discover two elementary constructions (with ruler and compasses only), namely that of a perpendicular to a straight line from an external point (Eucl. I. 12), and that of an angle equal to a given rectilineal angle (Eucl. I. 23). Vogt infers that geometry must have been in a very rudimentary condition at the time. I do not think this follows; the explanation would seem to be rather that, the restriction of the instruments used in constructions to the ruler and compasses not having been definitely established before the time when Oenopides wrote, it had not previously occurred to anyone to substitute new constructions based on that principle for others previously in vogue. In the case of the perpendicular, for example, the construction would no doubt, in earlier days, have been made by means of a set square.

EXCURSUS II.

POPULAR NAMES FOR EUCLIDEAN PROPOSITIONS.

Although some of these time-honoured names are familiar to most educated people, it seems to be impossible to trace them to their original sources, or to say who applied them for the first time respectively. It may be that they were handed down by oral tradition for long periods in each case before they found their way into written documents.

We begin with

I. 5.

1. This proposition is in this country universally known as the *Pons Asinorum*, "Asses' Bridge." Even in this case opinion is not unanimous as to the exact implication of the term. Perhaps the more general view is that taken in the *Stanford Dictionary of Anglicised Words and Phrases* (by C. A. M. Fennell) where the description is: "Name of the fifth proposition of the first Book of Euclid, suggested by the figure and the difficulties which poor geometricians find in mastering it." This is certainly the equivalent of what I gathered, in my early days at school, from a former Fellow of St John's, the Reverend Anthony Bower, who was a high Wrangler in 1846 and a friend of Todhunter's. The "ass" on this interpretation is a synonym for "fool." But there is another view (as I have learnt lately) which is more complimentary to the ass. It is that, the figure of the proposition being like that of a trestle-bridge, with a ramp at each end which is the more practicable the flatter the figure is drawn, the bridge is such that, while a horse could not surmount the ramp, an ass could; in other words, the term is meant to refer to the surefooted-ness of the ass rather than to any want of intelligence on his part. (I may perhaps mention that Sir George Greenhill is a strong supporter of this view.)

An epigram of 1780 is the earliest reference to the term in Murray's English Dictionary:

> "If this be rightly called the bridge of asses,
> He's not the fool that sticks but he that passes."

The writer's own view is not too clear. He seems to imply that, while the inventor of the name meant that only the fool finds the bridge difficult to pass, the more proper view would be that, since the ass can get over, and "ass" is synonymous with "fool," therefore it must be the fool who *can* get over; in other words, he seems to object to the phrase as being a contradiction in terms.

But we have also to take account of the fact that the French apply the term to I. 47. Now in Euclid's figure for I. 47 there is no suggestion of a bridge, and the reference can only be to the nature of the theorem, its diffi-culty or otherwise. It is curious that the French dictionaries give two different explanations of *Pont aux ânes*. Littré makes it "ce que personne ne doit ni ne peut ignorer; ce qui est si facile que tout le monde doit y réussir." Now no intelligent person could have applied the name to Eucl. I. 47 for this reason, namely that it was so easy that even a fool could not help knowing it. Larousse is better informed; there we find "*Pont aux ânes*, certaine difficulté, certaine question qui n'arrête que les ignorants, et qui sert de critérium pour juger l'intelligence de quelqu'un, et particulièrement d'un écolier. C'est ainsi que, dans les classes de mathématiques, on ne manque jamais de dire que le carré de l'hypoténuse est le *pont aux ânes* de la géométrie. La plupart des dictionnaires entendent par ce mot une chose si -simple, si facile, que personne ne doit l'ignorer : c'est une erreur évidente." Larousse is clearly right. But it will be observed that, so far as it goes, Larousse's interpretation rather supports the first of the two alternative explanations of the meaning of "Asses' Bridge" as applied to 1. 5, namely that it is difficult for the fool (= "ass") to master.

In the *Stanford Dictionary* it is added that "in logic the term was in the 16 c. applied to the conversion of propositions by the aid of a difficult diagram for finding middle terms"; and if the mathematicians borrowed the term from logic, this again would be rather in favour of the first explanation of its use for 1. 5.

If it is permitted *desipere in loco*, I would add for the benefit of future generations (in the hope that they will still be able to appreciate the joke or, in the alternative, will be tempted to discuss learnedly what could possibly have been meant) a very topical allusion in a recent *Punch* (14 Oct. 1925): "When they film Euclid, as is suggested, we shall no doubt see a very thrilling rescue over the burning Pons Asinorum."—And yet it is safe to prophesy that the "Asses' Bridge" will outlive the "film"!

2. *Elefuga.*

This name for Eucl. 1. 5 is mentioned by Roger Bacon (about 1250), who also gives an explanation of it (*Opus Tertium*, c. vi). He observes that in his day people in general, finding no utility in any science such as geometry, for example, recoiled from the idea of studying it unless they were boys forced to it by the rod, so that they would hardly learn so much as three or four propositions. Hence it is, he says, that the fifth proposition is called "Elefuga, id est, fuga miserorum ; elegia enim Graece dicitur, Latine miseria ; et elegi sunt miseri." That is, according to Roger Bacon, Elefuga is "flight of the miserable." This explanation no doubt accounts for the verses about *Dulcarnon* in Chaucer's *Troilus and Criseyde*, III, ll. 933–5 :

> "Dulcarnon called is 'fleminge of wrecches';
> It seemeth hard, for wrecches wol not lere
> For verray slouthe or othere wilful tecches";

since "fleminge of wrecches," "banishment of the miserable," is a translation of "fuga miserorum." Only Dulcarnon is there wrongly taken to be the same proposition as Elefuga, i.e. 1. 5, whereas, as we shall see, Dulcarnon was really the name for the Pythagorean theorem 1. 47.

Etymologically, Roger Bacon's explanation leaves something to be desired. The word would really seem to be an attempt to compound the two Greek words ἔλεος, pity (or the object of pity), and φυγή, flight (cf. note *ad loc.* in Skeat's edition of Chaucer). Notwithstanding the confusion of tongues, the object seems to be a play upon the two words *Elementa* and ἔλεος, which both begin with the same three letters, and the implication is that "escape from the Elements" (which normally came when Prop. 5 was reached) was equivalent to "escape from misery" or "trouble." A better form for the word would perhaps be Eleufuga ; and this form actually occurs in Alanus' *Anticlaudianus*, III, c. 6 (cited by Du Cange, *Glossarium*, s.v.). The word also occurs, according to Skeat's note, in Richard of Bury's *Philobiblon*, c. xiii, where it was somewhat oddly translated by J. B. Inglis in 1832 "How many scholars has the Helleflight of Euclid repelled !"

I. 47.

The Pythagorean proposition about the square on the hypotenuse has taken even a deeper hold of the minds of men, and has been known by a number of names.

1. *The Theorem of the Bride* (θεώρημα τῆς νύμφης).

This name is found in a MS. of Georgius Pachymeres (1242–1310) in the Bibliothèque Nationale at Paris; there is a note to this effect by Tannery (*La Géométrie grecque*, p. 105), but, as he says nothing more, it is probable that the passage gives the mere name without any explanation of it. We have, however, much earlier evidence of the supposed connexion of the proposition with marriage. Plutarch (born about 46 A.D.) says (*De Iside et Osiride* 56, p. 373 F) "We may imagine the Egyptians (thinking of) the most beautiful of triangles (and) likening the nature of the All to this triangle most particularly, for it is this same triangle which Plato is thought to have employed in the *Republic*, when he put together the Nuptial Figure (γαμήλιον διάγραμμα)"—διάγραμμα, though literally meaning "diagram" or "figure," was commonly used in the sense of "proposition"—"and in that triangle the perpendicular side is 3, the base 4, and the hypotenuse, the square on which is equal to the sum of the squares on the sides containing (the right angle), 5. We must, then, liken the perpendicular to the male, the base to the female and the hypotenuse to the offspring of both....For 3 is the first odd number and is perfect, 4 is the square on an even side, 2, while the 5 partly resembles the father and partly the mother, being the sum of 3 and 2."

Plato used the three numbers 3, 4, 5 of the Pythagorean triangle in the formation of his famous Geometrical Number; but Plato himself does not call the triangle the Nuptial Triangle nor the number the Nuptial Number. It is later writers, Plutarch, Nicomachus and Iamblichus, who connect the passage about the Geometrical Number with marriage; Nicomachus (*Introd. Ar.*, II, 24, 11) merely alludes to "the passage in the Republic connected with the so-called Marriage," while Iamblichus (*In Nicom.*, p. 82, 20 Pistelli) only speaks of "the Nuptial Number in the Republic."

It would appear, then, that the name "Nuptial Figure" or "Theorem of the Bride" was originally used of one particular right-angled triangle, namely (3, 4, 5). A late Arabian writer Behā-ad-dīn (1547–1622) seems to have applied the term "Figure of the Bride" to the same triangle; the Arabs therefore seemingly followed the Greeks. The idea underlying the use of the term, first for the triangle (3, 4, 5), and then for the general theorem of I. 47, seems to be roughly that of the two parties to a marriage becoming one, just as the two squares on the sides containing the right angle become the one square on the hypotenuse in the said theorem.

2. *The "Bride's Chair."*

The origin of this name is more obscure. It must presumably have been suggested by a supposed resemblance between the figure of the proposition and such a chair. D. E. Smith (*History of Mathematics*, II, pp. 289–90) remarks that the "Bride's Chair" may be so-called "because the Euclid figure is not unlike the chair which a slave carries on his back and in which the Eastern bride is sometimes transported to the ceremony," and he cites a note from Edouard Lucas' *Récréations Mathématiques*, II, p. 130: "La démonstration que nous venons de donner du théorème de Pythagore sur le carré de l'hypoténuse ne diffère pas essentiellement de la démonstration hindoue, connue sous le nom de la *Chaise de la petite mariée*, que l'on rencontre dans l ouvrage de Bhascara (Bija-Ganita, § 146)." The figure of Bhāskara is not that of Euclid but that shown at the top of p. 355 above ; I have however not been able to find the name "Bride's Chair" in Colebrooke's translation of the work of Bhāskara.

Notwithstanding the apparent frivolity of the setting, I venture to suggest that light may be thrown on the question by a very modern version of the "Bride's Chair" which appeared during or since the War in *La Vie Parisienne.* The illustration represents Euclid's figure for I. 47 and, drawn over it, as on a frame, a *poilu* in full fighting kit carrying on his back his bride and his household belongings. Roughly speaking, the soldier is standing (or rather walking) in the middle of the large square, his head and shoulders are bending to the right within the contour of one of the small squares, while the lady, with mirror and powder-puff in action, is sitting with her back to him in the right angle between the two smaller squares (*HAG* in the figure on p. 349 above)[1]. I am informed by Sir George Greenhill that there was also an earlier version "showing the chair as it is in use to-day in Cairo and Egypt, the earliest version of a taxi-chair, a pattern as early as Euclid and suggesting the nickname of the proposition." This recalls to my mind the remark of a friend to whom I mentioned the subject and showed the figure of the proposition ; he observed at once on seeing it "But I should have said it was more like a sedan chair," the large square suggesting to him the actual chair and the two smaller squares the two bearers.

3. *Dulcarnon.*

This name for I. 47 appears, as above mentioned, in Chaucer's *Troilus and Criseyde,* III, ll. 930-3, where Criseyde says :

> ' I am, til God me bettre minde sende,
> At dulcarnon, right at my wittes ende.'
> Quod Pandarus, ' ye, nece, wol ye here?
> Dulcarnon called is " fleminge of wrecches."' '

Billingsley, too, in his edition of Euclid (1570) observes of I. 47 that "it hath bene commonly called of barbarous writers of the latter time Dulcarnon."

Dulcarnon (see Skeat's note *ad loc.*) seems to represent the Persian and Arabic *du'lkarnayn,* lit. *two-horned,* from Pers. *du,* two, and *karn,* horn. The name was applied to I. 47 because the two smaller squares stick up like two horns and, as the proposition is difficult, the word here takes the sense of "puzzle"; hence Criseyde was "at dulcarnon" because she was perplexed and at her wit's end.

4. *Francisci tunica* = "Franciskaner Kutte," "Franciscan's cowl."

This name is quoted by Weissenborn (*Die Uebersetzungen des Euklid durch Campano und Zamberti,* p. 42) as given in a *Geometrie* by one Kunze. The name is quite appropriate, one of the squares representing the hood thrown back.

III. 7, 8.

I have already mentioned the names "Goose's Foot" (*Pes anseris*) and "Peacock's Tail" (*Cauda pavonis*) applied, suitably enough, to these propositions respectively. They come from Luca Paciuolo's edition of Euclid published in 1509 (*vide* Weissenborn, *ibid.*).

---

[1] Old Cambridge men will recall a picture in some respects not unlike, though less artistic than, the cartoon in *La Vie Parisienne,* I mean the painting of "The Man Loaded with Mischief" which used to be over the door of the former inn of that name on the St Neots Road, a short distance from Cambridge.

ENGLISH INDEX.

al-'Abbās b. Sa'īd al-Jauharī 85
"Abthiniathus" (or "Anthisathus") 203
Abū 'l 'Abbās al-Faḍl b. Ḥātim, *see* an-
Nairīzī
Abū 'Abdallāh Muḥ. b. Mu'ādh al-Jayyānī 90
Abū 'Alī al-Baṣrī 88
Abū 'Alī al-Ḥasan b. al-Ḥasan b. al-Haitham
88, 89
Abū Dā'ūd Sulaimān b. 'Uqba 85, 90
Abū Ja'far al-Khāzin 77, 85
Abū Ja'far Muḥ. b. Muḥ. b. al-Ḥasan
Naṣīraddīn aṭ-Ṭūsī, *see* Naṣīraddīn
Abū Muḥ. b. Abdalbāqī al-Baġdādī al-Faraḍī
8 *n.*, 90
Abū Muḥ. al-Ḥasan b. 'Ubaidallāh b. Sulai-
mān b. Wahb 87
Abū Naṣr Ġars al-Na'ma 90
Abū Naṣr Mansūr b. 'Alī b. 'Irāq 90
Abū Naṣr Muḥ. b. Muḥ. b. Tarkhān b.
Uzlaġ al-Fārābī 88
Abū Sahl Wījan b. Rustam al-Kūhī 88
Abū Sa'īd Sinān b. Thābit b. Qurra 88
Abū 'Uthmān ad-Dimashqī 25, 77
Abū 'l Wafā al-Būzjānī 77, 85, 86
Abū Yūsuf Ya'qūb b. Isḥāq b. aṣ-Ṣabbāḥ al-
Kindī 86
Abū Yūsuf Ya'qūb b. Muḥ. ar-Rāzī 86
Adjacent (ἐφεξῆς), meaning 181
Aenaeas (or Aigeias) of Hierapolis 28, 311
Aganis 27–8, 191
Aḥmad b. al-Ḥusain al-Ahwāzī al-Kātib 89
Aḥmad b. 'Umar al-Karābīsī 85
al-Ahwāzī 89
Aigeias (? Aenaeas) of Hierapolis 28, 311
Alexander Aphrodisiensis 7 *n.*, 29
Algebra, geometrical, 372–4: classical method
was that of Eucl. II. (cf. Apollonius) 373:
preferable to semi-algebraical method 377–
8: semi-algebraical method due to Heron
373, and favoured by Pappus 373: geome-
trical equivalents of algebraical operations
374: algebraical equivalents of propositions
in Book II. 372–3
Alī b. Aḥmad Abū 'l Qāsim al-Anṭākī 86
Allman, G. J. 135 *n.*, 318, 352
Alternate (angles) 308
Alternative proofs, interpolated, 58, 59
Amaldi 175, 179–80, 193, 201, 313, 328
Ambiguous case 306–7
Amphinomus 125, 128, 150 *n.*
Amyclas of Heraclea 117
Analysis (and synthesis) 18: alternative
proofs of XIII. 1–5 by, 137: definitions of,

interpolated, 138: described by Pappus
138–9: modern studies of Greek analysis
139: theoretical and problematical analysis
138: *Treasury of analysis* (τόπος ἀναλυόμενος) 8, 10, 11, 138: method of analysis
and precautions necessary to 139–40:
analysis and synthesis of problems 140–2:
two parts of analysis (a) *transformation*,
(b) *resolution*, and two parts of synthesis,
(a) *construction*, (b) *demonstration* 141:
example from Pappus 141–2: analysis
should also reveal διορισμός (conditions of
possibility) 142
Analytical method 36: supposed discovery
of, by Plato 134, 137
Anaximander 370
Anchor-ring 163
Andron 126
Angle. Curvilineal and rectilineal, Euclid's
definition of, 176 sq.: definition criticised
by Syrianus 176: Aristotle's notion of
angle as κλάσις 176: Apollonius' view of,
as *contraction* 176, 177: Plutarch and
Carpus on, 177: to which category does it
belong? *quantum*, Plutarch, Carpus, "A-
ganis" 177, Euclid 178; *quale*, Aristotle
and Eudemus 177–8: *relation*, Euclid 178:
Syrianus' compromise 178: treatise on *the
Angle* by Eudemus 34, 38, 177–8: classification of angles (Geminus) 178–9: curvilineal and "mixed" angles 26, 178–9,
horn-like (κερατοειδής) 177, 178, 182, 265,
lune-like (μηνοειδής) 26, 178–9, *scraper-like*
(ξυστροειδής) 178: angle *of* a segment 253:
angle *of* a semicircle 182, 253: definitions
of angle classified 179: recent Italian views
179–81: angle as cluster of straight lines
or rays 180–1, defined by Veronese 180:
as part of a plane ("angular sector") 179–
80: *flat* angle (Veronese etc.) 180–1, 269:
three kinds of angles, which is prior
(Aristotle)? 181–2: *adjacent* angles 181:
alternate 308: *similar* (=equal) 178, 182,
252: *vertical* 278: *exterior* and *interior*
(to a figure) 263, 280: *exterior* when re-
entrant 263: *interior and opposite* 280:
construction by Apollonius of angle equal
to angle 296: angle in a semicircle, treatment
of, 317–19: trisection of angle, by conchoid
of Nicomedes 265–6, by quadratrix of
Hippias 266, by spiral of Archimedes 267
al-Anṭākī 86
Antiphon 7 *n.*, 35

"Anthisathus" (or "Abthiniathus") 203
Āpastamba-Śulba-Sutra 352 : evidence in, as
to early discovery of Eucl. 1. 47 and use
of gnomon 360–4 : Bürk's claim that
Indians had discovered the irrational 363–
4 : approximation to √2 and Thibaut's
explanation 361, 363–4 : inaccurate values
of π in, 364
Apollodorus "Logisticus" 37, 319, 351
Apollonius : disparaged by Pappus in comparison with Euclid 3 : supposed by some
Arabians to be author of the *Elements* 5 :
a "carpenter" 5 : on elementary geometry
42 : on the *line* 159 : on the *angle* 176 :
general definition of *diameter* 325 : tried to
prove axioms 42, 62, 222–3 : his "general
treatise" 42 : constructions by, for bisection
of straight line 268, for an angle equal to
an angle 296 : for an angle equal to an angle 296 :
on parallel-axiom(?) 42–3 : adaptation to
conics of theory of application of areas
344–5 : geometrical algebra in, 373 : *Plane
Loci* 14, 259, 330 : *Plane νεύσεις* 151 : comparison of dodecahedron and icosahedron
6 : on the *cochlias* 34, 42, 162 : on unordered
irrationals 42, 115 : 138, 188, 221, 222, 246,
249, 259, 370, 373
Application of areas 36, 343–5 : contrasted
with *exceeding* and *falling-short* 343 :
complete method equivalent to geometric
solution of mixed quadratic equation 344–5,
383–5, 386–8 : adaptation to conics (Apollonius) 344–5 : *application* contrasted with
construction (Proclus) 343
"Aqaton" 88
Arabian editors and commentators 75–90
Arabic numerals in scholia to Book x.,
12th c., 71
Archibald, R. C. 9 *n.*, 10
Archimedes 20, 21, 116, 142 : "postulates"
in, 120, 123 : famous "lemma" (assumption)
known as Postulate of Archimedes 234 :
"Porisms" in, 11 *n.*, 13 : spiral of, 26, 267 :
on *straight line* 166 : on *plane* 171–2 : 225,
249, 370
Archytas 20
Areskong, M. E. 113
Arethas, Bishop of Caesarea 48 : owned
Bodleian MS. (B) 47–8 : had famous Plato
MS. of Patmos (Cod. Clarkianus) written 48
Argyrus, Isaak 74
Aristaeus 138 : on conics 3 : *Solid Loci* 16,
329 : comparison of five (regular solid)
figures 6
Aristotelian *Problems* 166, 182, 187
Aristotle : on nature of *elements* 116 : on
first principles 117 sqq. : on definitions 117,
119–20, 143–4, 146–50 : on distinction between hypotheses and definitions 119, 120,
between hypotheses and postulates 118,
119, between hypotheses and axioms 120 :
on axioms 119–21 : axioms indemonstrable 121 : on definition by negation
156–7 : on *points* 155–6, 165 : on *lines*,
definitions of 158–9, classification of 159–

60 : quotes Plato's definition of *straight
line* 166 : on definitions of *surface* 170 :
on the *angle* 176–8 : on priority as between
right and acute angles 181–2 : on *figure*
and definition of 182–3 : definitions of
"squaring" 149–50, 410 : on parallels 190–
2, 308–9 : on *gnomon*. 351, 355, 359 : on
attributes κατὰ παντός and πρῶτον καθόλου
319, 320, 325 : on the *objection* 135 : on
reductio 135 : on *reductio ad absurdum*
136 : on the *infinite* 232–4 : supposed postulate or axiom about divergent lines taken
by Proclus from, 45, 207 : gives pre-Euclidean proof of I. 5 252–3 : on theorem of
angle in a semicircle 149 : on sum of angles
of triangle 319–21 : on sum of exterior
angles of polygon 322 : 38, 45, 117, 150*n.*,
181, 184, 185, 187, 188, 195, 202, 203,
221, 222, 223, 226, 259, 262–3, 283, 411
Arithmetical calculations in scholia to Bk. x.
71, 74
al-Arjānī, Ibn Rāhawaihi 86
Ashkāl at-ta'sīs 5 *n.*
Ashraf Shamsaddīn as-Samarqandī, Muh. b.
5 *n.*, 89
Astaroff, Ivan 113
Asymptotic (non-secant) lines 40, 161, 203
Athelhard of Bath 78, 93–6
Athenaeus 20, 351
Athenaeus of Cyzicus 117
August, E. F. 103
Austin, W. 103, 111
Autolycus, *On the moving sphere* 17
Avicenna 77, 89
Axioms, distinguished from postulates by
Aristotle 118–9, by Proclus (Geminus and
"others") 40, 121–3 : Proclus on difficulties in distinctions 123–4 : distinguished
from hypotheses by Aristotle 120–1, by
Proclus 121–2 : indemonstrable 121 : attempt by Apollonius to prove 222–3 :
= "common (things)" or "common
opinions" in Aristotle 120, 221 : common
to all sciences 119, 120 : called "common
notions" in Euclid 121, 221 : which are
genuine? 221 sqq. : Proclus recognises five
222, Heron three 222 : interpolated axioms
224, 232 : Pappus' additions to axioms
25, 223, 224, 232 : axioms of congruence,
(1) Euclid's Common Notion 4, 224–7,
(2) modern systems (Pasch, Veronese and
Hilbert) 228–31 : "axiom" with Stoics =
every simple declaratory statement 41, 221

Bacon, Roger 94, 416
Balbus, *de mensuris* 91
Barbarin 219
Barlaam, arithmetical commentary on Eucl. II.
74
Barrow, 103, 105, 110, 111
Base, meaning 248–9
Basel, *editio princeps* of Eucl. 100–1
Basilides of Tyre 5, 6
Baudhāyana Sulba-Sūtra 360
Bayfius (Baïf, Lazare) 100